Environmental Modelling
Finding Simplicity in Complexity

Editors

John Wainwright
and
Mark Mulligan

Environmental Monitoring and Modelling Research Group,
Department of Geography,
King's College London,
Strand
London WC2R 2LS
UK

John Wiley & Sons, Ltd

Other Wiley Editorial Offices

John Wiley & Sons Inc., 111 River Street, Hoboken, NJ 07030, USA

Jossey-Bass, 989 Market Street, San Francisco, CA 94103-1741, USA

Wiley-VCH Verlag GmbH, Boschstr. 12, D-69469 Weinheim, Germany

John Wiley & Sons Australia Ltd, 33 Park Road, Milton, Queensland 4064, Australia

John Wiley & Sons (Asia) Pte Ltd, 2 Clementi Loop #02-01, Jin Xing Distripark, Singapore 129809

John Wiley & Sons Canada Ltd, 22 Worcester Road, Etobicoke, Ontario, Canada M9W 1L1

Wiley also publishes its books in a variety of electronic formats. Some content that appears
in print may not be available in electronic books.

Library of Congress Cataloging-in-Publication Data

Environmental modelling : finding simplicity in complexity / editors, John Wainwright and
  Mark Mulligan.
      p. cm.
    Includes bibliographical references and index.
    ISBN-13 978-0-471-49617-5 (acid-free paper) – 978-0-471-49618-2 (pbk. : acid-free paper)
    ISBN-10 0-471-49617-0 (acid-free paper) – 0-471-49618-9 (pbk. : acid-free paper)
    1. Environmental sciences – Mathematical models. I. Wainwright, John, 1967-II.
  Mulligan, Mark, Dr.

  GE45.M37E593 2004
  628–dc22

                                                          2003062751

British Library Cataloguing in Publication Data

A catalogue record for this book is available from the British Library

ISBN-13 978-0-471-49617-5 (Cloth)   978-0-471-49618-2 (Paper)
ISBN-10 0-471-49617-0 (Cloth)   0-471-49618-9 (Paper)

Typeset in 9/11pt Times by Laserwords Private Limited, Chennai, India
Printed and bound in Great Britain by Antony Rowe Ltd, Chippenham, Wiltshire
This book is printed on acid-free paper responsibly manufactured from sustainable forestry
in which at least two trees are planted for each one used for paper production.

For my parents, Betty and John, without whose quiet support over the years none of this would be possible. (JW)

To my parents, David and Filomena, who taught (and teach) me so much and my son/daughter-to-be whom I meet for the first time in a few weeks (hopefully *after* this book is complete). (MM)

Contents

List of Contributors

Andrew Baird, Department of Geography, University of Sheffield, Sheffield, S10 2TN, UK.
http://www.shef.ac.uk/geography/staff/baird_andrew.html

Chris J. Baker, School of Engineering, Mechanical Engineering, The University of Birmingham, Edgbaston, Birmingham, B15 2TT, UK.
http://www.eng.bham.ac.uk/civil/people/bakercj.htm

Keith J. Beven, Centre for Research on Environmental Systems and Statistics, Institute of Environmental and Natural Sciences, Lancaster University, Lancaster LA1 4YQ, UK.
http://www.es.lancs.ac.uk/hfdg/kjb.html

Nick R. Bond, School of Biological Sciences, Monash University (Clayton Campus), Victoria 3800, Australia.
http://biolsci.dbs.monash.edu.au/directory/labs/fellows/bond/

Sophia Burke, Environmental Monitoring and Modelling Research Group, Department of Geography, King's College London, Strand, London WC2R 2LS, UK.
http://www.kcl.ac.uk/kis/schools/hums/geog/smb.htm

Arun Chotai, Centre for Research on Environmental Systems and Statistics, Institute of Environmental and Natural Sciences, Lancaster University, Lancaster LA1 4YQ, UK.
http://www.es.lancs.ac.uk/cres/staff/achotai/

Andrew Collison, Senior Associate, Philip Williams and Associates, 720 California St, San Francisco, CA 94108, USA.
http://www.pwa-ltd.com

Nick A. Drake, Environmental Monitoring and Modelling Research Group, Department of Geography, King's College London, Strand, London WC2R 2LS, UK.
http://www.kcl.ac.uk/kis/schools/hums/geog/nd.htm

Guy Engelen, Research Institute for Knowledge Systems bv, P.O. Box 463, 6200 AL Maastricht, The Netherlands.
http://www.riks.nl/Projects/WadBOS

David Favis-Mortlock, School of Geography, Queen's University Belfast, Belfast BT7 1NN, Northern Ireland, UK.
http://www.qub.ac.uk/geog

Francesco Giannino, Facoltà di Agraria, Università di Napoli 'Federico II', Portici (NA), Italy.
http://www.ecoap.unina.it

James Griffiths, River Regimes Section, Centre for Ecology and Hydrology, Wallingford OX10 1BB, UK.
http://www.nwl.ac.uk/ih/

Hördur V. Haraldsson, Unit of Biogeochemistry, Chemical Engineering, Lund University, Box 124, 221 00 Lund, Sweden.
http://www2.chemeng.lth.se/staff/hordur/index.shtml

L.D. Danny Harvey, Department of Geography, University of Toronto, 100 St. George Street, Toronto, Ontario, M5S 3G3 Canada.
http://www.geog.utoronto.ca/info/faculty/Harvey.htm

Eric F. Lambin, Department of Geography, University of Louvain, 3, place Pasteur, B-1348 Louvain-la-Neuve, Belgium.
http://www.geo.ucl.ac.be/Recherche/Teledetection/index.html

Colin Legg, School of GeoSciences, The University of Edinburgh, Darwin Building, King's Buildings, Mayfield Road, Edinburgh EH9 3JU, Scotland, UK.
http://www.geos.ed.ac.uk/contacts/homes/clegg/

Stefano Mazzoleni, Facoltà di Agraria, Università di Napoli 'Federico II', Portici (NA), Italy.
http://www.ecoap.unina.it

Katerina Michaelides, School of Geographical Sciences, University of Bristol, University Road, Bristol, BS8 1SS, UK.
http://www.ggy.bris.ac.uk/staff/staff_michaelides.htm

Mark Mulligan, Environmental Monitoring and Modelling Research Group, Department of Geography, King's College London, Strand, London WC2R 2LS, UK.
http://www.kcl.ac.uk/kis/schools/hums/geog/mm.htm

Mark A. Nearing, United States Department of Agriculture, Agricultural Research Service, Southwest Watershed Research Center, 2000 E. Allen Road, Tucson, AZ 85719, USA.
http://www.tucson.ars.ag.gov/

Colin P. Osborne, Department of Animal and Plant Sciences, University of Sheffield, Sheffield, S10 2TN, UK.
http://www.shef.ac.uk/aps/staff-colin-osborne.html

George L.W. Perry, Environmental Monitoring and Modelling Research Group, Department of Geography, King's College London, Strand, London WC2R 2LS, UK.
http://www.kcl.ac.uk/kis/schools/hums/geog/gp.htm

John N. Quinton, Cranfield University, Silsoe, Bedford MK45 4DT, UK. (Now at Department of Environmental Science, Institute of Environmental and Natural Sciences, Lancaster University, Lancaster LA1 4YQ, UK.)
http://www.es.lancs.ac.uk/people/johnq/

Francisco Rego, Centro de Ecologia Aplicada Prof. Baeta Neves, Instituto Superior de Agronomia, Tapada da Ajuda 1349-017, Lisbon, Portugal.
http://www.isa.utl.pt/ceabn

Harald U. Sverdrup, Unit of Biogeochemistry, Chemical Engineering, Lund University, Box 124, 221 00 Lund, Sweden.
http://www2.chemeng.lth.se/staff/harald/

John B. Thornes, Environmental Monitoring and Modelling Research Group, Department of Geography, King's College London, Strand, London WC2R 2LS, UK.
rthornes@btinternet.co.uk

Mark J. Twery, Research Forester, USDA Forest Service, Northeastern Research Station, Aiken Forestry Sciences Laboratory, 705 Spear Street, PO Box 968, Burlington, VT 05402-0968, USA.
http://www.fs.fed.us/ne/burlington

John Wainwright, Environmental Monitoring and Modelling Research Group, Department of Geography, King's College London, Strand, London WC2R 2LS, UK.
http://www.kcl.ac.uk/kis/schools/hums/geog/jw.htm

Nigel G. Wright, School of Civil Engineering, The University of Nottingham, University Park, Nottingham NG7 2RD, UK.
http://www.nottingham.ac.uk/%7Eevzngw/

Peter C. Young, Centre for Research on Environmental Systems and Statistics, Systems and Control Group, Institute of Environmental and Natural Sciences, Lancaster University, Lancaster LA1 4YQ, UK; and CRES, Australian National University, Canberra, Australia.
http://www.es.lancs.ac.uk/cres/staff/pyoung/

Xiaoyang Zhang, Department of Geography, Boston University, 675 Commonwealth Avenue, Boston, MA 02215, USA.
http://crsa.bu.edu/~zhang

Preface

Attempting to understand the world around us has been a fascination for millennia. It is said to be part of the human condition. The development of the numerical models, which are largely the focus of this book, is a logical development of earlier descriptive tools used to analyse the environment such as drawings, classifications and maps. Models should be seen as a complement to other techniques used to arrive at an understanding, and they also, we believe uniquely, provide an important means of testing our understanding. This understanding is never complete, as we will see in many examples in the following pages. This statement is meant to be realistic rather than critical. By maintaining a healthy scepticism about our results and continuing to test and re-evaluate them, we strive to achieve a progressively better knowledge of the way the world works. Modelling should be carried out alongside field and laboratory studies and cannot exist without them. We would therefore encourage all environmental scientists not to build up artificial barriers between 'modellers' and 'non-modellers'. Such a viewpoint benefits no-one. It may be true that the peculiarities of mathematical notation and technical methods in modelling form a vocabulary which is difficult to penetrate for some but we believe that the fundamental basis of modelling is one which, like fieldwork and laboratory experimentation, can be used by any scientist who, as they would in the field or the laboratory, might work with others, more specialist in a particular technique to break this language barrier.

Complexity is an issue that is gaining much attention in the field of modelling. Some see new ways of tackling the modelling of highly diverse problems (the economy, wars, landscape evolution) within a common framework. Whether this optimism will go the way of other attempts to unify scientific methods remains to be seen. Our approach here has been to present as many ways as possible to deal with environmental complexity, and to encourage readers to make comparisons across these approaches and between different disciplines. If a unified science of the environment does exist, it will only be achieved by working across traditional disciplinary boundaries to find common ways of arriving at simple understandings. Often the simplest tools are the most effective and reliable, as anyone working in the field in remote locations will tell you!

We have tried to avoid the sensationalism of placing the book in the context of any ongoing environmental 'catastrophe'. However, the fact cannot be ignored that many environmental modelling research programmes are funded within the realms of work on potential impacts on the environment, particularly due to anthropic climate and land-use change. Indeed, the modelling approach – and particularly its propensity to be used in forecasting – has done much to bring potential environmental problems to light. It is impossible to say with any certainty as yet whether the alarm has been raised early enough and indeed which alarms are ringing loudest. Many models have been developed to evaluate what the optimal means of human interaction with the environment are, given the conflicting needs of different groups. Unfortunately, in many cases, the results of such models are often used to take environmental exploitation 'to the limit' that the environment will accept, if not beyond. Given the propensity for environments to drift and vary over time and our uncertain knowledge about complex, non-linear systems with threshold behaviour, we would argue that this is clearly not the right approach, and encourage modellers to ensure that their results are not misused. One of the values of modelling, especially within the context of decision-support systems (see Chapter 14) is that non-modellers and indeed non-scientists can use them. They can thus convey the opinion of the scientist and the thrust of scientific knowledge with the scientist *absent*. This gives modellers and scientists contributing to models (potentially) great influence over the decision-making process (where the political constraints to this process are not paramount). With this influence comes a great responsibility for the modeller to ensure that the models used are both accurate and comprehensive in terms of the driving forces and affected factors and that these models are not applied out of context or in ways for which they were not designed.

This book has developed from our work in environmental modelling as part of the Environmental Monitoring and Modelling Research Group in the Department of Geography, King's College London. It owes a great debt to the supportive research atmosphere we have found there, and not least to John Thornes who initiated the group over a decade ago. We are particularly pleased to be able to include a contribution from him (Chapter 18) relating to his more recent work in modelling land-degradation processes. We would also like to thank Andy Baird (Chapter 3), whose thought-provoking chapter on modelling in his book *Ecohydrology* (co-edited with Wilby) and the workshop from which it was derived provided one of the major stimuli for putting this overview together. Of course, the strength of this book rests on all the contributions, and we would like to thank all of the authors for providing

excellent overviews of their work and the state-of-the-art in their various fields, some at very short notice. We hope we have been able to do justice to your work.

We would also like to thank the numerous individuals who generously gave their time and expertise to assist in the review of the chapters in the book. Roma Beaumont re-drew a number of the figures in her usual cheerful manner. A number of the ideas presented have been tested on our students at King's over the last few years – we would like to thank them all for their inputs. Finally, we would like to thank Keily Larkins and Sally Wilkinson at John Wiley and Sons for bearing with us through the delays and helping out throughout the long process of putting this book together.

John Wainwright and Mark Mulligan
London
December 2002

Introduction

JOHN WAINWRIGHT AND MARK MULLIGAN

1 INTRODUCTION

We start in this introduction to provide a prologue for what follows (possibly following a tradition for books on complex systems, after Bar-Yam, 1997). The aim here is to provide a brief general rationale for the contents and approach taken within the book.

In one sense, everything in this book arises from the invention of the zero. Without this Hindu-Arabic invention, none of the mathematical manipulations required to formulate the relationships inherent within environmental processes would be possible. This point illustrates the need to develop abstract ideas and apply them. Abstraction is a fundamental part of the modelling process.

In another sense, we are never starting our investigations from zero. By the very definition of the environment as that which surrounds us, we always approach it with a number (nonzero!) of preconceptions. It is important not to let them get in the way of what we are trying to achieve. Our aim is to demonstrate how these preconceptions can be changed and applied to provide a fuller understanding of the processes that mould the world around us.

2 WHY MODEL THE ENVIRONMENT?

The context for much environmental modelling at present is the concern relating to human-induced climate change. Similarly, work is frequently carried out to evaluate the impacts of land degradation due to human impact. Such *application-driven* investigations provide an important means by which scientists can interact with and influence policy at local, regional, national and international levels. Models can be a means of ensuring environmental protection, as long as we are careful about how the results are used (Oreskes *et al.*, 1994; Rayner and Malone, 1998; Sarewitz and Pielke, 1999; Bair, 2001).

On the other hand, we may use models to develop our understanding of the processes that form the environment around us. As noted by Richards (1990), processes are not observable features, but their effects and outcomes are. In geomorphology, this is essentially the debate that attempts to link process to form (Richards *et al.*, 1997). Models can thus be used to evaluate whether the effects and outcomes are reproducible from the current knowledge of the processes. This approach is not straightforward, as it is often difficult to evaluate whether process or parameter estimates are incorrect, but it does at least provide a basis for investigation.

Of course, understanding-driven and applications-driven approaches are not mutually exclusive. It is not possible (at least consistently) to be successful in the latter without being successful in the former. We follow up these themes in much more detail in Chapter 1.

3 WHY SIMPLICITY AND COMPLEXITY?

In his short story 'The Library of Babel', Borges (1970) describes a library made up of a potentially infinite number of hexagonal rooms containing books that contain every permissible combination of letters and thus information about everything (or alternatively, a single book of infinitely thin pages, each one opening out into further pages of text). The library is a model of the Universe – but is it a useful one? Borges describes the endless searches for the book that might be the 'catalogue of catalogues'! Are our attempts to model the environment a similarly fruitless endeavour?

Environmental Modelling: Finding Simplicity in Complexity. Edited by J. Wainwright and M. Mulligan
© 2004 John Wiley & Sons, Ltd   ISBNs: 0-471-49617-0 (HB); 0-471-49618-9 (PB)

Compare the definition by Grand (2000: 140): 'Something is complex if it contains a great deal of information that has a high utility, while something that contains a lot of useless or meaningless information is simply complicated.' The environment, by this definition, is something that may initially appear complicated. Our aim is to render it merely complex! Any explanation, whether it is a qualitative description or a numerical simulation, is an attempt to use a model to achieve this aim. Although we will focus almost exclusively on numerical models, these models are themselves based on conceptual models that may be more-or-less complex (see discussions in Chapters 1 and 11). One of the main questions underlying this book is whether simple models are adequate explanations of complex phenomena. Can (or should) we include Ockham's Razor as one of the principal elements in our modeller's toolkit?

Bar-Yam (1997) points out that a dictionary definition of complex means 'consisting of interconnected or interwoven parts'. 'Loosely speaking, the complexity of a system is the amount of information needed in order to describe it' (ibid.: 12). The most complex systems are totally random, in that they cannot be described in shorter terms than by representing the system itself (Casti, 1994) – for this reason, Borges' Library of Babel is not a good model of the Universe, unless it is assumed that the Universe is totally random (or alternatively that the library *is* the Universe!). Complex systems will also exhibit *emergent* behaviour (Bar-Yam, 1997), in that characteristics of the whole are developed (emerge) from interactions of their components in a nonapparent way. For example, the properties of water are not obvious from those of its constituent components, hydrogen and oxygen molecules. Rivers emerge from the interaction of discrete quantities of water (ultimately from raindrops) and oceans from the interaction of rivers, so emergent phenomena may operate on a number of scales.

The optimal model is one that contains sufficient complexity to explain phenomena, but no more. This statement can be thought of as an information-theory rewording of Ockham's Razor. Because there is a definite cost to obtaining information about a system, for example by collecting field data (see discussion in Chapter 1 and elsewhere), there is a cost benefit to developing such an optimal model. In research terms there is a clear benefit because the simplest model will not require the clutter of complications that make it difficult to work with, and often difficult to evaluate (see the discussion of the Davisian cycle by Bishop, 1975, for a geomorphological example).

Opinions differ, however, on how to achieve this optimal model. The traditional view is essentially a reductionist one. The elements of the system are analysed and only those that are thought to be important in explaining the observed phenomena are retained within the model. Often this approach leads to increasingly complex (or possibly even complicated) models where additional process descriptions and corresponding parameters and variables are added. Generally, the law of diminishing returns applies to the extra benefit of additional variables in explaining observed variance. The modelling approach in this case is one of deciding what level of simplicity in model structure is required relative to the overall costs and the explanation or understanding achieved.

By contrast, a more holistic viewpoint is emerging. Its proponents suggest that the repetition of simple sets of rules or local interactions can produce the features of complex systems. Bak (1997), for example, demonstrates how simple models of sand piles can explain the size of and occurrence of avalanches on the pile, and how this approach relates to a series of other phenomena. Bar-Yam (1997) provides a thorough overview of techniques that can be used in this way to investigate complex systems. The limits of these approaches have tended to be related to computing power, as applications to real-world systems require the repetition of very large numbers of calculations. A possible advantage of this sort of approach is that it depends less on the interaction and interpretations of the modeller, in that emergence occurs through the interactions on a local scale. In most systems, these local interactions are more realistic representations of the process than the reductionist approach that tends to be conceptualized so that distant, disconnected features act together. The reductionist approach therefore tends to constrain the sorts of behaviour that can be produced by the model because of the constraints imposed by the conceptual structure of the model.

In our opinion, both approaches offer valuable means of approaching an understanding of environmental systems. The implementation and application of both are described through this book. The two different approaches may be best suited for different types of application in environmental models given the current state of the art. Thus, the presentations in this book will contribute to the debate and ultimately provide the basis for stronger environmental models.

4 HOW TO USE THIS BOOK

We do not propose here to teach you how to suck eggs (nor give scope for endless POMO discussion), but would like to offer some guidance based on the way we have structured the chapters. This book is divided into four parts. We do not anticipate that many readers will want (or need) to read it from cover to cover in one go. Instead, the different elements can be largely understood and followed separately, in almost any order. Part I provides an introduction to modelling approaches in general, with a specific focus on issues that commonly arise in dealing with the environment. We have attempted to cover the process of model development from initial concepts, through algorithm development and numerical techniques, to application and testing. Using the information provided in this chapter, you should be able to put together your own models. We have presented it as a single, large chapter, but the headings within it should allow simple navigation to the sections that are more relevant to you.

The twelve chapters of Part II form the core of the book, presenting a state of the art of environmental models in a number of fields. The authors of these chapters were invited to contribute their own viewpoints of current progress in their specialist areas using a series of common themes. However, we have not forced the resulting chapters back into a common format as this would have restricted the individuality of the different contributions and denied the fact that different topics might require different approaches. As much as we would have liked, the coverage here is by no means complete and we acknowledge that there are gaps in the material here. In part, this is due to space limitations and in part due to time limits on authors' contributions. We make no apology for the emphasis on hydrology and ecology in this part, not least because these are the areas that interest us most. However, we would also argue that these models are often the basis for other investigations and thus are relevant to a wide range of fields. For any particular application, you may find building blocks of relevance to your own interests across a range of different chapters here. Furthermore, it has become increasingly obvious to us while editing the book that there are a number of common themes and problems being tackled in environmental modelling that are currently being developed in parallel behind different disciplinary boundaries. One conclusion that we have come to is that if you cannot find a specific answer to a modelling problem relative to a particular type of model, then a look at the literature of a different discipline can often provide answers. Even more importantly, this can lead to the demonstration of different problems and new ways of dealing with issues. Cross-fertilization of modelling studies will lead to the development of stronger breeds of models!

In Part III, the focus moves to model applications. We invited a number of practitioners to give their viewpoints on how models can or should be used in their particular field of expertise. These chapters bring to light the different needs of models in a policy or management context and demonstrate how these needs might be different from those in a pure research context. This is another way in which modellers need to interface with the real world, and one that is often forgotten.

Part IV deals with a number of current approaches in modelling: approaches that we believe are fundamental to developing strong models in the future. Again, the inclusion of subjects here is less than complete, although some appropriate material on error, spatial models and validation is covered in Part I. However, we hope this part gives at least a flavour of the new methods being developed in a number of areas of modelling. In general, the examples used are relevant across a wide range of disciplines. One of the original reviewers of this book asked how we could possibly deal with future developments. In one sense this objection is correct, in the sense that we do not possess a crystal ball (and would probably not be writing this at all if we did!). In another, it forgets the fact that many developments in modelling await the technology to catch up for their successful conclusion. For example, the detailed spatial models of today are only possible because of the exponential growth in processing power over the past few decades. Fortunately the human mind is always one step ahead in posing more difficult questions. Whether this is a good thing is a question addressed at a number of points through the book!

Finally, a brief word about equations. Because the book is aimed at a range of audiences, we have tried to keep it as user-friendly as possible. In Parts II, III and IV we asked the contributors to present their ideas and results with the minimum of equations. In Part I, we decided that it was not possible to get the ideas across without the use of equations, but we have tried to explain as much as possible from first principles as space permits. Sooner or later, anyone wanting to build their own model will need to use these methods anyway. If you are unfamiliar with text including equations, we would simply like to pass on the following advice of the distinguished professor of mathematics and physics, Roger Penrose:

If you are a reader who finds any formula intimidating (and most people do), then I recommend a procedure I normally adopt myself when such an offending line presents itself. The procedure is, more or less, to ignore that line completely and to skip over to the next actual line of text! Well, not exactly this; one should spare the poor formula a perusing, rather than a comprehending glance, and then press onwards. After a little, if armed with new confidence, one may return to that neglected formula and try to pick out some salient features. The text itself may be helpful in letting one know what is important and what can be safely ignored about it. If not, then do not be afraid to leave a formula behind altogether.

(Penrose, 1989: vi)

5 THE BOOK'S WEBSITE

As a companion to the book, we have developed a related website to provide more information, links, examples and illustrations that are difficult to incorporate here (at least without having a CD in the back of the book that would tend to fall out annoyingly!). The structure of the site follows that of the book, and allows easy access to the materials relating to each of the specific chapters. The URL for the site is www.kcl.ac.uk/envmod. We will endeavour to keep the links and information as up to date as possible to provide a resource for students and researchers of environmental modelling. Please let us know if something does not work and, equally importantly, if you know of exciting new information and models to which we can provide links.

REFERENCES

Bair, E. (2001) Models in the courtroom, in M.G. Anderson and P.D. Bates (eds) *Model Validation: Perspectives in Hydrological Science*, John Wiley & Sons, Chichester, 57–76.

Bak, P. (1997) *How Nature Works: The Science of Self-Organized Criticality*, Oxford University Press, Oxford.

Bar-Yam, Y. (1997) *Dynamics of Complex Systems*, Perseus Books, Reading, MA.

Bishop, P. (1975) Popper's principle of falsifiability and the irrefutability of the Davisian cycle, *Professional Geographer* **32**, 310–315.

Borges, J.L. (1970) *Labyrinths*, Penguin Books, Harmondsworth.

Casti, J.L. (1994) *Complexification: Explaining a Paradoxical World Through the Science of Surprise*, Abacus, London.

Grand, S. (2000) *Creation: Life and How to Make It*, Phoenix, London.

Oreskes, N., Shrader-Frechette, K. and Bellitz, K. (1994) Verification, validation and confirmation of numerical models in the Earth Sciences, *Science* **263**, 641–646.

Penrose, R. (1989) *The Emperor's New Mind*, Oxford University Press, Oxford.

Rayner, S. and Malone, E.L. (1998) *Human Choice and Climate Change*, Batelle Press, Columbus, OH.

Richards, K.S. (1990) 'Real' geomorphology, *Earth Surface Processes and Landforms* **15**, 195–197.

Richards, K.S., Brooks, S.M., Clifford, N., Harris, T. and Lowe, S. (1997) Theory, measurement and testing in 'real' geomorphology and physical geography, in D.R. Stoddart (ed.) *Process and Form in Geomorphology*, Routledge, London, 265–292.

Sarewitz, D. and Pielke Jr, R.A. (1999) Prediction in science and society, *Technology in Society* **21**, 121–133.

Part I

Modelling and Model Building

1

Modelling and Model Building

MARK MULLIGAN AND JOHN WAINWRIGHT

Modelling is like sin. Once you begin with one form of it you are pushed to others. In fact, as with sin, once you begin with one form you ought to consider other forms. . . . But unlike sin – or at any rate unlike sin as a moral purist conceives of it – modelling is the best reaction to the situation in which we find ourselves. Given the meagreness of our intelligence in comparison with the complexity and subtlety of nature, if we want to say things which are true, as well as things which are useful and things which are testable, then we had better relate our bids for truth, application and testability in some fairly sophisticated ways. This is what modelling does.

(Morton and Suárez, 2001: 14)

1.1 THE ROLE OF MODELLING IN ENVIRONMENTAL RESEARCH

1.1.1 The nature of research

Research is a means of improvement through understanding. This improvement may be personal, but it may also be tied to broader human development. We may hope to improve human health and well-being through research into diseases such as cancer and heart disease. We may wish to improve the design of bridges or aircraft through research in materials science, which provides lighter, stronger, longer-lasting or cheaper bridge structures (in terms of building and of maintenance). We may wish to produce more or better crops with fewer adverse impacts on the environment through research in biotechnology. In all of these cases, research provides in the first instance better understanding of how things are and how they work, which can then contribute to the improvement or optimization of these systems through the development of new techniques, processes, materials and protocols.

Research is traditionally carried out through the accumulation of observations of systems and system behaviour under 'natural' circumstances and during experimental manipulation. These observations provide the evidence upon which hypotheses can be generated about the structure and operation (function) of the systems. These hypotheses can be tested against new observations and, where they prove to be reliable descriptors of the system or system behaviour, then they may eventually gain recognition as tested theory or general law.

The conditions, which are required to facilitate research, include:

1. a means of observation and comparative observation (measurement);
2. a means of controlling or forcing aspects of the system (experimentation);
3. an understanding of previous research and the state of knowledge (context);
4. a means of cross-referencing and connecting threads of 1, 2 and 3 (imagination).

1.1.2 A model for environmental research

What do we mean by the term *model*? A model is an abstraction of reality. This abstraction represents a complex reality in the simplest way that is adequate for the purpose of the modelling. The best model is always that which achieves the greatest realism (measured objectively as agreement between model outputs

Environmental Modelling: Finding Simplicity in Complexity. Edited by J. Wainwright and M. Mulligan
© 2004 John Wiley & Sons, Ltd ISBNs: 0-471-49617-0 (HB); 0-471-49618-9 (PB)

and real-world observations, or less objectively as the process insight gained from the model) with the least parameter complexity and the least model complexity.

Parsimony (using no more complex a model or representation of reality than is absolutely necessary) has been a guiding principle in scientific investigations since Aristotle who claimed: 'It is the mark of an instructed mind to rest satisfied with the degree of precision which the nature of the subject permits and not to seek an exactness where only an approximation of the truth is possible' though it was particularly strong in medieval times and was enunciated then by William of Ockham, in his famous 'razor' (Lark, 2001). Newton stated it as the first of his principles for fruitful scientific research in *Principia* as: 'We are to admit no more causes of natural things than such as are both true and sufficient to explain their appearances.'

Parsimony is a prerequisite for scientific explanation, not an indication that nature operates on the basis of parsimonious principles. It is an important principle in fields as far apart as taxonomy and biochemistry and is fundamental to likelihood and Bayesian approaches of statistical inference. In a modelling context, a parsimonious model is usually the one with the greatest explanation or predictive power and the least parameters or process complexity. It is a particularly important principle in modelling since our ability to model complexity is much greater than our ability to provide the data to parameterize, calibrate and validate those same models. Scientific explanations must be both relevant *and* testable. Unvalidated models are no better than untested hypotheses. If the application of the principle of parsimony facilitates validation, then it also facilitates utility of models.

1.1.3 The nature of modelling

Modelling is not an alternative to observation but, under certain circumstances, can be a powerful tool in understanding observations and in developing and testing theory. Observation will always be closer to truth and must remain the most important component of scientific investigation. Klemeš (1997: 48) describes the forces at work in putting the modelling 'cart' before the observation 'horse' as is sometimes apparent in modelling studies:

> *It is easier and more fun to play with a computer than to face the rigors of fieldwork especially hydrologic fieldwork, which is usually most intensive during the most adverse conditions. It is faster to get a result by modeling than through acquisition and analysis of*
> *more data, which suits managers and politicians as well as staff scientists and professors to whom it means more publications per unit time and thus an easier passage of the hurdles of annual evaluations and other paper-counting rituals. And it is more glamorous to polish mathematical equations (even bad ones) in the office than muddied boots (even good ones) in the field.*
>
> (Klemeš, 1997: 48)

A model is an abstraction of a real system, it is a simplification in which only those components which are seen to be significant to the problem at hand are represented in the model. In this, a model takes influence from aspects of the real system and aspects of the modeller's perception of the system and its importance to the problem at hand. Modelling supports in the conceptualization and exploration of the behaviour of objects or processes and their interaction as a means of better understanding these and generating hypotheses concerning them. Modelling also supports the development of (numerical) experiments in which hypotheses can be tested and outcomes predicted. In science, understanding is the goal and models serve as tools towards that end (Baker, 1998).

Cross and Moscardini (1985: 22) describe modelling as 'an art with a rational basis which requires the use of common sense at least as much as mathematical expertise'. Modelling is described as an art because it involves experience and intuition as well as the development of a set of (mathematical) skills. Cross and Moscardini argue that intuition and the resulting insight are the factors which distinguish good modellers from mediocre ones. Intuition (or imagination) cannot be taught and comes from the experience of designing, building and using models. Tackling some of the modelling problems presented on the website which complements this book will help in this.

1.1.4 Researching environmental systems

Modelling has grown significantly as a research activity since the 1950s, reflecting conceptual developments in the modelling techniques themselves, technological developments in computation, scientific developments in response to the increased need to study systems (especially environmental ones) in an integrated manner, and an increased demand for extrapolation (especially prediction) in space and time.

Modelling has become one of the most powerful tools in the workshop of environmental scientists who are charged with better understanding the interactions between the environment, ecosystems and the populations of humans and other animals. This understanding is increasingly important in environmental stewardship (monitoring and management) and the development of increasingly sustainable means of human dependency on environmental systems.

Environmental systems are, of course, the same systems as those studied by physicists, chemists and biologists but the level of abstraction of the environmental scientist is very different from many of these scientists. Whereas a physicist might study the behaviour of gases, liquids or solids under controlled conditions of temperature or pressure and a chemist might study the interaction of molecules in aqueous solution, a biologist must integrate what we know from these sciences to understand how a cell – or a plant or an animal – lives and functions. The environmental scientist or geographer or ecologist approaches their science at a much greater level of abstraction in which physical and chemical 'laws' provide the rule base for understanding the interaction between living organisms and their nonliving environments, the characteristics of each and the processes through which each functions.

Integrated environmental systems are different in many ways from the isolated objects of study in physics and chemistry though the integrated study of the environment cannot take place without the building blocks provided by research in physics and chemistry. The systems studied by environmental scientists are characteristically:

- *Large-scale, long-term.* Though the environmental scientist may only study a small time-scale and space-scale slice of the system, this slice invariably fits within the context of a system that has evolved over hundreds, thousands or millions of years and which will continue to evolve into the future. It is also a slice that takes in material and energy from a hierarchy of neighbours from the local, through regional, to global scale. It is this context which provides much of the complexity of environmental systems compared with the much more reductionist systems of the traditional 'hard' sciences. To the environmental scientist, models are a means of integrating across time and through space in order to understand how these contexts determine the nature and functioning of the system under study.
- *Multicomponent.* Environmental scientists rarely have the good fortune of studying a single component of their system in isolation. Most questions asked of environmental scientists require the understanding of interactions between multiple living (biotic)

and nonliving (abiotic) systems and their interaction. Complexity increases greatly as the number of components increases, where their interactions are also taken into account. Since the human mind has some considerable difficulty in dealing with chains of causality with more than a few links, to an environmental scientist models are an important means of breaking systems into intellectually manageable components and combining them and making explicit the interactions between them.

- *Nonlaboratory controllable.* The luxury of controlled conditions under which to test the impact of individual forcing factors on the behaviour of the study system is very rarely available to environmental scientists. Very few environmental systems can be re-built in the laboratory (laboratory-based physical modelling) with an appropriate level of sophistication to adequately represent them. Taking the laboratory to the field (field-based physical modelling) is an alternative, as has been shown by the Free Atmosphere CO_2 Enrichment (FACE) experiments (Hall, 2001), BIOSPHERE 2 (Cohn, 2002) and a range of other environmental manipulation experiments. Field-based physical models are very limited in the degree of control available to the scientist because of the enormous expense associated with this. They are also very limited in the scale at which they can be applied, again because of expense and engineering limitations. So, the fact remains that, at the scale at which environmental scientists work, their systems remain effectively noncontrollable with only small components capable of undergoing controlled experiments. However, some do argue that the environment itself is one large laboratory, which is sustaining global-scale experiments through, for example, greenhouse gas emissions (Govindasamy *et al.*, 2003). These are not the kind of experiments that enable us to predict (since they are real-time) nor which help us, in the short term at least, to better interact with or manage the environment (notwithstanding the moral implications of this activity!). Models provide an inexpensive laboratory in which mathematical descriptions of systems and processes can be forced in a controlled way.
- *Multiscale, multidisciplinary.* Environmental systems are multiscale with environmental scientists needing to understand or experiment at scales from the atom through the molecule to the cell, organism or object, population of objects, community or landscape through to the ecosystem and beyond. This presence of multiple scales means that environmental scientists are rarely just environmental scientists, they may be physicists, chemists, physical chemists, engineers,

biologists, botanists, zoologists, anthropologists, population geographers, physical geographers, ecologists, social geographers, political scientists, lawyers, environmental economists or indeed environmental scientists in their training but who later apply themselves to environmental science. Environmental science is thus an interdisciplinary science which cuts across the traditional boundaries of academic research. Tackling contemporary environmental problems often involves large multidisciplinary (and often multinational) teams working together on different aspects of the system. Modelling provides an integrative framework in which these disparate disciplines can work on individual aspects of the research problem and supply a module for integration within the modelling framework. Disciplinary and national boundaries, research 'cultures' and research 'languages' are thus no barrier.

- *Multivariate, nonlinear and complex.* It goes without saying that integrated systems such as those handled by environmental scientists are multivariate and as a result the relationships between individual variables are often nonlinear and complex. Models provide a means of deconstructing the complexity of environmental systems and, through experimentation, of understanding the univariate contribution to multivariate complexity.

In addition to these properties of environmental systems, the rationale behind much research in environmental systems is often a practical or applied one such that research in environmental science also has to incorporate the following needs:

- *The need to look into the future.* Environmental research often involves extrapolation into the future in order to understand the impacts of some current state or process. Such prediction is difficult, not least because predictions can only be tested in real time. Models are very often used as a tool for integration of understanding over time and thus are well suited for prediction and postdiction. As with any means of predicting the future, the prediction is only as good as the information and understanding upon which it is based. While this understanding may be sufficient where one is working within process domains that have already been experienced during the period in which the understanding was developed, when future conditions cross a process domain, the reality may be quite different to the expectation.
- *The need to understand the impact of events that have not happened (yet).* Environmental research

very often concerns developing scenarios for change and understanding the impacts of these scenarios on systems upon which humans depend. These changes may be developmental such as the building of houses, industrial units, bridges, ports or golf courses and thus require environmental impact assessments (EIAs). Alternatively, they may be more abstract events such as climate change or land use and cover change (LUCC). In either case where models have been developed on the basis of process understanding or a knowledge of the response of similar systems to similar or analogous change, they are often used as a means of understanding the impact of expected events.

- *The need to understand the impacts **of** human behaviour.* With global human populations continuing to increase and *per capita* resource use high and increasing in the developed world and low but increasing in much of the developing world, the need to achieve renewable and nonrenewable resource use that can be sustained into the distant future becomes more and more pressing. Better understanding the impacts of human resource use (fishing, forestry, hunting, agriculture, mining) on the environment and its ability to sustain these resources is thus an increasing thrust of environmental research. Models, for many of the reasons outlined above, are often employed to investigate the enhancement and degradation of resources through human impact.
- *The need to understand the impacts **on** human behaviour.* With human population levels so high and concentrated and with *per capita* resource needs so high and sites of production so disparate from sites of consumption, human society is increasingly sensitive to environmental change. Where environmental change affects resource supply, resource demand or the ease and cost of resource transportation, the impact on human populations is likely to be high. Therefore understanding the nature of variation and change in environmental systems and the feedback of human impacts on the environment to human populations is increasingly important. Environmental science increasingly needs to be a supplier of reliable forecasts and understanding to the world of human health and welfare, development, politics and peacekeeping.

1.2 MODELS WITH A PURPOSE (THE PURPOSE OF MODELLING)

Modelling is thus the canvas of scientists on which they can develop and test ideas, put a number of ideas together and view the outcome, integrate and communicate those ideas to others. Models can play one or more of many roles, but they are usually developed with one or two roles specifically in mind. The type of model built will, in some way, restrict the uses to which the model may be put. The following seven headings outline the purposes to which models are usually put:

1. *As an aid to research*. Models are now a fairly commonplace component of research activities. Through their use in assisting understanding, in simulation, as a virtual laboratory, as an integrator across disciplines and as a product and means of communicating ideas, models are an aid to research activities. Models also facilitate observation and theory. For example, understanding the sensitivity of model output to parameter uncertainty can guide field data-collection activities. Moreover, models allow us to infer information about unmeasurable or expensively measured properties through modelling them from more readily measured variables that are related in some way to the variable of interest.

2. *As a tool for understanding*. Models are a tool for understanding because they allow (indeed, require) abstraction and formalization of the scientific concepts being developed, because they help tie together ideas that would otherwise remain separate and because they allow exploration of the outcomes of particular 'experiments' in the form of parameter changes. In building a model, one is usually forced to think very clearly about the system under investigation and in using a model one usually learns more about the system than was obvious during model construction.

3. *As a tool for simulation and prediction*. Though there are benefits to be gained from building models, their real value becomes apparent when they are extensively used for system simulation and/or prediction. Simulation with models allows one to integrate the effects of simple processes over complex spaces (or complex processes over simple spaces) and to cumulate the effects of those same processes (and their variation) over time. This integration and cumulation can lead to the prediction of system behaviour outside the time or space domain for which data are available. This integration and cumulation are of value in converting a knowledge or hypothesis of process into an understanding of the outcome of this process over time and space – something that is very difficult to pursue objectively without modelling. Models are thus extensively employed in extrapolation beyond measured times and spaces, whether that means prediction (forecasting) or postdiction (hindcasting) or near-term casting (nowcasting) as is common in meteorology and hydrology. Prediction using models is also a commonly

used means of making better any data that we do have through the interpolation of data at points in which we have no samples, for example, through inverse distance weighting (IDW) or kriging techniques. Furthermore, an understanding of processes can help us to model high resolution data from lower resolution data as is common in climate model downscaling and weather generation (Bardossy, 1997; see also Chapters 2 and 19).

4. *As a virtual laboratory*. Models can also be rather inexpensive, low-hazard and space-saving laboratories in which a good understanding of processes can support model experiments. This approach can be particularly important where the building of hardware laboratories (or hardware models) would be too expensive, too hazardous or not possible (in the case of climate-system experiments, for example). Of course, the outcome of any model experiment is only as good as the understanding summarized within the model and thus using models as laboratories can be a risky business compared with hardware experimentation. The more physically based the model (i.e. the more based on proven physical principles), the better in this regard and indeed the most common applications of models as laboratories are in intensively studied fields in which the physics are fairly well understood such as computational fluid dynamics (CFD) or in areas where a laboratory could never be built to do the job (climate-system modelling, global vegetation modelling).

5. *As an integrator within and between disciplines*. As we will see in Chapters 14 and 15, models also have the ability to integrate the work of many research groups into a single working product which can summarize the understanding gained as well as, and sometimes much better than, traditional paper-based publications. Understanding environmental processes at the level of detail required to contribute to the management of changing environments requires intensive specialization by individual scientists at the same time as the need to approach environmental research in an increasingly multidisciplinary way. These two can be quite incompatible. Because of these two requirements, and because of funding pressures in this direction, scientific research is increasingly a collaborative process whereby large grants fund integrated analysis of a particular environmental issue by tens or hundreds of researchers from different scientific fields, departments, institutions, countries and continents working together and having to produce useful and consensus outcomes, sometimes after only three years. This approach of big science is particularly clear if we look at the scientific approach of the large UN conventions: climate change, biological diversity, desertification and is also evident

in the increasingly numerous authorship on individual scientific papers.

Where archaeologists work with hydrologists working with climate scientists working with ecologists and political scientists, the formal language of mathematics and the formal data and syntax requirements of models can provide a very useful language of communication. Where the research group can define a very clear picture of the system under study and its subcomponents, each contributing group can be held responsible for the production of algorithms and data for a number of subcomponents and a team of scientific integrators is charged with the task of making sure all the subcomponents or modules work together at the end of the day. A team of technical integrators is then charged with making this mathematical construct operate in software. In this way the knowledge and data gained by each group are tightly integrated where the worth of the sum becomes much more than the worth of its individual parts.

6. *As a research product.* Just as publications, websites, data sets, inventions and patents are valid research products, so are models, particularly when they can be used by others and thus either provide the basis for further research or act as a tool in practical environmental problem solving or consultancy. Equally, models can carry forward entrenched ideas and can set the agenda for future research, even if the models themselves have not been demonstrated to be sound. The power of models as research products can be seen in the wealth of low cost publicly available models especially on the online repositories of models held at the CAMASE Register of Agro-ecosystems Models (http://www.bib.wau.nl/camase/srch-cms.html) and the Register of Ecological Models (http://eco.wiz.uni-kassel.de/ecobas.html). Furthermore, a year-by-year analysis of the number of English language academic research papers using one prominent, publicly available hydrological model (TOPMODEL) indicates the amount of research that can stem from models. According to ISI, from 1991 to 2002 inclusive, some 143 scientific papers were published using TOPMODEL, amounting to more than 20 per year from 1997 to 2002 (this figure therefore does not include the rash of papers in conference proceedings and other nonpeer-reviewed publications). Models can also be very expensive 'inventions', marketed to specialist markets in government, consultancy and academia, sometimes paying all the research costs required to produce them, often paying part of the costs.

7. *As a means of communicating science and the results of science.* To write up science as an academic paper, in most cases, confines it to a small readership and to fewer still users. To add the science as a component to a working model can increase its use outside the research group that developed it. In this way, models can make science and the results of science more accessible both for research and for education. Models can be much more effective communicators of science because, unlike the written word, they can be interactive and their representation of results is very often graphical or moving graphical. If a picture saves a thousand words, then a movie may save a million and in this way very complex science can be hidden and its outcomes communicated easily (but see the discussion below on the disadvantages in this approach).

Points 1–4 above can be applied to most models while 5, 6 and 7 apply particularly to models that are interface-centred and focused on use by end users who are not the modellers themselves. In environmental science these types of model are usually applied within the context of policy and may be called 'policy models' that can be used by policy advisers during the decision-making or policy-making process. They thus support decisions and could also be called decision-support systems (DSS: see Chapters 14 and 15) and which perform the same task as do policy documents, which communicate research results. The hope is that policy models are capable of doing this better, particularly in the context of the reality of scientific uncertainty compared with the myth that policy can be confidently based on exact, hard science (Sarewitz, 1996). The worry is that the extra uncertainties involved in modelling processes (parameter uncertainty, model uncertainty, prediction uncertainty) mean that although models may be good communicators, what they have to communicate can be rather weak, and, worse still, these weaknesses may not be apparent to the user of the model output who may see them as prophecy, to the detriment of both science and policy. There is no clear boundary between policy models and nonpolicy (scientific models) but policy models or DSS in general tend to focus on 5, 6 and 7 much more than purely scientific models, which are usually only used by the model builder and few others. We will see later how the requirements of research and policy models differ.

1.2.1 Models with a poorly defined purpose

We have seen why modelling is important and how models may be used but before building or using a model we must clearly define its purpose. There are no generic models and models without a purpose are models that will find little use, or worse, if they do find use, they will often be inappropriate for the task at hand. In defining the purpose of a model, one must first clearly define the purpose of the research: what is the problem being addressed?; what are the processes involved?; who are the stakeholders?; what is the physical boundary of the problem and what flows cross it from outside of the modelling domain?; over which timescale should the problem be addressed?; and what are the appropriate levels of detail in time and space?

Here, the research question is the horse and the model the cart. The focus of all activities should be to answer the research question, not necessarily to parameterize or produce a better model. Once the research has been defined, one must ask whether modelling is the appropriate tool to use and, if so, what type of model. Then follows the process of abstracting reality and defining the model itself.

1.3 TYPES OF MODEL

Models are by no means a new tool in the scientists' toolbox. Environmental scientists have used spatial models of populations, environments, infrastructures, geologies and geographies in the form of maps and drawings for as long as science itself. Maps and drawings are abstractions of the *form* of nature in the same way that models are (usually) abstractions of the *process* of nature. Mathematics has its origins in the ancient Orient where it developed as a practical science to enable agriculture and agricultural engineering through the development of a usable calendar, a system of mensuration and tools for surveying and design. With the ancient Greeks, mathematics became more abstract and focused much more on deduction and reasoning. Mathematical models have been developed since the origin of mathematics, but there was a significant increase in modelling activity since the development of calculus by Newton and Leibniz working independently in the second half of the seventeenth century. Cross and Moscardini (1985) define three ages of modelling: (1) the 'Genius Age'; (2) the 'Transitional Age'; and (3) the 'Contemporary Age'. The 'Genius Age' followed the development of calculus and is characterized by the development of models of complex physical phenomena such as those of gravitation by Newton, of electromagnetic waves by Clerk Maxwell (of our own university) and of relativity by Einstein. Modelling in the 'Genius Age' was always limited by the need to produce analytical solutions to the set of equations developed. Cross and Moscardini's 'Transitional Age' was initiated by the availability of mechanical and then electromechanical aids to arithmetic but these devices were expensive, difficult to use

and slow. The development of increasingly inexpensive computer power, a bank of numerical techniques that can yield accurate solutions to most equation sets and the softening of the human–computer communication barrier through the development of personal computers (PCs) and high-level programming languages have moved the 'Transitional Age' into the 'Contemporary Age'. Thus, modern numerical computer models can be seen rather simply as the continuation of the relationship between science and mathematics with the greater sophistication afforded by advances in the power of computer input, processing and display. No-one knows what will come next but we make some educated guesses in the last chapters of this book.

Models can be classified hierarchically. The two top-level model types are the *mathematical* models and the physical or *hardware* models (not to be confused with physically based, mathematical models). Hardware models are scaled-down versions of real-world situations and are used where mathematical models would be too complex, too uncertain or not possible because of lack of knowledge. Examples include laboratory channel flumes, wind tunnels, free atmosphere CO_2 enrichment apparatus, rhizotrons and lysimeters, and the BIOSPHERE 2 laboratory (Cohn, 2002). Many instruments are also hardware models which allow control of some environmental conditions and the measurement of the response of some system to these controls. The Parkinson leaf chamber which forms the basis of most leaf photosynthesis systems is a good example. The chamber is a small chamber (usually $5\,cm^3$) which is clamped onto a leaf *in vivo* and controls the input humidity and CO_2 concentration and measures the output of the same so that photosynthesis and transpiration can be measured.

Hardware models are usually small-scale compared with the systems which they simulate and their results may be prone to uncertainty resulting from scale effects which have to be balanced against the increased cost of larger-scale hardware models. Hardware models are also expensive. The 1.27 ha BIOSPHERE 2 (BIOSPHERE 1 is, apparently, the Earth) experiment in the Santa Catalina Hills, near Oracle, Arizona, USA, is now used to test how tropical forest, ocean and desert ecosystems might respond to rising CO_2 and climate change and cost some US$150 million (UK-based readers may want to visit the Eden project near St Austell, Cornwall, to get an impression of a similar physical model [www.edenproject.com]). Most physical models are considerably less ambitious and thus much cheaper but this does, nevertheless, give an indication of the kind of costs that are required to 'simulate' ecosystems in hardware. Hardware models give a degree of control on the systems under investigation, in the case of BIO-SPHERE 2, CO_2, water and nutrient inputs can be controlled. There is, however, always the problem that the hardware representation of a process is only as good as our understanding of that process and our ability to replicate it: BIOSPHERE 2 cannot simulate storms with high winds or damaging downpours or the diversity of soil complexes that exist in the rainforest which it simulates. This may render the BIOSPHERE 2 rainforest response to climate change rather simplistic compared with the field reality. Furthermore, the structure of the hardware model may also interfere with natural processes in the environment, for example, the glass windows of BIO-SPHERE 2 cut out some 50% of the incoming light. Because of the cost involved, it is also usually difficult to replicate hardware experiments: there is only one BIOSPHERE 2 and laboratories usually only have one wind tunnel or channel flume. Nevertheless hardware models couple the scientific rigour of observation with the controllability of mathematical modelling. For many applications, it is only logistic difficulties and cost which keep them as a relatively uncommon approach to modelling: physical models require a great deal more set-up and maintenance costs than software and data.

Mathematical models are much more common and represent states and rates of change according to formally expressed mathematical rules. Mathematical models can range from simple equations through to complex software codes applying many equations and rules over time and space discretizations. One can further define mathematical models into different types but most models are actually mixtures of many types or are transitional between types. One might separate mathematical models into empirical, conceptual or physically based:

- Empirical models describe observed behaviour between variables on the basis of observations alone and say nothing of process. They are usually the simplest mathematical function, which adequately fits the observed relationship between variables. No physical laws or assumptions about the relationships between variables are required. Empirical models have high predictive power but low explanatory depth, they are thus rather specific to the conditions under which data were collected and cannot be generalized easily for application to other conditions (i.e. other catchments, other forests, other latitudes).
- Conceptual models explain the same behaviour on the basis of preconceived notions of how the system works in addition to the parameter values, which describe the observed relationship between the variables. A conceptual model of hillslope hydrology may separately model processes of surface runoff, subsurface matrix quickflow and subsurface pipeflow (see Chapter 4). While all these models are empirical, their separation incorporates some process understanding. Conceptual models have slightly greater explanatory depth but are as nongeneralizable as the empirical models which make them up.
- Physically based models should be derived deductively from established physical principles and produce results that are consistent with observations (Beven, 2002) but in reality physically based models often do one of these but rarely both. In general use, there is a continuum of models that falls broadly under the heading of physically based, but that might include some level of empirical generalization in order to allow them to operate at an appropriate environmental scale, or to fill gaps where the physics is not known. *Process* models emphasize the importance of the processes transforming input to output. In some respects, this tradition may arise from the link between the study of process and form within the discipline of geomorphology. Similarly, ecologists often talk of *mechanistic* models. It may be that the choice of terminology (relating to physics or mechanics) represents a desire to suggest the science underlying the model has a sound basis. Physically based models tend to have good explanatory depth (i.e. it is possible to interrogate the model to find out exactly why, in process terms, an outcome is as it is). On the other hand, physically based models are characterized by low predictive power: they often do not agree with observations. This lack of agreement often demonstrates a poor understanding of the physics of the system. These models thus often need to be calibrated against observations (see Chapter 4, for example). One has to think carefully about the explanatory depth of a model that does not replicate the observed reality well (Beven, 2001). Where they are not highly calibrated to observed data *and* if they are appropriately and flexibly structured, physically based models offer a greater degree of generality than empirical models.

According to the level of process detail and understanding within the model, it may be termed black box or white box. In a black box model, only the input and output are known and no details on the processes which transform input to output are specified, and the transformation is simulated as a parameter or parameters defining the relationship of output to input. On the other hand, in a white box model, all elements of the physical processes transforming input to output are known and specified. There are very few systems for which white box models can be built and so most models, being a mixture of physically based and empirical approaches, fall in between white and black to form various shades of grey boxes. Empirical models are usually closer to the black box end of the spectrum while physically based models fall in between this and the white box, depending on their detail and the extent to which they are calibrated to observed data.

There are no universally accepted typologies of models and, given the diversity of approaches apparent in even a single model code in the multi-process models, which are increasingly common, there is little point in specifying one. Nevertheless, it is useful to understand the properties according to which models may be classified. We have already discussed the different types of model (empirical, conceptual, physically based). Models can be further subdivided according to how the equations are integrated (either analytically solving the model equations as differential equations or numerically solving them within a computer as difference equations, see this chapter). Further subdivision can take place according to the mathematics of the model, for example, whether the equations are deterministic, that is, a single set of inputs always produces one (and the same) output. In the alternative, stochastic approach, a single set of inputs can produce very different outputs according to some random processes within the model. Up to this point most models are still mixtures of many of these types, though two further properties are still to be specified. Models are of different spatial types. Lumped models simulate a (potentially) spatially heterogeneous environment as a single – lumped – value. Semi-distributed models may have multiple lumps representing clearly identifiable units such as catchments. Distributed models break space into discrete units, usually square cells (rasters) or triangular irregular networks (TINs, e.g. Goodrich *et al.*, 1991) or irregular objects. The spatial realm of a model may be one-dimensional, two-dimensional (sometimes within the context of a geographical information system or GIS) and, sometimes, three-dimensional. All models are lumped at the scale of the cell or triangle and because the sophistication of modelling techniques is way ahead of the sophistication of measurement techniques, data limitations mean that most distributed models use lumped data. Finally, one has to consider the manner in which the model handles time. Static models exclude time whereas dynamic ones include it explicitly. A summary of the potential means of classifying models is given below:

Conceptual type: empirical, conceptual, physically based or mixed
Integration type: analytical, numerical or mixed
Mathematical type: deterministic or stochastic or mixed
Spatial type: lumped, semi-distributed, distributed, GIS, 2D, 3D or mixed
Temporal type: static, dynamic or mixed.

1.4 MODEL STRUCTURE AND FORMULATION

1.4.1 Systems

Contemporary mathematical modelling owes much to systems thinking and systems analysis. A system is a set of inter-related components and the relationships between them. Systems analysis is the 'study of the composition and functioning of systems' (Huggett, 1980). In practice, systems analysis involves the breaking down or modularization of complexity into simple manageable subsystems connected by flows of causality, matter, energy or information. The purpose of systems analysis is to make complex systems more easily understood. Systems usually comprise of *compartments* or *stores* that represent quantities such as height (m), mass (kg), volume (m^3), temperature ($^{\circ}$C), annual evaporation (mm) and which are added to or subtracted from by flows or fluxes such as height increment (m a^{-1}), weight gain (kg a^{-1}), volume increment (m^3 a^{-1}) and evaporation (mm month^{-1}). Further details on systems analysis can be found in Chorley and Kennedy (1971), Huggett (1980) and Hardisty *et al.* (1993).

In modelling a *variable* is a value that changes freely in time and space (a compartment or flow) and a state variable is one which represents a state (compartment). A *constant* is an entity that does not vary with the system under study, for example, acceleration due to gravity is a constant in most Earth-based environmental models (but not in geophysics models looking at gravitational anomalies, for example). A *parameter* is a value which is constant in the case concerned but may vary from case to case where a case can represent a different model run or different grid cells or objects within the same model.

1.4.2 Assumptions

In order to abstract a model from reality, a set of assumptions has to be made. Some of these assumptions will be wrong and known to be so but are necessary for the process of abstraction. The key to successful modelling is to know which assumptions are likely to be wrong and to ensure that they are not important for the purpose for which the model is intended. Further, one should only use the model for that purpose and ensure that no-one else uses the model for purposes which render incorrect assumptions significant or correct assumptions invalid. The value of a model depends totally on the validity and scope of these assumptions. These assumptions must be well understood and explicitly stated with reference to the conditions under which they are valid and, more importantly, the conditions under which they are invalidated. Abstraction should always be guided by the principle of parsimony. Perrin *et al.* (2001) indicate that simple models (with few optimized parameters) can achieve almost as high a level of performance as more complex models in terms of simulating their target variable. Although the addition of model parameters and more detailed process description may have benefits from a theoretical point of view, they are unlikely to add greatly to model predictive capability even if substantial data resources are available to keep the uncertainty in these parameters to a minimum (de Wit and Pebesma, 2001). The greater the number of parameters, the greater the likelihood that they will be cross-correlated and that each extra parameter will add relatively little to the explanation (van der Perk, 1997). Perrin *et al.* (2001) concur with Steefel and van Cappellen (1998) who indicate, for models with equal performance, that the best model is the simplest one. Simplicity must be strived for, but not at the cost of model performance. In this way building models is best achieved by starting with the simplest possible structure and gradually and accurately increasing the complexity as needed to improve model performance (see Nash and Sutcliffe, 1970). Figures 11.3–11.5 (in Chapter 11) indicate the manner in which model performance, model complexity, the costs (in time and resources) of model building and the uncertainty of results can be related.

1.4.3 Setting the boundaries

We have now defined the problem and agreed that modelling is part of the solution. Further, we know what the purpose of the model is and can thus define what kind of model is appropriate. The next task is to begin building the model. We must first set the boundaries in time and space to identify which times and spaces will be modelled and which must be supplied as data. We call these data the boundary conditions for data, representing processes outside the spatial domain of the model and the initial conditions for data, representing processes internal to the model spatial domain but external (before) the model temporal domain. Model results, from those of planetary rotation (Del Genio, 1996), through river flooding (e.g. Bates and Anderson,

1996) to moth distributions (Wilder, 2001) are usually sensitive to the initial and boundary conditions so these must be carefully specified. In the case of a general circulation model of the atmosphere (GCM), the boundary is usually the top of the atmosphere and thus one of the variables which must be supplied as a boundary condition, because it is not modelled within the GCM, is the incoming solar radiation flux. It is important to mention that boundary conditions may exist outside the conceptual space of a model even if they are inside the physical space of the same. For example, until recently, global vegetation cover was not simulated within GCMs and thus, despite the fact that it resides within the spatial domain of GCMs, it had to be specified as a boundary condition for all time steps (and also as an initial condition for the first time step). Nowadays many GCMs incorporate an interactive vegetation modelling scheme so this is no longer necessary. GCMs have to be supplied with an initial condition sea surface temperature (SST) field for each sea surface grid cell (GCMs are usually 3D raster distributed) for time zero after which the model will calculate the SST for each timestep.

Let us consider a simple model of soil erosion. The objective of the model is for us to understand more of the relative importance of climatic, landscape and plant factors on soil erosion. Specifically, we will simulate wash erosion (E, variable over space and time) which is simulated on the basis of runoff (Q, variable over space and time), slope gradient (S, variable over space constant over time), vegetation cover (V, state variable over space and time) and the erodibility (K, variable over space constant in time) of the soil and three parameters, m, n and i. Wash erosion is soil erosion by water running across the land surface and, after Thornes (1990) the model can be expressed as:

$$E = kQ^m S^n e^{-iV} \tag{1.1}$$

where
E = erosion (mm month^{-1})
k = soil erodibility
Q = overland flow (mm month^{-1})
m = flow power coefficient (1.66)
S = tangent of slope (mm^{-1})
n = slope constant (2.0)
V = vegetation cover (%)
i = vegetation erosion exponential function (dimensionless).

In this way, soil erosion is a function of the erodibility of the soil (usually controlled by its organic matter content, structure, texture and moisture), the 'stream power' of runoff (Q^m), the slope angle effect (S^n) and the protection afforded by vegetation cover ($e^{-0.07V}$). Let us say that it is a distributed model running over 25-metre raster cells and at a time resolution of one month for 50 years. Erosion and runoff are fluxes (flows), vegetation, slope gradient and erodibility are states (compartments). Since the model is of soil erosion, it does not simulate vegetation growth nor runoff generation, hence these are outside the boundary of the model and they must be specified for each timestep as a boundary condition. No initial conditions need be specified since the only state variable, vegetation cover, is already specified as a boundary condition. If we were to calculate the change in soil thickness (Z, state variable) according to erosion, then we must specify an initial condition for Z.

1.4.4 Conceptualizing the system

The understanding gained from being forced to rationalize one's conceptual view of a process or system and quantify the influence of each major factor is often the single most important benefit of building a mathematical model (Cross and Moscardini, 1985). If the model results match reality, then this conceptual view is supported and, if not, then the conceptualization may be, partly or wholly, at fault and should be reviewed. Different people will produce quite different conceptualizations of the same system depending upon their own background and experience: to a climate scientist a forest is a surface cover interacting with the atmosphere and land surface and affecting processes such as the CO_2 concentration of the atmosphere and the energy and water fluxes across the atmospheric boundary layer. To a forester a forest is a mass of wood-producing trees of different ages, sizes and monetary values. Cross and Moscardini (1985) specify five stages of mathematical modelling: problem identification, gestation, model building, simulation and pay-off. These stages are taken sequentially, although one may move back to earlier stages at any time as needed.

Problem identification is a fairly obvious first stage. If the problem is not properly identified, then it will not be possible to arrive at a solution through modelling or any other means. The gestation stage is an important, though often neglected, stage consisting of the gathering of background information, amassing an understanding of the system under investigation, separating relevant information from irrelevant and becoming familiar with the whole context of the problem. The two substages of gestation may be considered as modularization and reviewing the science. Modularization

breaks the problem down into solvable chunks in the manner of systems analysis. A clear mental map of the processes involved will help very much in the separation of big complex processes into families of small and solvable self-contained modules. Once modularized, the existing science and understanding of each process must be reviewed, allowing abstraction of the relevant from the irrelevant. On departing from this stage the modeller should have a good conceptual understanding of the problem, its context and how it will be solved. This is where the process of abstraction is most important and the modeller's intuition as to what is and is not important will be most valuable. Having devised an acceptable conceptual framework and a suitable data subset, the formulation of the mathematical model is usually fairly straightforward. The process of model building incorporates a further three substages: developing the modules, testing the modules and verifying the modules. Developing the modules will often involve some re-use of existing models, some routine model development and some flashes of inspiration. It is important that, even at this early stage, the modules are tested so that the solution developed has a reasonable chance of producing consistent results and not, for example, negative masses, humidities or heights. In addition to defining the model itself, one will also have to, at this stage, give some thought to the method of solution of the model equations. Before detailing the numerical methods available for this, we will outline in some detail the practical aspects of putting a simple model together.

1.4.5 Model building

Model building may consist of stringing together sets of equations to which an analytical solution will be derived, but more likely these days it will involve compartmentalization of the problem and its specification as either compartments and flows within a graphical model building environment such as STELLA (http://hps-inc.com), VENSIM (http://www.vensim.com), Power-Sim (http://www.powersim.com), ModelMaker (http://www.cherwell.com), SIMULINK, the graphical modelling environment of MATLAB or SIMILE (http://www.simulistics.com) or as routines and procedures in a high level computer programming language such as BASIC, FORTRAN, Pascal, C++ or Java, or a custom modelling language such as PCRASTER (http://www.pcraster.nl).

Some modellers prefer to build their models graphically by adding compartments and flows, linking them with dependencies and entering the appropriate equations into the relevant compartments, flows or variables. Indeed, this approach is fairly close to systems analysis and the way that many non-programmers conceptualize their models. Others, usually those who are well initiated in the art of computer programming, prefer to construct their models in code, or indeed even in the pseudo-code which modern spreadsheet programs allow for the solution of equations as reviewed extensively by Hardisty *et al.* (1994). By way of an introduction to model building let us look back on the soil-erosion equation we introduced a few pages back and examine the manner in which this could be constructed in (a) a graphical model building program, in this case SIMILE which is produced by the Edinburgh-based simulistics.com and is currently distributed free of charge for educational use from http://www.simulistics.com; (b) a spreadsheet program, in this case Microsoft Excel, available from http://www.microsoft.com; and (c) a spatial model building language called PCRASTER produced by the University of Utrecht in The Netherlands and distributed free of charge for educational use from http://www.pcraster.nl. For simplicity we will keep the model simple and nonspatial in the SIMILE and Excel implementation and make it spatial in the PCRASTER implementation.

Though the graphical user interface (GUI) and syntax of these specific software tools will not be identical to any others that you may find, the basic principles of working with them will be similar and so this exercise should also prepare you for work using other tools. There is not space here to enter into the complexities of high-level programming syntax but suffice to say that, for a modeller, knowing a programming language – and they all share the same basic constructs – is a very useful but not indispensable skill to have mastered. Coding in a high-level language does allow more efficient models to be developed in terms of the time they take to run (excluding BASIC and Java from the above list because they interpret commands just like PCRASTER) and the memory they occupy. Coding also sidesteps the limitations of software tools by giving the programmer full, unlimited access to the computer's functions.

1.4.6 Modelling in graphical model-building tools: using SIMILE

SIMILE is a new addition to the suite of graphical systems-modelling tools and is attractive because of its low cost, comprehensive online documentation and powerful modelling capabilities. The construction of a model in SIMILE is achieved first through the specification of compartments and the flow variables between them and subsequently by the specification of the parameters that affect the flow variables and the direction of the influence they have. The SIMILE interface is as shown in Figure 1.1. The modelling canvas is initially blank and the modeller adds compartments, flows, parameters, variables and influences using the toolbar short cuts. These are then given a label and are populated with the relevant values or equations through the equation bar above the model canvas, which becomes active when a compartment, flow or variable is clicked with the computer mouse or other pointing device. The model shown on the canvas is our soil-erosion model but do not look too closely at this yet as we will now attempt to construct it.

The first step is to define the compartment or state variable. In this case of soil erosion, this could be soil thickness so we will label it as such. Note that the symbols that SIMILE produces are standard systems-analysis symbols. Now let's add the flow (Figure 1.2). This is, of course, soil erosion that is a flow out of the soil-thickness compartment and not into it because, for simplicity, we are only simulating erosion and not deposition. The flow is given the label **E** and an influence arrow must be drawn from **E** to the compartment which represents soil thickness since erosion affects soil thickness (note that we use a **bold** symbol to denote the parameter as used in the model formalization compared to the *italicized* symbol when discussing the model form in an equation). Note that until all influences and equations are fully specified, the flow and its associated arrows remain red. On full parameterization all model components will be black in colour. We cannot now specify the equation for erosion until we have fully specified all of the parameters and variables that influence it. We therefore add the variables: **Q, k, s, m, n, i** and **V** (Figure 1.3) and draw influence arrows from each of them to soil erosion (Figure 1.4) since they all affect soil erosion. We can now specify the parameter values of each of **Q, k, s, m, n, i** and **V** either as (fixed) parameters or as variables which are time dependent and either calculated in SIMILE or read from an input file or is the outcome of another equation or submodel in SIMILE. To keep things simple we will enter these values as constants, $Q = 100$ (mm month^{-1}), $k = 0.2$, $s = 0.5$, $m = 1.66$, $n = 2.0$, $i = -0.07$ and $V = 30(\%)$. Finally, we can click on the flow, **E** and enter the equation (from Equation 1.1), which determines the value of **E** as a function of **Q, k, s, m, n, i, V** (see the equation bar in Figure 1.1). Note that in the SIMILE syntax we only enter the right-hand side of the equals sign in any equation and so Equation 1.1

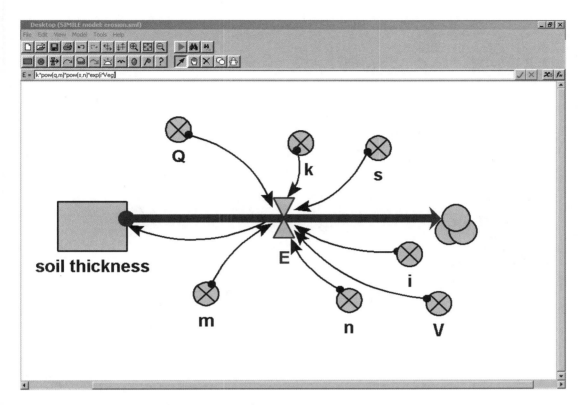

Figure 1.1   Interface of the SIMILE modelling software

Figure 1.2   Adding the flow component to the SIMILE model developed in the text

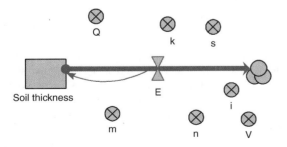

Figure 1.3   The SIMILE erosion model with the variables (**Q**, **k**, **s**, **m**, **n**, **i**, **V**) added

becomes k*pow(Q,m)*pow(s,n)*exp(i*V) with pow(**x**) representing to the power of **x** and exp(**x**) representing

e (the base of natural logarithms) to the power of **x**. We must also specify the initial condition for soil thickness, in mm because **E** is also in mm. Let's say the initial thickness is 10 m (10000 mm). All that remains is to build or compile the model allowing SIMILE to convert this graphic representation to a pseudo-code which is then interpreted into machine code at runtime.

To run the model we need to specify the number and frequency of timesteps in the runtime environment and then run the model. Output is logged for each timestep and the user can examine the model output variable by variable, graphically or as text. If checked and working, this model can be wrapped and used as a component of a larger, more complex model. Many of the graphical modelling tools also have helpers or wizards for performing model calibration, optimization and sensitivity analysis. The great advantage of graphical model-building tools is the rapidity with which they can be learned and the ease with which even very complex systems can be represented. Their disadvantages are that they are generally expensive and it can be rather difficult to do more advanced modelling since they are

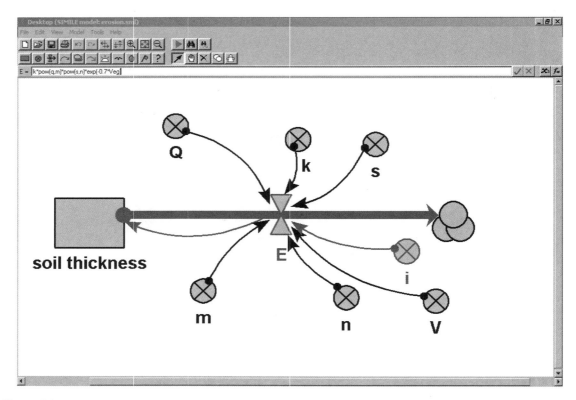

Figure 1.4   Influences between the variables and the flow for the SIMILE soil-erosion model

rather more constrained than modelling in code. If you want to look into it further, the online help and tutorial files of SIMILE provide much greater detail on the more powerful aspects of its functionality including its ability to deal with many (spatial) instances of a model representing for example, individual hillslopes or plants or catchments. You can download this erosion model from the companion website for this book.

1.4.7 Modelling in spreadsheets: using Excel

A very detailed training for modelling in Excel is given in Hardisty *et al.* (1994), so we only provide a simple example here. To build the same soil-erosion model in a spreadsheet such as Microsoft Excel requires the specification of the time interval and number of timesteps upfront since these will be entered directly into the first column of the spreadsheet. Open the spreadsheet and label column **A** as **Timestep** at position **A:2** in the spreadsheet (column A, row 2). Then populate the next 30 rows of column **A** with the numbers 1 to 30. This can be easily achieved by adding the number 1 to **A:3**. At position **A:4** type the following equation: **=A3+1**

and press enter. The spreadsheet will now calculate the results of this equation which will appear as the number in **A:4**. We can now highlight cells **A:4** through to **A:32** with our pointing device (left click and drag down) and go to the menu item **Edit** and the **Fill** and then **Down** (or Ctrl-D) to copy this equation to all the cells in the highlighted range. Note that in each cell, the cell identifier to the left of the plus sign in the equation (which for the active cell can be viewed in the equation bar of Excel) is changed to represent the cell above it so that in **A:4** it is **A:3** but in **A:5** it is **A:4** and so on. This is relative referencing which is a very useful feature of Excel as far as modelling is concerned because it allows us to define variables, i.e. parameters which change with time and time, which are usually represented along the rows of the spreadsheet or even on another spreadsheet. Model instances representing different locations or objects are usually represented across the columns of the spreadsheet, as are the different compartment, flow and variable equations.

If we want to avoid relative referencing in order to specify a constant or parameter (rather than a variable), we can enter it into a particular cell and use absolute referencing to access it from elsewhere on the spreadsheet. In absolute referencing a $ must be placed in front of the identifier which we do not want to change when the equation is filled down or across. To specify a parameter which remains constant in time (down the spreadsheet) we would place the $ before the row identifier, for example, =A$3+1, to specify a parameter which remains constant across a number of model instances (across the spreadsheet) we place the $ before the column identifier, =$A3+1. To specify a constant that changes nowhere on the spreadsheet, then we use for example =A3+1. This form always refers to a single position on the spreadsheet, whenever the equation is filled to.

Going back to our soil-erosion model, we have a number of model parameters which will not change in time or space for this implementation and these can be sensibly placed in row 1, which we have left uncluttered for this purpose. We can label each parameter using the cell to the left of it and enter its value

directly in the cell. So let's label **Q**, **k**, **s**, **m**, **n**, **i** and **V** and give them appropriate values, Q = 100 (mm month^{-1}), k = 0.2, s = 0.5, m = 1.66, n = 2.0, i = −0.07 and V = 30(%). See Figure 1.5 for the layout. Now that we have specified all of the parameters we can enter the soil erosion equation in cell **B:3** (we should place the label Soil Erosion (mm month^{-1}) in **B:2**). We will use absolute referencing to ensure that our parameters are properly referenced for all timesteps. In column **D** we can now specify soil thickness in metres so let us label **C:2** with 'Soil Thickness (m)' and specify an initial condition for soil thickness, of say 10 m, in cell **D:2**. In the syntax of Excel Equation 1.1 becomes =E$1*(C$1^I$1)*(G$1^K$1)*EXP(M$1*O$1) where E$1 holds the value of **k**, C$1 holds the value of **Q**, I$1 holds the value of **m**, G$1 holds the value of **s**, K$1 holds the value of **n**, M$1 holds the value of **i** and O$1 holds the value of **V**. In Excel a caret (^) represents 'to the power of' and EXP(**x**) represents **e** (the base of natural logarithms) to the power of **x**. On pressing enter, this equation will produce a value of

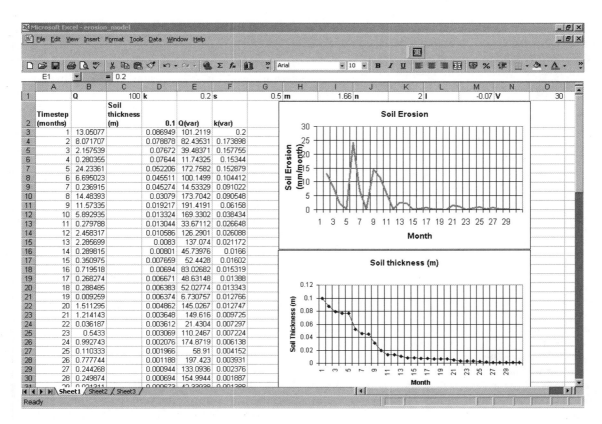

Figure 1.5 Layout of the soil-erosion model in Excel

12.7923 (mm month^{-1}) of erosion. If we fill this down all 30 months then we find that the value remains the same for all months since no inputs vary over time. We can now update the soil thickness for each timestep by removing the soil eroded each timestep from the soil thickness at the beginning of that timestep. To do this type **=D2-(B3/1000)** into **D:3**, where **D:2** is the initial soil thickness and **B:3** is the soil erosion during the first timestep. This produces a soil thickness at the end of the first month of 9.987 metres. Fill this down all 30 months and you will see the soil thickness after 30 months has fallen to 9.616 metres. Note that, because you have used relative referencing here, each month the new soil thickness is arrived at by taking the soil erosion for the month from the soil thickness at the end of the previous month (the row above). Why not now add a couple of line graphs indicating the evolution of soil erosion and soil thickness over time as in Figure 1.5? So, we have a model but not a very interesting one. Soil erosion is a constant 12.79 mm month^{-1} and as a result soil thickness decreases constantly over time (note that for illustration purposes this rate is very high and would rapidly lead to total soil impoverishment unless it were replaced by weathering at least at the same rate). We can now change the values of the model parameters to view the effect on soil erosion and soil thickness. Changing their value causes a shift in the magnitude of soil erosion and soil thickness but no change in its temporal pattern.

We might make the model a little more interesting by specifying **Q** as a variable over time. We can do this by creating a new column of Qs for each timestep. Put the label 'Qvar' in cell **E:2** and in E:3 type =rand()*100 and fill this down to **E:32**. This will generate a random Q between 0 and 100 for each timestep. Now go back to the erosion equation in **B:3** and replace the reference to C$1, the old Q with a relative reference to your new **Qvar** at **E:3**. Highlight **B3:B32** and fill the equation down. You will see immediately from the updated graphs that erosion is now much more variable in response to the changing Q and, as a result, soil thickness changes at different rates from month to month. We can also add a feedback process to the model by decreasing the erodibility as soil erosion proceeds. This feedback reflects the field observation that soil erosion takes away the fine particles first and leads to the armouring of soils thus reducing further erosion. To do this we will first reduce the soil thickness to 0.5 metres at **D:2**. We will then specify a new variable **k** in column **F**. Label **F:2** with the label 'k(var)' and in **F:3** add **=E$1*(D2/D$2)**. This modification will ensure that as soil thickness reaches zero so will erodibility, which is handy because it will also ensure that your model

does not produce negative erosion if left to run for long enough. Erodibility will decrease linearly with the ratio of current to original soil thickness. It is important to note that the erodibility at time t is calculated on the basis on the soil thickness at the end of timestep $t - 1$. For the first timestep that means that erodibility is calculated on the basis of the initial soil thickness. If this were not the case, then when we come to update soil erosion at **B:3** by changing the reference to **k** from **E$1** to **F:3**, we would produce a circularity because soil erosion at t_1 would depend upon erodibility at time t_1 which would, in turn, depend upon soil erosion at time t_1 and so the calculations would be circular and could not be solved using Excel (see the section on iterative methods below). When you have changed the reference to **k** at **B:3** from **E$1** to **F:3** and filled this down to **B:32**, you should notice that the rate of soil erosion, though still very much affected by the random rate of runoff, declines over time in response to the declining erodibility. This will be particularly apparent if you extend the time series by filling down **A32** through to **F32** until they reach **A92** through to **F92** (you will have to re-draw your graphs to see this). On the other hand, you could simply increase the lid on the random runoff generation to 200 at cell **E:3** and fill this down. The increased rate of erosion will mean that the negative feedback of decreased erodibility kicks in sooner. So now the model is showing some interesting dynamics, you could go on to make the vegetation cover respond to soil thickness and/or make the slope angle respond to erosion. You now have the basics of modelling in spreadsheets.

The great advantage of spreadsheet modelling is afforded by the ease with which equations can be entered and basic models created and integrated with data, which usually reside in spreadsheets anyway. Furthermore, modelling in spreadsheets allows very rapid production and review of results and very rapid analysis of model sensitivity to parameter changes. It is therefore very useful, at least at the exploratory stage of model development. Most computer users will have used spreadsheet software and it is fairly widely available since it is not specialist software like the graphical model building tools or the modelling languages. You can download this model from the companion website for this book.

1.4.8 Modelling in high-level modelling languages: using PCRASTER

'High-level' in this context does not mean that a high level of detailed knowledge is required. High-level computer languages are those which are closest to human language compared with low-level languages which are closer to machine code. High-level languages define a rigid syntax in which the user is able to develop software or a string of calculations without knowing too much about how the computer actually performs the calculations or functions that will be done. Examples of high level computer programming languages include BASIC, Pascal, FORTRAN, C++ and Java.

A number of researchers have designed generic computer modelling languages for the development of spatial and nonspatial models (van Deursen, 1995; Wesseling *et al.*, 1996; Maxwell and Costanza, 1997; Karssenberg *et al.*, 2000). None has been universally accepted and many have not been realized in software. The flexibility required of a generic modelling language is so great that few codes are robust enough to handle all potential applications. PCRASTER (http://www.pcraster.nl) is a good example of a high-level, spatial modelling language. Instead of working with parameters, constants and variables which point to single numbers, PCRASTER parameters, constants and variables point to raster grid maps which may contain categorical data, ordinal values, real numbers, a Boolean (true or false) value or one of a number of more specialist types of data. PCRASTER is a pseudo-code interpreter, which provides the modeller with high-level functions for cartographic modelling (GIS) and dynamic modelling (spatial modelling over time). These functions can be invoked at the MS-DOS command line to perform simple calculations or in model script files for the integration of dynamic spatial models. A number of utilities are available for map and time series display, import and export and map attribute conversion. A PCRASTER model consists of a simple ASCII text file in which commands are listed in the sequence in which calculation is required. PCRASTER reads and interprets this file line by line at runtime to realize the model simulation. A PCRASTER model consists of the following sections: binding, areamap, timer, initial and dynamic. In the binding section the user specifies a set of variable names that will be used in the model script and the corresponding disk file names from which the data can be read or written, in the format:

VariableName=mapname.map; for a variable which points to a single map and always reads or writes to that map.

VariableName=mapname; for a variable which points to a different map each timestep, in the first timestep the map is given the suffix mapname0.001, in the second timestep mapname0.002 and so on, with zeros being used to fill the file name to eight characters and also to fill the unused parts of the file extension.

VariableName=timeseriesname.tss; for a variable which reads from or to columns in a timeseries file.

VariableName=tablename.tbl; for a variable which reads or writes to a data table or matrix.

Note that like many programming and scripting languages PCRASTER is case sensitive. You must be consistent with the usage of lower and upper case in your scripts.

In the **areamap** section the user specifies a clone map of Boolean type that identifies the area for calculations with a TRUE (1) and the area in which no calculations will be performed with a FALSE (0). In the **timer** section the user specifies the initial timestep, the time interval and the number of timesteps. In the **initial** section the user specifies any initial conditions (calculations that will be performed only once at the start of the simulation). The bulk of a PCRASTER model usually falls within the dynamic section which specifies calculations that will be performed in sequence for every timestep. The text below represents a PCRASTER version of the soil-erosion model.

```
# Erosion Model.
binding
#input
Dem=dem.map;
#output
Erosion=eros;
#time series and tables
ErosTss=Erosion.tss;
OflowTss=oflow.tss;

#constants
m=1.66;
n=2.0;
```

```
e=2.718;
Q=100;
i=-0.07;
k=0.2;
V=30;
areamap
clone.map;
timer
1 92 1;
initial
SlopeDeg=scalar(slope(Dem));#slope tangent
Thickness=scalar(10);
dynamic
report Erosion=k*(Q**m)*(SlopeDeg**n)*(e**(i*V);#(mm)
report ErosTss=maptotal(Erosion);
Thickness=Thickness-(Erosion/1000);#(m)
```

Note that in this very simple model the only spatially variable quantity is slope angle (**SlopeDeg**), which is derived from the digital elevation model (Dem) using the PCRASTER **slope**() command which performs a window operation to define the slope on the basis of the change in altitude between each cell and its eight contiguous neighbours (see Chapter 4). All other quantities are defined as constants. Even the soil thickness, which is specified here as an initial condition equal to 100, is constant across the map though, as a state variable, it will vary over time. Equation 1.1 in PCRASTER syntax becomes:

report Erosion= k*(Q**m)*(SlopeDeg**n)

*(e**(i*V); #(mm)

where 'report' is the keyword which ensures that the results of the calculation are written to the output map on disk, ** represents 'to the power of' and * represents 'multiplied by'. All PCRASTER script command lines (except the section headings) must end in a :.. The # (hash) symbol indicates that all material to the right of the hash and on the same line is a comment which is not to be interpreted by PCRASTER. The spatial functionality of PCRASTER is used in the **maptotal** function that, in this case, sums erosion across the whole map for output as a single value per timestep in a time series file. The final line of the code updates the soil thickness according to the erosion which has taken place. It might seem unusual to have a statement such as **Thickness=Thickness-(Erosion/1000);** because the same variable is both sides of the equals sign. This is in fact a fairly common construct in programming and modelling languages and is equivalent to $Thickness_t = Thickness_{t-1} - (Erosion_t/1000)$;

in other words, it allows passage of information between timesteps. In order that soil thickness is never allowed to become negative we should use the syntax **Thickness=max(Thickness-(Erosion/1000),0);**.

Again this is not a particularly interesting model. When supplied with a DEM, in this case for a catchment in Spain, this model will produce a series of 92 maps of erosion, which will vary over space according to slope angle but will be constant in time (Figure 1.6). To make the model more interesting we may, for example, use some of the functionality of PCRASTER to (a) make the spatial distribution of Q more realistic; (b) produce a temporally variable Q; and (c) vary soil thickness according to landscape and provide a negative feedback to soil erosion to represent armouring. In order to distribute Q spatially we must define the local drainage direction (LDD) and then specify a precipitation and an infiltration rate and calculate the difference between them (the runoff) for each timestep. Further, we must accumulate this runoff down the stream network or LDD. The cumulated runoff will represent a much more realistic situation than a uniform runoff. In the initial section we use the PCRASTER function LDDCREATE to calculate the drainage direction for each cell based on the steepest drop to its eight contiguous neighbours (the so-called D8 algorithm; see O'Callaghan and Mark, 1984, or Endreny and Wood, 2001). The syntax is: **Ldd=lddcreate(Dem, elevation, outflowdepth, corevolume, corearea, catchmentprecipitation)** where the five latter parameters are thresholds used in the removal of local pits that can be produced by inaccuracies in the production of the DEM and can be safely set to 1e31 each here since the DEM to be used is

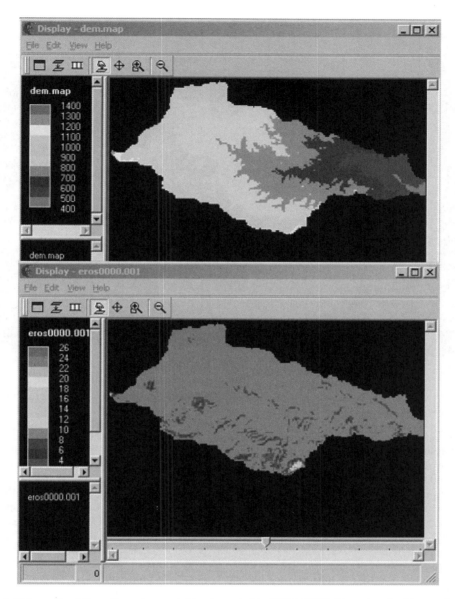

Figure 1.6   Initial results of the soil-erosion model implemented in PCRASTER. Reproduced with permission (see van Deursen 1995)

sound. The resulting LDD map has a drainage direction for each cell and this produces a clear drainage network (Figure 1.7). We can ensure that the map is written to disk by reporting it and adding a reference in the binding section. We can now specify a rainfall and soil infiltration rate, which could be spatially variable but, for simplicity, we will use a lumped approach. The infiltration rate can be added to the

binding section as a constant: **Infil=50;#(mm/month)** but in order to allow temporal variation we will specify the rainfall as a time series which we may have prepared from meteorological station data. In the binding section rainfall is defined as a time series using **Rainfall=rainfall.tss;#(mm/month)** and in the dynamic model section the rainfall is read from the rainfall series using the command **timeinputscalar** (because

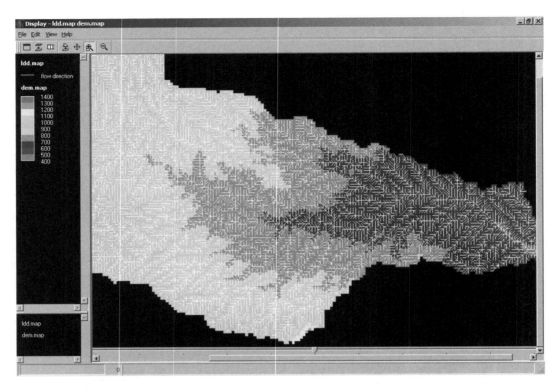

Figure 1.7   Drainage direction for each cell in the PCRASTER soil-erosion model and its corresponding altitude. Reproduced with permission (see van Deursen 1995)

rainfall is a scalar quantity that is a continuous, real, as opposed to integer, number). The command which must be added to the dynamic section before erosion is calculated is **Rainfallmm=timeinputscalar(Rainfall,1);**. The number 1 indicates that the first column of the rainfall file is read and applied to all cells, but a map of nominal values could have equally been supplied here allowing the reading of different columns of the time series file for different cells or areas of the map. As in the spreadsheet, time passes down the rows of time series files. The format for time series files is given in the PCRASTER online documentation. We can now specify the infiltration excess for each cell as: **InfilExcess=max(Rainfallmm-Infil,0);**. The max statement ensures that InfilExcess is never less than zero. In order to cumulate this excess along the flow network we use the **accuflux** command in the form: **report Q=accuflux(Ldd, InfilExcess);**. Now remove (or comment out using #) the reference to Q as a constant and the new Q will be distributed along the local drainage network. We can now specify soil thickness as a spatially variable entity instead of a constant by commenting out its definition as a constant in the initial section and

by specifying it as a function of landscape properties instead. Saulnier *et al.* (1997) model soil thickness as a decreasing function of slope on the basis of greater erosion on steep slopes and greater deposition on shallow ones. Here we will approximate a distribution of soil thickness according to a compound topographic index, the so-called TOPMODEL wetness index (WI), which is the ratio of upslope area to local slope angle and therefore expresses the potential for soil accumulation as well as erosion. We can calculate the WI in the initial section as **WI=accuflux(Ldd,25*25)/SlopeDeg;** where 25 is the width of the individual raster grid cells (in metres). We can scale soil thickness in a simple way as **Thickness=(WI/mapmaximum(WI))*MaxThickness;** where MaxThickness is the maximum potential soil thickness. Finally, we can replicate the armouring algorithm we developed in the spreadsheet model by storing the initial soil thickness using **InitialThickness= Thickness;** in the initial section, removing the definition of erodibility (**k**) as a constant in the binding section and replacing it with the following definition in the dynamic section of the model script: **k = 0.2*(Thickness/Initial-Thickness);**.

We now have a more sophisticated model in which erosion will respond to passing rainfall events over time and spatially to the cumulation of infiltration excess down the spatially variable stream network. Furthermore, the soil erosion will vary with slope gradient and a negative feedback mechanism reduces erosion as it progresses to simulate surface armouring. The model is still very simple and could be made more realistic in many ways (by making the vegetation respond to soil moisture and soil thickness, by making slope angle responsive to erosion, and so on). However, even this simple model can enhance our understanding of the spatial patterns of soil erosion. Figure 1.8 shows (a) the DEM; (b) the calculated slope angle; (c) the natural log of soil erosion for month 4 (72.4 mm rainfall); and (d) the natural log of soil erosion for month 32 (53.8 mm rainfall). The scale used is logarithmic in order that the spatial pattern can be more readily seen. The final PCRASTER code is given in the box:

```
# Erosion Model.
binding
#input
Dem=dem.map;
Rainfall=rainfall.tss;#(mm/month)
#output
Erosion=eros;
Ldd=ldd.map;
#time series and tables
ErosTss=Erosion.tss;
OflowTss=oflow.tss;
#constants
m=1.66;
n=2.0;
e=2.718;
#Q=100;
i=-0.07;
#k=0.2;
V=30;
Infil=50;#(mm/month)
areamap
clone.map;
timer
1 92 1;
initial
SlopeDeg=scalar(slope(Dem));#slope tangent
WI=accuflux(Ldd,25*25)/SlopeDeg;
Thickness=(WI/mapmaximum(WI))*10;
InitialThickness=Thickness;
report Ldd=lddcreate(dem.map,1e31,1e31,1e31,1e31);


dynamic
k=0.2*(Thickness/InitialThickness);
Rainfallmm=timeinputscalar(Rainfall,1);
InfilExcess=max(Rainfallmm-Infil,0);
report Q=accuflux(Ldd,InfilExcess);
Erosion=k*(Q**m)*(SlopeDeg**n)*(e**(i*V));#(mm)
report ErosTss=maptotal(Erosion);
Thickness=max(Thickness-(Erosion/1000),0);#(m)
```

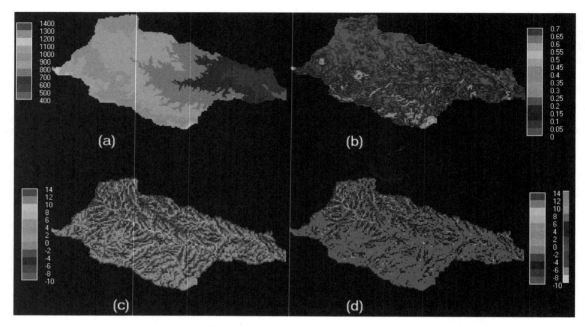

Figure 1.8   Final PCRASTER soil-erosion model results showing: (a) the DEM; (b) the calculated slope angle; (c) the natural log of soil erosion for month 4 (72.4 mm rainfall); and (d) the natural log of soil erosion for month 32 (53.8 mm rainfall). Reproduced with permission (see van Deursen 1995)

You can download this model from the companion website for this book.

The main advantage of scripting languages like PCRASTER is in their power and flexibility and in the rapidity with which even complex operations can be performed through the application of pre-coded high-level functions. The main disadvantage is that such languages require a considerable effort to learn and have steep learning curves, but, once mastered, significant modelling feats are relatively easy. Scripting languages are not, however, as flexible and powerful as high-level computer languages but these in turn can require large programming marathons to achieve the same as a few lines in a well-crafted scripting language. High-level computer languages offer the advantages of flexibility to implement new procedures and the ability to optimize complex procedures to run as quickly as possible. For detailed climate (Chapter 2) or computational fluid dynamics (Chapter 20) modelling, such optimization (perhaps by running a model in parallel on a number of processors at once) is an absolute necessity so that results may be obtained within a reasonable time frame. Even in general applications, a compiled language (e.g. FORTRAN, Pascal or C++) will be noticeably faster than an interpreted one (e.g. PCRASTER, BASIC and usually Java).

So, we have used three modelling approaches to produce essentially the same model. Each approach has its advantages and disadvantages and you will probably have made up your mind already which approach you feel more comfortable with for the kind of modelling that you wish to do. Whichever technique you use to build models, it is critically important to keep focused on the objective and not to fall into the technology trap that forces all of your mental power and effort into jumping the technological hurdles that may be needed to achieve a set of calculations and as a result leaves little energy and time after the model building stage for its testing, validation and application to solving the problem at hand. A simpler model which is thoroughly tested and intelligently applied will always be better than a technically advanced model without testing or sufficient application.

1.4.9 Hints and warnings for model building

Cross and Moscardini (1985) give a series of hints and warnings intended to assist novice modellers in building useful models. This list is expanded here according to our own environmentally biased perspective on modelling.

1. Remember that all models are partial and will never represent the entire system. Models are never finished, they always evolve as your understanding of the system improves (often with thanks to the previous model). In this way all models are wrong (Sterman, 2002)! Nevertheless, it is important to draw some clear lines in model development which represent 'finished' models which can be tested thoroughly and intensively used.

2. Models are usually built for a specific purpose so be careful of yourself or others using them for purposes other than those originally intended. This issue is particularly relevant where models are given a potentially wide user base through distribution via the Internet.

3. Do not fall in love with your model (it will not love you back!).

4. Take all model results with a grain of salt unless those *specific* results have been evaluated in some way.

5. Do not distort reality or the measurement of reality to fit with the model, however convincing your conceptualization of reality appears to be.

6. Reject properly discredited models but learn from their mistakes in the development of new ones.

7. Do not extrapolate beyond the region of validity of your model assumptions, however powerful it makes you feel!

8. Keep ever present the distinction between models and reality. The sensitivity of a model to climate change is not the sensitivity of reality to climate change. If you write about 'model sensitivity' then do not omit the word 'model' (Baker, 2000).

9. Be flexible and willing to modify your model as the need arises. Note that this is *modify* not expand upon.

10. Keep the objectives ever present by continually asking 'What is it that I am trying to do?'

11. As in all aspects of science, keep the model honest and realistic despite any short-term financial or status gain that may seem to accrue from doing the opposite.

1.5 DEVELOPMENT OF NUMERICAL ALGORITHMS

1.5.1 Defining algorithms

An algorithm is the procedure or set of rules that we use to solve a particular problem. Deciding on the algorithm to use in any particular model can be thought about in two stages. The first has already been covered in the previous section – that of conceptualizing the system and developing a mental model of the system. The second is the conversion of this conceptualization to a form that can be calculated. It may be that, at this stage, the problem as formalized is not calculable, and we must return to the former stage and try to re-evaluate and describe the system in question, so that an approximation can be arrived at without compromising the reliability of the model results.

1.5.2 Formalizing environmental systems

Environmental models are focused upon change. This change may be of properties in time or in space. Increasingly, models are being developed where both temporal and spatial variations are evaluated, so we need to have techniques that can assess both of these changes. The branch of mathematics that allows us to assess change is known as calculus, initially derived in the seventeenth century independently by Newton and Leibniz. A basic understanding of calculus is fundamental for all modelling, so we will provide a brief overview here. More detailed information can be found in numerous books, some of the more accessible being Bostock and Chandler (1981), De Sapio (1976), Ferrar (1967), Friedman (1968), Haeussler and Paul (1988), Lang (1978) and Thompson (1946).

Consider the case where we are interested in the transport of coarse sediment through a river. To investigate the process we have tagged rocks with radio transmitters that allow us to record the location of the rocks through time (e.g. Schmidt and Ergenzinger, 1992). In Figure 1.9 we have plotted the position along the river channel of a rock over a period of 24 hours. These results can also be plotted as a graph showing the location in space and time of the rock (Figure 1.9b). This graph can also be used to define the change in position through time. The change in position through time is effectively the velocity of the rock as it moves. Remember that velocity is the change in distance over the time to move that distance. Over the 24 hours, the rock moved a total of 8.57 m, so that its average velocity was $\frac{8.57\,\mathrm{m}}{24\,\mathrm{h}} = 0.36\,\mathrm{m\,h^{-1}}$. Notice, however, that the rock was not moving at a constant velocity through the day, as the graph is not a straight line. We can use the same technique to show how the velocity of the particle varied over time. In the second hour, the particle moved 0.72 m, so its velocity was $0.72\,\mathrm{m\,h^{-1}}$. Fourteen hours later, the velocity was $0.22\,\mathrm{m\,h^{-1}}$ (Figure 1.9c). These velocities are simply the slope of the graph at successive points in time. We now have enough information

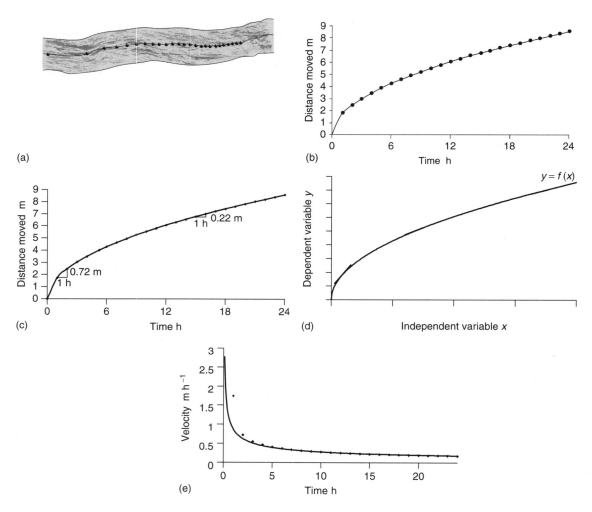

Figure 1.9 (a) Map showing position along a stream channel of a tagged rock over a period of 24 hours; (b) Graph showing the location in time and space of the tagged rock as shown in Figure 1.9a. The line interpolates the position of the rock at times between those measured; (c) Graph showing how the variation in the velocity of the measured rock can be calculated as the local slope of the graph; (d) Graph showing relationship between the two variables as a continuous function. The thick lines show examples of the tangent of the slope at specific points; (e) Graph showing the velocity of the rock particle through time. The line shows the continuous change in velocity calculated using the formula $\frac{dy}{dx} = y' = 0.875x^{-0.5}$, obtained by differentiation, whereas the points show the measured points at hourly intervals. Note the major difference in the early part of the graph, where the velocity is rapidly changing, due to the fact that the hourly measurements are based on an average rather than continuous measurements

to model the movement of the rock at a daily and hourly timestep.

However, if we need to model movements at shorter time intervals, we need to have some method for estimating the velocity of the rock, and thus the slope of the line, at these shorter intervals. To do this, we may be able to use a function that shows us how the two variables are related, either empirically or from physical principles. Typically, this function may be written in the form:

$$y = f(x) \tag{1.2}$$

In this case, we find empirically (for example, using regression analysis) that the relationship has a simple form:

$$y = 1.75x^{0.5} \tag{1.3}$$

where the variable y is the distance moved in metres and x is the time in hours. The velocity of the particle is still the slope at each point on the graph; we can now estimate these slopes continuously from our function, as the tangent to the curve at each successive point (Figure 1.9d). To calculate this tangent, we need to work out the derivative of the function, which has the standard form for power-law relationships such as Equation 1.3:

$$\frac{dy}{dx} = nx^{n-1} \tag{1.4}$$

How this standard form is obtained is shown in Figure 1.10.

The derivative $\frac{dy}{dx}$ (usually said as 'dy by dx') may also be written as y' or $f'(x)$. (The difference between these is essentially that the former is Newton's notation, whereas the latter is that of Leibniz.) Applying this rule to the example of Equation 1.3, we have:

$$\frac{dy}{dx} = y' = 0.875x^{-0.5} \tag{1.5}$$

Thus, we can now calculate the instantaneous velocity of our rock at any point in time. The process of calculating the derivative is known as differentiation. More complex functions require different rules for differentiation, and these can be found in the standard texts cited above and on the book's website.

Environmental models may also commonly use terms relating to acceleration or diffusion. Acceleration is the change in velocity through time, so we can also calculate it using differentiation. This time, we plot velocity against time (Figure 1.9e) and find the tangent at a point by differentiating this curve. Relative to our original measurement, we have calculated:

$$\frac{d\frac{dy}{dx}}{dx} = \frac{d^2y}{dx^2} = f''(x) = y'' \tag{1.6}$$

This term is called the second derivative. The standard form for power-law relationships means that we simply need to repeat the operation in Equation 1.4 twice to calculate the specific value of the second derivative:

$$\frac{d^2y}{dx^2} = n(n-1)x^{n-2} \tag{1.7}$$

Thus, our rock particle is accelerating at a rate of $y'' = -0.5 \times 0.875x^{-1.5} = -0.4375x^{-1.5}\,\mathrm{m\,h^{-1}}$, or in other words it is consistently decelerating.

Thornes (1985, 1988) uses derivatives in modelling the interaction between vegetation and erosion to show cases of stability in the landscape. Given a functional relationship for vegetation change, it is possible to find the conditions where vegetation cover becomes constant. This is simply where $\frac{dV}{dt} = 0$, where V is the vegetation amount and t is time. This example is examined further in Chapter 18. Another use for derivatives is in the process of optimization to find the optimal values of parameters in a model. These optimal values may fall at maximum or minimum points of the graph showing the relationship. At these points, the following conditions for the first and second derivatives will hold:

$$\text{Maximum:} \quad \frac{dy}{dx} = 0 \quad \frac{d^2y}{dx^2} < 0 \tag{1.8a}$$

$$\text{Minimum:} \quad \frac{dy}{dx} = 0 \quad \frac{d^2y}{dx^2} > 0 \tag{1.8b}$$

Care must, however, be taken that the graph is not simply at a local plateau, or inflexion point, where:

$$\text{Inflexion:} \quad \frac{dy}{dx} = 0 \quad \frac{d^2y}{dx^2} = 0. \tag{1.8c}$$

Integration is the opposite process to differentiation. It allows us to calculate the sum of a property over space or time. If we plot the function that represents the relationship we are interested in, then the sum of this property is the area under the curve. To see why this is so, consider a function that relates the change in depth of water along a section of river (Figure 1.11). If the depth (h) is constant along section of length y, then the sum of the values of y is simply $y \cdot h$. In other words, it is the area under water along the section length in which we are interested – in this case represented by a simple rectangle. If, however, the depth increases linearly along the section, then the area under the curve will be a trapezium with area $0.5(h_1 + h_2) \cdot y$, where h_1 is the depth at the start of the reach and h_2 is the depth at the end. We often use this approach to calculate integrals where the shape of the relationship is complex, or known only from data at certain points. The 'trapezium rule' involves dividing the curve or data into a set of known intervals, calculating the area of the trapezium that is a simple representation of each interval, and summing the total area. This approach is best when we know the change between points is linear, or the distance between points is sufficiently small. Note that we are not restricted to integration in one dimension. If we integrated the area of water by the width of the

Deriving the Standard Form for Differential Calculus

Let us take, as an example, the function $y = x^2 + 2x$. In tabular and graphical forms, this function is:

x =	y =
0	0
1	3
2	8
3	15
4	24
5	35
...	...

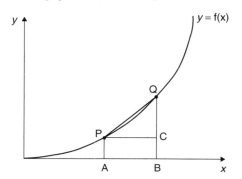

A line joining any two points on the curve is known as a *chord*. In the graph below, the line PQ is a chord.

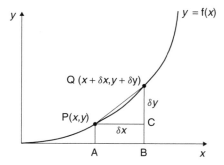

The slope of this line is:

$$\text{gradient of PQ} = \frac{QC}{PC}$$

To define the gradient at a point, let the length of PC in the graph above be h (an arbitrary length). If the point C is (2, 8), then the value of the coordinates of point Q are given by the function, by inserting the x value of Q, which is $(2 + h)$. The y value at Q is then:

$$y = (2 + h)^2 + 2(2 + h)$$

which is $8 + 6h + h^2$. The distance QC is then $6h + h^2$ (the 8 disappears because it is the value of y at C) and the

gradient of the chord PQ is:

$$\text{gradient of PQ} = \frac{QC}{PC} = \frac{6h + h^2}{h} = 6 + h$$

To find the value of the gradient at the point P, we let the value of h become vanishingly small, so that the points P and Q are in the same location. In other words, the value of h approaches zero. This process is known as taking the *limit* and is written: $h \xrightarrow{\text{lim}} 0$. Thus, we have:

$$\text{gradient at P} = h \xrightarrow{\text{lim}} 0 \, (6 + h) = 6$$

In more general terms, if P is any point (x, y), then Q is given by the function as $(x + h, (x + h)^2 + 2(x + h))$, or by expansion of terms $(x + h, x^2 + 2xh + h^2 + 2x + 2h)$ and the gradient of the chord PQ is:

$$\text{gradient of PQ} = \frac{2xh + h^2 + 2h}{h} = 2x + h + 2$$

In this case, the value of y at P (i.e. $x^2 + 2x$) disappears when calculating the distance from Q, leaving the remaining terms (i.e. $2xh + h^2 + 2h$). The gradient at P then becomes:

$$\text{gradient at P} = h \xrightarrow{\text{lim}} 0 \, (2x + h + 2) = 2x + 2$$

which we can verify by substitution, so that if $x = 2$, the gradient at P is 6. The derivative of $y = x^2 + 2x$ is thus demonstrated as being $\dfrac{dy}{dx} = 2x + 2$, which is the same as given by the standard form in the main text (i.e. we apply $\dfrac{dy}{dx} = nx^{n-1}$ to each of the terms which contain an x in the function).

It is more usual to use the incremental notation as illustrated below:

where δx means the increment in x and δy the increment in y. We can then re-write:

$$\text{gradient of PQ} = \frac{QC}{PC} = \frac{\delta y}{\delta x}$$

and by definition:

$$\frac{dy}{dx} = \delta x \xrightarrow{\text{lim}} 0 = \frac{\delta y}{\delta x}$$

Figure 1.10 Derivation of the standard form of differentiation

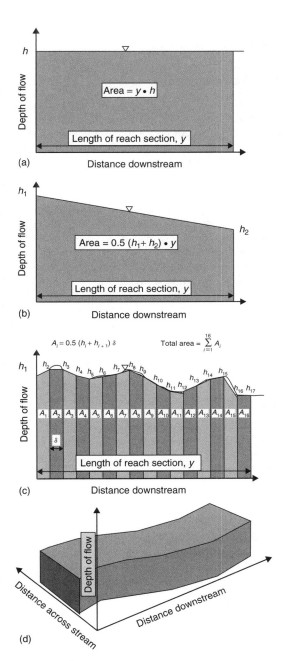

(a)

(b)

(c)

(d)

channel, then we would have a means of calculating the volume of water.

The trapezium rule *approximates* the integral in cases where change is nonlinear. It is also possible to calculate integrals directly if we know the form of the function. The relationship between differentiation and integration can be seen when considering the simple power-law relationship considered above as a standard form. The symbol $\int$ is used to mean 'integral of':

$$\int n x^{n-1}\, dx = x^n + c \qquad (1.9a)$$

or

$$\int x^n\, dx = \frac{1}{n+1} x^{n+1} + c \qquad (1.9b)$$

Thus, practically as well as conceptually, integration is the reverse of differentiation. The value c is known as the constant of integration, and arises because the derivative of a constant is zero. Other standard functions can be integrated by reading tables of derivatives in the reverse direction. The dx in Equation 1.9 simply tells us which variable we are integrating over. It does *not* mean that we multiply by the change in x. The general form of an integral is $\int f(x)\, dx$. In more complicated functions, there may be other variables as well as constants, so it is vital to know which of the variables we are considering.

$\int$ is known more specifically as the *indefinite integral*. It is used to define the general form of the integral we are interested in. Often, we will want to calculate the value of the integral over a specific range of values, as for example in the case above where we were interested in a specific reach of a river, rather than the entire length of the river. In this case, we write $\int_a^b f(x)\, dx$ where a is the starting value of x that we want to consider and b is the end value that we want to consider. To solve a definite integral, first derive the indefinite integral for the relationship. In the case of our rock moving in the river channel, let us consider the sum of its velocities by calculating the integral of Equation 1.4a:

$$\int 0.875 x^{-0.5}\, dx = 0.875 \frac{1}{-0.5+1} x^{-0.5+1} + c$$
$$= 1.75 x^{0.5} \qquad (1.10)$$

Notice two things here: first, the constant of integration has disappeared because by definition the distance moved at the start of the observations (i.e. when $x = 0$) was zero. Second, the result we have obtained is the same as Equation 1.3. In other words, the sum of the velocities of our particle movement is the same as the total distance moved. Obtaining this relationship in reverse should further emphasize the relationship

Figure 1.11 Derivation of integration in one and two dimensions using the example of changing flow depth along a river section: (a) the case where the depth is constant along the reach of interest; (b) linear change of depth along the reach; (c) use of the trapezium rule for a complex function representing change in water depth; and (d) integrating by the width at each point to give the volume of water in the reach

between integration and differentiation. The second stage of calculation involves substituting the start and end values into the indefinite integral. Third, we subtract the starting value from the end value to give us the final answer. Over the whole 24-hour period of observation, the rock therefore moved:

$$\int_0^{24} 0.875x^{-0.5}\, dx = (1.75 \cdot 24^{0.5}) - (1.75 \cdot 0^{0.5})$$

$$= 8.57\,\text{m} \qquad (1.11)$$

which agrees with our initial observations. Confirm for yourself that the rock moved 6.06 m during the first 12 hours and 2.51 m in the second 12 hours (note the total distance moved when calculated this way); between 14 hours and 17 hours 30 minutes it moved 0.77 m. What happens if you calculate the distance moved between hours 0 and 6, 6 and 12, 12 and 18 and 18 and 24? What would happen to the total distance moved if you used smaller intervals to calculate? Note that the constant of integration *always* vanishes when calculating a definite

integral, whether we know its value or not, as we subtract it at *b* from itself at *a*.

1.5.3 Analytical models

It is possible to use developments of the approaches outlined above to provide solutions to specific problems under specific sets of conditions. These we would term analytical solutions. Consider the simple population-growth model (Burden and Faires, 1997):

$$N_{t+1} = N_t + \lambda N_t \qquad (1.12)$$

This model states that the size of the population N at the next timestep we consider (the $t + 1$ subscript) is given by the current population, N_t, plus a growth rate that is a linear function of the current population. The parameter λ represents the net growth rate (number of births minus deaths) over the time interval. Versions of this model are commonly used for simulating vegetation growth and animal (including human) populations (see Chapters 6–8 and 12). How does this model fit

---

To demonstrate that the analytical solution of Equation 1.13 is Equation 1.14, we need to recognize that the population at time t is simply the integral of the changes in population to that point in time, or:

$$\int_0^t \frac{dN}{dt}\, dt \quad \int_0^t \lambda N\, dN$$

If we rearrange this to have similar terms on each side of the equation, we have:

$$\int_0^t \frac{dN}{N}\, dN \quad \int_0^t \lambda\, dt$$

These are both standard integral forms, so that we can evaluate the integrals as:

$$\log_e N \quad c_1 \quad \lambda t \quad c_2$$

where c_1 and c_2 are the respective coefficients of integration. If we move c_1 to the right-hand side and exponentiate to remove the logarithm, this becomes:

$$N \quad e^{\lambda t\ (c_2 - c_1)}$$

We can split the terms on the right-hand side to give:

$$N \quad e^{\lambda t} \quad e^{(c_2 - c_1)}$$

If we set t 0, we can evaluate the meaning of the coefficients of integration. The value of $e^{\lambda t}$ at this point will equal 1 (by definition) so that:

$$N(0) \quad e^{(c_2 - c_1)}$$

so we can see that the terms on the right-hand side are the size of the population at this time. Defining N_0 $N(0)$, we arrive at Equation 1.14:

$$N \quad N_0\, e^{\lambda t}$$

Figure 1.12   Derivation of the analytical solution to the population equation (Equation 1.13)

into what we have just been saying about differential equations? The model is expressing change in the population, so we can also specify how the population is changing. In this case, it is the second part on the right-hand side of Equation 1.11 that represents the change, so that we can also say:

$$\frac{\mathrm{d}N}{\mathrm{d}t} = \lambda N \qquad (1.13)$$

The solution (see Figure 1.12) to this equation is:

$$N_t = N_0 \, e^{\lambda t} \qquad (1.14)$$

where N_0 is the initial population. The solution shows that populations with a linear growth rate will grow in size exponentially through time. In reality, this (Malthusian) assumption only tends to be valid over short time periods because of resource constraints to sustained linear growth rates. We will consider how to make the model more valid below.

However, in many cases, environmental systems may be described by functions for which there are no analytical solutions. In other words, we cannot use algebra to reformulate the problem into a final form. Alternatively, we may be interested in spatial or temporal variability in our initial and/or boundary conditions that cannot be described in a functional form. For example, in rainfall-runoff modelling, we have to deal with rainfall rates that vary through space and time and irregular topography that is not a simple function of distance along a slope. In these cases, we need to develop alternative approaches that allow us to approximate solutions to the systems we are trying to model. These approaches require numerical solutions. Numerical solutions typically involve approximation and iterative attempts to provide an answer to the problem in hand. The use of iterative approaches is considered in more detail below.

1.5.4 Algorithm appropriateness

In most systems, there will be more than one way of representing the system components and their interlinkages. At the most fundamental level, we can contrast black box approaches with grey box or white box approaches. How do we evaluate whether an algorithm is appropriate?

One set of criteria to evaluate for appropriateness is to judge in terms of process representation. If we consider the example of population growth, we have already seen what is probably the simplest possible model, in Equations 1.12 to 1.14. The population is considered as a single, simple unity, and one parameter is used to control how it changes. This parameter is meaningful, in that it represents the difference between births and deaths through time, but assumes that this value is constant. In other words, there is always the same proportion of births and deaths relative to the size of the population. This assumption is reasonable *on average*, but might not be appropriate in all circumstances. We have also noted that the model predicts exponential growth, which is usually only true for short time periods, for example, because increased competition for resources means that continued growth is unsustainable. We might want to account for these limitations by modifying the model as follows:

$$\frac{\mathrm{d}N}{\mathrm{d}t} = \beta \left(1 - \frac{N}{N_{max}}\right) N - \theta N \qquad (1.15)$$

where β is the birth rate in individuals per unit time, θ is the death rate in individuals per unit time and N_{max} is the maximum number of individuals that can exist in the specific environment. N_{max} is often called the carrying capacity of the environment. In the first part of this model, we have essentially added a component, $\left(1 - \frac{N}{N_{max}}\right)$, which says that as the population approaches the carrying capacity, and thus the ratio $\frac{N}{N_{max}}$ approaches one, the birth rate is multiplied by an increasingly small number, and so the rate of increase gets smaller. This effect is illustrated in Figure 1.13. This S-shaped curve is known as the logistic growth curve. Note that in the second part of this model, we have simply assumed that death rates are again a constant proportion of the population. In terms of process representation, we may again feel that this is inadequate, because it fails to deal with different ways in which members of the population may die – for example, due to old age, disease or accidents – and want to add extra components to account for these processes and how they might change with population or resources. These components may be simple functions, or may depend on the population size or density. Alternatively, we may want to account for apparently random elements in these processes by using stochastic functions to represent them (see below).

One assumption we have still incorporated implicitly in this model is that the population grows in an isolated environment, without competition from other populations. For example, it is common to investigate population changes in terms of separate populations of predators and prey. These investigations form the basis of the Lotka–Volterra population models. These models

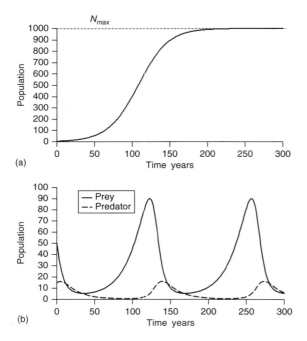

Figure 1.13 (a) Graph showing population simulated using Equation 1.12, with a limitation to a carrying capacity of 1000 individuals. The birth rate β is 0.05 individuals per year in this example; (b) Graph showing population simulated using Equation 1.16, with $\beta_H = 0.05$ individuals per year, $p = 0.01$ individuals per predator per prey per year, $\beta_H = 0.002$ individuals per year per prey eaten per year and $\theta = 0.06$ individuals per year

link the populations of a prey species, H, and a predator species, P, as:

$$\frac{dH}{dt} = \beta_H H - pHP \tag{1.16a}$$

$$\frac{dP}{dt} = \beta_P HP - \theta P \tag{1.16b}$$

where β_H is the birth rate of the prey species, p is the rate of predation, β_P is the birth rate of the predator per prey eaten, and θ is the death rate of the predator species. Figure 1.13b shows an example of the output of this model, showing cycles of increase and decrease of the prey and predator species. Note that in this formulation the death rate of the predator is given (compare Equation 1.15) but death of the prey is only considered to occur through predation. This model is still a simple one that assumes an isolated environment and an interaction of two species, but can still provide complex results.

All of these population models still make the assumption that 'population' is a useful level of generalization. It may, however, be more appropriate to deal with the population number as an *emergent property* of the actions of the individuals that make it up. In other words, we can assess the birth, death and interaction of individuals within an environment, and derive the population as the sum of all remaining individuals. Certainly, this is a more realistic model representation of what is actually happening. Berec (2002) provides an overview of the approach and shows the relationship between the average outputs of individual-based models and those based on population statistics. Pascual and Levin (1999) used a cellular automata approach (see below) to represent the predator–prey model and show how spatial patterns are important in considering outcomes. Bousquet *et al.* (2001) provide a more specific example of the hunting of antelope in Cameroon, showing that spatial patterns of trap sites are important as well as cooperation between hunters. Other overviews can be found in Epstein and Axtell (1996) and Ginot *et al.* (2002).

Another useful example of different process representations is the ways that river flows are modelled. In the simplest case, we might take a highly empirical form that relates the amount of rainfall in the catchment to the bulk or peak outflow from the catchment (see the review of lumped catchment models by Blackie and Eeles, 1985). This approach will allow us to make basic estimations of available water resources or of the potential for flooding. However, it is highly dependent on data availability and only works for a single point in the catchment. If we want to evaluate how flows change locally, for example, to investigate flooding in particular areas, or to understand how the flow dynamics interact with the channel shape, or how they control sediment movement, then clearly this level of simplification is not useful.

Flows are most often simulated using the St Venant approximations (named after the French hydrologist Jean-Claude Barré de St Venant) and first formulated in 1848. These approximations state that water flows are controlled by the continuity (mass balance) equation:

$$\frac{\partial h}{\partial t} + h\frac{\partial u}{\partial x} + u\frac{\partial h}{\partial x} - q = 0 \tag{1.17a}$$

and a conservation of momentum equation made up of the following components:

$$\underset{(1)}{\frac{1}{g}\frac{\partial u}{\partial t}} + \underset{(2)}{\frac{u}{g}\frac{\partial u}{\partial x}} + \underset{(3)}{\frac{\partial h}{\partial x}} - \underset{(4)}{(S - S_f)} = 0 \tag{1.17b}$$

where h is depth of flow (m), u is flow velocity (m s^{-1}), t is time (s), x is distance downstream (m), q is unit lateral inflow (m s^{-1}), g is acceleration due to gravity (m s^{-2}), S is the slope of the channel bed (m m^{-1}) and S_f is the slope of the water slope (also known as the friction slope) (m m^{-1}). If we use all of these components (1–4), then the model is known as a dynamic wave approximation (Graf and Altinakar, 1998; Singh, 1996; Dingman, 1984). Component (1) reflects the inertia of the flow; (2) the gravity effect on the flow; (3) the pressure effect; and (4) the friction effect. However, it may be possible to ignore the inertial term because its value is so small. In this case we omit the component (1) and produce the diffusive wave approximation:

$$\frac{u}{g}\frac{\partial u}{\partial x} + \frac{\partial h}{\partial x} - (S - S_f) = 0 \qquad (1.18a)$$

which, when combined with the continuity equation 1.15a, gives the diffusion-wave approximation:

$$\frac{\partial h}{\partial t} + u\frac{\partial h}{\partial x} - D\frac{\partial^2 h}{\partial x^2} - q = 0 \qquad (1.18b)$$

where D is the diffusion coefficient (equal to $\frac{Q}{2wS}$ where Q is the discharge [m^3 s^{-1}] and w the flow width [m]). This approach may be valid under conditions of negligible convective acceleration. In other cases we can omit the diffusive component of the flow, so that we assume the flows are simply controlled by a single relationship between depth and discharge. In this case, components 2 and 3 become negligible, $S = S_f$ and:

$$\frac{\partial h}{\partial t} + u\frac{\partial h}{\partial x} - q = 0 \qquad (1.19)$$

which is known as the kinematic wave approximation. This approximation is valid particularly in overland flows on hillslopes (see below) and where waves in the flow are of low amplitude relative to their wavelength. Where there are backwater effects, as at tributary junctions or upstream of lakes and estuaries, the assumptions of the kinematic wave approximation break down (Dingman, 1984) so that diffusive or dynamic wave algorithms should be used in these cases. Singh (1996) contains numerous references to and examples of the use of these equations. Further details are also given in Chapter 5.

In all these cases, we have simply made one-dimensional approximations to the flow process. In other words, we assume that it is valid to model average conditions at a series of cross-sections down the channel. If we want to estimate flooding, then this approach may be very reasonable (e.g. Bates and De Roo, 2000). However, we may want to know how conditions change

across a cross-section, or as a continuum along a channel. These approaches would be better for looking at the effects of flows on structures, for example, bridge supports, or on how the flow interacts with the morphology. In these cases, approaches using more complex, computational fluid dynamics (CFD) models will usually be more appropriate. Such models are covered in detail in Chapter 20.

Overall, we can summarize that the process representation is not a single definition, and different criteria will be used depending on our modelling objectives. Both bottom-up and top-down approaches may be useful for modelling the same system in different conditions. We can see how this difference can lead to different levels of complexity in terms of model components, equations representing their interactions, and expectations in terms of outputs.

Second, we might evaluate algorithm appropriateness in terms of computational resources. With the rapidly increasing rate of progress of microprocessor design and reduction of costs of microprocessors, memory and data storage, it is possible to design and implement larger and more complex models on desktop computers. Moore's 'law' of increasing computer power is based on the fact that more transistors can be crammed into successively smaller spaces on integrated circuits, allowing them to run progressively faster. In the past 35 years, transistor density has doubled about once every 18 months, allowing the corresponding exponential increase of model size (number of points simulated) and complexity (Voller and Porté-Agel, 2002). The design of supercomputers and parallel processing, where a number of calculations can be carried out simultaneously, is also proceeding apace. One criterion we might use is the time taken to execute the model. This time will include that required to read in all of the data and parameters required, carry out the calculations in our algorithm, and output all of the results. Algorithms are often evaluated in terms of the number of operations required, as this can often be related to the processing speed (measured in FLOPS or floating-point operations per second – modern machines are capable of megaflops [millions] or gigaflops [thousands of millions]). Modern computers are designed to be able to make calculations on the several memory addresses involved in a floating-point calculation at once.

For example, if we consider the population-growth models from this section, the models based on gross population characteristics have substantially fewer calculations than those based on individuals. Therefore, if either algorithm is appropriate from a process point of view, but we need the calculations to be carried out quickly (for example, if we then use these calculations to drive another part of a larger model) or if computer time has an associated cost which we need to minimize, then it would be better to choose the population-based approach. A second criterion in terms of computational resources is that of available memory while the model is running or data storage for inputs and outputs. Many environmental models are spatially distributed, requiring a large amount of spatial information to be stored to carry out calculations. This information is usually held in dynamic (volatile) memory, which is relatively expensive, although much less so than even a few years ago. If parts of the spatial information need to be read from and stored back onto other media, such as hard drives, model performance suffers significantly. Thus, it may be necessary to design large spatial models to store only absolutely necessary information, and to access other information as and when necessary. Specific techniques are currently being designed to implement these approaches. For example, in the analysis of large digital elevation models (DEMs), Arge *et al.* (2002) demonstrated that a parallel approach to flow routing is between twice and ten times as fast as a serial approach for large datasets (they used examples with $13\,500 \times 18\,200$ and $11\,000 \times 25\,500$ cells). It is also capable of processing larger datasets (e.g. $33\,454 \times 31\,866$ cells, occupying 2 Gb of memory) that are not possible to evaluate using a serial approach due to machine-memory limitations.

Third, we need to consider our models in relation to available data. Data are required as basic inputs, or parameters, and for use in testing the results that are obtained. It may be better to use available data rather than have to produce estimates or large-scale interpolations or extrapolations from those that are available. On the other hand, this restriction should not always be too strongly applied, as models can often be used in the design of field data-collection campaigns by suggesting where critical measurements could be made. Increasingly, techniques are becoming available for relating model results at coarse scale to field measurements, which are often taken at points. These techniques are investigated further in Chapter 19.

1.5.5 Simple iterative methods

The most straightforward numerical approaches involve simple repetition or iteration of a calculation to evaluate how it changes through time. As an example, we will use a simple 'bucket' storage model of how a lake level will behave through time. The model is illustrated in Figure 1.14a and is described by the equation:

$$\frac{dh}{dt} = r + i - e - kh \qquad (1.20a)$$

and the inequality:

$$0 \le h \le h_{max} \qquad (1.20b)$$

In other words, the depth in the lake changes each year by an amount equal to the rainfall, r (in m a^{-1}) plus the river and surface inflow, i (in m a^{-1}), less the annual evaporation from the lake, e (in m a^{-1}) less the seepage from the base of the lake, which is a linear function of the depth of water, h (in m) in the lake. This function is

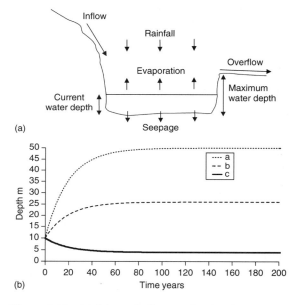

Figure 1.14 (a) Schematic diagram showing the representation of a lake water balance as a 'bucket' model with a series of inputs: rainfall and inflow from rivers; and outputs: evaporation, seepage and overflow once the maximum depth of the lake has been reached; and (b) graph showing the results of the 'bucket' model representing lake depth changes. Curve (a) represents rainfall of 1 m a^{-1}, an inflow rate of 0.8 m a^{-1}, evaporation rate of 0.05 m a^{-1} and seepage intensity of 0.05 a^{-1}; curve (b) has a lower inflow rate of 0.2 m a^{-1}, and higher evaporation rate of 1 m a^{-1}, conditions perhaps reflecting higher temperatures than in (a); curve (c) has a higher inflow of 2 m a^{-1}, compared to (a), perhaps representing vegetation removal, or the existence of a larger catchment area. In case (c), overflow from the lake occurs because the maximum depth threshold of 50 m is reached

controlled by a rate coefficient k, which is related to the permeability of the lake bed. There is a maximum level to which the lake can be filled, h_{max}, in m, above which overflow occurs. Similarly, the level cannot go below the ground surface. We need to use a numerical solution because of the inequalities constraining the minimum and maximum depths. To calculate iteratively, we use the following steps:

1. Define initial lake level $h_{t=0}$.
2. Calculate the annual change in level $\dfrac{dh}{dt}$ using Equation 1.20a.
3. Calculate the new level h_{t+1} by adding the annual change value.
4. If $h_{t+1} > h_{max}$ (Equation 1.20b), then calculate the overflow as $h_{t+1} - h_{max}$, and constrain $h_{t+1} = h_{max}$. Alternatively, if $h_{t+1} < 0$, constrain $h_{t+1} = 0$.
5. Go back to step (2) and repeat until the simulation time is complete.

Figure 1.14b illustrates the results for simple scenarios of environmental change, or for comparing lakes in different geomorphic settings. Similar sorts of models have been used for lake-level studies in palaeoenvironmental reconstructions (see discussion in Bradley, 1999). This type of storage model is also used in studies of hillslope hydrology, where the bucket is used to represent the ability of soil to hold water (e.g. Dunin, 1976). Note that this method is not necessarily the most accurate iterative solution, as it employs a backward-difference calculation (see below). More accurate solutions can be achieved using Runge–Kutta methods (see Burden and Faires, 1997; Engeln-Müllges and Uhlig, 1996).

Another example of where we would want to use iterative solutions is where we cannot calculate a closed algebraic form for a relationship that might be used in the model. For example, many erosion models use the settling velocity of particles to predict the amount of deposition that will occur (e.g. Beuselinck *et al.*, 2002; Morgan *et al.*, 1998). The settling velocity states that a sediment particle will move downwards (settle) through a body of water at a rate given by Stokes' Law:

$$v_s^2 = \frac{4}{3} \frac{g \left(\dfrac{\rho_s}{\rho} - 1 \right) d}{C_D} \qquad (1.21)$$

where v_s is the settling velocity (m s^{-1}), g is acceleration due to gravity (m s^{-2}), ρ_s is the density of the particle (kg m^{-3}), ρ is the density of water (kg m^{-3}), d is the particle diameter (m) and C_D is the drag coefficient of the particle. As noted by Woolhiser *et al.* (1990),

the drag coefficient is a function of the Reynolds number (Re) of the particle, a dimensionless relationship showing the balance between the gravitational and viscous forces on the particle:

$$C_D = \frac{24}{\text{Re}} + \frac{3}{\sqrt{\text{Re}}} + 0.34 \qquad (1.22)$$

However, the Reynolds number is itself a function of the particle settling velocity:

$$\text{Re} = \frac{v_s d}{\nu} \qquad (1.23)$$

where ν is the kinematic viscosity of the water (m^2 s^{-1}). Now we have the settling velocity on both sides of the equation, and no way of algebraically rearranging the equation to provide a simple solution in terms of v_s.

We can use a set of techniques for finding roots of equations (e.g. Burden and Faires, 1997) to tackle this problem. If we have an equation in the form:

$$\text{f}(x) = 0 \qquad (1.24)$$

then a root-solving technique allows us to find the value of x for which the function gives the result of zero. This approach can be implemented by combining Equations 1.21 to 1.23 and rearranging them to give:

$$v_s^2 \left(\frac{24}{\dfrac{v_s d}{\nu}} + \frac{3}{\sqrt{\dfrac{v_s d}{\nu}}} + 0.34 \right) - \frac{4}{3} g \left(\frac{\rho_s}{\rho} - 1 \right) d = 0.$$
$$(1.25)$$

The bisection method is a very robust technique for finding roots, as it will always converge to a solution. One problem that we will find with a number of numerical techniques is that they can fail to find the correct solution, or may even provide incorrect solutions to the model.

Bisection works by choosing an interval in which the solution is known to lie. Estimates are made of the lower and the upper value of this interval. This choice can easily be checked by calculating the function with the specific values. We can find the root by successively halving the interval, and comparing the midpoint with the two end points to find which subinterval the root lies. The computational steps are as follows (Figure 1.15):

1. Set an iteration counter i equal to 1.
2. Calculate the midpoint m_i of the lower (l_i) and upper (u_i) estimates of the interval:

$$m_i = \tfrac{1}{2}(l_i + u_i)$$

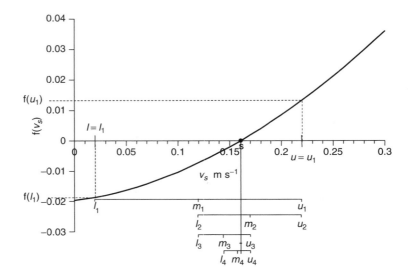

Figure 1.15   Graph to show how the bisection technique can be used to solve a root numerically by taking an interval about the solution, *s*, by subdivision of the interval it is known to be in. The lower value of the interval is called *l* and the upper value *u* in this diagram. The numbered subscripts show the successive iterations of these values. The function shown is Equation 1.25, used to calculate the settling velocity of a 1-mm diameter sediment particle with a density of $2500 \, \text{kg m}^{-3}$

3. Calculate $f(m_i)$. If the value is zero, then the root has been found. Otherwise . . .
4. Compare $f(m_i)$ with $f(l_i)$. If they have the same sign, then the root must lie between m_i and u_i, so set $l_{i+1} = m_i$ and $u_{i+1} = u_i$. If they have different signs, then the root must lie between l_i and m_i, so set $l_{i+1} = l_i$ and $u_{i+1} = m_i$.
5. If the interval between l_{i+1} and u_{i+1} is acceptably small (i.e. less than a predefined tolerance, ε), then accept the midpoint as the root.
6. Otherwise, increment the value of *i*. If *i* is within a predefined limit, return to step 2 and repeat the process.

The bisection method is useful because it *always* converges to a solution. It is thus useful even where quite sophisticated techniques will fail. The disadvantages of the technique are that it is very slow in converging and that in the process we can be close to a good approximation at an intermediate stage in the process without realizing it. The efficiency of the algorithm can be assessed because the number of iterations (*n*) required to achieve a result with a given tolerance ε is:

$$n = \log_2 \frac{u_i - l_i}{\varepsilon}. \quad (1.26)$$

The Newton–Raphson technique is far more efficient in that it converges much more rapidly. However, problems with the Newton–Raphson technique can arise (a) if the initial estimate is too far away from the root; (b) if the function has a local maximum or minimum; or (c) in the case of some functions that can cause the algorithm to enter an infinite loop without arriving at a solution. Numerous other techniques for root solution, together with discussion of their respective advantages and limitations, can be found in Burden and Faires (1997) and Press *et al.* (1992).

1.5.6 More versatile solution techniques

In this section, we consider a range of techniques that can be used to solve environmental models where analytical and simple iterative approaches are not appropriate. There are many variants of these techniques available, but space here only permits a brief overview. In all these cases, we assume that a model formalization is available in the form of a differential equation, starting with straightforward examples, but then considering more commonly used and more complicated forms.

1.5.6.1 Finite differences

The finite difference approach assumes that a solution to a differential equation can be arrived at by

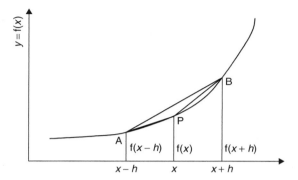

Figure 1.16 Graph showing the derivation of the finite difference technique for finding a solution to a function at point x. The distance h is the 'finite difference' of the name, and the entire function can be solved by repeating the calculations with x located at different places along the axis

approximation at a number of known, fixed points. These points are separated from each other by a known, fixed distance – the finite difference of the name of the technique (as compared with the infinitesimal distances used to calculate derivatives). In a simple case, we can consider how the technique works by looking at the function in Figure 1.16. To evaluate the rate of change in our model at point x, we need to find the slope of the function $f(x)$ $\left(\text{i.e. } \dfrac{dy}{dx}\right)$ at this point. If the distance (h) between our calculation points is small and/or the function does not change too dramatically, then the slope is best estimated by the line AB, in other words, the slope between the values of the function at $x + h$ and $x - h$.

This estimate is called the *central difference approximation*, and is given by the relationship:

$$f'(x) \approx \frac{f(x + h) - f(x - h)}{2h} \qquad (1.27)$$

Similarly, the line PB is also an estimate of the slope at P. This estimate is the *forward difference approximation* with:

$$f'(x) \approx \frac{f(x + h) - f(x)}{h} \qquad (1.28)$$

In the same way, the line AP is also an estimate, called the *backward difference approximation*, with:

$$f'(x) \approx \frac{f(x) - f(x - h)}{h} \qquad (1.29)$$

It is also possible to define the second derivative at x using a similar approach. The *central difference*

approximation is:

$$f''(x) \approx \frac{f(x + h) - 2f(x) + f(x - h)}{h^2} \qquad (1.30)$$

Note that in talking about the finite difference approach, the term approximation is fundamental (observe the use of $\approx$ meaning 'approximately equal to' in Equations 1.27 to 1.30 and in general through the rest of this section). We will see in a later section that evaluating the errors inherent in models is also fundamental. The error involved in these approximations can be assessed using a technique called Taylor expansion. For the central difference approximations the error is related to values of $h^2 f''(x) + h^3 f'''(x) + h^4 f''''(x) + \ldots$. This sequence is commonly written as $O(h^2)$, which means 'values of the order h^2 and greater'. Because the value of the higher order derivatives rapidly approaches zero, the approximation is reasonable. Similarly, the error in the backward difference approximations is $O(h)$. Clearly this is a much poorer approximation, but is valid in cases where the values change less rapidly, or if such error is within the bounds of acceptability.

We will commonly come across the case where the variable we are looking at is a function of more than one variable, e.g. $y = f(x, t)$, so that the variable changes through space and time. In this case it is useful to think of the value on a regularly spaced grid, as shown in Figure 1.17. Our finite differences are given by distances h, which is the increment in the x direction, and k, the increment in the t direction. We define *index parameters* to describe points on the grid – i in the x direction and j in the t direction – which have integer values. We would then denote the value at any point on the grid P(ih, jk) as:

$$y_P = y(ih, jk) = y_{i,j} \qquad (1.31)$$

The central difference approximation of $\dfrac{\partial y}{\partial t}$ at P is:

$$\frac{\partial y}{\partial t} \approx \frac{y_{i,j+1} - y_{i,j-1}}{2k} \qquad (1.32)$$

while the central difference approximation of $\dfrac{\partial y}{\partial x}$ at P is:

$$\frac{\partial y}{\partial x} \approx \frac{y_{i+1,j} - y_{i-1,j}}{2h} \qquad (1.33)$$

Other derivations are given in Figure 1.17. Note the change in notation from $\dfrac{dy}{dx}$ to $\dfrac{\partial y}{\partial x}$. The first case, with the normal d, denotes an ordinary differential equation (ODE); that is, the change we are interested in relates to one variable only. The curly ∂ denotes a partial

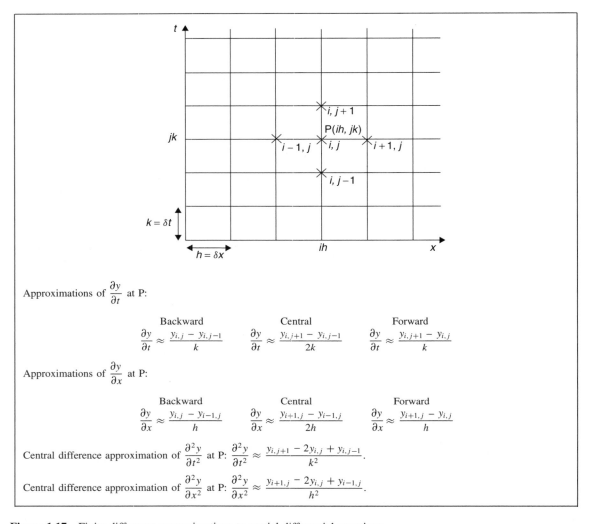

Figure 1.17   Finite difference approximations to partial differential equations

differential equation (PDE). Change in this example is partially related to two different variables – x and t – at the same time. There are three different types of partial differential equation – hyperbolic, parabolic and elliptic – as shown in Figure 1.18.

We will commonly want to solve two forms of problem. The first occurs as above where one of the variables we are looking at is time, and we want to solve change through time from a set of given conditions, known as *initial values*. These conditions will usually be prescribed for $t = t_0$. On our grid, the solution would appear as in Figure 1.19.

This is called an *initial value problem*. Note that we also have to specify the values immediately beyond our

grid so we can make statements about the solution of the problem at the edges. These values are called *boundary conditions*, and may be specific values, or functions of values at the edges of the grid.

A less frequent problem involves two (or more) variables which do not include time and require the solution of the function in terms of the variable at a single point in time over a finite are. The variables are commonly spatial coordinates, but may define another parameter space. A heat-flow problem might be a common example of the former – either in two or three dimensions. In this case our grid appears as in Figure 1.20, and we have a *boundary-value problem*.

Partial Differential Equations

PDEs come in three different varieties, and the solution methods for each vary. The three types are:

1. *hyperbolic* – for which the one-dimensional wave equation is the prototype form:

$$\frac{\partial^2 u}{\partial t^2} = v^2 \frac{\partial^2 u}{\partial x^2}$$

where v is the velocity of wave propagation.

2. *parabolic* – for which the diffusion equation is the prototype form:

$$\frac{\partial u}{\partial t} = \frac{\partial}{\partial t}\left(D \frac{\partial u}{\partial x}\right)$$

where D is the diffusion rate.

3. *elliptic* – for which the Poisson equation is the prototype form:

$$\frac{\partial^2 u}{\partial t^2} + \frac{\partial^2 u}{\partial y^2} = \rho(x, y)$$

where ρ is a given source term. If $\rho = 0$, this is called Laplace's equation.

Figure 1.18 Types of differential equation

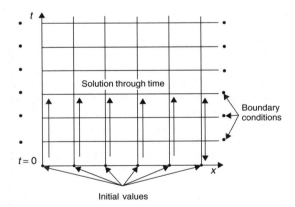

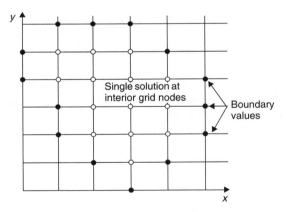

Figure 1.19 Illustration of an initial value problem. Given starting conditions at time $t = 0$ and boundary conditions beyond the domain of interest, solution is progressively carried out for each spatial location at successive time intervals on the finite difference grid

Figure 1.20 Illustration of a boundary-value problem. A single (steady-state) solution is calculated across the grid, given known boundary values or boundary conditions

The most straightforward way of solving a finite difference approximation to a partial differential equation is to use the *backward difference* technique (also called Euler's method in some texts). As an example, consider the kinematic wave approximation to the flow of water over a hillslope on which runoff is being generated at a

known rate:

$$\frac{\partial d}{\partial t} + v \frac{\partial d}{\partial x} = e(x, t) \tag{1.34}$$

where d is depth of water (L), t is time (T), v is flow velocity (L T^{-1}), x is horizontal distance (L) and $e(x, t)$ is the unit runoff rate at a given location and time (L T^{-1}) (Singh, 1996: compare the general derivation of

Equation 1.19 above which is identical apart from the source term has changed from q to e (x, t) to make it explicitly relate to runoff from infiltration excess). This is an example of a *hyperbolic equation*. If we know the initial condition:

$$d(x, 0) = g(x), \quad x \geq 0 \qquad (1.35a)$$

and the boundary condition

$$d(0, t) = b(t), \quad t > 0 \qquad (1.35b)$$

then we can solve the equation using backward differences. The backward difference expression for the first term is:

$$\frac{\partial d}{\partial t} \approx \frac{d(x, t) - d(x, t - k)}{k} + O(k) \qquad (1.36)$$

and for the second term:

$$\frac{\partial d}{\partial x} \approx \frac{d(x, t - k) - d(x - h, t - k)}{h} + O(h) \quad (1.37)$$

Putting these together and ignoring the error terms, we have a full solution, which is:

$$\frac{d(x, t) - d(x, t - k)}{k}$$
$$+ v \frac{d(x, t - k) - d(x - h, t - k)}{h} = e(x, t) \quad (1.38)$$

As we know the runoff rate (possibly using a simple bucket model of infiltration as described above), and are trying to solve for the new water depth, we rearrange this equation to give:

$$d(x, t) = d(x, t - k) + \frac{kv}{h} \{d(x, t - k)$$
$$- d(x - h, t - k)\} + \{e(x, t)k\} \quad (1.39)$$

Thus, to calculate the solution for a given surface, we define the grid of values in x and t, and carry out an iterative solution for $x = ih, i = 1, 2, \ldots, n$, for each value $t = jk, j = 1, 2, \ldots, N$. The solution works because at each time t, the solution depends only on known values of d at the previous timestep $t - k$. The solution then follows from the initial condition. This type of solution is known as an *explicit scheme*. All the values needed for the calculations at a specific timestep are known explicitly because we are using a backward difference solution. Other details required for the full solution of this equation as applied to desert hillslopes are given in Scoging (1992) and Scoging *et al.* (1992).

Implications of its use are discussed in Parsons *et al.* (1997; see also Chapter 3).

An advantage of the backward difference scheme given above is its simplicity, but it has two disadvantages which can lead to problems in certain circumstances. First, as noted above, the error terms are relatively high, being $O(h) + O(k)$. Second, the solution can become unstable under specific conditions. Such conditions are known as the Courant–Friedrichs–Lewy stability criterion, which states that stable solutions will only be produced if:

$$\frac{vk}{h} \leq 1 \qquad (1.40)$$

The practical implication of this is that for cases where the velocity, v, is high, the timestep k must be reduced to maintain stability. Very short timesteps may mean that the computer time required becomes excessively high to solve the problem, and thus the algorithm may not be appropriate in computer-resource terms, as discussed above. Alternatively, the spatial resolution would have to be compromised by increasing h, which may not be viable or useful in certain applications.

There are a number of other explicit techniques which can be used to solve similar equations, which improve the accuracy of the solution by reducing the error terms. Alternatively, higher accuracy and reduced error may be obtained by using an *implicit* solution to the problem. In this case, there are several unknowns during the calculations at any one timestep. Solution of simultaneous equations and/or iterative calculations are thus required to arrive at a final solution. Further details can be found in a number of texts, such as Reece (1976), Smith (1985), Ames (1992) or Press *et al.* (1992).

Boundary conditions are a critical element of any numerical solution. Basically, we must define what happens at the edge of the solution space so that it interacts realistically with the space around it. In some cases, the boundary conditions can be very simple. For example, when simulating water flows on hillslopes, the boundary at the top of a catchment will have no water flux entering the solution domain because there is nowhere for it to come from. At the base of the hillslope, flow may occur freely out of the system. We may set the gradient that controls the speed of this flow as equal to the gradient at the edge cell. Other ways of controlling the behaviour at the edges of the solution domain are to define the values of boundary cells as a function of time (Dirichelet boundary condition) or to specify the gradient at the boundary (Neumann boundary condition) (Press *et al.*, 1992). For example, in our hillslope-hydrology case, a Dirichelet boundary condition might be used if the channel flow is known at the base of the slope as a function of time (perhaps using an approximation such as the instantaneous unit hydrograph [see Chapter 5]); the Neumann boundary condition might be appropriate where there is a flood control structure where the slope is fixed. Correct definition of boundary conditions is often fundamental in controlling the correct solution of simulations, for example by preventing the development of numerical oscillations or exponentially increasing values.

1.5.6.2 Finite elements

An alternative method for solving models of this type is the finite element method. Rather than dividing the area in which we are interested into a regular grid, finite elements allow us to use regular shapes (triangles, quadrilaterals) to represent the area of interest. It is also possible to use cubes, cuboids and hexahedrons to represent volumes in three-dimensional models, and to have curved boundaries to the solution domain. Although it is possible to use a regular grid, it is much more common to vary the size of the elements so that locations where change occurs rapidly are represented by much smaller areas. In this way, the effects of errors in approximations may be minimized. This compares with the finite difference approach, where changes in grid size must take place uniformly across the same row or same column of the grid, allowing much less flexibility. A second advantage comes with the way in which boundary conditions are addressed. Because the boundary conditions are directly included in the formulation in the finite element approach, it is not necessary to approximate gradients at the boundaries. On the other hand, finite element calculations can require much more computer memory and computing power (see Smith and Griffiths, 1998, for example).

The finite element method is based on the fact that any integral over a specific range can be evaluated by adding the integrals of subintervals making up that range (see the discussion of definite integrals above) (Henwood and Bonet, 1996). In one dimension, we have:

$$\int_{x_1}^{x_n} f(x)\, dx = \int_{x_1}^{x_2} f(x)\, dx + \int_{x_2}^{x_3} f(x)\, dx$$
$$+ \cdots + \int_{x_{n-1}}^{x_n} f(x)\, dx \qquad (1.41)$$

In other words, we can provide a solution over the whole domain by splitting it into subdomains (elements) and using local solutions, or approximations to them, to evaluate the whole. We need to solve the overall integral in a way that provides the best approximations to the local conditions. One way of doing this is to use the weighted residuals (R) of the function:

$$\int R \cdot w\, dx = 0 \qquad (1.42)$$

where w are a series of weights, to provide a spatially averaged best solution.

The simplest approach is to use approximations to local solutions of the elements using linear interpolation, using the 'hat' function (Figure 1.21). At any point i, the value of the hat function, N_i, using this approach is:

$$N_i = \begin{cases} 0 & if \quad x \le x_{i-1} \\ \dfrac{x - x_{i-1}}{x_i - x_{i-1}} & if \quad x_{i-1} \le x \le x_i \\ \dfrac{x_{i+1} - x}{x_{i+1} - x_i} & if \quad x_i \le x \le x_{i+1} \\ 0 & if \quad x \ge x_{i+1} \end{cases} \qquad (1.43)$$

We therefore have an approximation to function $f(x)$ at any point within an element, based on known values at any point i as:

$$f(x) \approx N_i f(x_i) + N_{i+1} f(x_{i+1}) \qquad (1.44)$$

which is called the linear basis function for the approximation. In the Galerkin approach to finite elements, the weights in Equation 1.42 are set as equal to the values of the basis function, i.e. $w = N_i$.

As an example, consider the case where we are interested in the distribution of heat in a soil, for which we have a temperature measurement at the surface and at a known depth. Under steady-state conditions, the temperature distribution will be described by a function of the form:

$$\frac{\partial}{\partial x}\left(-k\frac{\partial \tau}{\partial x}\right) + \tau = 0 \qquad (1.45)$$

where τ is the temperature in °C and k the rate of heat transmission or thermal conductivity (W m^{-1} °C^{-1}). In the finite element method, this function becomes the residual term to be minimized, so that we insert this equation into Equation 1.42:

$$\int_0^{x_n} \left\{ \frac{\partial}{\partial x}\left(-k\frac{\partial \tau}{\partial x}\right) w + \tau w \right\} dx = 0 \qquad (1.46)$$

Integration by parts (see Wilson and Kirkby, 1980) can be used to transform this relationship into a set of

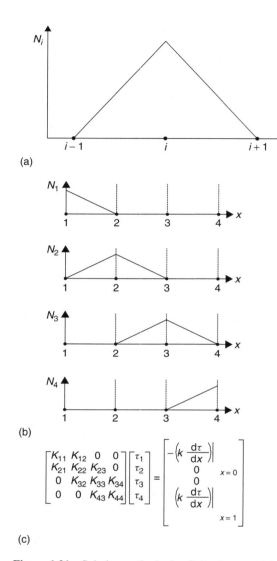

(a)

(b)

(c)

Figure 1.21 Solution method using finite elements: (a) hat function; (b) solution of by piece-wise interpolation using sliding hat function; and (c) finite element matrix

ordinary differential equations:

$$\int_0^{x_n} \left(-k\frac{d\tau}{dx}\frac{dw}{dx} + \tau w \right) dx = \left[k\frac{d\tau}{dx}w \right]_0^{x_n} \quad (1.47)$$

Next, we split the whole domain into a number of elements – in this example, three – and apply the Galerkin estimate of the weights using the linear basis functions and estimate the local value of τ using Equation 1.44.

Taking the element as having unit length, and multiplying by a scaling factor $\left| \dfrac{dx}{d\xi} \right|$, we have at node i:

$$\tau_i \int_0^1 \left(k\frac{dN_{i-1}}{d\xi}\frac{d\xi}{dx}\frac{dN_i}{d\xi}\frac{d\xi}{dx} + N_{i-1}N_i \right) \left| \frac{dx}{d\xi} \right| d\xi$$

$$= \left[k\frac{d\tau}{dx}N_i \right]_0^1 \quad (1.48)$$

where ξ (the Greek letter 'xi') refers to the fact that we are integrating along the element rather than the solution domain as a whole. For each node in the area of interest (Figure 1.21b) we now have a relationship based on the values of N_i and N_{i-1} that can be used to solve within the elements. As $\dfrac{dN_{i-1}}{d\xi} = -1$, $\dfrac{dN_{i-1}}{d\xi} = 1$ and $\dfrac{d\xi}{dx} = 3$ (the number of elements), we can now define directly the values of the integral used to multiply the values τ_i on the left-hand side of Equation 1.48. This provides a matrix of values (Figure 1.21c) that is usually known as the 'stiffness' matrix (because of the historical origins of the technique in structural analysis). It now remains to solve the right-hand side of Equation 1.48 by inserting the relevant weights. In all but the first and last cases, these cancel, providing source and sink terms at each of the boundaries of interest (the soil surface and the location of the temperature probe in our case). Using our known boundary conditions at these points (T_1 is the temperature at the surface and T_4 is the temperature at the probe), we produce a set of simultaneous equations:

$$\begin{aligned} \tau_1 &= T_1 \\ K_{21}\tau_1 + K_{22}\tau_2 + K_{23}\tau_3 &= 0 \\ K_{32}\tau_2 + K_{33}\tau_3 + K_{34}\tau_4 &= 0 \\ \tau_4 &= T_4 \end{aligned} \quad (1.49)$$

that can be solved using standard techniques (see Press *et al.*, 1992, for details). Finally, the values of τ_1 can be substituted into Equation 1.48 to give the heat fluxes into the top of the soil and out of the base of the soil (Equation 1.49).

Finite elements are widely used in structural and engineering applications, including slope stability (see Chapter 10), fluid and heat-flow studies. More details on the technique can be found in Henwood and Bonet (1996), Wait and Mitchell (1985), Smith and Griffiths (1998), Gershenfield (1999), Burden and Faires (1997) and Zienkiewicz and Taylor (2000), including information on how to model time-dependent conditions. Flow modelling examples can be found in the articles by Muñoz-Carpena *et al.* (1993, 1999), Motha and Wigham (1995) and Jaber and Mohtar (2002). Beven (1977) provides a subsurface flow example as applied to hillslope hydrology.

1.5.6.3 Finite volumes

The finite volume or control volume approach is similar to the finite element method in that it uses integration over known areas, but differs in the solution methods used. Versteeg and Malalasekera (1995) provide a very readable introduction to the technique, which we largely follow here. The area under study is first divided into a series of control volumes by locating a number of nodes throughout it. The boundaries of the control volumes are located midway between each pair of nodes (Figure 1.22). The process studied is investigated by integrating its behaviour over the control volume as:

$$\int_0^{\Delta V} \mathrm{f}'(x)dV = A_{i+\frac{1}{2}\delta x}\mathrm{f}(x_{i+\frac{1}{2}\delta x}) - A_{i-\frac{1}{2}\delta x}\mathrm{f}(x_{i-\frac{1}{2}\delta x})$$

(1.50)

which is the same as looking at the mass balance over the control volume.

Using the soil heat-flow example of Equation 1.45 again, the control-volume equation becomes:

$$\int_0^{\Delta V} \frac{\partial}{\partial x}\left(-k\frac{\partial\tau}{\partial x}\right) + \tau\,dV = \left(-kA\frac{d\tau}{dx}\right)_{i+\frac{1}{2}\delta x}$$

$$- \left(-kA\frac{d\tau}{dx}\right)_{i-\frac{1}{2}\delta x} + \overline{\tau}_i\Delta V = 0 \qquad (1.51)$$

and we can use linear interpolation to calculate the appropriate values of k and finite differences to estimate the values of $\frac{d\tau}{dx}$ at the edges of the control volume.

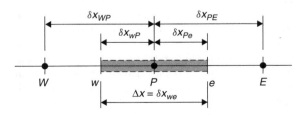

Figure 1.22   Discretization of the solution space in the finite volume technique
Source: Versteeg and Malalasekera, 1995

Assuming a uniform node spacing, these approximations give:

$$\left(-kA\frac{d\tau}{dx}\right)_{i+\frac{1}{2}\delta x} = -\frac{k_{i+1}+k_i}{2}A_{i+\frac{1}{2}\delta x}\left(\frac{\tau_{i+1}-\tau_i}{0.5\delta x}\right)$$

(1.52a)

$$\left(-kA\frac{d\tau}{dx}\right)_{i-\frac{1}{2}\delta x} = -\frac{k_{i-1}+k_i}{2}A_{i-\frac{1}{2}\delta x}\left(\frac{\tau_i-\tau_{i-1}}{0.5\delta x}\right)$$

(1.52b)

Substitution of these into Equation 1.51 gives:

$$-\frac{k_{i+1}+k_i}{2}A_{i+\frac{1}{2}\delta x}\left(\frac{\tau_{i+1}-\tau_i}{0.5\delta x}\right) + \frac{k_{i-1}+k_i}{2}A_{i-\frac{1}{2}\delta x}$$

$$\left(\frac{\tau_i-\tau_{i-1}}{0.5\delta x}\right) + \tau_i\Delta V = 0 \qquad (1.53)$$

Rearranging to give common terms in τ_{i-1}, τ_i and τ_{i+1}, we have:

$$\left(\frac{k_{i+1}+k_i}{\delta x}A_{i+\frac{1}{2}\delta x} + \frac{k_{i-1}+k_i}{\delta x}A_{i-\frac{1}{2}\delta x} + \Delta V\right)\tau_i$$

$$= \left(\frac{k_{i-1}+k_i}{\delta x}A_{i-\frac{1}{2}\delta x}\right)\tau_{i-1} + \left(\frac{k_{i+1}+k_i}{\delta x}A_{i+\frac{1}{2}\delta x}\right)\tau_{i+1}$$

(1.54)

Over the whole solution domain, this equation is again a set of simultaneous equations that can be solved in exactly the same ways as mentioned before.

Finite volume approaches are used commonly in computational fluid dynamics approaches (see Chapter 20). More detailed applications in this area can be found in the books by Versteeg and Malalasekera (1995) and Ferziger and Perić (1999). Applications relating to simpler flow models (such as the kinematic wave approximation used in the finite difference example) can be found in the articles by Gottardi and Venutelli (1997) and Di Giammarco *et al.* (1996).

1.5.6.4 Alternative approaches

There is increasing interest in models that do not depend on numerical discretization as described in the sections above. Of these, the most common is the cellular automata approach. This method is dealt with in detail in Chapter 21, so only a summary is given here. The advantages of cellular automata are that they can represent discrete entities directly and can reproduce emergent properties of behaviour. However, because the number of entities simulated in realistic models is high, they can have a large computational load, although special computer architectures that take advantage of the structure of cellular automata have been used to speed the process considerably (Gershenfield, 1999).

Cellular automata use a large number of cells or nodes that can take a number of states. Each cell or node is progressively updated using a fixed set of rules that depend only on local conditions. Thus at no stage is information on the whole solution space required. The most commonly cited example is John Conway's 'Game of Life'. A cell can either be 'on' or 'off', representing life or no life in that location. There are three rules that determine the state of a cell in the next iteration of the model based on counting the number of occupied cells in the immediate eight neighbours (Figure 1.23a):

1. 'Birth' – a cell will become alive if exactly three neighbours are already alive.
2. 'Stasis' – a cell will remain alive if exactly two neighbours are alive.
3. 'Death' – a cell will stop living under any other conditions (0 or 1 neighbours reflecting isolation; or ≥ 4 neighbours reflecting overcrowding).

Compare these rules to the population models discussed above. One of the interesting outcomes of this approach is that patterns can be generated that are self-reproducing.

Cellular automata have been used in a wide variety of applications, in particular fluid flows. It is possible for example to demonstrate the equivalence of the Navier–Stokes equations of fluid flow to cellular automata approximations (Wolfram, 1986; Gershenfield, 1999). More details of how to apply the method can be found in Ladd (1995), Bar-Yam (2000) and Wolfram (2002). See Chapter 21 for a discussion of environmental applications in more detail.

A further approach is the use of artificial neural networks. An artificial neural network is a computer representation of the functioning of a human brain. It does so by representing the neurons that are the fundamental building blocks of the brain structure. A number of inputs are passed through weighting functions representing synapses (Figure 1.24). The weighted values are summed and then affected by a local bias value. The result is passed to a function that decides whether the current neurone should be activated, and thus controls its output. The number of inputs and the number of neurons and ways in which they are combined control the function of the neural network. An important feature of neural networks is their ability to capture non-linear behaviour.

A common use of artificial neural networks is in rainfall-runoff modelling. A feed-forward model structure is usually used, with a single hidden layer of neurons (Figure 1.24). For example, Xu and Li (2002) used 29 input variables to estimate the next inflow into a reservoir in Japan. These inputs were hourly average upstream rainfall over seven hours, the seven hours of previous releases from two upstream reservoirs, the four previous hours of tributary inflow and the four previous hours of inflow into the reservoir. Xu and Li found reasonable predictions for small and medium-sized flows up to seven hours in advance. It is also possible to interpret the weights on the different inputs and their combinations to suggest the relative importance of the different input parameters. Essentially, though the issues relating to the use of artificial neural networks are those relating to any data-based approach (see Chapter 22). There are limits to what can be captured based on the data used to train the neural network. It is unlikely that significant numbers of extreme events will be available for forecasting of major events using this approach. Imrie *et al.* (2000) discuss these issues and suggest potential improvements relating to restricting the values of the weighting function during the learning process, and ensuring that the structure of the artificial neural network is appropriate for all cases.

Haykin (1999), Bar-Yam (2000), Blum (1992), Welstead (1994) and Rogers (1997) provide extensive details of how to apply the approach. Other applications in rainfall-runoff and flow modelling can be found in Hsu *et al.* (1995), Sajikumar and Thandaveswara (1999), Shamseldin (1997) and Campolo *et al.* (1999). French *et al.* (1992) present an application to rainfall forecasting; Dekker *et al.* (2001) use the approach in a forest transpiration model; Tang and Hsieh (2002) use it in atmospheric modelling; and De Falco *et al.* (2002) have used it to predict volcanic eruptions.

Stochastic approaches can provide important simplifications, particularly when key elements of the system are incompletely known. In a stochastic model, one or more probability distributions are employed to represent either deviations from known values of parameters or specific processes that have an apparently random behaviour (at least at the scale of observation). As an example of the former, consider the population-growth model discussed above in Equation 1.15. It has three parameters – β the birth rate, θ the death rate and N_{max} the environmental carrying capacity. Measurements of each of these variables may show that they are not constant through time, but exhibit a natural variability. Birth and death rates may vary according to the incidence of disease, whereas the carrying capacity may vary according to climatic conditions (e.g. rainfall controlling crop production). Assuming that the deviations are normally

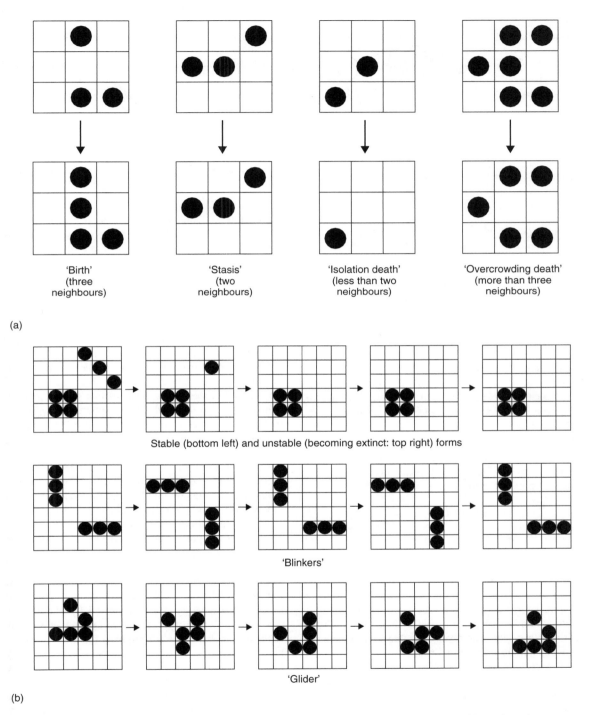

Figure 1.23   Conway's 'Game of Life' cellular automata model: (a) rules; and (b) examples of how the rules operate on the local scale to produce stable, unstable and self-reproducing features

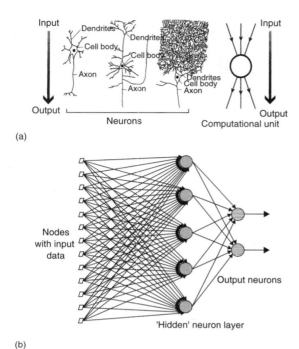

(a)

(b)

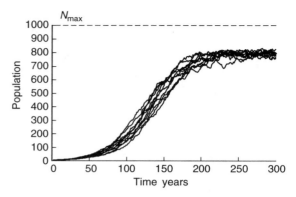

Figure 1.25 Examples of ten stochastic simulations of population growth using Equation 1.55 (the stochastic version of Equation 1.15), with $\beta = 0.05$ individuals per year, $\sigma_\beta = 0.025$ individuals per year, $\theta = 0.01$ individuals per year, $\sigma_\theta = 0.005$ individuals per year, $N_{max} = 1000$ individuals and $\sigma_{N_{max}} = 100$ individuals

Figure 1.24 (a) Structure of typical human neurons compared to that of a neuron used in an artificial neural net (from McLeod *et al.*, 1998); and (b) example of an artificial neural network model – a feed-forward model with a single hidden layer of neurons

distributed, with standard deviations σ_β, σ_θ and $\sigma_{N_{max}}$, respectively, we could modify Equation 1.15 to give a stochastic model with:

$$\frac{dN}{dt} = (\beta + n[\sigma_\beta]) \left(1 - \frac{N}{N_{max} + n[\sigma_{N_{max}}]}\right)$$
$$N - (\theta + n[\sigma_\theta])N \qquad (1.55)$$

where $n[\sigma]$ is a normally distributed random number with mean 0 and standard deviation σ. As the values of σ_β, σ_θ and σ_{Nmax} increase, so does the variability in output. Example simulations are shown in Figure 1.25. The model is very sensitive to high variability in carrying capacity, where very variable environments are often subject to population crashes.

The second type of stochastic model assumes that the best process description at a particular scale of observation is as a random process. This description may arise because there are too many different variables to measure, or because it is impossible to provide deterministic values for all of them because they require very detailed and/or impractical measurements.

This approach is commonly used when modelling climatic inputs into other models. For example, if the daily rainfall input is required over 100 years for a site that only has a short data record, the stochastic characteristics of the data record can be used to provide a model. The rainfall process at this scale can be simply approximated by stating that there is a distribution function for the length of time without rain. This distribution function is commonly a Poisson distribution, representing the random occurrence of dry spells of a specific, discrete length of time. Once a day is defined as having rainfall on it, a second distribution function is used to define the amount of rainfall that is produced. Commonly used distribution functions in this case are negative exponential or gamma distributions, with a strong positive skew representing the fact that most storms are small, with very few large events (see Wainwright *et al.*, 1999a; Wilks and Wilby, 1999). Another approach to simulating the length of dry spells is to use Markov transition probabilities. These probabilities define the likelihood that a dry day is followed by another dry day and that a wet day is followed by another wet day (only two probabilities are required for the four possible combinations: if p_{dd} is the probability that a dry day is followed by a dry day, then $p_{dw} = 1 - p_{dd}$ is the probability that a wet day follows a dry day; similarly, if pww is the probability that a wet day is followed by another wet day, then $p_{wd} = 1 - p_{ww}$ is the probability that a dry day follows a wet day). The Markov approach can be extended to define probabilities that a dry day is followed by two dry days, or a wet day is followed by two wet days and so on. This approach is useful if the climate system can be thought to have a 'memory'. For example, in regions dominated by frontal rainfall, it is likely that there will be periods of continuously wet weather, whereas thunderstorms in drylands events generally occur on particular days and change atmospheric conditions such that a further thunderstorm is unlikely. Note that it is still possible to interpret the meaning of such stochastic models in a physically meaningful way. Further details on stochastic rainfall generators can be found in Wilks and Wilby (1999). Markov models are described in more detail in Collins (1975) and Guttorp (1995), with other applications in sedimentology (e.g. Parks *et al.*, 2000), mapping (e.g. Kovalevskaya and Pavlov, 2002) and ecology (e.g. Hill *et al.*, 2002) among others.

Another commonly employed example of a stochastic model is related to sediment transport. Although the balance of forces on individual particles is generally well understood (e.g. Middleton and Southard, 1984), there are a number of factors that make it impractical to measure all the necessary controls in an operational way. For example, the likelihood of movement depends on the way in which a stone sits on the bed of a channel. If there is a variable particle size (which there is in all but the best-controlled laboratory experiments), then the angle at which the stone sits will vary according to the size of the stones in the bed and their orientation. Similarly, river flows exhibit turbulence, so that the force exerted on the stone is not exactly predictable. Einstein (1942) defined an approach that defines the sediment-transport rate as a function of the time periods over which sediment particles do not move and a function of the travel distance once a particle starts moving (see Raudkivi, 1990, for a derivation). Naden (1987) used this approach successfully to simulate the development of bedforms in gravel-bed rivers. A modified version of this approach was used by Wainwright *et al.* (1999b) to simulate the development of desert pavements.

It is a common misconception that stochastic models will always be faster than corresponding deterministic models. This situation is not necessarily the case because there is often a high computational overhead in generating good random numbers with equal likelihood that all values appear and that successive values are not correlated or repeated over short time intervals. In fact, it is impossible for a computer to generate a true random number. Most random-number generators depend on arithmetic manipulations that ensure that there are cycles of numbers that repeat but that the number over which they repeat is very large. The outputs are called pseudo-random numbers. Generating and evaluating good pseudo-random numbers are dealt with in detail in Press *et al.* (1992) and Kloeden *et al.* (1994). Problems in certain packages are presented in Sawitzki (1994) and McCullough and Wilson (2002).

1.6 MODEL PARAMETERIZATION, CALIBRATION AND VALIDATION

1.6.1 Chickens, eggs, models and parameters?

Should a model be designed around available measurements or should data collection be carried out only once the model structure has been fully developed? Many hardened modellers would specify the latter choice as the most appropriate. The parameters that are required to carry out specific model applications are clearly best defined by the model structure that best represents the processes at work. Indeed, modelling can be used in this way to design the larger research programme. Only by taking the measurements that can demonstrate that the operation of the model conforms to the 'real world' is it possible to decide whether we have truly understood the processes and their interactions.

However, actual model applications may not be so simple. We may be interested in trying to reconstruct past environments, or the conditions that led to catastrophic slope collapse or major flooding. In such cases, it is not possible to measure all of the parameters of a model that has a reasonable process basis, as the conditions we are interested in no longer exist. In such cases, we may have to make reasonable guesses based on indirect evidence. The modelling procedure may be carried out iteratively to investigate which of a number of reconstructions may be most feasible.

Our optimal model structure may also produce parameters that it is not possible to measure in the field setting, especially at the scales in which they are represented in the model. The limitations may be due to cost, or the lack of appropriate techniques. It may be necessary to derive transfer functions from (surrogate) parameters that are simpler to measure. For example, in the case of infiltration into hillslopes, the most realistic results are generally obtained using rainfall simulation, as this approach best represents the process we are trying to parameterize (although simulated rain is never exactly the same as real rain – see Wainwright *et al.*, 2000, for implications). However, rainfall simulation is relatively difficult and expensive to carry out, and generally requires large volumes of water. It may not be feasible to obtain or transport such quantities, particularly in remote locations – and most catchments contain some remote locations. Thus, it may be better to parameterize using an alternative measurement such as cylinder infiltration, or pedo-transfer functions that only require information about soil texture. Such measurements may not give exactly the same values as would occur under real rainfall, so it may be necessary to use some form of calibration or tuning for such parameters to ensure agreement between model output and observations. In extreme cases, it may be necessary to attempt to calibrate the model parameter relative to a known output if information is not available. We will return to the problems with this approach later.

Parameterization is also costly. Work in the field requires considerable investment of time and generally also money. Indeed, some sceptics suggest that the research focus on modelling is driven by the need to keep costs down and PhDs finished within three years (Klemeš, 1997). Equipment may also be expensive and, if it is providing a continuous monitored record, will need periodic attention to download data and carry out repairs. Therefore, it will generally never be possible to obtain as many measurements as might be desirable in any particular application. As a general rule of thumb, we should invest in parameter measurement according to how great an effect the parameter has on the model output of interest. The magnitude of the effect of parameters on model output is known as the sensitivity of a model to its parameters. This important stage of analysis will be dealt with in more detail below.

1.6.2 Defining the sampling strategy

Like models, measurements are also abstractions of reality, and the results of a measurement campaign will depend as much upon the timing, technique, spatial distribution, scale and density of sampling as on the reality of the data being measured. As in modelling, it is imperative that careful thought is put into the conceptualization and design of a sampling strategy appropriate to the parameter being measured and the objective of the measurement. This is particularly true when the sampled data are to be used to parameterize or to validate models. If a model under-performs in terms of predictive or explanatory power, this can be the result of inappropriate sampling for parameterization or validation as much as model performance itself. It is often assumed implicitly that data represents reality better than a model does (or indeed that data is reality). Both are models and it is important to be critical of both.

We can think of the sampling strategy in terms of (a) the variables and parameters to be measured for parameterization, calibration and validation; (b) the

direct or indirect techniques to be used in measurement and their inherent scale of representation; (c) the spatial sampling scheme (distributed, semi-distributed, lumped) and its density; and (d) the temporal sampling scheme (duration and temporal resolution). Choosing which variables will be measured for parameterization and the intensity of measurement will depend very much on the sensitivity of the significant model outputs to those (see below). Highly sensitive parameters should be high on the agenda of monitoring programmes but since model sensitivity to a parameter is usually dependent also on the value of other parameters, this is not always as straightforward as it might at first appear. Where variables are insensitive, either they should not be in the model in the first place or their measurement can be simplified to reflect this. Calibration parameters should be as much as possible those without physical meaning so as not to compromise the physical basis of the model and their measurement will be necessary for the application of models to new environments or epochs. Validation parameters and variables should be those which are the critical model outputs in terms of the purpose of the model. A robust validation of the key model output would tend to indicate that the model has performed well in a predictive sense. This outcome does not mean that the results have been obtained for the correct reasons, in other words, good prediction is no guarantee of good explanation. In this way, if one were to validate the output of a catchment hydrological model using measured discharge data and obtain good agreement between model and data, this success can come about as the result of many different configurations of the driving variables for discharge. It is thus important to validate the output required but also some internal variable which would indicate whether that output has been arrived at for the correct reasons, in this case the spatial distribution of soil moisture around the catchment.

The techniques used for measurement will depend upon a number of logistic constraints such as availability, cost, dependence on power supplies, training required for use and safety, but must also depend upon the spatial and temporal structure of the model for which these techniques will provide data since it is important that the model and the data are representing the same thing. A good example is soil moisture. Soil is a three-phase medium consisting of the soil matrix, rock fragments greater than 2 mm in diameter and porosity. Soil moisture occupies the porosity which is usually around half of the soil volume. In many soils, rock-fragment content can be in excess of

30% (van Wesemael *et al.*, 2000) and while rock fragments sometimes have a small porosity, it is usually pretty insignificant for the purposes of moisture retention. Volumetric measurement of soil moisture usually provides an output of m^3 water per m^3 soil fine fraction which does not usually contain rock fragments. These tend to be avoided in the installation of electronic sensors of soil moisture and not accounted for in calibration, and which tend to be avoided or sieved out of gravimetric samples. Soil-moisture measurements are usually an aggregate of small sample measurements of the fine soil fraction. However, soil tends to be represented as large blocks with dimensions of tens to hundreds of metres in hydrological models. This must therefore incorporate a significant loss of available porespace because of the presence of rock fragments and thus the nature of soil moisture at this scale is quite different to that at the point scale of measurement. The need to balance data and model attributes is particularly clear where indirect measurements, in particular remote sensing, are used for model parameterization.

Over recent decades there has been a move away from lumped models in which spatial heterogeneity is not represented towards distributed models in which it is. Advances in computing power and GIS technologies have enabled the development of complex spatial models based on the discretization of landscapes into vector polygons, triangular irregular networks, objects of complex form or simple raster grids. Despite recent advances in remote sensing, there are still very many parameters that cannot be measured using electromagnetic radiation and thus remote sensing. The sophistication of spatial models has rapidly outgrown our ability to parameterize them spatially and they thus remain conceptually lumped (Beven, 1992). The appropriate scale of distribution and the optimum configuration of measurements for model parameterization or calibration are the subject of much debate. For example, Musters and Bouten (2000) used their model of root-water uptake to determine the optimal sampling strategy for the soil-moisture probes used to parameterize it. Fieldwork is an expensive, labour-intensive, time-consuming and sometimes uncomfortable or even hazardous activity. Traditional random or structured sampling procedures usually require that a very large number of samples be collected in order to fulfil the assumptions of statistical inference. In order to reduce the sampling effort, prior knowledge about the system under study may be used to guide convenience or nonrandom sampling which is still statistically viable, with the appropriate method depending on the type of prior knowledge available (Mode *et al.*, 2002). Ranked set sampling (Mode *et al.*, 1999)

reduces the cost of sampling by using 'rough but cheap' quantitative or qualitative information to guide the sampling process for the real, more expensive sampling process. Chao and Thompson (2001) and others indicate the value of optimal adaptive sampling strategies in which the spatial or temporal sampling evolves over time according to the values of sites or times already sampled. A number of authors indicate how optimal sampling can be achieved by algorithmic approaches which maximize entropy in the results obtained (e.g. Bueso *et al.*, 1998; Schaetzen *et al.*, 2000). The luxury of optimizing your sampling scheme in this way is, however, not always available to the modeller, especially within the context of policy models which are applied using existing datasets generated by government agencies, for example, where 'you get what you are given' and which may not be collected with uniform or standard protocols (e.g. as outlined for soils data in Spain by Barahona and Iriarte, 2001) or where the protocol may evolve over time, affecting the legitimacy of time-series analysis. Usually the spatial sampling scheme chosen is a compromise between that which best represents the system under review and the computational resources and data available. This compromise is most clearly seen in the extensive discussions on the problem of grid size and subgrid variability in general circulation models (GCMs). May and Roeckner (2001), among others, indicate the importance of grid resolution in affecting the results of GCMs. Smaller grid sizes produce more realistic results, especially in highly mountainous areas, but smaller grids also have substantially higher computational and data costs.

Wainwright *et al.* (1999a) indicated the importance of the temporal detail of climate data for accurate hydrological modelling. The calculation of evapotranspiration using the Penman–Monteith formula for hourly data and then the same data aggregated to a single value for each day and then separately for each day and night indicates that the day–night aggregation produces much closer results to the original hourly data than does the daily aggregation because of the domain change in net radiation values from daylight hours when they are positive to night-time hours when they are negative. The error induced by aggregation to daily timestep is of the order of 100% and varies with the month of the year too. This indicates that one must pay attention to the natural scales and boundaries of the processes being modelled when devising the time (or space) scale for sampling. Similarly, Mulligan (1998) demonstrated the importance of high temporal resolution rainfall intensity data in understanding the partitioning between infiltration and overland flow. Where soil infiltration rates fall within the range of measured instantaneous rainfall intensities (as they often do), it is important to understand rainfall intensities as the distribution function of instantaneous intensities. The greater the timescale over which these intensities are aggregated, the lower the measured intensity would be. Such aggregation can have major effects on the predicted levels of Hortonian or infiltration-excess overland-flow production – which is, after all, a threshold process (see Wainwright and Parsons, 2002, for spatial implications). Hansen *et al.* (1996) suggested the importance of data quality in determining streamflow prediction for the lumped IHACRES rainfall-runoff model to conclude that rain-gauge density and the sampling interval of rainfall are the most critical across a range of catchments. Understanding these sensitivities is critical to designing an appropriate sampling scheme.

1.6.3 What happens when the parameters don't work?

It is frequently the case that initial parameter estimates will produce model outputs that are incompatible with the known behaviour of the system. There are usually good reasons for this outcome, so do not give up! Given that the parameter base for distributed models is generally small relative to the detail simulated, it is perhaps not surprising. Similarly, lumped models have a sparse parameter representation relative to natural variability in the system. Point measurements or spatial averages are often poor representations of the parameter interactions in this case. Evaluating errors from these sources will be dealt with later.

However, in terms of model parameterization, it may be impossible to return to the field to carry out more measurements, but we still need to obtain results for our model application. Thus, we need to adopt an iterative approach to the evaluation of the correct model parameters. This procedure is generally known as model calibration.

1.6.4 Calibration and its limitations

Kirkby *et al.* (1992) distinguish between physical parameters, which define the physical structure of the system under study, and process parameters, which define the order of magnitude of processes. Most models will contain both types of parameter. Definition of these process parameters is known as calibration or model tuning. Where they are physically based, this can be achieved by their measurement, where not, they are calibrated using a

process of optimization (optimized) against a measure of the agreement between model results and a set of observations used for calibration. The calibration dataset must be independent from any dataset which is used later to validate the model, if the same dataset is used for both it should be no surprise that the model is a perfect predictor! Split sample approaches, in which the available data is separated into a calibration set and a separate validation set, are usually the solution to this problem.

Calibration should pay particular attention to the sensitivity of parameters with sensitive parameters being calibrated carefully against high quality datasets to ensure that the resulting model will produce reliable outcomes. The simplest form of optimization is trial and error whereby model parameters are altered and a measure of goodness of fit between model results and the calibration dataset is noted. This process is repeated iteratively to obtain the best possible fit of observed against predicted. Of course the calibration will be specific to the model results calibrated against and will produce a model which should forecast this result well at the expense of other model outputs not involved in the calibration procedure. The choice of calibration parameters, measures and techniques will thus depend upon the purpose to which the model will be put. Moreover, a model calibration by one user with a particular understanding of its function may be quite different from that of another (Botterweg, 1995) and a model calibrated to a particular objective such as the prediction of peak runoff may then be useless in the prediction of total annual runoff. Some prior knowledge of, for example, the reasonable ranges of parameter values, will also be necessary and calibration will usually follow a preliminary sensitivity or uncertainty analysis which is performed to test the validity of the model. The relationship between the range of values for a parameter and the model agreement is known as the calibration curve for that parameter. A parameter that shows a significant change in error with a change in its value (with all other parameters held constant) is known as a sensitive parameter. If a model has only one parameter, it is usually fairly straightforward to find the optimal value for that parameter. This procedure becomes only marginally more difficult for models with more than one parameter where the parameters are independent. In most models, parameters are highly interdependent and this will confound the definition of an optimum parameterization. In these cases other – automated – techniques are used to define the optimal parameter set. These techniques include genetic algorithms and fuzzy logic approaches as used to calibrate a rainfall-runoff model to multiple objectives (peak discharge, peak time and total runoff volume) by Cheng *et al.* (2002). Calibration is particularly challenging in distributed models which tend to have a large number of parameters. Stochastic or evolutionary genetic algorithms seem to be the most successful approaches under these circumstances (Eckhardt and Arnold, 2001) and have been applied widely where there are multiple objectives of the calibration (Madsen, 2000; 2003). In distributed models, there may also be advantages of calibrating different areas, such as subcatchments, separately and independently rather than as an integrated whole (e.g. Seibert *et al.*, 2000). Ratto *et al.* (2001) highlight the utility of the global sensitivity analysis (GSA) and generalized likelihood uncertainty estimation (GLUE) approaches (see below) in the calibration of over-parameterized models with strong parameter interaction. GSA is a model-independent approach, which is based on estimating the fractional contribution of each input factor to the variance in the model output, accounting also for interaction terms. GLUE allows model runs to be classified according to the likelihood of their being a good simulator of the system, recognizing that many different combinations of parameter values can produce accurate model results (the issue of equifinality). By applying GSA to the GLUE likelihoods, the parameter sets driving model simulations with a good fit to observations are identified along with the basic features of the parameter interaction structure. Extensive model calibration tends to remove the physical basis of a model. Part of the objective in building a physically based model should be to produce a sufficiently good model structure and conceptualization to avoid the need for substantial calibration.

1.6.5 Testing models

The terms verification and validation have very specific meanings in computer science. Verification is used to denote the process of checking that the computer code (program) does exactly what the algorithm is designed to do. As well as a formal confirmation, the procedure also involves the removal of bugs that have crept into the program during its writing (due to typing mistakes as well as mis-conceptualizations). Validation, on the other hand, refers to the testing of the model output to confirm the results that should be produced in reality (Fishman and Kiviat, 1968). One common method of validation is the comparison of a numerical model against the analytical solution for specific boundary conditions or against field measured data for the period and place of the model simulation.

However, Oreskes *et al.* (1994) point out that the difference between these specific uses and the common usage of the same terms can often lead to confusion, particularly when model results are being presented to nonmodellers. Rykiel (1996) suggests that the terms are essentially synonymous in everyday language, so the distinction is hard to see to a nonuser. Furthermore, the roots of the words may imply that a model is better than was actually intended when the author of a paper noted that the model was verified and validated. The root meaning of verify comes from the Latin *verus*, meaning truth, while the Latin *validare* means to declare or give legal authority to something. Thus Oreskes *et al.* (1994) suggest that the nonmodeller may tend to feel a verified model presents the truth, and one that is validated can have legal authority, or is at least 'does not contain known or detectable flaws and is internally consistent'. They suggest that benchmarking is a more appropriate term for verification and model evaluation should be used in place of validation. However, it could be argued that these are equally value-laden terms. In reality, most model output actually seems to generate a healthy dose of scepticism in nonmodellers (see the debate in Aber, 1997, 1998; Dale and Van Winkle, 1998; Van Winkle and Dale, 1998, for example). Lane and Richards (2001), on the other hand, suggest that validation is used as a linguistic means of hiding from such criticism.

Much more fundamentally in this debate, the nature of environment systems and scientific practice means that whatever term is used for validation/model evaluation, it will always tend to overstate the case for belief in the model results. There are six reasons stated by Oreskes *et al.* (1994) for this problem. First, all environmental systems are open. Logically, it is only possible to demonstrate the truth of a closed system (although even this proposition is called into question by Gödel's theorem – see the excellent overview by Hofstadter, 1979). Second, there are problems due to the presence of unknown parameters and the scaling of nonadditive parameters (see below and Chapter 19). Third, inferences and embedded assumptions underlie all stages of the observation and measurement of variables – dependent and independent alike. Fourth, most scientific theories are developed by the addition of 'auxiliary hypotheses', that is, those not central to the principal theory, but fundamental in a specific context for putting it into action. Thus, it is impossible to tell whether the principal or an auxiliary hypothesis is incorrect should deductive verification fail. Fifth, as we have seen, more than one model formulation can provide the same output. This property is known formally as nonuniqueness or underdetermination (the Duhem–Quine thesis: Harding, 1976). Sixth, errors in auxiliary hypotheses may cancel, causing incorrect acceptance of the model. Many modellers would now accept that full validation is a logical impossibility (e.g. Refsgaard and Storm, 1996; Senarath *et al.*, 2000). Morton and Suárez (2001) suggest that in most practical contexts the term 'model' can be thought of as being synonymous with 'theory' or 'hypothesis', with the added implication that they are being confronted and evaluated with data. Often the models represent simplifications of complex, physically based theories, analogies of other systems, summaries of data, or representations of the theories themselves. It is this set of approaches that allows the provisional nature of scientific knowledge to be tested. Conversely, it is possible for models to continue being used for a range of reasons relating to the social, economic and political contexts of science (Oreskes and Bellitz, 2001).

Rykiel (1996) provides an overview of how validation has been employed in modelling, and distinguishes (a) operational or whole-model validation (correspondence of model output with real-world observations); (b) conceptual validation (evaluation of the underlying theories and assumptions); and (c) data validation (evaluation of the data used to test the model). He suggests that there are at least 13 different sorts of validation procedure that are commonly employed, explicitly or implicitly. These procedures are:

(a) face validation – the evaluation of whether model logic and outputs appear reasonable;
(b) Turing tests – where 'experts' are asked to distinguish between real-world and model output (by analogy with the test for artificial intelligence);
(c) visualization techniques – often associated with a statement that declares how well the modelled results match the observed data;
(d) comparison with other models – used for example in general circulation model evaluations (although note the high likelihood of developing an argument based on circular logic here!);
(e) internal validity – e.g. using the same data set repeatedly in a stochastic model to evaluate whether the distribution of outcomes is always reasonable;
(f) event validity – i.e. whether the occurrence and pattern of a specific event are reproduced by the model;
(g) historical data validation – using split-sample techniques to provide a subset of data to build a model and a second subset against which to test the model results (see also Klemeš, 1983);

(h) extreme-condition tests – whether the model behaves 'reasonably' under extreme combinations of inputs;
(i) traces – whether the changes of a variable through time in the model are realistic;
(j) sensitivity analyses – to evaluate whether changes in parameter values produce 'reasonable' changes in model output (see below);
(k) multistage validation (corresponding to the stages a, b and c noted above);
(l) predictive validation – comparison of model output with actual behaviour of the system in question;
(m) statistical validation – whether the range of model behaviour and its error structure match that of the observed system (but see the discussion on error propagation below).

Clearly, all these tests provide some support for the acceptance of a model, although some are more rigorous than others. The more tests a model can successfully pass, the more confidence we might have in it, although this is still no reason to believe it absolutely for the reasons discussed above. But in complex models, validation is certainly a nontrivial procedure – Brown and Kulasiri (1996: 132) note, for example, that 'a model can be considered to be successfully validated if all available techniques fail to distinguish between field and model data'. Any model test will in part be evaluating the simplifications upon which the model is based, and in part the reasonableness of its parameterization. However, if a number of parameterizations fail for a specific model, we might seriously reconsider its conceptual basis. As with other aspects of modelling, evaluation is an iterative process.

1.6.6 Measurements of model goodness-of-fit

Calibration and validation or model evaluation as outlined above all require some measurement of how well the model represents actual measurements. These measurements are often known as objective functions, or goodness-of-fit statistics. It should be recognized that there are a number of different goodness-of-fit measures, and they will each be sensitive to different aspects of model (mis.)behaviour. The choice of an appropriate measure is therefore vital in a robust model evaluation. In the following discussion, we use M_i as a sequence of model outputs to be compared to O_i observed system values, $\overline{M}$ is the mean model output and $\overline{O}$ the mean observed value.

We can use the coefficient of determination, r^2, to represent the proportion of variance in the observed data explained by the model results. The value is calculated as:

$$r^2 = \left(\frac{\sum\limits_{i=1}^{n} [O_i - \overline{O}] \cdot [M_i - \overline{M}]}{\sqrt{\sum\limits_{i=1}^{n} [O_i - \overline{O}]^2} \cdot \sqrt{\sum\limits_{i=1}^{n} [M_i - \overline{M}]^2}} \right)^2 \quad (1.56)$$

or the ratio of the covariance in the observed and modelled values to the product of the individual variances. The value of r^2 varies from 1, which means all of the variance in the data is explained by the model, to 0, where none of the variance is explained. As noted by Legates and McCabe (1999), this measure is insensitive to constant, proportional deviations. In other words, perfect agreement occurs if the model consistently underestimates or overestimates by a specific proportion, so that there is a linear relationship between M_i and O_i which does not lie on the 1:1 line (compare the use of the coefficient of determination in evaluating regression models). As correlation is not a robust statistical measure, the value of r^2 will be sensitive to outliers (values that lie a large distance away from the mean), which will tend to increase its value spuriously if there are extreme events in the observed dataset.

The Nash and Sutcliffe (1970) model-efficiency measure is commonly used:

$$NS = 1 - \frac{\sum\limits_{i=1}^{n} (O_i - M_i)^2}{\sum\limits_{i=1}^{n} (O_i - \overline{O})^2} \quad (1.57)$$

which is a measure of the mean square error to the observed variance. If the error is zero, then $NS = 1$, and the model represents a perfect fit. If the error is the same magnitude as the observed variance, then $NS = 0$ and the observed mean value is as good a representation as the model (and thus the time invested in the model is clearly wasted). As the error continues to increase, the values of the index become more negative, with a theoretical worst-case scenario of $NS = -\infty$. NS does not suffer from proportional effects as in the case of r^2, but it is still sensitive to outliers (Legate and McCabe, 1999).

A modified 'index of agreement' was defined as the ratio of mean-square error to total potential error

by Willmott (1981):

$$W = 1 - \frac{\sum\limits_{i=1}^{n} (O_i - M_i)^2}{\sum\limits_{i=1}^{n} (|M_i - \overline{O}| + |O_i - \overline{O}|)^2} \quad (1.58)$$

where 1 again means a perfect model fit, but the worst model fit is represented by a value of 0. As with the two previous measures, the squared terms make it sensitive to outliers. Legates and McCabe suggest that it may be better to adjust the comparison in the evaluation of NS and W to account for changing conditions, so that rather than using the observed mean, a value that is a function of time (e.g. a running mean) is more appropriate if the model covers highly variable situations. It may also be desirable to calculate absolute errors such as the root-mean square ($RMSE$: see below) and the mean absolute error:

$$MAE = \frac{\sum\limits_{i=1}^{n} |O_i - M_i|}{n} \quad (1.59)$$

and to use the extent to which $\frac{RMSE}{MAE} > 1$ as an indicator of the extent to which outliers are affecting the model evaluation. In a comparison of potential evapotranspiration and runoff models, Legates and McCabe found the best way of evaluating the results was to ignore r^2, and to use baseline-adjusted NS and W indices, together with an evaluation of $RMSE$ and MAE. Other authors have suggested that it is also useful to provide a normalized MAE, by dividing by the observed mean (thus giving a ratio measurement – compare the normalized $RMSE$ discussed below) (Alewell and Manderscheid, 1998). The effects of outliers can be minimized by making the comparison on values below a specific threshold (e.g. Madsen, 2000) or by using log-transformed values.

The choice of appropriate objective functions and model-evaluation procedures is a critical one. Using Bayesian analysis, Kirchner *et al.* (1996) demonstrate how insufficiently strict tests will cause a reasonable sceptic to continue to disbelieve a model, while the model developer may continue regardless. Only if a model has a chance of ≫90% of being rejected when it is a poor representation of observed values will the testing procedure be found convincing. As with all aspects of science, we must endeavour to use the most rigorous tests available that do not wrongly reject the model when the data are at fault.

1.7 SENSITIVITY ANALYSIS AND ITS ROLE

Sensitivity analysis is the process of defining model output sensitivity to changes in its input parameters. Sensitivity analysis is usually carried out as soon as model coding is complete and at this stage it has two benefits: to act as a check on the model logic and the robustness of the simulation and to define the importance of model parameters and thus the effort which must be invested in data acquisition for different parameters. The measurement of the sensitivity of a model to a parameter can also be viewed relative to the uncertainty involved in the measurement of that parameter in order to understand how important this uncertainty will be in terms of its impact on the model outcomes. If sensitivity analysis at this stage indicates that the model has a number of parameters to which the model is insensitive, then this may indicate over-parameterization and the need for further simplification of the model.

Sensitivity analysis is usually also carried out when a model has been fully parameterized and is often used as a means of learning from the model by understanding the impact of parameter forcing, and its cascade through model processes to impact upon model outputs (see, for example, Mulligan, 1996; Burke *et al.*, 1998; Michaelides and Wainwright, 2002). In this way the behaviour of aggregate processes and the nature of their interaction can better be understood. After calibration and validation, sensitivity analysis can also be used as a means of model experiment and this is very common in GCM studies where sensitivity experiments of global temperature to greenhouse forcing, to large-scale deforestation or to large-scale desertification are common experiments. Sensitivity analysis is also used in this way to examine the impacts of changes to the model structure itself, its boundary or initial conditions or the quality or quantity of data on its output (for example, May and Roeckner, 2001).

The sensitivity of model parameters is determined by their role in the model structure and if this role is a reasonable representation of their role in the system under study, then there should be similarities between the sensitivity of model output to parameter change and the sensitivity of the real system response to physical or process manipulation. Nevertheless one must beware of attributing the model sensitivity to parameter change as the sensitivity of the real system to similar changes in input (see Baker, 2000).

The methods of sensitivity analysis are covered in some detail by Hamby (1994) and, more recently, by Saltelli *et al.* (2000) and will not be outlined in detail here. In most sensitivity analyses a single parameter is varied incrementally around its normal value, keeping all other parameters unaltered. The model outputs of interest are monitored in response to these changes and the model sensitivity is usually expressed as the proportional change in the model output per unit change in the model input. In Figure 1.26 we show an example sensitivity analysis of the simple soil-erosion model (Equation 1.1), first in terms of single parameters and then as a multivariate sensitivity analysis. The former demonstrates the relative importance of vegetation cover, then slope, runoff and finally soil erodibility in controlling the amount of erosion according to the model. The multivariate analysis suggests that spatially variable parameters can have significant and sometimes counter-intuitive impacts on the sensitivity of the overall system. A sensitive parameter is one that changes the model outputs of interest significantly per unit change in its value and an insensitive parameter is one which has little effect on the model outputs of interest (though it may have effects on other aspects of the model). Model sensitivity to a parameter will also depend on the value of other model parameters, especially in systems where thresholds operate, even where these remain the same between model runs. It is important to recognize the propensity for parameter change in sensitivity analysis, that is a model can be highly sensitive to changes in a particular parameter but if changes of that magnitude are unlikely ever to be realized, then the model sensitivity to them will be of little relevance. In this way, some careful judgement is required of the modeller to set the appropriate bounds for parameter variation and the appropriate values of varying or nonvarying parameters during the process of sensitivity analysis.

Sensitivity analysis is a very powerful tool for interacting with simple or complex models. Sensitivity analysis is used to do the following:

1. better understand the behaviour of the model, particularly in terms of the ways in which parameters interact;
2. verify (in the computer-science sense) multi-component models;
3. ensure model parsimony by the rejection of parameters or processes to which the model is not sensitive;
4. target field parameterization and validation programmes for optimal data collection;
5. provide a means of better understanding parts of or the whole of the system being modelled.

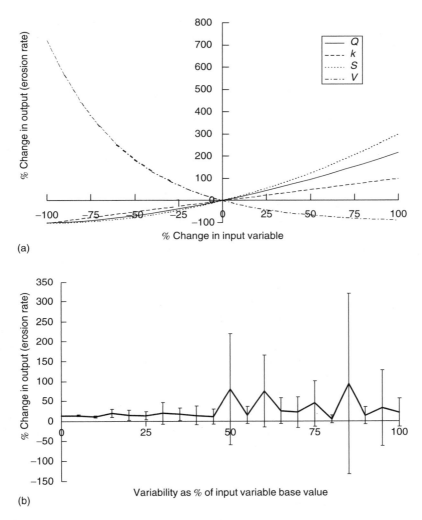

Figure 1.26 Example sensitivity analysis of the simple erosion model given in Equation 1.1: (a) univariate sensitivity analysis of the simple erosion model given in Equation 1.1. Base values are $Q = 100\,\text{mm}\,\text{month}^{-1}$, $k = 0.2$, $S = 0.5$, $m = 1.66$, $n = 2.0$, $i = 0.07$ and $V = 30\%$. The variables Q, k, S and V are varied individually from -100% to $+100\%$ of their base values and the output compared. Note that k has a positive linear response; Q a nonlinear response faster than k; S a nonlinear response faster than Q (because Q is raised to the power $m = 1.66$ while S is raised to the power $n = 2$); and V a negative exponential response. The order of parameter sensitivity is therefore $V > S > Q > k$.; and (b) Multivariate sensitivity analysis of the same model, where normally distributed variability is randomly added to each of the parameters as a proportion of the base value. Note the large fluctuations for large amounts of variability, suggesting that the model is highly sensitive where variability of parameters is $>50\%$ of the mean parameter value. No interactions or autocorrelations between parameter variations have been taken into account

1.8 ERRORS AND UNCERTAINTY

1.8.1 Error

No measurement can be made without error. (If you doubt this statement, get ten different people to write down the dimensions in mm of this page, without telling each other their measurements, and compare the results.) Although Heisenberg's uncertainty principle properly deals with phenomena at the quantum scale, there is always an element of interference when making an observation. Thus, the act of observation perturbs what we are measuring. Some systems may be particularly sensitive to these perturbations, for example, when we introduce devices into a river to measure patterns of turbulence. The very act of placing a flow meter into the flow causes the local structure of flow to change. If we were interested in the section discharge rather than the local changes in velocity, our single measuring device would have less significant impacts by perturbation, but the point measurement would be a very bad measurement of the cross-section flow. To counter this problem, we may repeat the measurement at a number of positions across the cross-section and provide an average discharge (usually weighting by the width of flow represented by each measurement). But this average will only be as good as the choice of positions taken to represent the flow. Sampling theory suggests that a greater number of measurements will provide a better representation, with the standard error decreasing with the square root of the number of measurements made. However, a larger number of samples will take a longer time to make, and thus we have possible temporal changes to contend with in giving added error. Clearly, this approach is impractical when flows are rapidly changing. If we require continuous measurements, we may build a structure into the flow, such as a flume or weir (e.g. French, 1986) which again perturbs the system being measured (possibly with major effects if the local gradient is modified significantly, or if sediment is removed from the system). Furthermore, the devices used to measure flow through these structures will have their own uncertainties of measurement, even if they are state-of-the-art electronic devices. In the flume case, the measurement is usually of a depth, which is then calibrated to a discharge by use of a rating curve. The rating curve itself will have an uncertainty element, and is usually proportional to depth to a power greater than one. Any error in measurement of depth will therefore be amplified in the estimation of discharge. Such measurements also tend to be costly, and the cost and disturbance therefore prevent a large number of measurements being taken in a specific area, which is problematic if we are interested in spatial variability.

Other environmental modelling questions might require even further perturbation of the system. For example, soil depth is a critical control of water flow into and through the soil and thus has an impact on other systems too, such as vegetation growth or mass movement. In a single setting, we might dig a soil pit. Even if we try to replace the soil as closely as possible in the order in which we removed it, there will clearly be a major modification to the local conditions

(most notably through bulk-density changes, which often mean a mound is left after refilling; a lower bulk density means more pore space for water to flow through, so we have significantly altered what we intended to observe). Even if we were foolish enough to want to use this approach at the landscape scale, it is clearly not feasible (or legal!), so we might use a spatial sample of auger holes. However, the auger might hit a stone and thus have difficulty penetrating to the soil base – in stony soils, we usually only reach into the upper part of the C horizon. We might therefore try a noninvasive technique such as ground-penetrating radar (GPR). GPR uses the measured reflections of transmitted radio frequency waves (usually in the 25–1200 MHz range) to provide 'observations' of the changes in dielectric constant, in engineering structures such as building and bridges or in the ground. Dielectric constant is controlled by changes in material density and water content. Where transmitted waves encounter a change in dielectric constant, some energy passes through the interface and some is reflected. The reflected energy from a GPR transmitter is recorded on a nearby receiver with the time delay (in nanoseconds) between the transmission of the pulse and its receipt indicating the distance of the reflecting object from the transmitter-receiver array. In this way GPR can image the subsurface and has found applications in archaeology (Imai *et al.*, 1987), hydrology (van Overmeeren *et al.*, 1997), glaciology (Nicollin and Kofman, 1994) and geology (Mellett, 1995). The difficulty with electromagnetic noninvasive techniques is that while dielectric discontinuities can be fairly obviously seen, the techniques provide little information on what these discontinuities are (rocks, roots or moisture, for example). Thus, noninvasive techniques are also subject to significant potential error. The implication is that all measurements should have their associated error cited so that the implications can be considered and due care be taken in interpreting results. Care is particularly necessary in the use of secondary data, where you may have very little idea about how the data were collected and quality controlled.

Field measurements are often particularly prone to error, because of the difficulty of collecting data. We may choose to use techniques that provide rapid results but which perhaps provide less precise measurements, because of the high costs involved in obtaining field data. Note the difference between error and precision (sometimes called the tolerance of a measurement) – the latter relates only to the exactness with which a measurement is taken. A lack of precision may give very specific problems when measuring features with fractal characteristics, or when dealing with systems that are sensitive to initial conditions. Thus, a consideration of the modelling requirements is often important when deciding the precision of a measurement.

Specification errors can arise when what is being measured does not correspond to the conceptualization of a parameter in the model. This problem may arise if a number of processes are incorporated into a single parameter. For example, if erosion is being considered as a diffusion process in a particular model, diffusion may occur by a number of processes, including rainsplash, ploughing, animal activity and soil creep. The first two might be relatively easy to measure, albeit with their own inherent problems (e.g. Torri and Poesen, 1988), while the latter may be more difficult to quantify either because of inherent variability in the case of bioturbation or because of the slow rate of the process in the case of creep. Interactions between the different processes may make the measurement of a compound diffusion parameter unreliable. It is also possible that different ways of measurement, apparently of the same process, can give very different results. Wainwright *et al.* (2000) illustrate how a number of measurements in rainfall-runoff modelling can be problematic, including how apparent differences in infiltration rate can be generated in very similar rainfall-simulation experiments. Using pumps to remove water from the surface of the plot led to significant over-estimation of final infiltration because of incomplete recovery of ponded water, when compared to direct measurement of runoff from the base of the plot (which itself incorporates ponding into the amount of infiltration and thus also over-estimates the real rate). The pumping technique also introduces a significant time delay to measurements, so that unsaturated infiltration is very poorly represented by this method. Differences between infiltration measured using rainfall simulation, cylinder infiltration and the falling-head technique from soil cores for the same location can be orders of magnitude (e.g. Wainwright, 1996) because each are representing infiltration in different ways. Different infiltration models may be better able to use measurements using one technique rather than another. Such specification errors can be very difficult to quantify, and may in fact only become apparent when problems arise during the modelling process. It should always be borne in mind that errors in model output may be due to incorrect parameter specification. When errors occur, it is an important part of the modelling process to return to the parameters and evaluate whether the errors could be caused in this way.

Environmental models operate in a defined space in the real world. However, the representation of that space will always be some form of simplification. At the

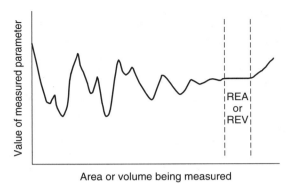

Figure 1.27 Definition of the representative elemental area (REA) or volume (REV) concept

extreme case, the system will be completely lumped, with single values for parameters and each input and output. Such models can be a useful generalization, for example in the forecasting of flooding or reservoir filling (e.g. Blackie and Eeles, 1985 – another example is the population models discussed above, although these can be spatialized as shown by Thornes, 1990). However, the definition of each parameter may be non-trivial for all but the simplest of catchments. Wood *et al.* (1988) used the term representative elemental area (REA) to evaluate the scale at which a model parameter might be appropriate (Figure 1.27). At the opposite end of the spectrum is the fully distributed model, in which all parameters are spatialized. There still remains the issue of the REA in relation to the grid size used (distributed applications may still have grid sizes of kilometres or hundreds of kilometres in the case of General Circulation Models). However, in addition, there is the issue of how to estimate parameters spatially. Field measurements are costly so that extensive data collection may be impossible financially, even if the people and equipment were available on a sufficient scale. Therefore, it is usual to use some sort of estimation technique to relate parameters to some easily measured property. For example, Parsons *et al.* (1997) used surface stone cover to estimate infiltration rates, finding that the spatial structure provided by this approach gave a better solution than simply assigning spatial values based on a distribution function of measured infiltration rates, despite the relatively high error in the calibration between stone cover and infiltration rate. Model sensitivity to different parameters (see above) may mean that different techniques of spatialization are appropriate for the parameters of the model in question. It is important that distributed models are tested with spatially distributed data, otherwise it is possible to arrive at completely different conclusions about their reliability (Grayson *et al.*, 1992a,b). Similarly, if the scale of the output is not related to the scale of the test data, errors in interpretation can arise. It is for this reason that techniques of upscaling or downscaling of model results are important (see Chapter 19).

Engeln-Müllges and Uhlig (1996) classify all of the above problems as input errors. They may be present as errors in model parameterization, or in data against which the model is tested. It is possible that a model could be rejected because of the latter errors, so we should be careful to specify the magnitude of errors as far as possible at every stage in our approach. There are three further types of error associated with the modelling process itself. These are procedural, computational and propagation errors (Engeln–Müllges and Uhlig 1996).

Procedural errors arise from using a numerical approximation to a problem where there is no known analytical solution. These errors are the difference between the exact and approximate solutions. Indeed, a useful test for model behaviour is a comparison of a numerical approximation against a known analytical solution. We saw above how the backward-difference solution to the kinematic wave equation for water flow (Equations 1.34 to 1.39) had errors associated with $O(h)$ and $O(k)$ – that is of the same order of magnitude of the timestep and grid size. In terms of the procedural error, a central difference solution is better in that its errors would be $O(h^2)$ and $O(k^2)$ and thus much smaller

format. However, we may encounter two common problems when using integer values. First, overflow (or underflow) may occur if the space (memory) allocated to storing the number is not sufficient. For a two-byte, signed integer (i.e. 15 binary bits used to store the number and one bit to store the sign), the range of values possible is $\pm 32\,767$ (i.e. $\pm 2^{15}$). Using four bytes, the range increases to $\pm 2\,147\,483\,647$ (i.e. $\pm 2^{31}$ as 31 bits are used to store the number and one bit for the sign). Second, if we divide one integer by another, there may be a remainder, which will be discarded if the result is also stored in integer format.

Floating-point numbers are stored in a three-part format, made up of a single bit representing the sign, eight bits representing the exponent and 23 bits representing the mantissa. These make up any number as sign $\times$ Mantissa $\times 2^{\text{exponent}-\text{bias}}$, where the bias of the exponent is a fixed integer for any given computer (Press *et al.*, 1992). Some numbers cannot be represented exactly in this format – either because they are irrational (e.g. π, e, $\sqrt{2}$), or because the binary form is not precise enough to accommodate them. In the form above, numbers with more than six or seven decimal digits cannot be represented. The resulting number may either be chopped (truncated) or rounded depending on the computer, and the associated error is called round-off error. Press *et al.* (1992) give the example of calculating $3 + 10^{-7}$. With a bias of 151, the two values are stored as shown in this table:

Sign	Exponent	Mantissa	Decimal equivalent
		Binary representation of number	
0	1 0 0 0 0 0 1 0	1 1 0 0 0 0 0 0 0 0 0 0 0 0 0 0 0 0 0 0 0 0 0	3
0	0 1 1 0 1 0 0 1	1 1 0 1 0 1 1 0 1 0 1 1 1 1 1 1 1 0 0 1 0 1 0	10^{-7}

because of the higher order differences involved. Thus, some balance must be made between the amount of procedural error, the computational resources required and the potential for side effects such as numerical diffusion.

Computational errors arise during the computer representation of the algorithm that operationalizes the model. Accumulated computational error is the sum of local computational errors incurred during the carrying out of the algorithm. Local computational errors are typically due to rounding or truncation. To see how this might be a problem, let us look at how a computer (or calculator, for that matter) actually stores a number.

All computers store numbers in binary format. All integer numbers can be exactly represented in this

To carry out the addition in binary arithmetic, the exponent must be identical. To maintain the identical number, every time the exponent is increased, the binary number in the mantissa must be right-shifted (all the digits are moved one space to the right and a zero added in the empty space at the beginning). To convert 10^{-7} (exponent 01101001 binary $= 105$ decimal) into the same form as 3 (exponent 10000010 binary $= 130$ decimal), we require 25 right shifts of the mantissa. As the length of the mantissa is only 23 bits, it will now be full of zeroes, so the result of $3 + 10^{-7}$ would simply be 3, and a truncation error of 10^{-7} would have occurred.

Many programming languages allow the possibility of storing numbers with 64 bits (called either long real or double precision). This increase of storage space extends the range of values that can be represented from $c. \pm 1 \times 10^{\pm 38}$ to $c. \pm 1 \times 10^{\pm 308}$ (but also increases the amount of memory required to run a program and store the results). In the example given above, it would be possible to right-shift the mantissa of 10^{-7} 25 times while still storing all the information present, so $3 + 10^{-7}$ would be correctly represented. Truncation errors will still occur, but will be of a lower order of magnitude than storage at single precision. We should not assume that truncation errors will cancel. It is possible that repeated calculations of this sort will be carried out, so that the total error due to truncation will increase through time.

Propagation errors are the errors generated in the solution of a problem due to input, procedural and computational errors. At each stage a calculation is made based directly or indirectly on an input value that has an associated error, that error affects the subsequent calculation, and thus is said to propagate through the model. As more model iterations are carried out, the magnitude of the propagated errors will tend to increase, and affect more of the solution space (Figure 1.28). Similarly, procedural errors and rounding errors will also affect subsequent calculations, and thus propagate through the model. Analysis of error propagation can be carried out analytically using the approach outlined in Engeln-Müllges and Uhlig (1996: 7–9). If x is the actual value and $\hat{x}$ the value measured with error, then the relative error of the measurement is:

$$\varepsilon_x = \frac{|x - \hat{x}|}{|x|}, \quad x \neq 0 \qquad (1.60a)$$

or

$$\varepsilon_x = \frac{|x - \hat{x}|}{|\hat{x}|}, \quad \hat{x} \neq 0, \qquad (1.60b)$$

so that for basic arithmetic operations, propagation errors are as follows:

1. A sum is well conditioned (reasonably bounded) if both inputs have the same sign. If $y = (x_1 \pm \varepsilon_{x_1}) + (x_2 \pm \varepsilon_{x_2})$ then:

$$\varepsilon_y \approx \frac{x_1}{x_1 + x_2} \varepsilon_{x_1} + \frac{x_2}{x_1 + x_2} \varepsilon_{x_2}. \qquad (1.61)$$

Because the fractions $\frac{x_1}{x_1 + x_2} + \frac{x_2}{x_1 + x_2} \equiv 1$, and the relative errors are always positive, only a fraction of each of the errors is propagated. This process can be repeated for the addition of any number, n, of variables x, using the fraction $\frac{x_i}{\sum_{j=1}^{n} x_j}$ to multiply each error term. If the errors are known as standard

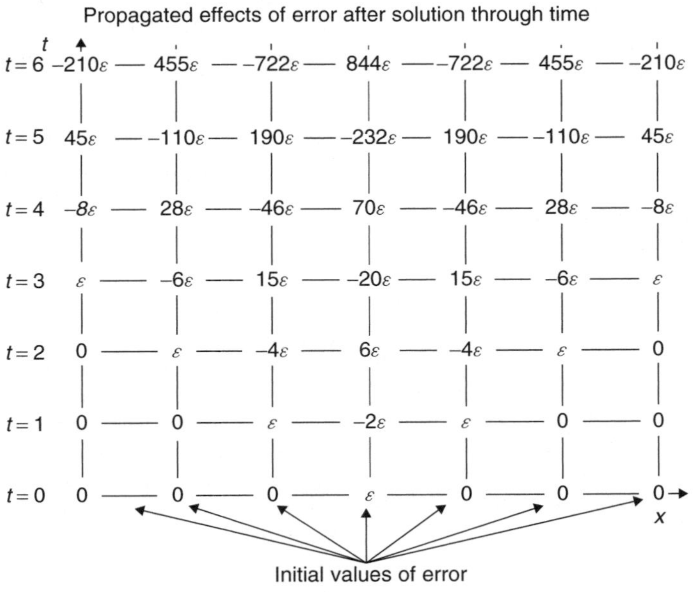

Propagated effects of error after solution through time

Figure 1.28 Effects of using a forward difference scheme in propagating errors through the solution domain of a model. Note that the errors spread spatially and increase through time

deviations σ_{x_i}, then the propagated error has a slightly different form:

$$\varepsilon_y \approx \sqrt{\sum_{i=1}^{n} \sigma_{x_i}^2} \qquad (1.62)$$

2. However, if the values of x_1 and x_2 have opposite signs and are nearly equal, amplification of errors occurs, because even though the fractions above still sum to one, at least one will be < -1 or > 1. Thus, the magnitude at least one of the errors will increase through propagation. This form of model should therefore be avoided. (But note that if both x_1 and x_2 are negative, the first rule applies.)
3. In the case of a product ($y = (x_1 \pm \varepsilon_{x_1}) \cdot (x_2 \pm \varepsilon_{x_2})$), error propagation is well conditioned, and related to the sum of the individual errors of the inputs:

$$\varepsilon_y \approx \varepsilon_{x_1} + \varepsilon_{x_2} \qquad (1.63)$$

Again, any number of multiplications can be propagated by successive adding of the error terms. Where the errors are known as standard deviations, we use:

$$\varepsilon_y \approx \sqrt{\sum_{i=1}^{n} \frac{\sigma_{x_i}^2}{x_{x_i}}} \qquad (1.64)$$

4. For a quotient $\left(y = \dfrac{x_1 \pm \varepsilon_{x_1}}{x_2 \pm \varepsilon_{x_2}}, \text{ assuming } x_2 \neq 0 \right)$, error propagation is well conditioned, and related to the difference in the error of the numerator and denominator in the function:

$$\varepsilon_y \approx \varepsilon_{x_1} - \varepsilon_{x_2} \qquad (1.65)$$

5. Propagation error of a power ($y = (x_1 \pm \varepsilon_{x_1})^p$) is related to the product of the power and the input error:

$$\varepsilon_y \approx p\varepsilon_{x_1} \qquad (1.66)$$

Thus roots (powers less than one) are well conditioned, and powers (>1) are ill conditioned. As the value of the power term increases, so does the error propagation. Wainwright and Parsons (1998) note the implications of this results in soil-erosion modelling, where certain model formulations will be significantly more susceptible to error than others. For an error defined as a standard deviation, we use:

$$\varepsilon_y \approx \sqrt{\frac{p\sigma_{x_1}^2}{x_{x_1}}} \qquad (1.67)$$

6. Combinations of the above operations can be carried out by combination of the relevant rules in the appropriate order.
7. Other functions require differentiation to be carried out, so that if $y = f(x)$, then:

$$\varepsilon_y \approx \frac{df(x)}{dx}\varepsilon_x \qquad (1.68)$$

so that if $y = \sin(x \pm \varepsilon_x)$, $\varepsilon_y \approx (\cos x) \cdot \varepsilon_x$ or $\varepsilon_y \approx \sqrt{[(\cos x) \cdot \sigma_x]^2}$. For multiple functions $y = f(x_i)$, $i = 1, \ldots, n$, partial differentiation with respect to each variable x_i must be carried out and summed:

$$\varepsilon_y \approx \sum_{i=1}^{n} \left| \frac{\partial f(x_i)}{\partial x_i} \right| \varepsilon_{x_i} \qquad (1.69)$$

For the example, if $y = \sin(x_1 \pm \varepsilon_x) \cdot e^{(x_2 \pm \varepsilon_{x_2})}$, the associated error will be $\varepsilon_y \approx (\cos x_1) \cdot \varepsilon_{x_1} + e^{x_2}\varepsilon_{x_2}$ or $\varepsilon_y \approx \sqrt{[(\cos x_1) \cdot \sigma_{x_1}]^2 + [e^{x_2}\sigma_{x_2}]^2}$.

An alternative approach to looking at error propagation is to use Monte Carlo analysis. This approach uses multiple realizations of a model using distribution functions to generate errors in inputs (e.g. De Roo *et al.*, 1992; Phillips and Marks, 1996; Fisher *et al.*, 1997; Parsons *et al.*, 1997; Nitzsche *et al.*, 2000). From these multiple realizations, error bands can be defined to give the range of possible outputs or even confidence intervals on outputs (e.g. Håkanson, 1996). The advantages of Monte Carlo analysis are that it can deal with the effects of spatially distributed parameters, whereas the analytical approach outlined above is more limited in this case, and that it can deal with relatively complex interactions of parameter error, which is impossible using the analytical approach. As a disadvantage, reliable results often require many tens or hundreds of simulations (Heuvelink, 1992), which are extremely costly in terms of computer resources for large simulations.

Beven and Binley (1992) developed the Generalized Likelihood Uncertainty Estimation (GLUE) technique to evaluate error propagation. The technique uses Bayesian estimators to evaluate the likelihood that specific combinations of model parameters are good predictors of the system in question. Monte Carlo simulations are used to provide the various parameter combinations, and the likelihood measured against a specified objective function (Figure 1.29). The choice of parameter combinations may be subjective, based on the experience of the user of the technique, and often fails to account for necessary relationships between parameter values. An advantage of the technique is that it requires no assumptions to be made about the distribution functions

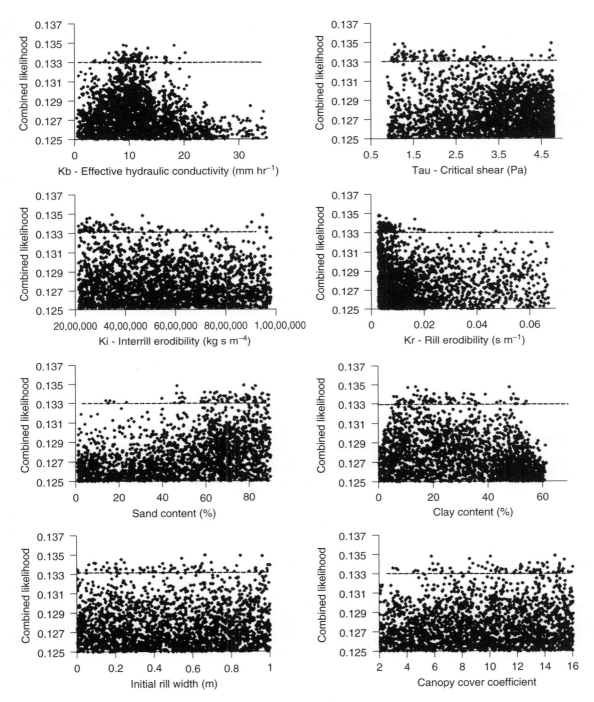

Figure 1.29 Example of error analysis using the GLUE approach: the study by Brazier *et al.* (2000) looking at the WEPP soil-erosion model showing the joint likelihood of goodness-of-fit for different combinations of eight parameters

of the input data (Beven, 1993). Choice of the objective function requires care that the key outputs of the model simulations are being tested reliably, as noted by Legates and McCabe (1999). Other techniques that can be used to estimate model uncertainty include Latin hypercube simulation, Rosenbleuth's and Harr's point-estimation methods, all of which can be used to sample the input-parameter space efficiently. Yu *et al.* (2001) found that Latin hypercube simulation provided a very efficient means of replicating more time-intensive Monte Carlo techniques (see Press *et al.*, 1992, for more detail on Latin hypercube sampling).

Hall and Anderson (2002) criticize error-estimation techniques based on objective measures because the level of data and detailed measurements required are lacking in many environmental modelling applications. For example, in the case of an extreme flood, the effects of the flood might be measurable (water depth from trash lines, aerial photographs of flood extent) after the event, but there may be little in the way of detailed measurements during the event. Rainfall fields are particularly difficult to reconstruct in high resolution in space and time; even if flow monitoring equipment were present, it may have been damaged during the passage of the flood wave (Reid and Frostick, 1996). Techniques therefore need to be developed that incorporate all information appropriate to model evaluation, including qualitative indications of permissible model behaviour.

1.8.2 Reporting error

We noted how the relative error of a measurement can be made (Equation 1.60). Different types of measurements of error also exist (Engeln-Müllges and Uhlig, 1996). Using the same notation as before, the true error of $\hat{x}$ is:

$$\Delta_x = x - \hat{x} \qquad (1.70)$$

while the absolute error of $\hat{x}$ is:

$$|\Delta_x| = |x - \hat{x}| \qquad (1.71)$$

and the percentage error of $\hat{x}$ is:

$$\frac{|x - \hat{x}|}{|x|} \cdot 100 \quad x \neq 0 \qquad (1.72)$$

If we have a series of values, $\hat{x}_i$, representing measurements with error on true values x_i, it is common to combine them as a root-mean square (RMS) error as follows:

$$RMS_x = \sqrt{\sum_{i=1}^{n} (x_i - \hat{x}_i)^2} \qquad (1.73)$$

RMS errors are in the same units as the original measurements, so are easy to interpret on a case-by-case basis. When comparing between cases with very different values, it may be useful to calculate a normalized RMS error:

$$NRMS_x = \frac{\sqrt{\sum_{i=1}^{n} (x_i - \hat{x}_i)^2}}{\sum_{i=1}^{n} x_i} \qquad (1.74)$$

which gives the proportion of error relative to the original value, and can thus be easily converted into a percentage error. In other cases, the goodness-of-fit measures described above may also be appropriate.

1.8.3 From error to uncertainty

Zimmermann (2000) defines six causes of uncertainty in the modelling process: lack of information, abundance of information, conflicting evidence, ambiguity, measurement uncertainty and belief. A *lack of information* requires us to collect more information, but it is important to recognize that the quality of the information also needs to be appropriate. It must be directed towards the modelling aims and may require the modification of the ways in which parameters are conceived of and collected. Information abundance relates to the complexity of environmental systems and our inability to perceive large amounts of complex information. Rather than collecting new data, this cause of uncertainty requires the simplification of information, perhaps using statistical and data-mining techniques. *Conflicting evidence* requires the application of quality control to evaluate whether conflicts are due to errors or are really present. Conflicts may also point to the fact that the model being used is itself wrong, so re-evaluation of the model structure and interaction of components may be necessary. *Ambiguity* relates to the reporting of information in a way that may provide confusion. Uncertainty can be removed by questioning the original informant, although this approach may not be possible in all cases. *Measurement uncertainty* may be reduced by invoking more precise techniques, although it must be done in an appropriate way. There is often a tendency to assume that modern gadgetry will allow measurement with fewer errors. It must be noted that other errors can be introduced (e.g. misrecording of electronic data if a data logger gets wet during a storm event) or that the new measurement may not be measuring exactly the same property as before. *Beliefs* about how data are to be interpreted can also cause uncertainty because different outcomes can result from the same starting point. Overcoming this uncertainty is a matter of iterative testing of the different belief structures. Qualitative assessment is thus as much an aspect of uncertainty assessment as qualitative analysis.

The quantitative evaluation of uncertainty has been discussed above in detail. Error-propagation techniques can be used in relatively simple models (or their subcomponents) to evaluate the impact of an input error (measurement uncertainty) on the outcome. In more complex scenarios, Monte Carlo analysis is almost certainly necessary. If sufficient runs are performed, then probability estimates can be made about the outputs. Carrying out this approach on a model with a large number of parameters is a nontrivial exercise, and requires the development of appropriate sampling designs (Parysow *et al.*, 2000). The use of sensitivity analysis can also be used to optimize this process (Klepper, 1997; Bärlund and Tattari, 2001). Hall and Anderson (2002) note that some applications may involve so much uncertainty that it is better to talk about possible outcomes rather than give specific probabilities. Future scenarios of climate change evaluated using General Circulation Models (see Chapters 2 and 12) are a specific case in point here. Another approach that can be used to evaluate the uncertainty in outcome as a function of uncertain input data is fuzzy set theory. Torri *et al.* (1997) applied this approach to the estimation of soil erodibility in a global dataset and Özelkan and Duckstein (2001) have applied it to rainfall-runoff modelling. Because all measurements are uncertain, the data used for model testing will also include errors. It is important to beware of rejecting models because the evaluation data are not sufficiently strong to test it. Monte *et al.* (1996) presented a technique for incorporating such uncertainty into the model evaluation process.

Distributed models may require sophisticated visualization techniques to evaluate the uncertainty of the spatial data used as input (Wingle *et al.*, 1999). An important consideration is the development of appropriate spatial and spatio-temporal indices for model evaluation, based on the fact that spatial data and their associated errors will have autocorrelation structures to a greater or lesser extent. Autocorrelation of errors can introduce significant nonlinear effects on the model uncertainty (Henebry, 1995).

Certain systems may be much more sensitive to the impacts of uncertainty. Tainaka (1996) discusses the problem of spatially distributed predator–prey systems, where there is a phase transition between the occurrence of both predator and prey, and the extinction of the predator species. Such transitions can occur paradoxically when there is a rapid increase in the number of the prey species triggered by instability in nutrient availability, for example. Because the phase transition represents a large (catastrophic) change, the model will be very sensitive to uncertainty in the local region of the parameter space, and it can thus become difficult or impossible to interpret the cause of the change.

1.8.4 Coming to terms with error

Error is an important part of the modelling process (as with any scientific endeavour). It must therefore be incorporated within the framework of any approach taken, and any corresponding uncertainty evaluated as far as possible. A realistic approach and a healthy scepticism to model results are fundamental. It is at best misleading to present results without corresponding uncertainties. Such uncertainties have significant impacts on model applications, particularly the use of models in decision-making. Large uncertainties inevitably lead to the rejection of modelling as an appropriate technique in this context (Beck, 1987). Recent debates on possible future climate change reinforce this conclusion (see the excellent discussion in Rayner and Malone, 1998).

In terms of modelling practice, it is here that we come full circle. The implication of error and uncertainty is that we need to improve the basic inputs into our models. As we have seen, this improvement does not necessarily just mean collecting more data. It may mean that it is better to collect fewer samples, but with better control. Alternatively, it may be necessary to collect the same number of samples, but with a more appropriate spatial and/or temporal distribution. Ultimately, the iterations involved in modelling should not just be within computer code, but also between field and model application and testing.

1.9 CONCLUSION

Modelling provides a variety of tools with which we can increase our understanding of environmental systems. In many cases, this understanding is then practically applied to real-world problems. It is thus a powerful tool for tackling scientific questions and answering (green!) engineering problems. But its practice is also something of an art that requires intuition and imagination to achieve appropriate levels of abstraction from the real world to our ability to represent it in practical terms. As a research tool, it provides an important link between theory and observation, and provides a means of testing our ideas of how the world works. This link is important in that environmental scientists generally deal with temporal and spatial scales that are well beyond the limits of observation of the individual. It is important to recognize that modelling is not itself a 'black box' – it forms a continuum of techniques that may be appropriately applied in different contexts. Part of the art of modelling is the recognition of the appropriate context for the appropriate technique.

Hopefully, we have demonstrated that there is no single way of implementing a model in a specific context. As we will see in the following chapters, there are many different ways to use models in similar settings. How the implementation is carried out depends on a range of factors including the aims and objectives of the study, the spatial and temporal detail required, and the resources available to carry it out. This chapter has presented the range of currently available techniques that may be employed in model building and testing. More details of these techniques are found on the book's website and in the numerous references provided. We have also addressed a number of the important conceptual issues involved in the modelling process, again to suggest that the modelling process is not as uniform as is often assumed, and that debate is often fruitful within and between disciplines in order to tackle some of these issues.

REFERENCES

Aber, J. (1997) Why don't we believe the models?, *Bulletin of the Ecological Society of America* **78**, 232–233.

Aber, J. (1998) Mostly a misunderstanding, I believe, *Bulletin of the Ecological Society of America* **79**, 256–257.

Alewell, C. and Manderscheid, B. (1998) Use of objective criteria for the assessment of biogeochemical ecosystem models, *Ecological Modelling* **107**, 213–224.

Ames, W.F. (1992) *Numerical Methods for Partial Differential Equations*, Academic Press, London.

Arge, L., Chase, J.S., Halpin, P., Vitter, J.S., Urban, D. and Wickremesinghe, R. (2002) Flow computation on massive grid terrains, http://www.cs.duke.edu/geo*/terrflow/.

Baker, V.R. (1998) Paleohydrology and the hydrological sciences, in G. Benito, V.R. Baker and K.J. Gregory (eds) *Palaeohydrology and Environmental Change*, John Wiley & Sons, Chichester, 1–10.

Baker, V.R. (2000) South American palaeohydrology: future prospects and global perspective, *Quaternary International* **72**, 3–5.

Barahona, E. and Iriarte, U.A. (2001) An overview of the present state of standardization of soil sampling in Spain, *The Science of the Total Environment* **264**, 169–174.

Bardossy, A. (1997) Downscaling from GCMs to local climate through stochastic linkages, *Journal of Environmental Management* **49**, 7–17.

Bärlund, I. and Tattari, S. (2001) Ranking of parameters on the basis of their contribution to model uncertainty, *Ecological Modelling* **142**, 11–23.

Bar-Yam, Y. (2000) *Dynamics of Complex Systems*, Perseus Books, Reading, MA.

Bates, P.D. and Anderson, M.G. (1996) A preliminary investigation into the impact of initial conditions on flood inundation predictions using a time/space distributed sensitivity analysis, *Catena* **26**, 115–134.

Bates, P.D. and De Roo, A.P.J. (2000) A simple raster-based model for flood inundation simulation, *Journal of Hydrology* **236**, 54–77.

Beck, M.B. (1987) Water quality modelling: a review of the analysis of uncertainty, *Water Resources Research* **23**, 1393–1442.

Berec, L. (2002) Techniques of spatially explicit individual-based models: construction, simulation and mean-field analysis, *Ecological Modelling* **150**, 55–81.

Beuselinck, L., Hairsine, P.B., Sander, G.C. and Govers, G. (2002) Evaluating a multiclass net deposition equation in overland flow conditions, *Water Resources Research* **38** (7), 1109 [DOI: 10.1029/2001WR000250].

Beven, K. (1977) Hillslope hydrographs by the finite element method, *Earth Surface Processes* **2**, 13–28.

Beven, K. (1992) Future of distributed modelling (editorial), *Hydrological Processes* **6**, 279–298.

Beven, K. (1993) Prophecy, reality and uncertainty in distributed hydrological modelling, *Advances in Water Resources* **16**, 41–51.

Beven, K. (2001) On explanatory depth and predictive power, *Hydrological Processes* **15**, 3069–3072.

Beven, K. (2002) Towards an alternative blueprint for a physically based digitally simulated hydrologic response modelling system, *Hydrological Processes* **16**, 189–206.

Beven, K. and Binley, A.M. (1992) The future of distributed models: model calibration and predictive uncertainty, *Hydrological Processes* **6**, 279–298.

Blackie, J.R. and Eeles, C.W.O. (1985) Lumped catchment models, in M.G. Anderson and T.P. Burt (eds) *Hydrological Forecasting*, John Wiley & Sons, Chichester, 311–345.

Blum, A. (1992) *Neural Networks in C++: An Object-Oriented Framework for Building Connectionist Systems*, John Wiley & Sons, Chichester.

Bostock, L. and Chandler, S. (1981) *Mathematics: The Core Course for A-Level*, ST(P) Ltd., Cheltenham.

Botterweg, P. (1995) The user's influence on model calibration results: an example of the model SOIL, independently calibrated by two users, *Ecological Modelling*, **81**, 71–81.

Bousquet, F., Le Page, C., Bakam, I. and Takforyan, A. (2001) Multiagent simulations of hunting wild meat in a village in eastern Cameroon, *Ecological Modelling* **138**, 331–346.

Bradley, R.S. (1999) *Palaeoclimatology: Reconstructing Climates of the Quaternary*. 2nd edn, Academic Press, London.

Brazier, R.E., Beven, K.J., Freer, J. and Rowan, J.S. (2000) Equifinality and uncertainty in physically based soil erosion models: Application of the GLUE methodology to WEPP – the Water Erosion Prediction Project – for sites in the UK and USA, *Earth Surface Processes and Landforms* **25**, 825–845.

Brown, T.N. and Kulasiri, D. (1996) Validating models of complex, stochastic, biological systems, *Ecological Modelling* **86**, 129–134.

Bueso, M.C., Angulo, J.M. and Alonso, F.J. (1998) A state-space model approach to optimum spatial sampling design based on entropy, *Environmental and Ecological Statistics* **5**, 29–44.

Burden, R.L. and Faires, J.D. (1997) *Numerical Analysis*, 7th edn, Brooks/Cole, Pacific Grove, CA.

Burke, S., Mulligan, M. and Thornes, J.B. (1998) Key parameters that control recharge at a catchment scale: using sensitivity analysis and path analysis to define homogenous recharge classes, in H.S. Wheater and C. Kirby (eds) *Hydrology in a Changing Environment*, Vol. II, John Wiley & Sons, Chichester, 229–243.

Campolo, M., Andreussi, P. and Soldati, A. (1999) River flood forecasting with a neural network model, *Water Resources Research* **35**, 1191–1197.

Chao, C.T. and Thompson, S.K. (2001) Optimal adaptive selection of sampling sites, *Environmetrics* **12**, 517–538.

Cheng, C.T., Ou, C.P. and Chau, K.W. (2002) Combining a fuzzy optimal model with a genetic algorithm to solve multi-objective rainfall-runoff model calibration, *Journal of Hydrology* **268**, 72–86.

Chorley, R.J. and Kennedy, B.A. (1971) *Physical Geography: A Systems Approach*, Prentice Hall, London.

Cohn, J.P. (2002) Biosphere 2: turning an experiment into a research station, *BioScience* **52**, 218–223.

Collins, L. (1975) *An Introduction to Markov Chain Analysis*. GeoAbstracts, Norwich.

Cross, M. and Moscardini, A.O. (1985) *Learning the Art of Mathematical Modelling* John Wiley & Sons, Chichester.

Dale, V.H. and Van Winkle, W. (1998) Models provide understanding, not belief, *Bulletin of the Ecological Society of America* **79**, 169–170.

De Falco, I., Giordanoa, A., Luongob, G., Mazzarella, B. and Tarantino, E. (2002) The eruptive activity of Vesuvius and its neural architecture, *Journal of Volcanology and Geothermal Research* **113**, 111–118.

Dekker, S.C., Bouten, W. and Schaap, M.G. (2001) Analysing forest transpiration model errors with artificial neural networks, *Journal of Hydrology* **246**, 197–208.

Del Genio, A.D. (1996) Simulations of superrotation on slowly rotating planets: sensitivity to rotation and initial condition, *Icarus* **120**, 332–343.

De Roo, A.P.J., Hazelhoff, L. and Heuvelink, G.B.M. (1992) The use of Monte Carlo simulations to estimate the effects of spatial variability of infiltration on the output of a distributed hydrological and erosion model, *Hydrological Processes* **6**, 127–143.

De Sapio, R. (1976) *Calculus for the Life Sciences*, Freeman, San Francisco.

De Wit, M.J.M. and Pebesma, E.J. (2001) Nutrient fluxes at the river basin scale. II: the balance between data availability and model complexity, *Hydrological Processes* **15**, 761–775.

Di Giammarco, P., Todini, E. and Lamberti, P. (1996) A conservative finite elements approach to overland flow: the control volume finite element formulation, *Journal of Hydrology* **175**, 267–291.

Dingman, S.L. (1984) *Fluvial Hydrology*, W.H. Freeman, New York.

Dunin, F.X. (1976) Infiltration: its simulation for field conditions, in J.C. Rhodda (ed.) *Facets of Hydrology*, John Wiley & Sons, Chichester, 199–227.

Eckhardt, K. and Arnold, J.G. (2001) Automatic calibration of a distributed catchment model, *Journal of Hydrology* **251**, 103–109.

Einstein, H.A. (1942) Formulae for transportation of bed-load, *Transactions of the American Society of Civil Engineers* **107**, 561–577.

Endreny, T.A. and Wood, E.F. (2001) Representing elevation uncertainty in runoff modelling and flowpath mapping, *Hydrological Processes* **15**, 2223–2236.

Engeln-Müllges and Uhlig, F. (1996) *Numerical Algorithms with FORTRAN*, Springer Verlag, Berlin.

Epstein, J.M. and Axtell, R. (1996) *Growing Artificial Societies: Social Science from the Bottom Up*, Brookings Institution Press, Washington, DC. See also http://www.brookings.edu/sugarscape/

Ferrar, W.L. (1967) *Calculus for Beginners*, Clarendon Press, Oxford.

Ferziger, J.H. and Perić, M. (1999) *Computational Methods for Fluid Dynamics*, Springer Verlag, Berlin.

Fisher, P., Abrahart, R.J. and Herbinger, W. (1997) The sensitivity of two distributed non-point source pollution models to the spatial arrangement of the landscape, *Hydrological Processes* **11**, 241–252.

Fishman, G.S. and Kiviat, P.J. (1968) The statistics of discrete-event simulation, *Simulation* **10**, 185–191.

Freer, J., Beven, K. and Ambroise, B. (1996) Bayesian-estimation of uncertainty in runoff prediction and the value of data – an application of the GLUE approach, *Water Resources Research* **32**, 2161–2173.

French, M.N., Krajewski, W.F. and Cuykendall, R.R. (1992) Rainfall forecasting in space and time using a neural network, *Journal of Hydrology* **137**, 1–31.

French, R.H. (1986) *Open-Channel Hydraulics*, McGraw-Hill, New York.

Friedman, N.A (1968) *Basic Calculus*, Scott, Foresman, Glenview, IL.

Gershenfield, N. (1999) *The Nature of Mathematical Modelling*, Cambridge University Press, Cambridge.

Ginot, V., Le Page, C. and Souissi, S. (2002) A multi-agent's architecture to enhance end-user individual-based modelling, *Ecological Modelling* **157**, 23–41. See also http://www.avignon.inra.fr/mobidyc

Goodrich, D.C., Woolhiser, D.A. and Keefer, T.O. (1991) Kinematic routing using finite-elements on a triangular irregular network, *Water Resources Research*, **27**, 995–1003.

Gottardi, G. and Venutelli, M. (1997) LANDFLOW: computer program for the numerical simulation of two-dimensional overland flow, *Computers and Geosciences* **23**, 77–89.

Govindasamy, B., Caldeira, K. and Duffy, P.B. (2003) Geoengineering Earth's radiation balance to mitigate climate change from a quadrupling of CO_2, *Global and Planetary Change* **771**, 1–12.

Graf, W.H. and Altinakar, M.S. (1998) *Fluvial Hydraulics: Flow and Transport Processes in Channels of Simple Geometry*, John Wiley & Sons, Chichester.

Grayson, R.B., Moore, I.D. and McMahon, T.A. (1992a) Physically based hydrologic modelling. 1. A terrain-based model for investigative purposes, *Water Resources Research* **28**, 2639–2658.

Grayson, R.B., Moore, I.D. and McMahon, T.A. (1992b) Physically based hydrologic modelling. 2. Is the concept realistic?, *Water Resources Research* **28**, 2659–2666.

Guttorp, P. (1995) *Stochastic Modeling of Scientific Data*, Chapman & Hall, London.

Haeussler, E.F. and Paul, R.S. (1988) *Calculus for the Managerial, Life and Social Sciences*. Prentice Hall International, London.

Håkanson, L. (1996) A general method to define confidence limits for model predictions based on validations, *Ecological Modelling* **91**, 153–168.

Hall, G.W. (2001) Elevated atmospheric CO_2 alleviates drought stress in wheat, *Agriculture, Ecosystems & Environment* **87**, 261–271.

Hall, J. and Anderson, M.G. (2002) Handling uncertainty in extreme or unrepeatable hydrological processes – the need for an alternative paradigm, *Hydrological Processes* **16**, 1867–1870.

Hamby, D.M. (1994) A review of techniques for parameter sensitivity analysis of environmental models, *Environmental Monitoring and Assessment* **32**, 135–154.

Hansen, D.P., Ye, W., Jakeman, A.J., Cooke, R. and Sharma, P. (1996) Analysis of the effect of rainfall and streamflow data quality and catchment dynamics on streamflow prediction using the rainfall-runoff model IHACRES, *Environmental Software* **11**, 193–202.

Harding, S. (1976) *Can Theories Be Refuted? Essays on the Duhem–Quine Thesis*, Reidel, Dordrecht.

Hardisty, J., Taylor, D.M. and Metcalfe, S.E. (1993) *Computerised Environmental Modelling: A Practical Introduction Using Excel*, John Wiley & Sons, Chichester.

Haykin, S. (1999) *Neural Networks: A Comprehensive Foundation*, 2nd edn, Prentice Hall, Upper Saddle River, NJ.

Henebry, G.M. (1995) Spatial model error analysis using autocorrelation indices, *Ecological Modelling* **82**, 75–91.

Henwood, D. and Bonet, J. (1996) *Finite Elements: A Gentle Introduction*, Macmillan, Basingstoke.

Heuvelink, G.B.M. (1992) *Error Propagation in Quantitative Spatial Modelling: Applications in Geographical Information Systems*, Netherlands Geographical Studies, Knag/Faculteit Ruimtelijke Wetenschappen Universiteit Utrecht, Utrecht.

Hill, M.F., Witman, J.D. and Caswell, H. (2002) Spatio-temporal variation in Markov chain models of subtidal community succession, *Ecology Letters* **5**, 665–675.

Hofstadter, D.R. (1979) *Gödel, Escher, Bach: An Eternal Golden Braid*, Penguin Books, Harmondsworth.

Hsu, K.-L., Gupta, H.V. and Sorooshian, S. (1995) Artificial neural network modeling of the rainfall-runoff process, *Water Resources Research* **31**, 2517–2530.

Huggett, R.J. (1980) *Systems Analysis in Geography*, Clarendon Press, Oxford.

Imai, T., Sakayama, T. and Kanemori, T. (1987) Use of ground-probing radar and resistivity surveys for archaeological investigations, *Geophysics* **52**, 137–150.

Imrie, C.E., Durucan, S. and Korre, S. (2000) River flow prediction using artificial neural networks: generalisation beyond the calibration range, *Journal of Hydrology* **233**, 138–153.

Jaber, F. and Mohtar, R.H. (2002) Stability and accuracy of finite element schemes for the one-dimensional kinematic wave solution, *Advances in Water Resources* **25**, 427–438.

Karssenberg, D., De Jong, K. and Burrough, P.A. (2000) A prototype computer language for environmental modeling in the temporal, 3D spatial and stochastic dimension, *Proceedings of the 4th International Conference on Integrating GIS and Environmental Modelling (GIS/EM4): Problems, Prospects and Research Needs*, Banff, Alberta, Canada, September 2–8.

Kirchner, J.W., Hooper, R.P., Kendall, C., Neal, C. and Leavesley, G. (1996) Testing and validating environmental models, *The Science of the Total Environment* **183**, 33–47.

Kirkby, M.J., Naden, P.S., Burt, T.P. and Butcher, D.P. (1992) *Computer Simulation in Physical Geography*, John Wiley & Sons, Chichester.

Klemeš, V. (1983) Conceptualization and scale in hydrology, *Journal of Hydrology* **65**, 1–23.

Klemeš, V. (1997) Of carts and horses in hydrological modelling, *Journal of Hydrologic Engineering* **1**, 43–49.

Klepper, O. (1997) Multivariate aspects of model uncertainty analysis: tools for sensitivity analysis and calibration, *Ecological Modelling* **101**, 1–13.

Kloeden, P.E., Platen, E. and Schurz, H. (1994) *Numerical Solution of SDE Through Computer Experiments*, Springer Verlag, Berlin.

Kovalevskaya, N. and Pavlov, V. (2002) Environmental mapping based on spatial variability, *Journal of Environmental Quality* **31**, 1462–1470.

Ladd, S.R. (1995) *C++ Simulations and Cellular Automata*. Hungry Minds, Inc., New York.

Lane, S.N. and Richards, K.S. (2001) The 'validation' of hydrodynamic models: some critical perspectives, in M.G. Anderson and P.D. Bates (eds) *Model Validation: Perspectives in Hydrological Science*, John Wiley & Sons, Chichester, 413–438.

Lang, S. (1978) *A First Course in Calculus*, Addison-Wesley, Reading, MA.

Lark, R.M. (2001) Some tools for parsimonious modelling and interpretation of within-field variation of soil and crop systems, *Soil and Tillage Research* **58**, 99–111.

Legates, D.R. and McCabe, Jr, G.J. (1999) Evaluating the use of 'goodness-of-fit' measures in hydrologic and hydroclimatic model validation, *Water Resources Research* **35**, 233–241.

Madsen, H. (2000) Automatic calibration of a conceptual rainfall-runoff model using multiple objectives, *Journal of Hydrology* **235**, 276–288.

Madsen, H. (2003) Parameter estimation in distributed hydrological catchment modelling using automatic calibration with multiple objectives, *Advances in Water Resources* **26**, 205–216.

Maxwell, T. and Costanza, R. (1997) A language for modular spatio-temporal simulation, *Ecological Modelling* **103**, 105–113.

May, W. and Roeckner, E. (2001) A time-slice experiment with the ECHAM 4 AGCM at high resolution: the impact of horizontal resolution on annual mean climate change, *Climate Dynamics* **17**, 407–420.

McCullough, B.D. and Wilson, B. (2002) On the accuracy of statistical procedures in Microsoft Excel 2000 and Excel XP, *Computational Statistics and Data Analysis* **40**, 713–721.

McLeod, P., Plunkett, K. and Rolls, E.T. (1998) *Introduction to Connectionist Modelling of Cognitive Processes*, Oxford University Press, Oxford.

Mellett, J.S. (1995) Ground-penetrating radar applications in engineering, environmental management and geology, *Journal of Applied Geophysics* **33**, 157–166.

Michaelides, K. and Wainwright, J. (2002) Modelling the effects of hillslope-channel coupling on catchment hydrological response, *Earth Surface Processes and Landforms* **27**, 1441–1458.

Middleton, G.V. and Southard, J.B. (1984) *Mechanics of Sediment Movement*, 2nd edn, Society of Economic Palaeontologists and Mineralogists, Short Course 3, Tulsa, OK.

Mode, N.A., Conquest, L.L. and Marker, D.A. (1999) Ranked set sampling for ecological research: accounting for the total costs of sampling, *Environmetrics* **10**, 179–194.

Mode, N.A., Conquest, L.L. and Marker, D.A. (2002) Incorporating prior knowledge in environmental sampling: ranked set sampling and other double sampling procedures, *Environmetrics* **13**, 513–521.

Monte, L., Håkanson, L., Bergström, U., Brittain, J. and Heling, R. (1996) Uncertainty analysis and validation of environmental models: the empirically based uncertainty analysis, *Ecological Modelling* **91**, 139–152.

Morgan, R.P.C., Quinton, J.N., Smith, R.E., Govers, G., Poesen, J.W.A., Auerswald, K., Chisci, G., Torri, D. and Styczen, M.E. (1998) The European soil erosion model (EUROSEM): a process-based approach for predicting sediment transport from fields and small catchments, *Earth Surface Processes and Landforms* **23**, 527–544.

Morton, A. and Suárez, M. (2001) Kinds of models, in Anderson, M.G. and Bates, P.D. (eds) *Model Validation: Perspectives in Hydrological Science*, John Wiley & Sons, Chichester, 11–21.

Motha, J.A. and Wigham, J.M. (1995) Modelling overland flow with seepage, *Journal of Hydrology* **169**, 265–280.

Mulligan, M. (1996) Modelling the complexity of landscape response to climatic variability in semi arid environments, in M.G. Anderson and S.M. Brooks (eds) *Advances in Hillslope Processes*, John Wiley & Sons, Chichester, 1099–1149.

Mulligan, M. (1998) Hydrology in a varying as well as changing environment, in H.S. Wheater and C. Kirby (eds) *Hydrology in a Changing Environment*, Vol. II, John Wiley & Sons, Chichester, 591–604.

Muñoz-Carpena, R., Miller, C.T. and Parsons, J.E. (1993) A quadratic Petrov–Galerkin solution for kinematic wave overland flow, *Water Resources Research* **29**, 2615–2627.

Muñoz-Carpena, R., Parsons, J.E. and Gilliam, J.W. (1999) Modelling hydrology and sediment transport in vegetative filter strips, *Journal of Hydrology* **214**, 111–129.

Musters, P.A.D. and Bouten, W. (2000) A method for identifying optimum strategies of measuring soil water contents for calibrating a root water uptake model, *Journal of Hydrology* **227**, 273–286.

Naden, P.S. (1987) Modelling gravel-bed topography from sediment transport, *Earth Surface Processes and Landforms* **12**, 353–367.

Nash, J.E. and Sutcliffe, J.V. (1970) River flow forecasting through conceptual models. Part I – a discussion of principles, *Journal of Hydrology* **27**, 282–290.

Nicollin, F. and Kofman, W. (1994) Ground-penetrating radar sounding of a temperate glacier – modelling of a multilayered medium, *Geophysical Prospecting* **42**, 715–734.

Nitzsche, O., Meinrath, G. and Merkel, B. (2000) Database uncertainty as a limiting factor in reactive transport prognosis, *Journal of Contaminant Hydrology* **44**, 223–237.

O'Callaghan, J.F. and Mark, D.M. (1984) The extraction of drainage networks from digital elevation data, *Computer Vision, Graphics and Image Processing* **28**, 323–344.

Oreskes, N. and Bellitz, K. (2001) Philosophical issues in model assessment, in M.G. Anderson and P.D. Bates (eds) *Model Validation: Perspectives in Hydrological Science*, John Wiley & Sons, Chichester, 23–41.

Oreskes, N., Shrader-Frechette, K. and Bellitz, K. (1994) Verification, validation and confirmation of numerical models in the Earth Sciences, *Science* **263**, 641–646.

Özelkan, E.C. and Duckstein, L. (2001) Fuzzy-conceptual rainfall-runoff models, *Journal of Hydrology* **253**, 41–68.

Parks, K.P., Bentley, L.R. and Crowe, A.S. (2000) Capturing geological realism in stochastic simulations of rock systems with Markov statistics and simulated annealing, *Journal of Sedimentary Research B* **70**, 803–813.

Parsons, A.J., Wainwright, J., Abrahams, A.D. and Simanton, J.R. (1997) Distributed dynamic modelling of interrill overland flow, *Hydrological Processes* **11**, 1833–1859.

Parysow, P., Gertner, G. and Westervelt, J. (2000) Efficient approximation for building error budgets for process models, *Ecological Modelling* **135**, 111–125.

Pascual, M. and Levin, S.A. (1999) From individuals to population densities: searching for the intermediate scale of nontrivial determinism, *Ecology* **80**, 2225–2236.

Perrin, C., Michel, C. and Andreassian, V. (2001) Does a large number of parameters enhance model performance? Comparative assessment of common catchment model structures on 429 catchments, *Journal of Hydrology* **242**, 275–301.

Phillips, D.L. and Marks, D.G. (1996) Spatial uncertainty analysis: propagation of interpolation errors in spatially distributed models, *Ecological Modelling* **91**, 213–229.

Press, W.H., Teukolsky, S.A., Vetterling, W.T. and Flannery, B.P. (1992) *Numerical Recipes in* FORTRAN: *The Art of Scientific Computing*, Cambridge University Press, Cambridge [or the equivalent versions in C++ or PASCAL].

Ratto, M., Tarantola, S. and Saltelli, A. (2001) Sensitivity analysis in model calibration: GSA-GLUE approach, *Computer Physics Communications* **136**, 212–224.

Raudkivi, A.J. (1990) *Loose Boundary Hydraulics*, Pergamon Press, Oxford.

Rayner, S. and Malone, E.L. (1998) *Human Choice and Climate Change*, Vol. 1, Batelle Press, Columbus, OH.

Reece, G. (1976) *Microcomputer Modelling by Finite Differences*. MacMillan, Basingstoke.

Refsgaard, J.C. and Storm, B. (1996) Construction, calibration and validation of hydrological models, in M.B. Abbott and J.C. Refsgaard (eds) *Distributed Hydrological Modelling*, Reidel, Dordrecht, 41–45.

Reid, I. and Frostick, L.E. (1996) Channel form, flows and sediments in deserts, in D.S.G. Thomas (ed.) *Arid Zone Geomorphology, Form and Change in Drylands*, John Wiley & Sons, Chichester, 205–229.

Rogers, J. (1997) *Object-Oriented Neural Networks in* C++, Academic Press, London.

Rykiel Jr, E.J. (1996) Testing ecological models: the meaning of validation, *Ecological Modelling* **90**, 229–244.

Sajikumar, N. and Thandaveswara, B.S. (1999) A non-linear rainfall-runoff model using an artificial neural network, *Journal of Hydrology* **216**, 32–55.

Saltelli, A., Chan, K. and Scott, M. (2000) *Sensitivity Analysis*, John Wiley & Sons, Chichester.

Sarewitz, D. (1996) *Frontiers of Illusion*, Temple University Press, Philadelphia, PA.

Saulnier, G.M., Beven, K. and Obled, C. (1997) Including spatially effective soil depths in TOPMODEL, *Journal of Hydrology* **202**, 158–172.

Sawitzki, G. (1994) Report on the numerical reliability of data-analysis systems, *Computational Statistics and Data Analysis* **18**, 289–301.

Schaetzen, W.B.F. de, Walters, G.A. and Savic, D.A. (2000) Optimal sampling design for model calibration using shortest path, genetic and entropy algorithms, *Urban Water* **2**, 141–152.

Schmidt, K.H. and Ergenzinger, P. (1992) Bedload entrainment, travel lengths, step lengths, rest periods – studied with passive (iron, magnetic) and active (radio) tracer techniques, *Earth Surface Processes and Landforms* **17**, 147–165.

Scoging, H.M. (1992) Modelling overland-flow hydrology for dynamic hydraulics, in A.J. Parsons and A.D. Abrahams (eds) *Overland Flow: Hydraulics and Erosion Mechanics*, UCL Press, London, 89–103.

Scoging, H.M., Parsons, A.J. and Abrahams, A.D. (1992) Application of a dynamic overland-flow hydraulic model to a semi-arid hillslope, Walnut Gulch, Arizona, in A.J. Parsons and A.D. Abrahams (eds) *Overland Flow: Hydraulics and Erosion Mechanics*, UCL Press, London, 105–145.

Seibert, J., Uhlenbrook, S. Leibundgut, C. and Halldin, S. (2000) Multiscale calibration and validation of a conceptual rainfall-runoff model, *Physics and Chemistry of the Earth B* **25**, 59–64.

Senarath, S.U.S., Ogden, F., Downer, C.W. and Sharif, H.O. (2000) On the calibration and verification of two-dimensional, distributed, Hortonian, continuous watershed models, *Water Resources Research* **36**, 1495–1510.

Shamseldin, A.Y. (1997) Application of a neural network technique to rainfall-runoff modelling, *Journal of Hydrology*, **199**, 272–294.

Singh, V.P. (1996) *Kinematic Wave Modelling in Water Resources: Surface–Water Hydrology*, John Wiley & Sons, Chichester.

Smith, G.D. (1985) *Numerical Solution of Partial Differential Equations: Finite Difference Methods*, Clarendon Press, Oxford.

Smith, I.M. and Griffiths, D.V. (1998) *Programming the Finite Element Method*, John Wiley & Sons, Chichester.

Steefel, C.I. and van Cappellen, P. (1998) Reactive transport modelling of natural systems, *Journal of Hydrology* **209**, 1–7.

Sterman, J.D. (2002) All models are wrong: reflections on becoming a systems scientist, *System Dynamics Review* **18**, 501–531.

Tainaka, K.-I. (1996) Uncertainty in ecological catastrophe, *Ecological Modelling* **86**, 125–128.

Tang, Y. and Hsieh, W.W. (2002) Hybrid coupled models of the tropical Pacific: II – ENSO prediction, *Climate Dynamics* **19**, 343–353.

Thompson, S.P. (1946) *Calculus Made Easy: Being a Very-Simplest Introduction to Those Beautiful Methods of Reckoning which are Generally Called by the Terrifying Names of the Differential Calculus and the Integral Calculus*, Macmillan, London. [Difficult to get hold of, but a wonderful title!]

Thornes, J.B. (1985) The ecology of erosion, *Geography* **70**, 222–236.

Thornes, J.B. (1988) Erosional equilibria under grazing, in J. Bintliff, D. Davidson and E. Grant (eds) *Conceptual Issues in Environmental Archaeology*, Edinburgh University Press, Edinburgh, 193–211.

Thornes, J.B. (1990) The interaction of erosional and vegetational dynamics in land degradation: spatial outcomes, in J.B. Thornes (ed.) *Vegetation and Erosion*, John Wiley & Sons, Chichester, 41–55.

Torri, D. and Poesen, J. (1988) The effect of cup size on splash detachment and transport measurements. Part II: theoretical approach, in A. Imeson and M. Sala (eds) *Geomorphic Processes in Environments with Strong Seasonal Contrasts*: Vol. 1 *Hillslope Processes, Catena* Supplement 12 *Catena*, Cremlingen, 127–137.

Torri, D., Poesen, J. and Borselli, L. (1997) Predictability and uncertainty of the soil erodibility factor using a global dataset, *Catena* **31**, 1–22.

Van Deursen, W.P.A. (1995) *A geographical information system and dynamic model development and application of a prototype spatial modelling language*, PhD thesis, Utrecht Knag/Faculteit Ruimtelijke Wetenschappen Universiteit Utrecht, Netherlands.

Van Overmeeren, R.A., Sariowan, S.V. and Gehrels, J.C. (1997) Ground-penetrating radar for determining volumetric water content; results of comparative measurements at two test sites, *Journal of Hydrology* **197**, 316–338.

Van der Perk, M. (1997) Effect of model structure on the accuracy and uncertainty of results from water quality models, *Hydrological Processes* **11**, 227–239.

Van Wesemael, B., Mulligan, M. and Poesen, J. (2000) Spatial patterns of soil water balance on intensively cultivated hillslopes in a semi-arid environment: the impact of rock fragments and soil thickness, *Hydrological Processes* **14**, 1811–1828.

Van Winkle, W. and Dale, V.H. (1998) Model interactions: a reply to Aber, *Bulletin of the Ecological Society of America* **79**, 257–259.

Versteeg, H.K. and Malalasekera, W. (1995) *An Introduction to Computational Fluid Dynamics: The Finite Volume Method*, Longman, Harlow.

Voller, V.R. and Porté-Agel, F. (2002) Moore's law and numerical modelling, *Journal of Computational Physics* **179**, 698–703.

Wainwright, J. (1996) A comparison of the infiltration, runoff and erosion characteristics of two contrasting 'badland' areas in S. France, *Zeitschrift für Geomorphologie Supplementband* **106**, 183–198.

Wainwright, J., Mulligan, M. and Thornes, J.B. (1999) Plants and water in drylands, in A.J. Baird and R.L. Wilby (eds) *Ecohydrology*, Routledge, London, 78–126.

Wainwright, J. and Parsons, A.J. (1998) Sensitivity of sediment-transport equations to errors in hydraulic models of overland flow, in J. Boardman and D. Favis-Mortlock (eds) *Modelling Soil Erosion by Water*, Springer-Verlag, Berlin, 271–284.

Wainwright, J. and Parsons, A.J. (2002) The effect of temporal variations in rainfall on scale dependency in runoff coefficients, *Water Resources Research* **38** (12), 1271 [Doi: 10.1029/2000WR000188].

Wainwright, J., Parsons, A.J. and Abrahams, A.D. (1999) Field and computer simulation experiments on the formation of desert pavement, *Earth Surface Processes and Landforms* **24**, 1025–1037.

Wainwright, J., Parsons, A.J. and Abrahams, A.D. (2000) Plot-scale studies of vegetation, overland flow and erosion interactions: case studies from Arizona and New Mexico, *Hydrological Processes* **14**, 2921–2943.

Wait, R. and Mitchell, A.R. (1985) *Finite Element Analysis and Applications*, John Wiley & Sons, Chichester.

Welstead, S.T. (1994) *Neural Network and Fuzzy Logic Applications in C/C++*. John Wiley & Sons, Chichester.

Wesseling, C.G., Karssenberg, D. van Deursen, W.P.A. and Burrough, P.A. (1996) Integrating dynamic environmental models in GIS: the development of a dynamic modelling language, *Transactions in GIS* **1**, 40–48.

Wilder, J.W. (2001) Effect of initial condition sensitivity and chaotic transients on predicting future outbreaks of gypsy moths, *Ecological Modelling* **136**, 49–66.

Wilks, D.S. and Wilby, R.L. (1999) The weather generation game: a review of stochastic weather models, *Progress in Physical Geography* **23**, 329–357.

Willmott, C.J. (1981) On the validation of models, *Physical Geography* **2**, 184–194.

Wilson, A.G. and Kirkby, M.J. (1980) *Mathematics for Geographers and Planners*. Oxford University Press, Oxford.

Wingle, W.L., Poeter, E.P. and McKenna, S.A. (1999) UNCERT: geostatistics, uncertainty analysis and visualization software applied to groundwater flow and contaminant transport modelling, *Computers and Geosciences* **25**, 365–376.

Wolfram, S. (1986) Cellular automaton fluids 1: basic theory, *Journal of Statistical Physics* **45**, 471–526.

Wolfram, S. (2002) *A New Kind of Science*. Wolfram Media, Champaign, IL.

Wood, E.F., Sivapalan, M., Beven, K. and Band, L. (1988) Effects of spatial variability and scale with implications to hydrologic modelling, *Journal of Hydrology* **102**, 29–47.

Woolhiser, D.A., Smith, R.E. and Goodrich, D.C. (1990) *KINEROS, a Kinematic Runoff and Erosion Model: Documentation and User Manual*. US Department of Agriculture, Agricultural Research Service, ARS-77, Washington, DC.

Xu, Z.X. and Li, J.Y. (2002) Short-term inflow forecasting using an artificial neural network model, *Hydrological Processes* **16**, 2423–2439.

Yu, P.-S., Yang, T.-C. and Chen, S.-J. (2001) Comparison of uncertainty analysis methods for a distributed rainfall-runoff model, *Journal of Hydrology* **244**, 43–59.

Zienkiewicz, O.C. and Taylor, R.L. (2000) *The Finite Element Method*, Vol. 3 *Fluid Dynamics*, Butterworth-Heinemann, Oxford.

Zimmermann, H.-J. (2000) An application-oriented view of modelling uncertainty, *European Journal of Operational Research* **122**, 190–198.

Part II

The State of the Art in Environmental Modelling

2

Climate and Climate-System Modelling

L.D. DANNY HARVEY

2.1 THE COMPLEXITY

Climate and climatic change are particularly difficult to model because of the large number of individual components involved in the climate system, the large number of processes occurring within each component, and the multiplicity of interactions between components. The identification of separate, interacting components of the climate system is to some extent arbitrary and a matter of convenience in order to facilitate human analysis and thinking. Here, the climate system will be thought of as consisting of the following components: the atmosphere, the oceans, the cryosphere (glaciers, sea ice, seasonal snow cover), the biosphere and the lithosphere (the Earth's crust). All of these components affect, and are affected by, the other components, so they form a single system. For example, the atmosphere and oceans influence each other through the transfer of momentum (by winds), heat and moisture. The growth and decay of glaciers, and the formation of sea ice and land-snow cover, depend on atmospheric temperature and the supply of moisture, but ice and snow surfaces in turn influence the absorption of solar energy and hence influence temperature, and modulate the flow of water vapour between the land or ocean and the atmosphere. The biosphere is clearly affected by atmospheric and/or oceanic conditions, but also influences climate through the effect of land vegetation on surface roughness and hence on winds, and through its effect on evaporation and the reflection or absorption of solar energy. Micro-organisms in the upper ocean affect the reflectivity of the ocean surface to sunlight. Both the marine and terrestrial biosphere play an important role in determining the atmospheric concentration of a number of climatically important trace gases, the most important being carbon

dioxide (CO_2). Finally, the Earth's crust affects the climate through the influence of land topography on winds and the distribution of rain, through the role of the ocean-continent configuration and the shape of the ocean basin on ocean currents, and through the role of chemical weathering at the Earth's surface and the deposition of marine carbonates in modulating the atmospheric concentration of CO_2 at geological time scales.

Even this thumbnail sketch of the climate system leaves out many important considerations. The occurrence of clouds, which are part of the atmosphere, is strongly influenced by conditions at the land surface. Their optical properties are influenced in part by micro-organisms in the upper few metres of the ocean, which emit sulphur compounds that ultimately become cloud-condensation nuclei. The occurrence of clouds dramatically affects the flow of both solar and infrared radiation between the atmosphere, land surface, and space. Many chemical reactions occur in the atmosphere which determine the concentrations of climatically important trace gases and aerosols; many of the important reactants in atmospheric chemistry are released from the terrestrial or marine biosphere. Cloud droplets serve as important sites for many important chemical reactions, and the optical properties of clouds are themselves influenced to some extent by the chemical reactions that they modulate. A more thorough discussion of the interactions between and within the various components of the climate system can be found in Chapter 2 of Harvey (2000a).

In building computer models of the climate system, as with any system, there are a number of basic considerations. The first involves the number of components to be included. The various components of the climate system change at different time scales. For example, at

Environmental Modelling: Finding Simplicity in Complexity. Edited by J. Wainwright and M. Mulligan
© 2004 John Wiley & Sons, Ltd ISBNs: 0-471-49617-0 (HB); 0-471-49618-9 (PB)

the decadal to century time scale, changes in the extent of the large ice caps (Greenland, Antarctica) can be ignored. There would thus be no need to include models of these components for a simulation of 100 years or less; rather, the observed present-day distribution can be simply prescribed as a lower boundary condition on the atmosphere. Similarly, changes in the geographical distribution of the major terrestrial biomes can be ignored at this time scale. However, at longer time scales, changes in biomes and in ice sheets, and their feedback effects on the atmosphere and oceans, would need to be considered. Thus, the comprehensiveness of a climate model (the number of components retained) depends in part on the time scale under consideration. The comprehensiveness is also dictated in part by the particular purpose for which one is building the climate model.

The flipside to comprehensiveness is model complexity. Generally, more comprehensive models tend to be less complex – that is, each component represents fewer of the processes that occur in reality, or represents them in a more simplified manner. This is because the more comprehensive models tend to be used for longer time scales, so limitations in computing power require that less detailed calculations be performed for a given period of simulated time. Furthermore, the addition of more climate-system components also tends to increase the overall computational requirements, which can be offset by treating each component in less detail.

Nature varies continuously in all three spatial dimensions, thus comprising an infinite number of infinitesimally close points (at least according to classical physics!). However, due to the finite memory capacity of computers, it is necessary to represent variables at a finite number of points, laid out on a grid of some sort. The calculations are performed only at the grid points. The spacing between the grid points is called the model resolution. In global atmospheric models the typical horizontal resolution is a few hundred kilometres. In ocean models the resolution can be as fine as a few tens of kilometres. Many important elements of the climate system (e.g. clouds, land-surface variation) have scales much smaller than this. Detailed models at high resolution are available for such processes by themselves, but these are computationally too expensive to be included in a climate model, and the climate model has to represent the effect of these subgrid-scale processes on the climate system at its coarse grid-scale. The formulation of the effect of a small-scale process on the large scale is called a *parameterization*. All climate models use parameterizations to some extent. Another kind of simplification used in climate models is to average over a complete spatial dimension. Instead of, for

instance, a three-dimensional longitude–latitude–height grid, one might use a two-dimensional latitude–height grid, with each point being an average over all longitudes at its latitude and height. When the dimensionality is reduced, more processes have to be parameterized but less computer time is required.

The equations used in climate models are a mixture of fundamental principles which are known to be correct (such as Newton's Laws of Motion and the First Law of Thermodynamics), and parameterizations. Parameterizations are empirical relationships between model-resolved variables and subgrid-scale processes that are not rigorously derived from physical principles (although the overall form of the parameterization might be derived from physical considerations), but are derived from observed correlations. An example of a parameterization would be to compute the percentage of cloud cover in a given atmospheric grid box based on the grid-averaged relative humidity and whether the vertical motion is upward or downward. However, because parameterizations are not rigorously derived, the observed correlation might not be applicable if the underlying conditions change – that is, as the climate changes.

Another critical issue in the formulation of climate models is the problem of *upscaling*. This arises due to the fact that the mathematical formulation of a particular process that is valid at a point is, in general, not valid when applied over an area as large as a model grid square. There are a number of reasons why upscaling problems arise, including but not limited to: the occurrence of spatial heterogeneity combined with nonlinear processes (this is especially important for land–air fluxes of heat, water vapour and momentum); feedbacks between different scales (especially important for the interaction between an ensemble of cumulus clouds and the larger-scale atmosphere); and the development of emergent system properties as more elements are added with increasing scale (this is important for the properties of sea ice). These and other conceptually distinct causes of upscaling problems, and the range of solutions adopted across a number of disciplines involved in global-change research, are thoroughly reviewed in Harvey (2000b). Closely related to the upscaling problem is the question of how best to incorporate the effects of *subgrid-scale variability* in nature. The mirror image of the upscaling problem is that of *downscaling* – taking model output that is averaged over a grid cell and deriving specific output values for a specific point within the grid cell. However, this is a problem of determining how the climate of specific surface types or slopes differs from the grid-average climate, rather than a problem of how to formulate the model equations (see also Chapter 19).

2.2 FINDING THE SIMPLICITY

In this section the major kinds of models used for each of the climate system components identified above are outlined. A comprehensive treatment of the most complex models used to simulate the various climate-system components can be found in the volume edited by Trenberth (1992), while a recent assessment of model performance can be found in the Third Assessment Report of the Intergovernmental Panel on Climate Change (IPCC, 2002).

2.2.1 Models of the atmosphere and oceans

2.2.1.1 One-dimensional radiative-convective atmospheric models

These models are globally averaged horizontally but contain many layers within the atmosphere. They treat processes related to the transfer of solar and infrared radiation within the atmosphere in considerable detail, and are particularly useful for computing the changes in net radiation – one of the possible drivers of climatic change – associated with changes in the composition of the atmosphere (e.g. Lal and Ramanathan, 1984; Ko *et al.*, 1993). The change in atmospheric water vapour amount as climate changes must be prescribed (based on observations), but the impact on radiation associated with a given change in water vapour can be accurately computed. The increase in water vapour in the atmosphere is thought to be the single most important feedback that amplifies an initial radiative perturbation, and radiative-convective models provide one means for assessing this feedback through a combination of observations and well-established physical processes.

2.2.1.2 One-dimensional upwelling-diffusion ocean models

In this type of model, which was first applied to questions of climatic change by Hoffert *et al.* (1980), the atmosphere is treated as a single well-mixed box that exchanges heat with the underlying ocean and land surface. The absorption of solar radiation by the atmosphere and surface depends on the specified surface reflectivity and the atmospheric transmissivity and reflectivity. The emission of infrared radiation to space is a linearly increasing function of atmospheric temperature

in this model; this increase serves to 'dampen' temperature changes and thus limits the final temperature response for a given radiative perturbation. Through an appropriate choice of the constant of proportionality in the parameterization of infrared radiation to space, the model can be forced to have any desired climate responsiveness to a given radiative perturbation. This process is in turn useful if one is trying to replicate the behaviour of more complex models in a computationally efficient way for the purposes of scenario analysis and testing the interactions between different model components. The ocean is treated as a one-dimensional column that represents a horizontal average over the real ocean, excluding the limited regions where deep water forms and sinks to the ocean bottom, which are treated separately. Figure 2.1 illustrates this model. The sinking in polar regions is represented by the pipe to the side of the column. This sinking and the compensating upwelling within the column represent the global-scale advective overturning (the thermohaline circulation).

2.2.1.3 One-dimensional energy balance models

In these models the only dimension that is represented is the variation with latitude; the atmosphere is averaged vertically and in the east–west direction, and often combined with the surface to form a single layer. The multiple processes of meridional (north–south) heat transport by the atmosphere and oceans are usually represented as diffusion, while infrared emission to space is represented in the same way as in the upwelling diffusion model. Figure 2.2 illustrates a single east–west zone in the energy balance model of Harvey (1988), which separately resolves land and ocean, surface and air, and snow- or ice-free and snow- or ice-covered regions within each land or ocean surface box. These models have provided a number of useful insights concerning the interaction of horizontal heat transport feedbacks and high-latitude feedbacks involving ice and snow (e.g. Held and Suarez, 1974).

2.2.1.4 Two-dimensional atmosphere and ocean models

Several different two-dimensional (latitude–height or latitude–depth) models of the atmosphere and oceans have been developed (e.g. Peng *et al.*, 1982, for the atmosphere; Wright and Stocker, 1991, for the ocean). The two-dimensional models permit a more physically based computation of horizontal heat transport than in one-dimensional energy balance models. In some two-dimensional ocean models (e.g. Wright and Stocker,

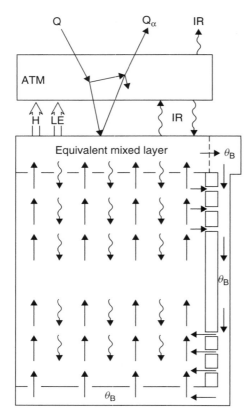

Figure 2.1 Illustration of the one-dimensional upwelling diffusion model. The model consists of boxes representing the global atmosphere and ocean mixed layer, underlain by a deep ocean. Bottom water forms at temperature θ_B, sinks to the bottom of the ocean, and upwells with an upwelling velocity w. The temperature profile in the deep ocean is governed by the balance between the upwelling of cold water (straight arrows) and downward diffusion of heat from the mixed layer (wavy arrows). Latent (LE) and sensible (H) heat and infrared radiation (IR) are transferred between the atmosphere and the surface. Q is solar radiation and α is albedo. The physical mixed layer depth is reduced to account for the fact that part of the Earth's surface consists of land, which responds more quickly than the ocean mixed layer to changes in the surface energy balance

1991), the intensity of the thermohaline overturning is determined by the model itself, while in others (e.g. de Wolde *et al.*, 1995), it is prescribed, as in the one-dimensional upwelling-diffusion model. The one-dimensional energy balance atmosphere–surface climate model has also been coupled to a two-dimensional ocean model (Harvey, 1992; Bintanja, 1995; de Wolde

et al., 1995). It is relatively easy to run separate two-dimensional ocean models for each of the Atlantic, Pacific and Indian Ocean basins, with a connection at their southern boundaries (representing the Antarctic Ocean) and interaction with a single, zonally-averaged atmosphere (as in de Wolde *et al.*, 1995).

2.2.1.5 Three-dimensional atmosphere and ocean general circulation models

The most complex atmosphere and ocean models are the three-dimensional atmospheric general circulation models (AGCMs) and ocean general circulation models (OGCMs), both of which are extensively reviewed in Gates *et al.* (1996) and in various chapters of Trenberth (1992). These models divide the atmosphere or ocean into a horizontal grid with a typical resolution of 2–4° latitude by 2–4° longitude in the latest models, and typically 10 to 20 layers in the vertical. They directly simulate winds, ocean currents, and many other features and processes of the atmosphere and oceans. Figure 2.3 provides a schematic illustration of the major processes that occur within a single horizontal grid cell of an AGCM. Both AGCMs and OGCMs have been used extensively in a stand-alone mode, with prescribed ocean–surface temperatures in the case of AGCMs and with prescribed surface temperatures and salinities, or the corresponding heat and freshwater fluxes, in the case of OGCMs. Only when the two models are coupled together do we have what can be considered to be a climate model, in which all the temperatures are freely determined. Coupled AOGCMs automatically compute the feedback processes associated with water vapour, clouds, seasonal snow and ice, as well as the uptake of heat by the oceans when driven by a prescribed, anthropogenic increase of atmospheric CO_2. The uptake of heat by the oceans delays and distorts the surface temperature response but contributes to sea-level rise through expansion of ocean water as it warms. AOGCMs compute radiative transfer through the atmosphere (explicitly modelling clouds, water vapour and other atmospheric components), snow and sea-ice, surface fluxes, transport of heat and water by the atmosphere and ocean, and storage of heat in the ocean. Because of computational constraints, the majority of these processes are parameterized to some extent (see Dickinson *et al.*, 1996, concerning processes in atmospheric and oceanic GCMs). More detailed representations are not practical, or have not been developed, for use in a global model. Some parameterizations inevitably include constants which have been tuned to observations of the current climate.

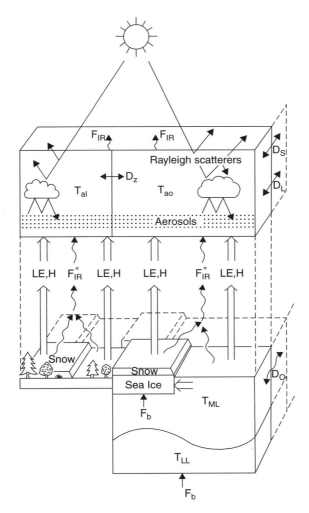

Figure 2.2 Illustration of a single latitude zone in the energy balance climate model of Harvey (1988). In each latitude zone, a distinction is made between land and ocean and the surface and atmosphere. Partial snowcover on land and sea ice in the ocean box can also occur. Vertical heat transfer occurs as latent (LE) and sensible (H) heat, infrared radiation (F_{IR}^*), and multiple scattering of solar radiation, while horizontal heat transfer is represented by diffusion (D_S for atmospheric sensible heat, D_L for atmospheric latent heat, and D_O for oceanic latent heat – all in the meridional direction – and D_Z for zonal (east–west) atmospheric transfers). T_{al}, T_{ao}, T_{ML} and T_{LL} are the atmospheric temperature in the land sector, atmospheric temperature in the oceanic sector, oceanic mixed layer temperature, and the temperature of a lower layer, respectively. F_{IR} is the IR emission to space, while F_b is a vertical heat transfer in the ocean at high latitudes (which is implicitly supplied from other zones)

2.2.2 Models of the carbon cycle

The upwelling-diffusion model that was described above can be used as the oceanic part of the carbon cycle, as in the work of Hoffert *et al*. (1981). The global mean atmosphere–ocean exchange of CO_2, the vertical mixing of total dissolved carbon by thermohaline overturning and diffusion, and the sinking of particulate material produced by biological activity can all be represented in this model. A two-dimensional ocean model has been used as the oceanic component of the global carbon cycle (Stocker *et al*., 1994). Finally, OGCMs can be used as the oceanic component of the global carbon cycle, in which the model-computed ocean currents

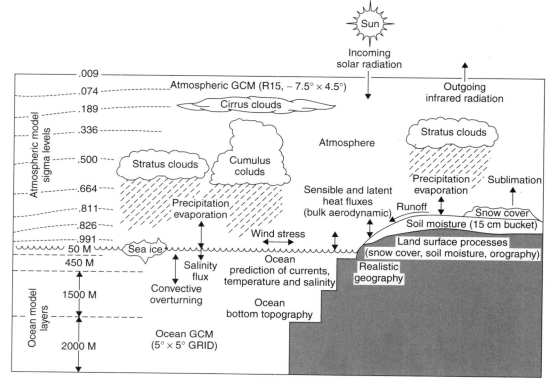

Figure 2.3 Illustration of the major processes occurring inside a typical horizontal grid cell of an atmospheric general circulation model (AGCM).
Source: Reproduced from Washington and Meehl (1989), where a full explanation can be found.
Note: 'Sigma level' refers to the ratio of pressure at a give height to the surface pressure

and other mixing processes are used, in combination with simple representations of biological processes and air–sea exchange (e.g., Bacastow and Maier-Reimer, 1990; Najjar *et al.*, 1992). CO_2 uptake calculations using 3-D models have so far been published only for stand-alone OGCMs, in which the circulation field and surface temperatures have been fixed. In a coupled simulation, changes in both of these variables in response to increasing greenhouse gas concentrations would alter the subsequent uptake of CO_2 to some extent.

The terrestrial biosphere can be represented by a series of interconnected boxes, where the boxes represent components such as leafy material, woody material, roots, detritus, and one or more pools of soil carbon. Each box can be globally aggregated such that, for example, the detrital box represents all the surface detritus in the world. The commonly used, globally aggregated box models are quantitatively compared in Harvey (1989). Figure 2.4 gives an example of a

globally aggregated terrestrial biosphere model, where numbers inside boxes represent the steady state amounts of carbon (Gt) prior to human disturbances, and the numbers between boxes represent the annual rates of carbon transfer (Gt a^{-1}). In globally aggregated box models it is not possible to simulate separate responses in different latitude zones (e.g. net release of carbon through temperature effects at high latitudes, net uptake of carbon in the tropics due to CO_2 fertilization). Since regional responses vary nonlinearly with temperature and atmospheric CO_2 concentration, extrapolation into the future using globally aggregated models undoubtedly introduces errors. An alternative is to run separate box models for major regions, as in van Minnen *et al.* (1996), rather than lumping everything together.

The role of the terrestrial biosphere in global climatic change has also been simulated using relatively simple models of vegetation on a global grid with resolution as fine as 0.5° latitude × 0.5° longitude. These models

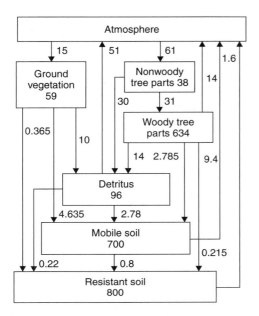

Figure 2.4 The six-box, globally aggregated terrestrial biosphere model of Harvey (1989). The numbers in each box are the steady state amounts of carbon (in Gt), while the arrows between the boxes represent the annual transfers (Gt C yr^{-1})

have been used to evaluate the impact on net ecosystem productivity of higher atmospheric CO_2 (which tends to stimulate photosynthesis and improve the efficiency of water use by plants) and higher temperatures (which can increase or decrease photosynthesis and increase decay processes). These models distinguish, as a minimum, standing biomass from soil organic matter. The more sophisticated varieties track the flows of both carbon and nitrogen (taken to be the limiting nutrient), and include feedbacks between nitrogen and the rates of both photosynthesis and decay of soil carbon (e.g. Rastetter *et al.*, 1991, 1992; Melillo *et al.*, 1993).

Grid-point models of the terrestrial biosphere have been used to assess the effect on the net biosphere–atmosphere CO_2 flux of hypothetical (or GCM-generated) changes in temperature and/or atmospheric CO_2 concentration, but generally without allowing for shifts in the ecosystem type at a given grid point as climate changes. More advanced ecosystem models are being developed and tested that link biome models (which predict changing ecosystem types) and ecophysiological models (which predict carbon fluxes) (e.g. Plöchl and Cramer, 1995). Different biomes vary in the proportion of 'plant functional types', examples of which are tropical evergreen forest, cool-temperate evergreen

trees, and cool grasses. An alternative to simulating the location of specific ecosystems is to predict the proportions of different plant functional types at each grid cell based on the carbon balance for each functional type, as determined from an ecophysiological model (Foley *et al.*, 1996). This effectively allows for the simulation of both rapid biophysical processes and the long-term adjustment of ecosystems to changing climate and atmospheric CO_2 concentration. Simulations with these and earlier models demonstrate the potential importance of feedbacks involving the nutrient cycle and indicate the potential magnitude of climate-induced terrestrial biosphere–atmosphere CO_2 fluxes. However, individual models still differ considerably in their responses (VEMAP Members, 1995). As with models of the oceanic part of the carbon cycle, such simulations have yet to be carried out interactively with coupled AOGCMs. These models also have not yet been combined with ocean carbon uptake OGCMs.

Rather detailed models of the marine biosphere, involving a number of species and interactions, have also been developed and applied to specific sites or regions (e.g. Gregg and Walsh, 1992; Sarmiento *et al.*, 1993; Antoine and Morel, 1995; Oschlies and Garçon, 1999; Oschlies *et al.*, 2000).

2.2.3 Models of atmospheric chemistry and aerosols

Atmospheric chemistry is central to the distribution and amount of ozone in the atmosphere. The dominant chemical reactions and sensitivities are significantly different for the stratosphere and troposphere. These processes can be adequately modelled only with three-dimensional atmospheric models (in the case of the troposphere) or with two-dimensional (latitude–height) models (in the case of the stratosphere). Atmospheric chemistry is also critical to the removal of CH_4 from the atmosphere and, to a lesser extent, all other greenhouse gases except H_2O and CO_2. In the case of CH_4, a change in its concentration affects its own removal rate and, hence, subsequent concentration changes. An accurate simulation of changes in the removal rate of CH_4 requires specification of the concurrent concentrations of other reactive species, in particular NO_x (nitrogen oxides), CO (carbon monoxide) and the VOCs (volatile organic compounds); and use of a model with latitudinal and vertical resolution. However, simple globally averaged models of chemistry–climate interactions have been developed. These models treat the global CH_4–CO–OH cycle in a manner which takes into account the effects of the heterogeneity of the chemical and transport processes, and provide estimates of future global or hemispheric mean changes in the chemistry of the Earth's atmosphere. Some of the models also simulate halocarbon concentrations and the resulting atmospheric chlorine concentration, as well as radiative effects due to halocarbons (Prather *et al.*, 1992). An even simpler approach, adopted by Osborn and Wigley (1994), is to treat the atmosphere as a single well-mixed box but to account for the effects of atmospheric chemistry by making the CH_4 lifetime depend on CH_4 concentration in a way that roughly mimics the behaviour of the above-mentioned globally averaged models or of models with explicit spatial resolution. Atmospheric O_3 and CH_4 chemistry has not yet been incorporated in AGCMs used for climate simulation purposes, although two-dimensional interactive chemistry-climate models have been built (Wang *et al.*, 1998).

Atmospheric chemistry is also central to the distribution and radiative properties of small suspended particles in the atmosphere referred to as aerosols, although chemistry is only part of what is required in order to simulate the effects of aerosols on climate. The primary aerosols that are affected by atmospheric chemistry are sulphate aerosols (produced from the emission of SO_2 and other S-containing gases), nitrate aerosols (produced from emission of nitrogen oxides), and organic carbon aerosols (produced from the emission of a variety of organic compounds from plants and gasoline). The key processes that need to be represented are the source emissions of aerosols or aerosol precursors; atmospheric transport, mixing, and chemical and physical transformation; and removal processes (primarily deposition in rainwater and direct dry deposition onto the Earth's surface). Since part of the effect of aerosols on climate arises because they serve as cloud condensation nuclei, it is also important to be able to represent the relationship between changes in the aerosol mass input to the atmosphere and, ultimately, the radiative properties of clouds. Establishing the link between aerosol emissions and cloud properties, however, involves several poorly understood steps and is highly uncertain.

Geographically distributed sulphur-aerosol emissions have been used as the input to AGCMs and, in combination with representations of aerosol chemical and physical processes, have been used to compute the geographical distributions of sulphur-aerosol mass using only natural emission sources and using natural + anthropic emission sources (e.g. Langner and Rodhe, 1991; Chin *et al.*, 1996; Pham *et al.*, 1996). Given the differences in the GCM-simulated aerosol distributions for these two sets of emission sources and a variety of assumptions concerning the aerosol optical properties, other studies have estimated the possible range of direct (cloud-free) effects on radiative forcing (e.g. Haywood *et al.*, 1997). With yet further assumptions concerning how clouds respond to sulphur aerosols, a range of indirect (cloud-induced) effects can also be computed (e.g. Boucher and Lohmann, 1995; Jones and Slingo, 1996; Kogan *et al.*, 1996; Lohmann and Feichter, 1997). The results of these and other calculations are extensively reviewed in Harvey (2000a: Chapter 7).

2.2.4 Models of ice sheets

High resolution (20 km × 20 km horizontal grid), two- and three-dimensional models of the polar ice sheets have been developed and used to assess the impact on global mean sea level of various idealized scenarios for temperature and precipitation changes over the ice sheets (e.g. Huybrechts and Oerlemans, 1990; Huybrechts *et al.*, 1991). AGCM output has also recently been used to drive a three-dimensional model of the East Antarctic ice sheet (Verbitsky and Saltzman, 1995), but has not yet been used to assess the possible contribution of changes in mountain glaciers to future sea-level rise. Output from high resolution ice-sheet models can be used to develop simple relationships in which the contribution of ice-sheet changes to future sea level is scaled with changes in global mean temperature.

2.2.5 The roles of simple and complex climate-system models

Climate-system models have been used to understand the role of different processes in determining the present climate and atmospheric composition, in trying to understand the causes of past climatic change, and in trying to estimate the range of possible future climate changes due to human activities (emissions of greenhouse gases, emissions of aerosols and aerosol precursors), and changes in the land vegetation (particularly past and ongoing large-scale deforestation). Atmospheric GCMs are also used for day-to-day weather prediction (the problem of weather prediction and of predicting climatic change are quite different, however, since weather prediction is concerned with specific events at specific places and times, whereas climate prediction is concerned with long-term changes and trends, averaged over a period of several decades). Simple and complex models have different but complementary roles to play in climate research.

Simple models allow investigation of the basic relationships between the components of the climate system, the factors driving climatic change, and the overall response of the system. They are useful mainly for exploring global-scale questions. The upwelling-diffusion model, for example, has been used to investigate the role of the oceans in delaying the climatic response to increasing greenhouse-gas concentrations and the role of ocean mixing–climate feedbacks in modifying the transient response (e.g. Hoffert *et al.*, 1980; Harvey and Schneider, 1985; Morantine and Watts, 1990), in exploring the importance of natural variability in observed global mean temperature variations during the past century (Wigley and Raper, 1990; Schlesinger and Ramankutty, 1994), in setting constraints on the magnitude of the global mean aerosol cooling effect (Wigley, 1989), and in assessing the relative roles of greenhouse gases, aerosols, and solar variability in explaining global mean temperature variations during the past century (Kelly and Wigley, 1992; Schlesinger and Ramankutty, 1992). Because simple models are relatively easy to analyse (due to their low dimensionality and the fact that only the most basic processes are included), and because multiple diagnostic tests can be performed (due to the fact that they require little computer time), simple models have provided a number of important insights into the climate system that would not have been easily discovered using complex models.

Three-dimensional models, on the other hand, allow for the explicit simulation of many of the key processes that determine climate sensitivity (the change in global mean temperature for a given radiative perturbation) and the longer-term feedbacks involving the terrestrial and marine biosphere. Although the strengths of many of the feedbacks in these models ultimately depend on the way in which subgrid-scale processes are parameterized (and can sometimes change dramatically when the parameterization is changed), more of the processes that determine climate sensitivity are explicitly represented. This in turn allows the investigation of interactions, and the development of insights, that could not be obtained using simple models.

One of the most important applications of climate system models is in projecting the global and regional-scale effects on climate of anthropic emissions of greenhouse gases and aerosols (or aerosol precursors). Three-dimensional AOGCMs are the only models that can make predictions at the regional scale (due to their regional resolution), but it is widely agreed that the regional-scale results are less reliable than the global-scale results (which themselves have a factor of three uncertainty for temperature). Nevertheless, three-dimensional AOGCMs present an internally consistent picture of regionally varying climatic change, and indicate the magnitude of climatic change that could occur

in specific regions in association with a given global-average change. They therefore indicate the risks associated with a given global-average climatic change. The globally averaged temperature change can therefore be thought of as an index of the risks posed by a particular scenario of global GHG emissions. Simple models, on the other hand, generally only predict the global mean temperature change. They can be directly and easily driven by scenarios of anthropogenic emissions, the build-up of greenhouse-gas concentrations can be computed, and the computed concentration changes can be used to project the temporal variation in global mean temperature. This projection can be carried out for a variety of assumptions concerning the response of the terrestrial biosphere carbon sink to changes in temperature and atmospheric composition, concerning climate sensitivity and the intensity of ocean mixing, and concerning the role of aerosols. Simple models have the additional advantage that they do not generate internal variability ('noise') that can obscure the interpretation of AOGCM output.

There is no scientific justification for developing policy responses at the national level based on the climatic changes and impacts within the jurisdiction under consideration, as given by any one model. Rather, the climate models can be used to support a collective (i.e. global-scale) policy response based on generalized risks, rather than nation-by-nation responses based on expected nation-by-nation impacts. Simple but comprehensive models provide an easy means for generating scenarios of global mean temperature change (the risk index), while complex, regionally resolved models – through the range of responses at the regional level given by different models – serve to illustrate the specific risks associated with a given value of the risk index (global mean temperature). Thus, simple and complex models play complementary roles in the development of a policy response to the threat of global warming.

2.3 THE RESEARCH FRONTIER

A long-term goal of the climate-research community is the development of increasingly sophisticated models that couple more and more components of the climate system. A large number of modelling groups have created three-dimensional models that couple the atmospheric and oceanic components of the climate system, and that include increasingly realistic representations of sea-ice and land-surface processes (in particular, the build-up and melting of snow cover, runoff generation, and the modulation of the fluxes of heat, moisture and momentum by vegetation).

Until very recently, land-surface modules in AGCMs represented the vegetation canopy (if at all) as a passive vegetated surface that modulated the fluxes of heat and moisture between the atmosphere and surface. The most recent schemes include the vegetation–atmosphere carbon flux, and its coupling to the heat and moisture fluxes through changes in the plant stomata. Consistent modelling of the surface–air carbon flux requires representing the processes of photosynthesis and respiration, and hence of plant growth. Hence the incorporation of simple plant biophysical models is required for the simulation of the climatically important fluxes of heat and moisture in a way that most closely mimics nature. The need for this inclusion is because these fluxes depend in part on the stomatal resistance, which in turn depends on the photosynthetic demand for carbon, as well as on the simulated leaf temperature and soil moisture content. The incorporation of such biophysical modules in the land-surface scheme allows simulation of the climatic effects of higher atmospheric CO_2 through its effect on surface–air moisture fluxes, as well as through its radiative effect (Sellers *et al.*, 1996). The next step will be to incorporate ecosystem models that allow for changes in the vegetation species composition at a given grid cell, as in the stand-alone model of Foley *et al.* (1996). These models are based on a small number of plant functional types (e.g. moist tropical forest, temperate broadleaf forest, boreal forest). Once joint ecosystem–biophysical models are coupled to AGCMs, on a latitude–longitude grid, it will be possible to simulate the effect of climatic change on vegetation and the CO_2 flux to the atmosphere through shifts in biome zones, as well as through changes in the rates of photosynthesis and respiration. This advance will in turn allow a full coupling between the atmosphere and terrestrial biosphere in a regionally resolved model. The effect of altered stomatal conductance as climate and the atmospheric CO_2 concentration change can be quite important to the simulated climatic change on a regional scale, although the effects are much less important when averaged globally (Sellers *et al.*, 1996).

Another major area of intensive research is in the representation of clouds in AGCMs. Differences in the way in which clouds are modelled are the main reason for the wide uncertainty in long-term warming resulting from a doubling atmospheric CO_2 concentration (with global mean responses ranging from 2.1°C to 4.8°C in recent models, as summarized by Harvey, 2000a: Chapter 9). The climatic effects of clouds depend on the details: exactly when, where and how high clouds are; their detailed microphysical properties, such as cloud droplet radius, ice-crystal radius and shape, and the extent of impurities (such as soot) in cloud droplets. As the climate changes, the various cloud properties will change in hard-to-predict ways, with a range of strong but competing effects on climatic change (some cloud changes will tend to amplify the change, thereby serving as a positive feedback, while others will serve to diminish the change, thereby serving as a negative feedback). Increasingly sophisticated cloud schemes are being developed for and implemented in AGCMs (e.g. Del Genio *et al.*, 1996; Kao and Smith, 1999), that explicitly compute more and more of the cloud processes of climatic importance. However, greater sophistication and detail do not guarantee that the predicted changes in climate will be more reliable. This problem arises because even the most detailed schemes still require parameterizations to represent the effects of subgrid-scale processes, and the nature of the parameterizations can strongly influence the results.

Other areas of ongoing research include improved treatment of sea ice, the realistic routing of runoff to the oceans, and the incorporation of atmospheric chemistry within AGCMs used for climate simulation purposes. Better determination of the magnitude of the various direct and indirect effects of different aerosols on solar and infrared radiation, which are computed from but affect the behaviour of AGCMs, is also an important priority, since it is a necessary step in being able to infer the climate sensitivity from observed temperature changes during the past century or so.

2.4 CASE STUDY

As a case study to illustrate the use of climate models in global warming policy analysis, and as an example of the constructive interplay between simple and complex climate models, I will discuss the MAGICC/SCENGEN model system of Hulme *et al.* (2000). This model was developed to be used in developing assessments of regional or national vulnerability to anthropogenic climatic change. MAGICC (**M**odel for the **A**ssessment of **G**reenhouse-gas **I**nduced **C**limate **C**hange) consists of a suite of simple models of each of the climate system components discussed under 'Finding the Simplicity'. MAGICC contains a simple representation of the carbon cycle and other trace gas cycles needed to compute the build-up in their concentration, given emissions scenarios for each of the gases. Simple relationships are then used to relate the change in greenhouse gas concentrations to the trapping of heat in the atmosphere. Simple relationships are also used to directly relate aerosol emissions to the heating or cooling effects of aerosols. Schematic ice-melt models for mountain glaciers, Antarctica and Greenland are used to calculate the contribution of small and large glaciers to sea-level rise. The climate component of the model is similar to the upwelling-diffusion model that was illustrated earlier; it is used to compute the time-dependent, globally averaged change in surface temperature, as well as the vertical variation of warming with depth in the ocean, from which the component of sea level rise due to thermal expansion of sea water can be computed.

The globally averaged surface temperature produced by MAGICC is used in conjunction with the spatial patterns of surface temperature change as simulated by a variety of atmospheric or coupled atmosphere-ocean GCMs. These patterns are provided by the SCEN-GEN (SCENario GENerator) part of the model. This part of the model is based on the observation that the time–space pattern of climatic change when GCMs are driven by increasing greenhouse gas (GHG) concentrations can be well represented by a fixed spatial pattern, which can be uniformly scaled up or down in proportion to the global mean warming (Mitchell *et al.*, 1999). This approach has been referred to as the *pattern-scaling* method, and was first used by Santer *et al.* (1990). Even though the cooling effect of sulphate aerosols (which must also be taken into account when assessing future climatic change) is concentrated over and downwind of polluted regions, the spatial patterns of climatic change with GHGs only or with GHGs and aerosols are remarkably similar. This result is due to the fact that the regional pattern of climatic change is to a large extent controlled by the large-scale climate feedbacks that are triggered by overall climatic change, rather than by the local forcing (see Chapter 3.6 of Harvey, 2000a, for a more thorough discussion of this property of the climate system). Since simple models, such as MAGICC, are capable of quickly computing the time variation in global mean warming for a wide variety of GHG and aerosol-precursor emission scenarios, one can easily generate scenarios of regional climatic change by combining the global mean temperature change from MAGICC with the regional pattern of temperature change as provided by one or more GCMs.

Thus, rather than running a computationally expensive AOGCM for every conceivable scenario of future GHG and aerosol-precursor emissions, one need run the AOGCM for only one scenario, determine the spatial pattern produced by the AOGCM in question, and then scale this spatial pattern up or down in proportion to global mean warming as computed by a suite of coupled climate–gas cycle–aerosol forcing models. This then makes it practical to consider a wide range of emission scenarios. Furthermore, the global mean temperature

responsiveness (or *climate sensitivity*) to a given radiative heating perturbation can easily be adjusted in simple models to match that of any given AOGCM, as can the rate of mixing of heat into the deep ocean. This approach ensures that the global mean temperature response of the simple model will match that which would have been produced by the AOGCM with the altered scenarios (Raper and Cubasch, 1996, have demonstrated the ability of simple models to closely replicate the global mean behaviour of AOGCMs).

Different GCMs can give quite different spatial patterns of temperature change. MAGICC/SCENGEN is set up to allow the user to specify the spatial response pattern from one of 16 GCMs. The user also selects one of several pre-programmed emission scenarios, or can choose his or her own scenario. The climate sensitivity can be prescribed independently of the spatial pattern used to generate regional climatic change. Figures 2.5 and 2.6 give an example of the global mean and regional temperature output that are produced by the model.

The pattern-scaling approach has also been applied to precipitation in MAGICC/SCENGEN. In this case, the spatially varying percentage change in precipitation associated with a particular global mean warming in a GCM is computed. This percentage is scaled up or down in proportion to the global mean temperature at any given time as computed by MAGICC; the resulting percentage change is applied to the observed present-day precipitation. However, whereas over 90% of the time–space variation in surface temperature in a GCM simulation can be accounted for by scaling a fixed temperature pattern, much less of the precipitation variation can be captured in this way. Thus, regional precipitation changes deduced with the pattern-scaling method can be expected to differ, often substantially, from that which would have been obtained from the AOGCM. Compounding this uncertainty in the regional precipitation change is the fact that there is substantial yearly and decadal variability ('noise') in the precipitation response of any given GCM to global warming, and the differences in the response between different GCMs in many critical regions and times (such as the Southeast Asian summer monsoon) can be substantial.

Note on web links

The website associated with this chapter (www.kcl.ac.uk/envmod) contains a link to websites devoted to the detailed description and comparison of high-resolution models of many of the components of the climate system (namely, the atmosphere, oceans, oceanic part of the

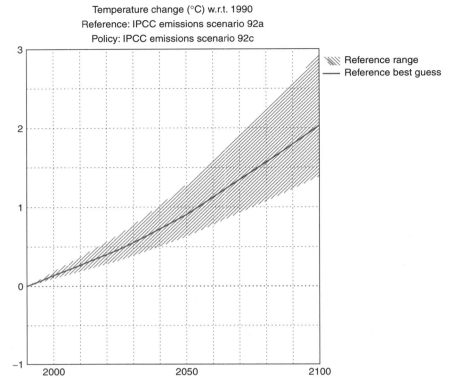

Figure 2.5 Illustration of the global mean temperature for a CO_2 doubling climate sensitivity of 2.5°C, and the range in global mean temperature for sensitivities ranging from 1.5°C to 4.5°C, produced by MAGICC

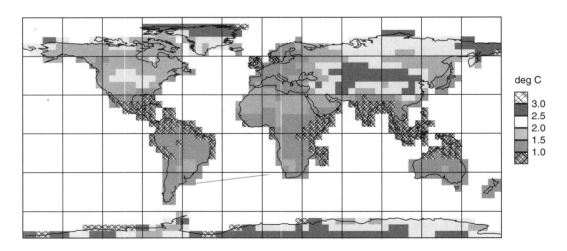

Figure 2.6 Illustration of the regional temperature change for 2050, generated by SCENGEN based on the temperature pattern of the Canadian Climate Centre atmosphere–ocean GCM

carbon cycle and the terrestrial biosphere). These sites contain numerous tables and colour figures comparing the models with one another and with observations, and also contain comprehensive lists of references. Some also contain links to the research centres where the models were developed. The website associated with this chapter also contains a link to the Climate Research Unit of the University of East Anglia, where a demonstration of the MAGGIC/SCENGEN schematic model can be found. A copy of the model can be downloaded, so that those who are interested (and who sign a licence agreement) can run the model on their own personal computers.

REFERENCES

Antoine, D. and Morel, A. (1995) Modelling the seasonal course of the upper ocean pCO_2 (i): development of a one-dimensional model, *Tellus* **47B**, 103–121.

Bacastow, R. and Maier-Reimer, E. (1990) Ocean-circulation model of the carbon cycle, *Climate Dynamics* **4**, 95–125.

Bintanja, R. (1995) The Antarctic ice sheet and climate, PhD thesis, Utrecht University.

Boucher, O. and Lohmann, U. (1995) The sulfate-CCN-cloud albedo effect: a sensitivity study with two general circulation models, *Tellus* **47B**, 281–300.

Chin, M., Jacob, D.J., Gardner, G.M., Foreman-Fowler, M.S. and Spiro, P.A. (1996) A global three-dimensional model of tropospheric sulfate, *Journal of Geophysical Research* **101**, 18667–18690.

Del Genio, A.D., Yao, M.-S., Kovari, W. and Lo, K.K. (1996) A prognostic cloud water parameterization for global climate models, *Journal of Climate* **9**, 270–304.

De Wolde, J.R., Bintanja, R. and Oerlemans, J. (1995) On thermal expansion over the last one hundred years, *Journal of Climate* **8**, 2881–2891.

Dickinson, R.E., Meleshko, V., Randall, D., Sarachik, E., Silva-Dias, P. and Slingo, A. (1996) Climate processes, in J.T. Houghton, L.G.F. Filho, B.A. Callander, N. Harris, A. Kattenberg and K. Maskell (eds) *Climate Change 1995: The Science of Climate Change*, Cambridge University Press, Cambridge, 193–227.

Foley, J.A., Prentice, C., Ramankutty, N., Levis, S., Pollard, D., Sitch, S. and Haxeltine, A. (1996) An integrated biosphere model of land surface processes, terrestrial carbon balance, and vegetation dynamics, *Global Biogeochemical Cycles* **10**, 603–628.

Gates, W.L. *et al.* (1996) Climate models – evaluation, in J.T. Houghton, L.G.F. Filho, B.A. Callander, N. Harris, A. Kattenberg and K. Maskell (eds) *Climate Change 1995: The Science of Climate Change*, Cambridge University Press, Cambridge, 229–284.

Gregg, W.W. and Walsh, J.J. (1992) Simulation of the 1979 spring bloom in the mid-Atlantic bight: a coupled physical/biological/optical model, *Journal of Geophysical Research* **97**, 5723–5743.

Harvey, L.D.D. (1988) A semi-analytic energy balance climate model with explicit sea ice and snow physics, *Journal of Climate* **1**, 1065–1085.

Harvey, L.D.D. (1989) Effect of model structure on the response of terrestrial biosphere models to CO_2 and temperature increases, *Global Biogeochemical Cycles* **3**, 137–153.

Harvey, L.D.D. (1992) A two-dimensional ocean model for long-term climatic simulations: stability and coupling to atmospheric and sea ice models, *Journal of Geophysical Research* **97**, 9435–9453.

Harvey, L.D.D. (2000a) *Global Warming: The Hard Science*, Prentice Hall, Harlow.

Harvey, L.D.D. (2000b) Upscaling in global change research, *Climatic Change* **44**, 225–263.

Harvey, L.D.D., Gregory, J., Hoffert, M., Jain, A., Lal, M., Leemans, R., Raper, S., Wigley, T. and de Wolde, J. (1997) *An Introduction to Simple Climate Models Used in the IPCC Second Assessment Report*, Intergovernmental Panel on Climate Change, Technical Paper II.

Harvey, L.D.D. and Huang, Z. (1995) An evaluation of the potential impact of methane-clathrate destabilization on future global warming, *Journal of Geophysical Research* **100**, 2905–2926.

Harvey, L.D.D. and Schneider, S.H. (1985) Transient climatic response to external forcing on $10^0–10^4$ year time scales, 1, Experiments with globally averaged coupled atmosphere and ocean energy balance models, *Journal of Geophysical Research* **90**, 2191–2205.

Haywood, J.M., Roberts, D.L., Slingo, A., Edwards, J.M. and Shine, K.P. (1997) General circulation model calculations of the direct radiative forcing by anthropogenic sulfate and fossil-fuel soot aerosol, *Journal of Climate* **10**, 1562–1577.

Held, I.M. and Suarez, M.J. (1974) Simple albedo feedback models of the icecaps, *Tellus* **26**, 613–630.

Hoffert, M.I., Callegari, A.J. and Hseih, C.-T. (1980) The role of deep sea heat storage in the secular response to climatic forcing, *Journal of Geophysical Research* **85**, 6667–6679.

Hoffert, M.I., Callegari, A.J. and Hseih, C.-T. (1981) A box-diffusion carbon cycle model with upwelling, polar bottom water formation and a marine biosphere, in B. Bolin (ed.) *Carbon Cycle Modeling, SCOPE 16*, John Wiley & Sons, New York, 287–305.

Houghton, J.T., Ding, Y., Griggs, D.J., Noguer, M., van der Linden, P.J., Dai, X., Maskell, K. and Johnson, C.A. (eds) (2001) *Climate Change 2001: The Scientific Basis*, Cambridge University Press, Cambridge.

Hulme, M., Wigley, T.M.L., Barrow, E.M., Raper, S.C.B., Centella, A., Smith, S. and Chipanshi, A. (2000) *Using a Climate Scenario Generator in Vulnerability and Adaptation Assessments: MAGICC and SCENGEN Workbook*, Climatic Research Unit, UEA, Norwich.

Huybrechts, P. and Oerlemans, J. (1990) Response of the Antarctic ice sheet to future greenhouse warming, *Climate Dynamics* **5**, 93–102.

Huybrechts, P., Letreguilly, A. and Reeh, N. (1991) The Greenland ice sheet and greenhouse warming, *Palaeogeography, Palaeoclimatology, Palaeoecology* **89**, 399–412.

Jones, A. and Slingo, A. (1996) Predicting cloud-droplet effective radius and indirect sulphate aerosol forcing using a general circulation model, *Quarterly Journal of the Royal Meteorological Society* **122**, 1573–1595.

Kao, C.Y.J. (1999) Sensitivity of a cloud parameterization package in the National Center for Atmospheric Research Community Climate model, *Journal of Geophysical Research* **104**, 11961–11983.

Kao, C.Y.J. and Smith, W.S. (1999) Sensitivity of a cloud parameterization package in the National Center for Atmospheric Research Community Climate Model, *Journal of Geophysical Research-Atmospheres* **104**, 11961–11983.

Kelly, P.M. and Wigley, T.M.L. (1992) Solar cycle length, greenhouse forcing and global climate, *Nature* **360**, 328–330.

Ko, M.K.W., Sze, N.D., Wang, W.-C., Shia, G., Goldman, A., Murcray, F.J., Murcray, D.G. and Rinsland, C.P. (1993) Atmospheric sulfur hexaflouride: sources, sinks, and greenhouse warming, *Journal of Geophysical Research* **98**, 10499–10507.

Kogan, Z.N., Kogan, Y.L. and Lilly, D.K. (1996) Evaluation of sulfate aerosol's indirect effect in marine stratocumulus clouds using observation-derived cloud climatology, *Geophysical Research Letters* **23**, 1937–1940.

Lal, M. and Ramanathan, V. (1984) The effects of moist convection and water-vapor radiative processes on climate sensitivity, *Journal of the Atmospheric Sciences* **41**, 2238–2249.

Langner, J. and Rodhe, H. (1991) A global three-dimensional model of the tropospheric sulfur cycle, *Journal of Atmospheric Chemistry* **13**, 225–263.

Lohmann, U. and Feichter, J. (1997) Impact of sulfate aerosols on albedo and lifetime of clouds: a sensitivity study with the ECHAM4 GCM, *Journal of Geophysical Research* **102**, 13685–13700.

Melillo, J.M., McGuire, A.D., Kicklighter, D.W., Moore III, B., Vorosmarty C.J. and Schloss, A.L. (1993) Global climate change and terrestrial net primary production, *Nature* **363**, 234–240.

Mitchell, J.F.B., Johns, T.C., Eagles, M., Ingram, W.J. and Davis, R.A. (1999) Towards the construction of climate change scenarios, *Climatic Change* **41**, 547–581.

Morantine, M. and Watts, R.G. (1990) Upwelling diffusion climate models: analytical solutions for radiative and upwelling forcing, *Journal of Geophysical Research* **95**, 7563–7571.

Najjar, R.G., Sarmiento, J.L. and Toggweiler, J.R. (1992) Downward transport and fate of organic matter in the ocean: simulations with a general circulation model, *Global Biogeochemical Cycles* **6**, 45–76.

Osborn, T.J. and Wigley, T.M.L. (1994) A simple model for estimating methane concentrations and lifetime variations, *Climate Dynamics* **9**, 181–193.

Oschlies, A. and Garçon, V. (1999) An eddy-permitting coupled physical-biological model of the North Atlantic, 1, Sensitivity to physics and numerics, *Global Biogeochemical Cycles* **13**, 135–160.

Oschlies, A., Koeve, W. and Garçon, V. (2000) An eddy-permitting coupled physical-biological model of the North Atlantic 2. Ecosystem dynamics and comparison with satellite and JGOFS local studies data, *Global Biogeochemical Cycles* **14**, 499–523.

Peng, L., Chou, M.-D. and Arking, A. (1982) Climate studies with a multi-layer energy balance model. Part I: Model description and sensitivity to the solar constant, *Journal of the Atmospheric Sciences* **39**, 2639–2656.

Pham, M., Müller, J.-F., Brasseur, G.P., Granier, C. and Mégie, G. (1996) A 3D model study of the global sulphur cycle: contributions of anthropogenic and biogenic sources, *Atmospheric Environment* **30**, 1815–1822.

Plöchl, M. and Cramer, W. (1995) Coupling global models of vegetation structure and ecosystem processes, *Tellus* **47B**, 240–250.

Prather, M., Ibrahim, A.M., Sasaki, T. and Stordal, F. (1992) Future chlorine-bromine loading and ozone depletion, in United Nations Environment Programme Staff (eds) *Scientific Assessment of Ozone Depletion: 1991*, World Meteorological Organization, Geneva.

Raper, S.C.B. and Cubasch, U. (1996) Emulation of the results from a coupled GCM using a simple climate model, *Geophysical Research Letters* **23**, 1107–1110.

Rastetter, E.B., McKane, R.B., Shaver, G.R. and Melillo, J.M. (1992) Changes in C storage by terrestrial ecosystems: how C-N interactions restrict responses to CO_2 and temperature, *Water, Air, and Soil Pollution* **64**, 327–344.

Rastetter, E.B., Ryan, M.G., Shaver, G.R., Melillo, J.M., Nadelhoffer, K.J., Hobbie, J. and Aber, J.D. (1991) A general biogeochemical model describing the responses of the C and N cycles in terrestrial ecosystems to changes in CO_2, climate, and N deposition, *Tree Physiology* **9**, 101–126.

Santer, B.D., Wigley, T.M.L., Schlesinger, M.E. and Mitchell, J.F.B. (1990) *Developing Climate Scenarios from Equilibrium GCM Results*, Max Planck Institute für Meteorologie, Report No. 47, Hamburg.

Sarmiento, J.L., Slater, R.D., Fasham, M.J.R., Ducklow, H.W., Toggweiler, J.R. and Evans, G.T. (1993) A seasonal three-dimensional ecosystem model of nitrogen cycling in the North Atlantic euphotic zone, *Global Biogeochemical Cycles* **7**, 417–450.

Schlesinger, M.E. and Ramankutty, N. (1992) Implications for global warming of intercycle solar irradiance variations, *Nature* **360**, 330–333.

Schlesinger, M.E. and Ramankutty, N. (1994) An oscillation in the global climate system of period 65–70 years, *Nature* **367**, 723–726.

Sellers, P.J. *et al.* (1996) Comparison of radiative and physiological effects of doubled atmospheric CO_2 on climate, *Science* **271**, 1402–1406.

Stocker, T.F., Broecker, W.S. and Wright, D.G. (1994) Carbon uptake experiment with a zonally averaged global ocean circulation model, *Tellus* **46B**, 103–122.

Trenberth, K.E. (ed.) (1992) *Climate System Modeling*, Cambridge University Press, Cambridge.

Van Minnen, J.G., Goldewijk, K.K. and Leemans, R. (1996) The importance of feedback processes and vegetation transition in the terrestrial carbon cycle, *Journal of Biogeography* **22**, 805–814.

VEMAP Members (1995) Vegetation ecosystem modeling and analysis project: comparing biogeography and biogeochemistry models in a continental-scale study of terrestrial ecosystem responses to climate change and CO2 doubling, *Global Biogeochemical Cycles* **9**, 407–437.

Verbitsky, M. and Saltzman, B. (1995) Behavior of the East Antarctic ice sheet as deduced from a coupled GCM/Ice-sheet model, *Geophysical Research Letters* **22**, 2913–2916.

Wang, C., Prinn, R.G. and Sokolov, A. (1998) A global interactive chemistry and climate model: formulation and testing, *Journal of Geophysical Research* **103**, 3399–3417.

Wigley, T.M.L. (1989) Possible climate change due to SO_2-derived cloud condensation nuclei, *Nature* **339**, 365–367.

Wigley, T.M.L. and Raper, S.C.B. (1990) Natural variability of the climate system and detection of the greenhouse effect, *Nature* **344**, 324–327.

Wright, D.G. and Stocker, T.F. (1991) A zonally averaged ocean model for the thermohaline circulation, I, Model development and flow dynamics, *Journal of Physical Oceanography* **21**, 1713–1724.

3

Soil and Hillslope Hydrology

ANDREW BAIRD

3.1 IT'S DOWNHILL FROM HERE

Any child can tell you 'water flows downhill'. However, it would be truer to say that water flows down an energy gradient from areas where the water has greater energy to areas where it has lower energy. This is true of water flowing over the ground surface and water flowing through a porous medium such as soil. Indeed, water can flow uphill if a suitable energy gradient exists. Hydrologists, ecologists and engineers, among others, are not just interested in the direction of flow, but also in the speed of flow, the amount (mass or volume) involved in the flow, and how flows cause variability in the amount of water on and in hillslopes both in space and time. Part of this interest involves a desire to model or predict water flows on and in hillslopes. With advancements in high-speed digital computing, the 1970s seemed to promise the development of sophisticated and accurate models of hillslope hydrology. The reality has been rather different (Beven, 1996, 2000). Some researchers would argue that our models are too complicated while others would suggest that they are too simple (cf. Baird, 1999). Certainly few, if any, models incorporate all known water-transfer processes on and in hillslopes. The rest of this chapter considers whether this is a good thing and also reviews current ideas on how we should model hillslope hydrology.

3.2 EASY DOES IT: THE LABORATORY SLOPE

Before looking at the real world, it is perhaps better to consider how water flows over and through simple artificial hillslopes in the laboratory. Such a consideration will allow us to readily identify the problems of describing and modelling flow on and in real slopes. Let us consider two artificial hillslopes. The first, Hillslope 1, consists simply of a relatively smooth, inclined plane constructed from impermeable material (Figure 3.1a). The second, Hillslope 2, also consists of an inclined plane of impermeable material but this time a homogeneous soil (say, a sandy loam) of constant thickness (say, 50 cm) has been spread across the plane (Figure 3.1b). The soil does not contain organic matter and is unvegetated. Water can be added to both slopes from above at any intensity commonly found in nature using a rainfall simulator. How does water flow on and through these idealized/artificial hillslopes?

3.2.1 Plane simple: overland flow on Hillslope 1

Almost immediately after the start of simulated rainfall, water will start to flow over the surface of Hillslope 1. How will water flow over the surface? A number of models exist to describe such flow. One of the most commonly used is the so-called kinematic wave. A kinematic wave model consists of an equation that describes flow coupled with a balance equation. Flow velocity is usually expressed as a power function of the depth of water on the slope, the gradient of the slope, and a friction factor describing the effects of frictional drag on water flow (see Baird, 1997a). To simulate changes in velocity and flow depth over time, this equation is combined with the balance equation describing time-variant inputs and outputs along the slope surface (Chow, 1959; Baird, 1997a,b). Because this approach ignores the effect of changes in momentum, it is called the kinematic approximation or kinematic wave approach (Ponce *et al.*, 1978; Dingman, 1984; Ponce, 1991) ('kinematic' means the study of motion without reference to mass or force; Hanks, 1986).

Environmental Modelling: Finding Simplicity in Complexity. Edited by J. Wainwright and M. Mulligan
© 2004 John Wiley & Sons, Ltd ISBNs: 0-471-49617-0 (HB); 0-471-49618-9 (PB)

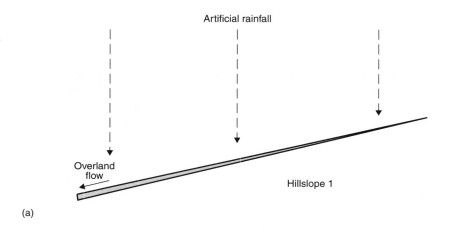

(a)

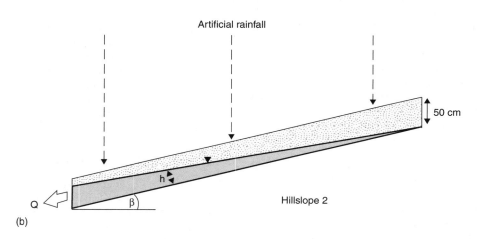

(b)

Figure 3.1   (a) Hillslope 1 – an inclined impermeable plane. (b) Hillslope 2 – an inclined impermeable plane topped with a sandy loam 50 cm thick

In essence, this approach implies that unsteady overland flow can be considered as a succession of steady uniform flows, with the water surface slope remaining constant at all times. The suitability and success of the kinematic wave approach in simulating overland flow on simple surfaces have been discussed by, among others, Henderson (1966), Woolhiser and Liggett (1967), Kibler and Woolhiser (1970), Woolhiser (1975), Ponce *et al.* (1978), Wu *et al.* (1978), Rose *et al.* (1983), Miller (1984), Cundy and Tento (1985), Bell *et al.* (1989) and Wheater *et al.* (1989).

For laboratory slopes like Hillslope 1 it seems the kinematic wave model provides very accurate descriptions of discharge from the base of the slope. This has been demonstrated by Bell *et al.* (1989), Hall *et al.*

(1989) and Wheater *et al.* (1989); who compared predictions from a simple kinematic wave model with measurements from a laboratory slope 1 m wide and 9.925 m long. They carried out 96 sets of measurements on the laboratory slope, varying both rainfall intensity and duration and slope gradient. For all sets of measurements they compared the fit of the model to the observed data using the following criterion:

$$R^2 = \frac{F_0^2 - F^2}{F_0^2} \tag{3.1}$$

where

$$F_0^2 = \sum_{i=1}^{n} (q_i - \overline{q})^2, \quad F^2 = \sum_{i=1}^{n} (q_i - q_i')^2$$

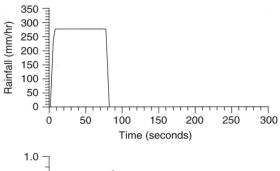

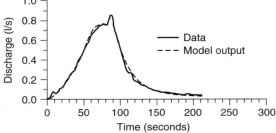

Figure 3.2 Comparison of measured and modelled (using a kinematic wave) overland flow on an artificial hillslope. In the example shown the slope gradient was 0.005
Source: Wheater *et al.* (1989)

and q_i is the observed discharge, q_i' the modelled discharge, n the number of discharge values and $\overline{q}$ the mean of the observed discharges. An R^2 of 1 represents a perfect correspondence between predicted and observed. Wheater *et al.* (1989) found that the mean R^2 of the 96 comparisons was very high at 0.967. An example of one of their comparisons is given in Figure 3.2.

3.2.2 Getting a head: soil-water/ground-water flow in Hillslope 2

Rainfall on Hillslope 2 will enter the soil and will then flow vertically and downslope through the soil profile. How the water enters the soil (infiltration) is dealt with in Chapter 4. Before we consider flow throughout the soil on Hillslope 2, it is useful to consider flow through a relatively small cube (say $50 \times 50 \times 50$ cm) of the soil. Saturated flow in such a soil is described well by Darcy's law, which, for steady flow in one dimension, is given by:

$$q = -K \frac{dH}{dx} \qquad (3.2)$$

where q is the discharge per unit area ($L^3 T^{-1} L^{-2} = L T^{-1}$), K is the hydraulic conductivity ($L T^{-1}$), H is the

hydraulic head (L) and x is the distance in the direction of flow (L). q is often called the specific discharge and K is the specific discharge under a unit hydraulic gradient (i.e. $dH/dx = 1$). H is a measure of the energy of the soil water and has units of length because it is expressed as energy per unit weight (i.e. $(M L^2 T^{-2})/(M L T^{-2}) = L$) (Fetter, 1994). To make q positive, a minus sign appears before K. This 'correction' is necessary because water flows down the hydraulic gradient, i.e. it flows along a negative gradient from areas of higher H to areas of lower H.

Flow is rarely one-dimensional. Sometimes it is necessary to consider dH/dz and dH/dy where z is depth and y the other horizontal direction. Sometimes, also, hydraulic conductivity varies according to direction, i.e. we need to consider K_x, K_y and K_v (Freeze and Cherry, 1979). When this occurs, the medium is said to be anisotropic (as opposed to isotropic). Flow is also rarely steady. To consider time-variant flows we would need to consider changes in soil water storage. In our cube this means that the water table will rise and fall depending on the difference between inflows and outflows of water. Of course, it is also possible for the water table to rise to the surface and for surface ponding to occur which might result in overland flow (see above). An exposition of the derivation of the transient flow equations for unconfined flow is beyond the scope of this chapter. Good descriptions can be found in Freeze and Cherry (1979) and Fetter (1994).

Darcy's law can also be modified to describe flow through unsaturated soil, where K is assumed to be a function of water content (θ), i.e.

$$q = -K(\theta) \frac{dH}{dx}. \qquad (3.3)$$

Where flow is unsteady, i.e. in most natural conditions, water content will change and so will K. Thus in unsaturated flow, unlike saturated flow, K is time-variant (except for some very simple cases). The unsaturated form of Darcy's law is usually called the Richards equation (especially when combined with the balance equation for time-variant flows) after L.A. Richards who was the first to realize that Darcy's law could be applied to unsaturated flows (Richards, 1931, cited in Hillel, 1980).

Darcy's law is the single most important relation used to describe saturated water flow through soils. Indeed, in a review of methods used to characterize the hydraulic properties of porous media, Narasimhan (1998) suggests:

It is reasonable to state that all the hydraulic characterization methods in use today have two themes in common: an empirical equation of motion, familiarly known as Darcy's law ... and the equation of transient heat conduction, originally proposed by Fourier in 1807, which has established itself as the working model for diffusion-type processes in physical sciences.

Provided the pores in our soil are not too large or too small, the hydraulic gradient not too large, and the soil fully saturated, it appears Darcy's law provides a good description of flow in laboratory soils. This has been demonstrated in numerous studies where a linear relationship between specific discharge and the hydraulic gradient has been observed. The physical basis of the law and its more general applicability to soils can be found in a classic paper by Hubbert (1940) and in standard hydrogeology text books (e.g. Freeze and Cherry, 1979; Fetter, 1994).

So, how can Darcy's law be used to model flow in our simple artificial slope? Various models based on Darcy's law have been developed to describe hillslope flow. The most sophisticated (at least mathematically) can model two- and three-dimensional time-variant flow, allow for spatial variability in soil physical properties and can simulate both saturated and unsaturated flow in a dense grid of points or nodes representing the hillslope. Models have even been developed to consider, for example, the effect of heat on water flow. For example, as water warms, its viscosity is reduced and it flows more readily. Thus K increases with water temperature (Klute, 1965; de Marsily, 1986). These sophisticated models are usually solved using numerical methods such as the finite-element and finite-difference method (see Chapter 1 and Wang and Anderson, 1982; Beven, 1985; Baird, 1999). Examples of these state-of-the-art models include the United States Geological Survey's MODFLOW and VS2D, the US Salinity Laboratory's HYDRUS, the German-based Institute for Water Resources Planning and Systems Research Ltd.'s FEFLOW, and Analytic and Computational Research Inc.'s PORFLOW (website details for all of these are given at the end of the list of references).

Some of the sophisticated models would be relatively easy to apply to our hillslope and many of their features would not be required for our simple flow case. For example, we could ignore variability and, perhaps, anisotropy of K. We could also ignore cross-slope variations in water table elevation and soil moisture content. An example of the application of this type of model to a relatively simple artificial slope can be found in Sloan and Moore (1984). Sloan and Moore used a finite-element model based on the 2-D Darcy–Richards

equation to model saturated–unsaturated flow in an experimental hillslope in the Coweeta Hydrological Laboratory in the southern Appalachian mountains, USA (Hewlett and Hibbert, 1963). The artificial hillslope consisted of a $0.92 \times 0.92 \times 13.77$ m concrete-lined trough constructed on a 40% slope and filled with recompacted C horizon forest soil. Measurements reported by Hewlett (1961) and Hewlett and Hibbert (1963) were used to assign values to the model's parameters, i.e. the model was not optimized to give the best possible predictions. Model predictions of discharge from the base of the slope were found to compare quite well with measurements. However, due to limitations of the experimental data it was not possible to compare measurements and predictions of water-table dynamics within the hillslope.

Although sophisticated Darcy-type models could be used, it is also possible to simplify our description of processes occurring in our artificial soil-covered slope (e.g. Sloan and Moore, 1984; Steenhuis *et al.*, 1988; Pi and Hjelmfelt, 1994; Fan and Bras, 1998; Wigmosta and Lettenmaier, 1999). A popular group of simple models for subsurface flow are, like the overland flow models described above, kinematic wave models. The simplest of this class of model ignore or greatly simplify flow in the unsaturated zone and (a) assume that flow in the soil is parallel with the soil surface; and (b) that the hydraulic gradient at any point within the saturated zone is equal to the slope (Henderson and Wooding, 1964; Beven, 1981). Thus, Darcy's law now becomes

$$Q = K h \sin \beta \qquad (3.4)$$

where Q is the discharge per unit width of hillslope in the downslope direction ($L^2 T^{-1}$) (i.e. specific discharge multiplied by depth of flow), h is the depth below the water table measured orthogonally to the impermeable plane at the base of the soil, and β is the slope of the plane (and soil surface) (see Figure 3.1b).

Equation 3.4 can be combined with a balance equation and solved numerically (finite-difference or finite-element method) or analytically (cf. Baird, 1999) to simulate the level of the water table at different points along a hillslope under both transient and steady rainfall. Subsurface flow models based on Equation 3.4 are linear kinematic wave models, whereas surface flow models based on a power law flow description produce nonlinear kinematic waves. For simple cases, the subsurface kinematic wave model has been shown by Beven (1981) to accord quite closely with models based on more rigorous Darcian assumptions. The model has also been extended to include one-dimensional unsaturated flow

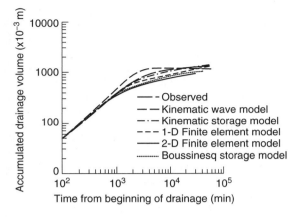

Figure 3.3 Comparison of measured and modelled discharge from the base of the Coweeta experimental soil trough
Source: Sloan and Moore (1984)

above the water table, time-variant rainfall, and variation of the hydraulic properties of the soil (Beven, 1982 – see also Sloan and Moore, 1984) and has even been used as a standard against which other models of hillslope hydrology are compared (e.g. Wigmosta and Lettenmaier, 1999). Even simpler models of hillslope hydrology are possible in which the whole hillslope segment is treated as a single control volume (see below and Sloan and Moore, 1984).

How good are these simplified models? This is a difficult question to answer because very few studies exist where model results have been compared to laboratory or field measurements. Sloan and Moore (1984) compared the complex 2-D model mentioned above with four other models of varying complexity, including a kinematic wave model and two very simple store and flow models. They found that some of the simpler models predicted flow from the base of the Coweeta laboratory slope as reliably as the 2-D model (see Figure 3.3). Moreover, the same simple models gave similar water-table predictions as the complex model.

3.3 VARIETY IS THE SPICE OF LIFE: ARE REAL SLOPES TOO HOT TO HANDLE?

Real hillslopes are rarely, if ever, as simple as our laboratory slopes. In fact, the more we think about hydrological processes on and within real slopes, the harder they become to conceptualize. For example, in the real world, hillslopes rarely have a constant gradient. They rarely consist of homogenous and isotropic soils (e.g. Quinton *et al.* 2000). Their soils usually vary in thickness and

they can be underlain by both permeable and impermeable strata. Stones within the soil might alter patterns and rates of surface and subsurface flow (Poesen, 1992; Mulligan, 1996; van Wesemael *et al.*, 2000). Subsurface flow might also be dominated by a few large pores (macropores – see Beven and Germann, 1982; McCoy *et al.*, 1994). Some pores may contain trapped gas bubbles that reduce flow through them. These bubbles may grow or shrink through time depending on their origin, and as they change size so rates of flow vary (Fayer and Hillel, 1986a; Faybishenko, 1995). The slope surface may not be a simple plane but instead consist of a complex (and sometimes living) microtopography. Water flow on this surface will tend to occur as different-sized threads of water moving at a range of velocities. Plant cover on the hillslope might also be highly variable, and different growth stages and species of plants will affect the storage and movement of water on and in the hillslope differently (Sánchez and Puigdefábregas, 1994; Verrecchia *et al.*, 1995). The soil surface might be susceptible to sealing and crusting before and after rainfall, and this, in turn, may alter the infiltration potential of the soil (Romkens *et al.*, 1990).

Although apparently reasonable for laboratory slopes, we might expect our kinematic wave models to be quite inadequate at simulating the real world as described above, and even our so-called sophisticated Darcy–Richards models might prove poor descriptors of subsurface flow processes in real soils. Attempting to model the complexity described above would seem to be a formidable task. Indeed, we might ask, can we model such variability and complexity? Or even, is it *desirable* to model such variability and complexity?

A pragmatic and perhaps mordant answer to the first question would be that Nature is generally too complex to model successfully and that we should pay more attention to runoff from entire hillslopes or at least large blocks of hillslope rather than the fine detail of runoff. That is, we should ignore the pattern and detail of the processes responsible for water movement and storage on and within the hillslope and instead view an entire hillslope as our smallest modelling unit. Such a view is not as radical as it first appears. Many models ignore fine detail for the very reason that the detail makes modelling too difficult or because the detail is *unimportant*. A good example of such a model is already under our noses.

3.3.1 Little or large?: Complexity and scale

It was suggested above that saturated flow through a block of laboratory soil is relatively simple and thus easy to describe using Darcy's law. However, the description was somewhat misleading. Flow at the scale of a few cubic decimetres of a fully saturated laboratory soil is easy to describe using Darcy's law. It is a simple matter to investigate whether flow in relatively small volumes of homogenous soil is Darcian (see the *Environmental Modelling* book website). However, flow through *individual* pores was not considered. At this smaller scale, flow is strongly heterogeneous and probably impossible to model. Even within the sandy loam, there will be pores varying in size by orders of magnitude. Rates of flow through the pores will also show great variability. Very few would argue that we should attempt to model flow in each and every pore of a soil. Instead we should consider larger, more manageable volumes of soil and describe the bulk or macroscopic scale flow through these volumes. This is exactly what Darcy's law does. Flow through individual pores is ignored in the law; it considers only the *macroscopic* flow of the medium. The scale at which this macroscopic flow is defined is termed the Representative Elementary Volume (REV) (Figure 3.4). The concept of the REV was first elucidated by Hubbert (1956) and was defined by Bear (1972) as a volume of sufficient size that there are no longer any statistical variations in the size of a particular property (in this case porosity) with the size of the element. The REV depends on the textural and structural properties of the soil and in Figure 3.4 is equal to V_3.

The arguments in the paragraph above apply also to larger-scale flows within and on hillslopes. Rather than describing all the elements of flow within a hillslope, would it not be better to describe flow at the scale of the hillslope, or at least some large hillslope elements? Over this larger scale, smaller-scale variability becomes less important; in other words, complexity at a small scale becomes simplicity at a larger scale. An

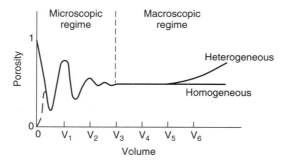

Figure 3.4 Definition of the representative elementary volume (REV). The REV in this case is V3
Source: Hubbert (1956) as drawn in Domenico and Schwartz (1990)

argument to this effect has recently been presented in the hydrological literature by Michel (1999) who suggested that it is better to consider the hillslope as a single quadratic reservoir rather than try to represent the details of flow within the hillslope. Michel was responding to a paper by Steenhuis *et al.* (1999a) in which two commonly used equations of water flow in hillslopes were shown, when simplified, to yield the same analytical description of flow through a sloping shallow soil. Steenhuis *et al.* (1999a) compared their analytical descriptions to data collected from the artificial hillslope (a soil-filled trough) at the Coweeta Hydrological laboratory described above (Hewlett and Hibbert, 1963). A key aspect of their study was that they needed to use so-called 'effective' values of hydraulic conductivity and water storage in order to achieve a good fit between modelled and measured discharge from the artificial hillslope. Such effective values are needed, it seems, to account for flow in the slope that is non-Darcian (i.e. relatively rapid flow in larger soil pores known as macropores – but see below). Michel (1999) suggests that a simple quadratic reservoir capable of describing flow from the soil-filled trough is:

$$Q = \frac{V^2}{\tau V_0} \quad (3.5)$$

where Q is the discharge from the slope ($L^3\ T^{-1}$), V is the volume of drainable water remaining in the hillslope (L^3), V_0 is the total volume of water that can drain from the hillslope (L^3), and τ is a system parameter (T). Michel shows that Equation 3.5 is a better description of flow from the hillslope than the equations presented by Steenhuis *et al.* (1999a). A key point of his argument is that many plausible models can be developed to describe flow from hillslopes. With this in mind he suggests that we should opt for the simplest plausible model that provides accurate predictions of whole hillslope behaviour. A similar, although more detailed, argument has been presented by Baird (1999). Of course, both Michel (1999) and Steenhuis *et al.* (1999a) were modelling flow through a laboratory slope and the discussion so far in this section has been on real slopes. However, implicit in Michel's argument is that, because even simplified slopes present modelling problems, lumped or simple models are needed for real (complex) slopes.

Michel's suggesting of lumping at the scale of an entire hillslope is, perhaps, rather extreme. Nevertheless, many other authors have investigated how suitable similar lumping might be. Relatively recently, Binley *et al.* (1989a,b) reviewed previous work on lumped hillslope models and also performed a series of numerical (computer) experiments designed to investigate whether lumped models can reproduce the behaviour of a heterogeneous hillslope. The first stage of their analysis was to simulate flow in a heterogenous hillslope using a 3-D saturated–unsaturated finite-element model based on Darcy's law. The model simulated a hillslope 150 m wide and 100 m long comprising 3024 finite-element nodes. Values of saturated hydraulic conductivity (from a known distribution) were assigned randomly across the model's nodes. Binley *et al.* (1989b) then attempted to reproduce the runoff from the modelled heterogeneous slope with a modelled slope that had a single 'effective' value of saturated hydraulic conductivity. They found that both the surface and subsurface runoff response of heterogeneous hillslopes with a generally high hydraulic conductivity could be described quite well by homogeneous slopes with single effective values of saturated hydraulic conductivity. However, in low hydraulic conductivity slopes, where the amount of surface runoff was quite high, the lumped models were unable to reproduce accurately the two components of runoff. Thus it appears that lumped models based on Darcy's law might have limited usefulness. A key difference between lumped models of the sort tested by Binley *et al.* (1989b) and that suggested by Michel (1999) is that those of Binley *et al.* (1989b) were lumped only in terms of a single parameter. That is, the model hillslopes used by Binley *et al.* were still divided up spatially, rather than being considered as single hillslope-sized stores. More on the problems of scale in hydrological modelling can be found in special issues of *Hydrological Processes* (see Blöschl and Sivapalan, 1995) and *Water Resources Research* (see Blöschl *et al.*, 1997).

Single-store representations of hillslope behaviour are not new, and many lumped conceptual-mathematical models of catchment runoff response to rainfall assume hillslope soils can be represented by single linear or non-linear 'reservoirs'. An example of a model containing such stores is HYRROM (Hydrological Rainfall-Runoff Model – Blackie and Eeles, 1985) which represents a catchment using only four stores: the interception store, the soil store, the ground water store, and the routing (channel) store (Figure 3.5). Details of more physically based single store models can be found in Sloan and Moore (1984; see also Watts, 1997).

Similar arguments to those presented above have been made in the context of modelling catchment runoff where it has been suggested that it is possible to define Representative Elementary Areas (REAs) at which runoff behaviour is stable (see Wood *et al.*, 1988, and Chapter 4), i.e. not affected by small-scale changes in, say, rainfall delivery or soil-hydraulic properties.

3.3.2 A fitting end: the physical veracity of models

An objection to models such as Equation 3.5 is that they are not physically meaningful (Steenhuis *et al.*, 1999b). However, care has to be taken in deciding what we mean by 'physically meaningful'. It is widely recognized that a good fit between an observed data set and part of a

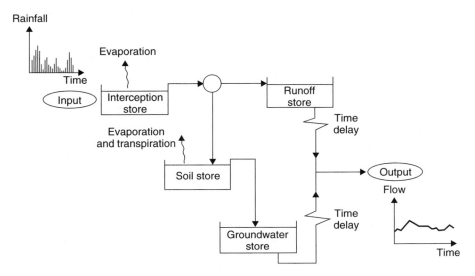

Figure 3.5 The structure of HYRROM (Hydrological Rainfall-runoff Model)
Source: Based on an original diagram by Blackie and Eeles (1985)

model's output is not a guarantee that the physical basis of the model has been validated (Beven, 1996; Baird, 1999; Gupta, 2001). For example, a model capable of predicting both water-table level and discharge from the base of a slope, is not validated if only its predictions of discharge have been shown to correspond closely with measurements (see Sloan and Moore, 1984). Clearly, it is important to check that the model also predicts water table elevations (in both space and time) accurately. Checking the veracity of different components of a model in this way is often termed internal validation.

A good example of internal validation is found in Parsons *et al.* (1997) who tested the detailed predictions of a distributed kinematic wave overland flow model. Remarkably, except for Scoging *et al.* (1992), it appears Parsons *et al.* (1997) are almost unique in testing the ability of such a model to predict both the slope hydrograph and patterns of flow on the slope. By performing a thorough internal validation of their model, Parsons *et al.* (1997) were able to reveal the limitations of a class of model that hitherto had been regarded as a good or satisfactory descriptor of hillslope overland flow. In particular, they found that such models only give acceptable predictions of overland flow hydraulics if they are finely tuned for each surface to which they are applied. Such fine tuning requires huge, and for many applications unrealistic, amounts of field data. This conclusion led Parsons *et al.* to suggest that process-based modelling of soil erosion may not be a realistic goal for hillslope researchers.

Even when a good fit is observed, it is important to bear in mind that a similarly good fit could be produced by a different suite of processes, i.e. it is possible that in Nature different groups of processes produce similar hydrological behaviour at certain scales of integration such as the hillslope or large hillslope segments ($10 s - 1000 s m^2$) (see Grayson *et al.*, 1992a,b). This problem is similar to, but not quite the same as, the problem of 'model equifinality' as defined and discussed by Beven (1996, 2001) (see also Baird, 1999) where different combinations (or values) of rate parameters applied to a *fixed set* of modelled processes can result in similar model predictions. It is important to realize that we are still very uncertain, and lack fundamental knowledge, about some of the processes controlling water flow in soils (despite the implicit claims of some modellers that we now know the suite of factors affecting water-transfer processes, with the only remaining challenge being that of identifying which of the factors are important in certain circumstances). This point is explored in more detail below. It seems we tend to

dismiss a soil-water model as lacking in physical meaning if it does not incorporate an explicit Darcy-based description of flow. However, to dismiss a model as not being physically meaningful on this basis may be misguided. A lumped model may not look physically based, but it could actually represent very well the *integrated* behaviour (at certain scales) of a number of key processes occurring within a soil. These processes may not be Darcian.

3.3.3 Perceptual models and reality: 'Of course, it's those bloody macropores again!'

Notwithstanding the problems of model equifinality, the need for checking a model's predictions with data is almost universally recognized (but see Fagerström, 1987; see also the discussion in Baird, 1999, p. 330). If we accept Popper's classical model of falsification (Popper, 1968; see also Wilby, 1997) in the sciences, then we might expect a mismatch between modelled and observed to be treated as evidence that a model is false. It might seem surprising, therefore, to find examples in the literature where authors have found mismatches between observed and modelled behaviour but still assume that their model provides a reliable description of part, even a substantial part, of the system being modelled. Moreover, we might be even more surprised to find that a lack of fit between model output and data is presented as evidence in its own right that some other named process is occurring in a soil. For example, there are examples in the literature where models based on the Darcy–Richards equation give relatively poor fits to observed data. When this happens, some authors suggest that the lack of a good correspondence between the model and the data is due to macropore flow. Furthermore, the authors suggest that the model is still a good description of flow in that part of the soil without macropores. Such statements might seem quite reasonable until it is realized that the authors have based their conclusions on an *assumption* that macropores occur in the soil. In other words they have no measurements of macropore flow, or indeed observations of macropores on which to base their conclusions about the importance of macropore flow. So why do researchers invoke macropores as an explanation of a lack of fit between Darcy–Richards models and observed data? And, are they right to do so?

Macropores are known to be very important controls of both patterns and rates of saturated and unsaturated water flow through certain soils (e.g. Beven and Germann, 1982; McCoy *et al.*, 1994; Buttle and Leigh, 1997). Macropores are relatively large pores present within a matrix of smaller pores. Exact definitions of macropores vary, but most agree they include features such as desiccation cracks, faunal (e.g. earthworm) burrows, and root channels in which capillary forces are negligible. Water flow in macropores (if it occurs) tends to be relatively rapid and bypasses more slowly moving water in the smaller pores of the matrix of the soil. Although often comprising a small proportion of the total pore space within a soil, macropores can be the dominant route of soil water movement. In addition, water flow through systems of macropores can be non-Darcian. The flow is often not strictly laminar as required by Darcy's law, and large discontinuities in the energy status of soil water over scales smaller than the REV (see above) can occur in the vicinity of macropores. A huge research effort on macropore effects began in the early 1980s and continues today, and much observed non-Darcian behaviour is now attributed to macropores. However, there are other phenomena of soil water flow, which are often overlooked, that could explain non-Darcian behaviour. An example is quasi-saturation. Quasi-saturation is the term given to soils that are not fully water-saturated below the water table (Faybishenko, 1995). Quasi-saturation exists for a variety of reasons. Pockets of soil air can be trapped during infiltration and subsequent water-table rise (Bond and Collis-George, 1981; Constantz *et al.*, 1988; Fayer and Hillel, 1986a,b; Faybishenko, 1995). Gases can also be generated *in situ* by soil bacteria (Reynolds *et al.*, 1992; Walter *et al.*, 1996; Beckwith and Baird, 2001). Surprisingly little work has been done on the effects of entrapped gas bubbles below the water table on flow. Yet these effects could be very important (e.g. Fayer and Hillel, 1986a,b; Reynolds *et al.*, 1992; Delozada *et al.*, 1994; Faybishenko, 1995; Beckwith and Baird, 2001), resulting in large reductions in hydraulic conductivity from the fully water-saturated condition. It is interesting to note how different branches of science make advances at different times. For example, in plant physiology the effect of gas bubbles on water flow through plant tissue is now widely recognized as being fundamental in affecting water movement in plants (see, for example, Tyree and Sperry, 1989; Tyree and Ewers, 1991; Huang and Nobel, 1993; Kolb *et al.*, 1996; Sobrado, 1997; Williams *et al.*, 1997). In soils, the same cannot be said. Although useful advances have been made in our understanding of quasi-saturation, most soil hydrologists are unaware of its potential importance. A search of the Web of Science database for journals published between 1996 and 2000 revealed only 17 papers dealing with quasi-saturation/air entrapment in soils. Fewer than half of these papers considered the effects of trapped gas on soil hydraulics. In contrast, in the same five-year period, 213 papers contained the words 'macropore' and 'soil' in their titles or abstracts. Macropores are the current fashion and have been for over 15 years.

Let us return to the example of a Darcy–Richards model being applied to a hillslope soil. It would be invidious to name specific authors who have rashly invoked macropore flow when a Darcy–Richards model does not provide a good description of observed data. So let us take a theoretical example. First, assume that we are comparing modelled and measured discharge from the base of the slope after a rainfall event. Second, assume that the model underpredicts the initially high rate of discharge from the slope but can be fitted quite well to the later recession. One explanation of the discrepancy between the Darcy–Richards model and observations is that the initially high rate of discharge from the slope is due to the rapid flow of water in macropores in the unsaturated zone, when these empty, Darcian flow processes in the saturated zone start to become dominant. The above conceptual model for the hillslope is certainly plausible and could be the truth. But we have to be careful. Other equally good and plausible models could be developed to explain what is going on. The problem is that these other plausible explanations do not exist in our 'perceptual model', i.e. we are unaware of the effects of quasi-saturation on the hydraulics of flow below the water table or we assume that it is unimportant. We are surrounded by papers telling us that macropores are important and do not feel the need to look elsewhere. As Beven (1996) noted: 'The perceptual model is inherently qualitative, but conditions both responses to experimental evidence and decisions about the dominant processes and representations of those processes to be included in quantitative models.'

It is quite possible that in our real system macropore flow is unimportant and the Darcy–Richards model is a modest or even poor descriptor of flow through the remainder of the soil. What really happens might be as follows. During and after a rain storm, the water table rises. With this increase in water table there is a general increase in the hydraulic conductivity of the soil (i.e. as dry soil becomes wetter) and outflow from the base of the hillslope increases sharply. However, during water-table rise, some pockets of air become trapped in the soil. The flow of water through soil pores entrains some of the trapped air bubbles. These soon start to coalesce at pore necks causing a decline in K and, therefore, flow from the soil (see Figure 3.6). Now we have another, quite reasonable, explanation for the high initial flow rates in the soil, followed by a steep decline. The pattern is the same but the processes are quite different.

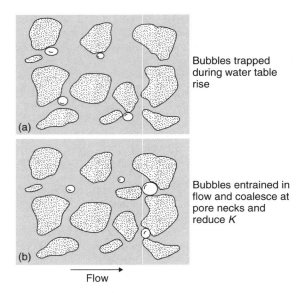

Bubbles trapped during water table rise

Bubbles entrained in flow and coalesce at pore necks and reduce *K*

Flow

Figure 3.6 (a) Trapped pockets of air in a hillslope soil after water table rise. (b) Coalescence of bubbles at pore necks within the soil causing a decline in hydraulic conductivity (*K*)

So we have to be extremely careful about fashions. Because our perceptual model is *partly* just that: a fashion. Macropores are still in fashion; quasi-saturation has yet to (and may never) hit the catwalk. Unconsciously, we tend to look to the process research that is in fashion in order to explain what we think is going on; in other words we look to popular 'perceptual' models. This is, perhaps, an unsurprising point, but nevertheless an important one to state. Fashionable research is likely to be important research; there is no doubt that macropores are important controls on water flow in some soils. However, just because something is important does not mean that it is the only important 'thing'. If we fail to think more deeply about other possibilities, we are being intellectually lazy. As Gupta (2001) has noted: 'Every assumption that goes into the modelling process should (if at all possible) be tested for violations . . . this requires a look at one's belief structure, which is surprisingly difficult to do.'

Of course, it is easy to be over-critical of previous work. Often field data are difficult to interpret and the processes occurring in a soil are far from clear. When Darcian models appear to fail, there is often *some* evidence that macropore flow might be the reason for a model's failure. A good example of the problems of comparing field data with model predictions is the study of Koide and Wheater (1992) who instrumented a small

(12 × 18 m) forested hillslope plot (0.02 ha) in Scotland with 21 tensiometers and a throughflow pit. They compared measurements of water movement with simulations from a 2-D saturated–unsaturated finite-element model based on Darcy's law and found large discrepancies between modelled and measured behaviour for some rainfall events. They suggested that this might be due to macropore flow; their soil contained numerous tree roots and these might have formed macropores through which short-circuiting flow occurred. However, it is also clear from their paper (and this is acknowledged by Koide and Wheater) that, individually or in combination, instrument errors, difficulties in parameterizing their model and difficulties in describing the boundary conditions of their study plot could also have been responsible for the apparent deficiencies of the model. What the study of Koide and Wheater shows is that differences between modelled and measured behaviour can be very difficult to attribute to a single mechanism and that the results of Darcian models should be used with great caution in this respect.

3.4 EENIE, MEENIE, MINIE, MO: CHOOSING MODELS AND IDENTIFYING PROCESSES

We have seen how simple laboratory slopes are quite easy to model. Nevertheless, even these present problems to modellers as the debate between Michel (1999) and Steenhuis *et al.* (1999a) demonstrates. Despite these problems, Darcian models are used very widely in studies of water flow through real soils and still dominate the research literature. Darcy's law still forms the core of most people's perceptual model of soil hydrology. Therefore, if confronted with the need to model water flow through a soil, most will opt for Darcy's law in some form.

It was suggested that real hillslopes might be too complex to model in detail. Indeed, it was noted that it might be undesirable to model all the known processes operating in and on a hillslope. Whether it is desirable always to consider the hillslope as a lumped reservoir such as in Equation 3.5 is another matter. An advantage of many lumped models is that they usually contain fewer parameters than more complex models. In addition, they usually contain greatly simplified descriptions of the processes taking place within a hillslope. There is also the possibility that lumped descriptions of flow at larger scales are physically meaningful. In this respect, it is worth noting that, strictly, Darcy's law is empirical, although throughout this chapter it has been suggested that the law is physically based. Darcy formulated his equation using data collected from experiments which involved measuring water flow through beds of sand. It was also noted above that Darcy's law ignores the fine-scale physics of water flow through individual pores. Despite its empirical origins and scale dependency, it has a physical basis as has been demonstrated by Hubbert (1940) and Gray and O'Neil (1976). Darcy's law, therefore, serves as an example of a lumped model having a clear physical basis. Therefore, there is every possibility that empirical equations such as Equation 3.5 which lump at larger scales also have a physical basis.

Yet good modelling practice should not just boil down to a search for the simplest model that describes hillslope hydrological behaviour as implied by Baird's (1999) complexity–simplicity dichotomy. In order to know whether a larger-scale description of system behaviour is physically meaningful, it is important to know what processes at smaller scales are contributing in an integrated fashion to the larger behaviour of the system. Thus, although it is impossible and quite undesirable to include every known hillslope hydrological process in a model, choosing a model is not just a question of choosing the simplest equation or set of equations that 'does the job' (cf. Baird, 1999). A further point in this respect is that many scientists are curious (although not as many as we might hope); they want to know how things work (knowing *why* is another matter). As such, they want to know what are the most important processes within a system. Darcy's law has dominated soil- and groundwater hydrology for well over 50 years, although many now recognize that macropores may control rates of flow in certain circumstances. However, even if we include macropores within our perceptual models, we are in danger of being too restrictive. We need to look beyond fashion and work hard in considering the range of possible factors that may be important in affecting soil water flow.

Finally, there needs to be a wider debate on how we use models and the relationship between modelling, fieldwork and laboratory work. As noted by Baird (1999), there is a tendency to assume, based on our existing perceptual models, that a certain suite of processes is important and to build a model that includes only these processes. Field work is then seen purely in terms of model verification. Even then, there have been remarkably few attempts to test a model thoroughly (Parsons *et al.*, 1997, being a notable exception). Field and laboratory work should also be seen in terms of driving the development of new models. Equally, it is important to realize that innovative model use can mean using a model to indicate possible phenomena worthy of further field investigation, i.e. models can be used as very valuable exploratory conceptual tools. Woolhiser

et al. (1996), Melone *et al.* (1998) and Singh (1998), for example, show how simple models can be used to give clues about what processes we should investigate in more detail in Nature. Thus, numerical exploration can be used as a precursor to field work.

ACKNOWLEDGEMENTS

Mark Mulligan, John Wainwright and an anonymous referee offered perceptive and helpful comments on an earlier draft of this chapter. The diagrams were drawn by Paul Coles and Graham Allsopp of Cartographic Services, Department of Geography, University of Sheffield.

REFERENCES

Baird, A.J. (1997a) Continuity in hydrological systems, in R.L. Wilby (ed.) *Contemporary Hydrology*, John Wiley & Sons, Chichester, 25–58.

Baird, A.J. (1997b) Runoff generation and sediment mobilisation by water, in D.S.G. Thomas (ed.) *Arid Zone Geomorphology*, John Wiley & Sons, Chichester, 165–184.

Baird, A.J. (1999) Modelling, in A.J. Baird and R.L. Wilby (eds) *Eco-Hydrology: Plants and Water in Terrestrial and Aquatic Environments*, Routledge, London, 300–345.

Bear, J. (1972) *Dynamics of Fluids in Porous Media*, Elsevier, New York.

Beckwith, C.W. and Baird, A.J. (2001) Effect of biogenic gas bubbles on water flow through poorly decomposed blanket peat, *Water Resources Research* **37** (3), 551–558.

Bell, N.C., Wheater, H.S. and Johnston, P.M. (1989) Evaluation of overland flow models using laboratory catchment data. II. Parameter identification of physically based (kinematic wave) models, *Hydrological Sciences Journal* **34** (3), 289–317.

Beven, K. (1981) Kinematic subsurface stormflow, *Water Resources Research* **17**, 1419–1424.

Beven, K. (1982) On subsurface stormflow: predictions with simple kinematic theory for saturated and unsaturated flows, *Water Resources Research* **18**, 1627–1633.

Beven, K. (1985) Distributed models, in M.G. Anderson and T.P. Burt (eds) *Hydrological Forecasting*, John Wiley & Sons, Chichester, 405–435.

Beven, K. (1996) Equifinality and uncertainty in geomorphological modelling, in B.L. Rhoads and C.E. Thorne (eds) *The Scientific Nature of Geomorphology: Proceedings of the 27th Binghampton Symposium in Geomorphology held 27–29 September 1996*, John Wiley & Sons, Chichester, 289–313.

Beven, K. (2000) On the future of distributed modelling in hydrology, *Hydrological Processes* **14**, 3183–3184.

Beven, K. (2001) On an acceptable criterion for model rejection, *Circulation* (The Newsletter of the British Hydrological Society) **68**, 9–11.

Beven, K. and Germann, P. (1982) Macropores and water flow in soils, *Water Resources Research* **18**, 1311–1325.

Binley, A., Beven, K. and Elgy, J. (1989a) A physically based model of heterogeneous hillslopes. 2. Effective hydraulic conductivities, *Water Resources Research* **25**, 1227–1233.

Binley, A., Elgy, J. and Beven, K. (1989b) A physically based model of heterogeneous hillslopes. 1. Runoff production. *Water Resources Research* **25**, 1219–1226.

Blackie, J.R. and Eeles, C.W.O. (1985) Lumped catchment models in M.G. Anderson and T.P. Burt (eds) *Hydrological Forecasting*, John Wiley & Sons, Chichester, 311–345.

Blöschl, G. and Sivapalan, M. (1995) Scale issues in hydrological modelling: a review, *Hydrological Processes* **9**, 251–290.

Blöschl, G., Sivapalan, M., Gupta, V., Beven, K. and Lettenmaier, D. (1997) Preface to the special section on scale problems in hydrology, *Water Resources Research* **33** (12), 2881–2881.

Bond, W.J. and Collis-George, N. (1981) Ponded infiltration into simple soil systems. 1. The saturation and transmission zones in the moisture content profile, *Soil Science* **131**, 202–209.

Buttle, J.M. and Leigh, D.G. (1997) The influence of artificial macropores on water and solute transport in laboratory soil columns, *Journal of Hydrology* **191**, 290–314.

Chow, V.T. (1959) *Open-Channel Hydraulics*, McGraw-Hill, London.

Constantz, J., Herkelrath, W.N. and Murphy, F. (1988) Air encapsulation during infiltration, *Soil Science Society of America Journal* **52**, 10–16.

Cundy, T.W. and Tento, S.W. (1985) Solution to the kinematic wave approach to overland flow routing with rainfall excess given by Philip's equation, *Water Resources Research* **21**, 1132–1140.

Delozada, D.S., Vandevivere, P., Baveye, P. and Zinder, S. (1994) Decrease of the hydraulic conductivity of sand columns by *Methanosarcina barkeri*, *World Journal of Microbiology and Biotechnology* **10**, 325–333.

de Marsily, C. (1986) *Quantitative Hydrogeology: Groundwater Hydrology for Engineers*. Academic Press, London.

Dingman, S.L. (1984) *Fluvial Hydrology*, Freeman, New York.

Domenico, P.A. and Schwartz, F.W. (1990) *Physical and Chemical Hydrogeology*, John Wiley & Sons, New York.

Fagerström, T. (1987) On theory, data and mathematics in ecology, *Oikos* **50**, 258–261.

Fan, Y. and Bras, R.L. (1998) Analytical solutions to hillslope subsurface storm flow and saturation overland flow, *Water Resources Research* **34**, 921–927.

Faybishenko, B.A. (1995) Hydraulic behavior of quasi-saturated soils in the presence of entrapped air: laboratory experiments, *Water Resources Research* **31**, 2421–2435.

Fayer, M.J. and Hillel, D. (1986a) Air encapsulation. 1. Measurement in a field soil, *Soil Science Society of America Journal* **50**, 568–572.

Fayer, M.J. and Hillel, D. (1986b) Air encapsulation. 2. Profile water storage and shallow water table fluctuations, *Soil Science Society of America Journal* **50**, 572–577.

Fetter, C.W. (1994) *Applied Hydrogeology*, 3rd edn, Macmillan, New York.

Freeze, R.A. and Cherry, J.A. (1979) *Groundwater*, Prentice-Hall, Englewood Cliffs, NJ.

Gray, W.G. and O'Neil, K. (1976) On the general equations for flow in porous media and their reduction to Darcy's law, *Water Resources Research* **12** (2), 148–154.

Grayson, R.B., Moore, I.D. and McMahon, T.A. (1992a) Physically based hydrologic modeling. 1. A terrain-based model for investigative purposes, *Water Resources Research* **28** (10), 2639–2658.

Grayson, R.B., Moore, I.D. and McMahon, T.A. (1992b) Physically based hydrologic modeling. 2. Is the concept realistic? *Water Resources Research* **28** (10), 2659–2666.

Gupta, H. (2001) The devilish Dr. M or some comments on the identification of hydrologic models, *Circulation* (The Newsletter of the British Hydrological Society) **68**, 1–4.

Hall, M.J., Johnston, P.M. and Wheater, H.S. (1989) Evaluation of overland flow models using laboratory catchment data. I. An apparatus for laboratory catchment studies, *Hydrological Sciences Journal* **34** (3), 277–288.

Hanks, P. (ed.) (1986) *Collins English Dictionary*, 2nd edn, Collins, Glasgow.

Henderson, F.M. (1966) *Open Channel Flow*, Macmillan, New York.

Henderson, F.M. and Wooding, R.A. (1964) Overland flow and groundwater flow from a steady rainfall of finite duration, *Journal of Geophysical Research* **69**, 1531–1540.

Hewlett, J.D. (1961) *Soil Moisture as a Source of Base Flow from Steep Mountain Watersheds*, USDA Forest Service Research Paper SE, **132**, 1–10, Southeastern Forest Experiment Station, Asheville, NC.

Hewlett, J.D. and Hibbert, A.R. (1963) Moisture and energy conditions within a sloping mass during drainage, *Journal of Geophysical Research* **68**, 1081–1087.

Hillel, D. (1980) *Fundamentals of Soil Physics*, Academic Press, San Diego.

Huang, B. and Nobel, P.S. (1993) Hydraulic conductivity and anatomy along lateral roots of cacti – changes with soil water status, *New Phytologist* **123** (3), 499–507.

Hubbert, M.K. (1940) The theory of groundwater motion, *Journal of Geology* **48**, 25–184.

Hubbert, M.K. (1956) Darcy's law and the field equations of the flow of underground fluids, *Transactions of the American Institute of Mining, Metallurgical and Petroleum Engineers* **207**, 222–239.

Kibler, D.F. and Woolhiser, D.A. (1970) *The Kinematic Cascade as a Hydrologic Model*, Colorado State University Hydrology Paper 39, Colorado State University, Fort Collins.

Klute, A. (1965) Laboratory measurements of hydraulic conductivity of saturated soil, in C.A. Black (ed.) *Methods of Soil Analysis. Part 1: Physical and Mineralogical Properties*, American Society of Agronomy, Madison, WI, 210–221.

Koide, S. and Wheater, H.S. (1992) Subsurface flow simulation of a small plot at Loch Chon, Scotland, *Hydrological Processes* **6**, 299–326.

Kolb, K.J., Sperry, J.S. and Lamont, B.B. (1996) A method for measuring xylem hydraulic conductance and embolism in entire root and shoot systems, *Journal of Experimental Botany* **47**, 1805–1810.

McCoy, E.L., Boast, C.W., Stehouwer, R.C. and Kladivko, E.J. (1994) Macropore hydraulics: taking a sledgehammer to classical theory, in R. Lal and B.A. Stewart (eds) *Soil Processes and Water Quality*, Lewis Publishers, London, 303–349.

Melone, F., Corradini, C. and Singh, V.P. (1998) Simulation of the direct runoff hydrograph at basin outlet, *Hydrological Processes* **12**, 769–779.

Michel, C. (1999) Comment on 'Can we distinguish Richards' and Boussinesq's equations for hillslopes?: The Coweeta experiment revisited' by T.S. Steenhuis *et al.*, *Water Resources Research* **35**, 3573.

Miller, J.E. (1984) *Basic Concepts of Kinematic Wave Models*. United States Geological Survey Professional Paper 1302. Government Printing Office, Washington, DC.

Mulligan, M. (1996) *Modelling hydrology and vegetation change in a degraded semi-arid environment*, unpublished PhD thesis, King's College, University of London.

Narasimhan, T.N. (1998) Hydraulic characterization of aquifers, reservoir rocks, and soils: a history of ideas, *Water Resources Research* **34**, 33–46.

Parsons, A.J., Wainwright, J., Abrahams, A.D., and Simanton, J.R. (1997) Distributed dynamic modelling of interrill overland flow *Hydrological Processes* **11**, 1833–1859.

Pi, Z. and Hjelmfelt, A.T. Jr (1994) Hybrid finite analytic solution of lateral subsurface flow, *Water Resources Research* **30** (5), 1471–1478.

Poesen, J.W.A. (1992) Mechanisms of overland-flow generation and sediment production on loamy and sandy soils with and without rock fragments, in A.J. Parsons and A.D. Abrahams (eds) *Overland Flow: Hydraulics and Erosion Mechanics*, UCL Press, London, 275–305.

Ponce, V.M. (1991) The kinematic wave controversy, *Journal of Hydraulic Engineering, American Society of Civil Engineers* **117**, 511–525.

Ponce, V.M., Li, R.M. and Simons, D.B. (1978) Applicability of kinematic and diffusion models, *Journal of the Hydraulics Division of the American Society of Civil Engineers* **104**, 353–360.

Popper, K. (1968) *Conjectures and Refutations: The Growth of Scientific Knowledge*, Harper and Row, New York.

Quinton, W.L., Gray, D.M. and Marsh, P. (2000) Subsurface drainage from hummock-covered hillslopes in the Arctic tundra, *Journal of Hydrology* **237**, 113–125.

Reynolds, W.D., Brown, D.A., Mathur, S.P. and Overend, R.P. (1992) Effect of in-situ gas accumulation on the hydraulic conductivity of peat, *Soil Science* **153**, 397–408.

Richards, L.A. (1931) Capillary conduction of liquids in porous mediums, *Physics* **1**, 318–333.

Romkens, M.J.M., Prasad, S.N. and Whisler, F.D. (1990) Surface sealing and infiltration, in M.G. Anderson and T.P. Burt (eds) *Process Studies in Hillslope Hydrology*, John Wiley & Sons, Chichester, 127–172.

Rose, C.W., Parlange, J.-Y., Sander, G.L., Campbell, S.Y. and Barry, D.A. (1983) Kinematic flow approximation to runoff on a plane: an approximate analytic solution, *Journal of Hydrology* **62**, 363–369.

Sánchez, G. and Puigdefábregas, J. (1994) Interactions of plant growth and sediment movement on slopes in a semi-arid environment, *Geomorphology* **9**, 243–260.

Scoging, H., Parsons, A.J. and Abrahams, A.D. (1992) Application of a dynamic overland-flow hydraulic model to a semi-arid hillslope, Walnut Gulch, Arizona, in A.J. Parsons and A.D. Abrahams (eds) *Overland Flow Hydraulics and Erosion Mechanics*, UCL Press, London, 105–145.

Singh, V.P. (1998) Effect of the direction of storm movement on planar flow. *Hydrological Processes* **12**, 147–170.

Sloan, P.G. and Moore, I.D. (1984) Modeling subsurface stormflow on steeply sloping forested watersheds, *Water Resources Research* **20**, 1815–1822.

Sobrado, M.A. (1997) Embolism vulnerability in drought-deciduous and evergreen species of a tropical dry forest, *Acta Oecologica – International Journal of Ecology* **18** (4), 383–391.

Steenhuis, T.S., Parlange, J.-Y., Parlange, M.B. and Stagnitti, F. (1988) A simple model for flow on hillslopes, *Agricultural Water Management* **14**, 153–168.

Steenhuis, T.S., Parlange, J.-Y., Sanford, W.E., Heilig, A., Stagnitti, F. and Walter, M.F. (1999a) Can we distinguish Richards' and Boussinesq's equations for hillslopes?: The Coweeta experiment revisited, *Water Resources Research* **35**, 589–593.

Steenhuis, T.S., Parlange, J.-Y., Sanford, W.E., Heilig, A., Stagnitti, F. and Walter, M.F. (1999b) Reply (to Michel 1999), *Water Resources Research* **35**, 3575–3576.

Tyree, M.T. and Ewers, F.W. (1991) Tansley review number 34: the hydraulic architecture of trees and other woody plants, *New Phytologist* **119**, 345–360.

Tyree, M.T. and Sperry, J.S. (1989) The vulnerability of xylem to cavitation and embolism, *Annual Review of Plant Physiology and Molecular Biology* **40**, 19–38.

van Wesemael, B., Mulligan M. and Poesen J. (2000) Spatial patterns of soil water balance on intensively cultivated hillslopes in a semi-arid environment: the impact of rock fragments and soil thickness, *Hydrological Processes* **14** (10), 1811–1828.

Verrecchia, E., Yair, A., Kidron, G.J. and Verrecchia, K. (1995) Physical properties of the psammophile cryptogamic crust and their consequences to the water regime of sandy soils, north-western Negev Desert, Israel, *Journal of Arid Environments* **29**, 427–437.

Walter, B.P., Heimann, M., Shannon, R.D. and White, J.R. (1996) A process-based model to derive methane emissions from natural wetlands, *Geophysical Research Letters* **23**, 3731–3734.

Wang, H.F. and Anderson, M.P. (1982) *Introduction to Groundwater Modeling: Finite Difference and Finite Element Methods*, Academic Press: San Diego.

Watts, G. (1997) Hydrological modelling in practice, in R.L. Wilby (ed.) *Contemporary Hydrology: Towards Holistic*

Environmental Science, John Wiley & Sons, Chichester, 151–193.

Wheater, H.S., Bell, N.C. and Johnston, P.M. (1989) Evaluation of overland flow models using laboratory catchment data. III. Comparison of conceptual models, *Hydrological Sciences Journal* **34** (3), 319–337.

Wigmosta, M.S. and Lettenmaier, D.P. (1999) A comparison of simplified methods for routing topographically driven subsurface flow, *Water Resources Research* **35**, 255–264.

Wilby, R.L. (1997) Beyond the river catchment, in R.L. Wilby (ed.) *Contemporary Hydrology: Towards Holistic Environmental Science*, John Wiley & Sons, Chichester, 317–346.

Williams, J.E., Davis, S.D. and Portwood, K. (1997) Xylem embolism in seedlings and resprouts of *Adenostoma fasciculatum* after fire, *Australian Journal of Botany* **45** (2), 291–300.

Wood, E.F., Sivapalan, M., Beven, K. and Band, L. (1988) Effects of spatial variability and scale with implications to hydrologic modeling, *Journal of Hydrology* **102**, 29–47.

Woolhiser, D.A. (1975) Simulation of unsteady flows, in K. Mahmood and V. Yevjevich (eds) *Unsteady Flow in Open Channels*, Water Resources Publications, Fort Collins, 485–508.

Woolhiser, D.A. and Liggett, J.A. (1967) Unsteady, one-dimensional flow over a plane – the rising hydrograph, *Water Resources Research* **3**, 753–771.

Woolhiser, D.A., Smith, R.E. and Giraldez, J.-V. (1996) Effects of spatial variability of saturated hydraulic conductivity on Hortonian overland flow, *Water Resources Research* **32**, 671–678.

Wu, Y.H., Yevjevich, V. and Woolhiser, D.A. (1978) *Effects of Surface Roughness and its Spatial Distribution on Run-Off Hydrographs*. Hydrology Paper 96, Colorado State University, Fort Collins.

WWW addresses for subsurface flow models

FEFLOW (variably saturated 3-D water, solute and heat flow): follow the links from http://www.wasy.de/

HYDRUS (variably saturated 2-D water and solute flow): follow the links from http://www.ussl.ars.usda.gov/MODELS/MODELS.HTM

MODFLOW (3-D saturated water and solute movement): follow the links from http://water.usgs.gov/software/ground_water.html

PORFLOW (includes variably saturated 3-D water, solute, and heat flow): follow the links from http://acriCFD.com/

VS2D (variably saturated 2-D water, solute and heat movement): use the same link as MODFLOW

4

Modelling Catchment Hydrology

MARK MULLIGAN

4.1 INTRODUCTION: CONNECTANCE IN HYDROLOGY

In Chapter 3, Baird outlines the complexities of water flow in soils and hillslopes and the importance of their spatial variability. In Chapter 5, Michaelides and Wainwright outline the factors controlling the movement of the same water within channels and the model approaches available in which to simulate and better understand this movement. This chapter takes a landscape view of water movement and examines the aggregation of hillslope processes into nested catchments connected by stretches of flowing water. We will examine the complexity of catchments in process, time and space and highlight some of the simple rules that determine their behaviour. We start by analysing the hydrological and computational definition of a catchment before discussing the range of models applied for hydrological modelling at this scale. We then present some of the state of the art and future directions in catchment modelling before identifying some of the gaps in our understanding of catchments and some of the simplicities which have emerged from the modelling of complex catchments.

Studying the hydrology of catchments involves an understanding of the water balance of individual hydrological response units (HRUs) through the interaction of climate, vegetation and soils in the determination of the local water balance (precipitation minus evapotranspiration) and resulting soil moisture and the partitioning of this balance between more rapid surface pathways for flow such as runoff and slower subsurface pathways such as throughflow (interflow in the US) or groundwater recharge. In addition, one has to understand the spatial distribution and topological network for connectance of these response units such that the propagation of water

between them can be understood. This propagation may involve amplification or diminution of flows depending upon the characteristics of the HRU. Finally, the rapid flows less connected with their substrate provided by permanent channels are added as a further layer of complexity. The interaction of spatially varying climatic and human impacts must also be taken into account. Thus, the spatial complexity of catchment hydrology can be significant, not to mention the fact that many of these factors change also in time, for secular processes at scales from minutes (for the meteorological) to hundreds of thousands of years (for the topological) and for extreme events at all scales and in any time.

The types of questions asked of hydrologists at the catchment scale are diverse. Leaving aside those *related to* hydrology (such as the impact of land-use change on soil and nutrient erosion, the transport of point source and nonpoint source contaminants, the extent and distribution of salinization, variations and change in aquatic or wetland environments and associated species) and concentrating on the solely hydrological issues, scientists, policy-makers and citizens are mainly interested in the following:

(a) the reasons for a devastating flood event or events;
(b) the reasons for a devastating hydrological (as opposed to climatic) drought;
(c) the potential impact of climate change on flooding and water resources;
(d) the potential impacts of land cover and/or land-use change or housing development on flooding and water resources;
(e) the seasonal and long-term regime of a river or set of rivers (for engineering purposes);

Environmental Modelling: Finding Simplicity in Complexity. Edited by J. Wainwright and M. Mulligan
© 2004 John Wiley & Sons, Ltd ISBNs: 0-471-49617-0 (HB); 0-471-49618-9 (PB)

(f) the potential downstream impacts of channel engineering, flood alleviation or other river or near-river engineering developments.

Tackling these issues is not solely a matter of modelling but, at the catchment scale (because controlled experiments are not possible), modelling plays an increasing important role alongside field monitoring and (paired) catchment experimentation. At the catchment scale we tend to be interested in streamflow (surface runoff), its magnitude and timing.

4.2 THE COMPLEXITY

4.2.1 What are catchments?

The catchment of any point in the landscape is that area which provides water to the point through lateral flow over the surface and underground. Catchments are usually delineated on the basis of watersheds determined from surface topography. In this way the complex and long-term processes which determine landscape form (geomorphology) also determine the topological properties of catchments at all scales. A watershed is a positive topographic anomaly which defines the boundary between two catchments such that the water on either side of the watershed flows into different catchments (often to different outlets). Topographically defined watersheds may not always be an accurate representation of the actual watershed since subsurface lateral flow within the soil and particularly within the bedrock will not always follow *surface* topographic control. Indeed, subterranean flows between catchments can be significant. Nevertheless, topographically defined watersheds are appropriate for most surface hydrological studies and are widely used. The term watershed is usually reserved for the boundary of a catchment but is also used to represent the catchment itself. The term drainage basin is also used to represent the hydrological catchment and indicates the catchment as a system and component of the hydrological cycle transferring water from the atmosphere through the interception of rainfall (and cloud; Zadroga, 1981) by land and plant either back to the atmosphere or to the sea by river. The drainage basin is thus a cog in the machine that is the hydrological cycle.

As well as this conceptual separation, catchments are very clearly spatially separated through their watersheds and are, indeed, one of the more easily separated aspects of the natural world. Catchments are multiscale with many small subcatchments joining to make up the estimated 6152 global catchments with coastal outlets that can be defined using the Vörösmarty *et al.* (2000)

30-minute global topological network of rivers. These catchments are similar in many ways even though they vary enormously in size. Vörösmarty *et al.* define the Amazon as the largest catchment in the world with an area of $5.854 \times 10^6 \, \text{km}^2$ and a length of 4327 km followed by the Nile, Zaire and Mississippi catchments. The Amazon has 33 main tributary catchments and a multitude of tributary streams[www].

4.2.2 Representing the flow of water in landscapes

Flow paths for surface water are usually defined using a neighbourhood operation on a digital elevation model (DEM) of topography. DEMs are common in hydrological modelling and provide a raster (square cellular, grid-based) representation of the topography of an area (O'Callaghan and Mark, 1984; Tarboton *et al.*, 1992). Each cell has a single value for altitude and the cell size of DEMs varies according to the area under study and data availability, but those applied in hydrological modelling are usually between 5 and 50 metres in length. Many hydrological parameters have been shown to be sensitive to DEM resolution, coarser DEMs leading to reduced slope and more sparse river networks. Increased catchment wetness and peak flow in the TOPMODEL have been observed using coarser representations of topography (Zhang and Montgomery, 1994) and a number of other models have been shown to be sensitive though the sensitivity is greatest for models with a high temporal resolution (e.g. Yang *et al.*, 2001). Thus, there needs to be appropriate balance between the spatial and temporal resolution of a catchment scale model.

DEMs are usually constructed by the interpolation of a raster grid from point-based altitude data[www]. Such data are acquired over small areas using field survey with an EDM (electronic distance measurement) system or, over larger areas, directly from the computerized interpretation of stereo ortho-photographic surveys or, most commonly, through the digitization of contour lines on existing cartography (with the contours usually having been derived from the prior manual interpretation of ortho-photographic survey). Alternatives to a raster interpretation of spatial data include (a) triangular irregular networks (TINs: Palacios-Velez and Cuevas-Renaud, 1986) the cells of which vary in size according to the topographic complexity and (b) contour-based models (O'Loughlin, 1986; Moore *et al.*, 1988). Contour-based models discretize a topographic surface into irregular shaped cells made of two consecutive contours connected by two lines which represent the steepest slope

---

[www] indicates material covered in more detail in the website which accompanies this book: http://www.kcl.ac.uk/envmod

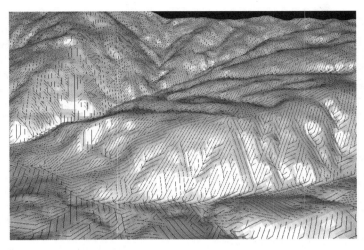

Figure 4.1   A digital elevation model (DEM) for the Tambito catchments in Colombia with cellular flow directions superimposed

connecting the two contours. While raster cells might not be the best representation for hydrological purposes, they are certainly the most practical for integration with pixel-based remotely sensed datasets and raster GIS calculation tools and are thus the most popular representation.

A DEM for the Tambito catchments in the Occidental Cordillera of the Colombian Andes is shown in Figure 4.1[WWW]. In order to define the catchments, one must first define the flow paths for water. Flow paths are defined according to the difference in altitude between the cell in which the flow originates and the altitude of neighbouring cells. Algorithms can be classified into those which flow into a single neighbouring cell and those which apportion flow to multiple neighbours. A number of algorithms have been used, including the so-called D8 algorithm (O'Callaghan and Mark, 1984), the MF (multiple flow direction) algorithm (Quinn *et al.*, 1991), the D-infinity algorithm (Tarboton, 1997), the digital elevation model network – DEMON (Costa-Cabral and Burges, 1994), the Rho 8 algorithm (Fairfield and Leymarie, 1991) and the kinematic routing algorithm (Lea, 1992). The D8 algorithm uses a nearest neighbour approach to define the relationship between a central cell and its eight contiguous neighbours (4 sides, 4 vertices) and defines the steepest downslope gradient along which all of the water is deemed to flow. There are only eight possible flow directions, each 45° apart and all water in the cell flows to only one of them. This algorithm is used in many geographical information

systems (GIS) and some hydrological models but does tend to produce rather linear flow networks when applied in areas of smoothed or shallow topography. An example of a D8 drainage network is superimposed on the Tambito DEM in Figure 4.1.

The Rho-8 algorithm randomly assigns the direction of flow to downslope neighbours weighted according to the degree of slope. The MF algorithm also uses the eight nearest neighbours but then apportions flow to *all* lower neighbouring cells based on the relative magnitude of their slopes. The aspect-based kinematic routing algorithm assigns flow direction according to the calculated aspect. A plane is fitted to the corners of the cell through interpolation of the altitudes of the pixel centres. Flow is routed as if it were a ball rolling along this plane. The DEMON and D-infinity algorithms use surface-fitting algorithms to determine a surface slope vector for the central pixel and the direction of this vector is used to proportion flow between the two neighbouring cells which fall either side of the vector (the two nearest cardinal directions in the case of DEMON and the cardinal and diagonal in the case of D-infinity). Multiple flow direction algorithms tend to produce flow dispersion while single flow direction algorithms lead to flow concentration.

Different algorithms will produce different definitions of a hydrological catchment and of many hydrologically important topographic indices such as a point's upslope area, A, and the specific catchment area, a. Upslope area is the total catchment area above a point or length of contour and the specific catchment

---

[WWW] indicates material covered in more detail in the website which accompanies this book: http://www.kcl.ac.uk/envmod

area is the upslope area per width of contour or cell size, L (Moore *et al.*, 1991). Specific catchment area is used to calculate saturation and saturation excess overland flow in hydrological models such as TOP-MODEL (Beven and Kirkby, 1979) and, along with other topographic indices, to calculate erosion and landsliding in many other models. Upslope area is commonly used for channel mapping on the basis of threshold upslope areas for channel initiation. Channels are usually estimated on the basis of either a constant threshold upslope area for channel initiation (O'Callaghan and Mark, 1984; Tarboton *et al.*, 1992) or a slope dependent one (Dietrich *et al.*, 1993). This threshold area can usually be obtained from a log, log plot of upslope area versus local slope (Tarboton *et al.*, 1992). Vogt *et al.* (2002) extended this concept to include the geological, soil, topographic and climatic characteristics which can affect the upslope area and local slope required for channel initiation.

Figure 4.2 shows a set of cumulative frequency distributions of upslope area calculated using these different algorithms for a random 65535 of the Tambito DEM cells. The MF and the D-infinity tend to produce a peak of subcatchments in the range 1500–5100 m² whereas the kinematic algorithm produces a sharp peak of subcatchments sized 1200 m² and the DEMON algorithm produces a broader peak centred at 2100 m². D-8 and rho-8 produce a rather stepped distribution of subcatchment sizes especially for small basins, reflecting the limited number of flow directions and thus basin aggregations that are possible. Depending on the routing algorithm used, individual cells can have very different upslope areas (and thus occupy hydrologically different zones). Figure 4.3 shows the spatial outcomes of these differences at the catchment scale. The upslope area

value of the D8 algorithm minus the equivalent value for each of the other algorithms is shown, in addition to the differences in subcatchment properties at the small scale outlined in Figure 4.2. Figure 4.3 indicates that, even at the catchment scale, large differences in the upslope area can occur between these algorithms.

The calculation of flow direction allows the delineation of catchments and subcatchments, the routing of lateral flows of water, the calculation of stream properties such as stream order and the calculation of trends in controlling variables, such as slope gradient, down the hydrological network. Figure 4.4 shows some of these properties calculated for the Tambito catchments. Slope gradient (the change in height per unit horizontal distance) is calculated using the method of Horn (1981) as implemented within the PCRASTER GIS (van Deursen, 1995) although other methods also exist, such as Zevenbergen and Thorne (1986). (Note the unrealistic slopes in the SE corner resulting from lack of data for interpolation of the topography because of cloud cover in the initial ortho-photography.) The method of Horn (1981) is also used to calculate aspect (the direction of the maximum rate of change in elevation). The catchment of any identified point may be calculated by following the drainage network upwards to the watershed. Figure 4.4 shows the catchments of the outflows of the Tambito DEM. Many of these are partial because the DEM boundaries are those of the map sheet but the largest catchment (that of the Tambito itself) is complete. Stream channels are calculated here as is commonly the case by initiating a channel when upslope area is greater than a threshold value, in this case 1 000 000 m² (1 km²), though this is, of course, rather arbitrary. The stream order is calculated directly from

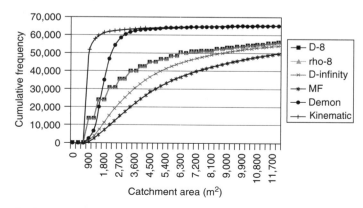

Figure 4.2   Differences in the distribution of micro-catchments within the Tambito catchment for six different routing algorithms

Figure 4.3   Spatial differences in accumulated upslope area between the D8 and four other algorithms for flow routing

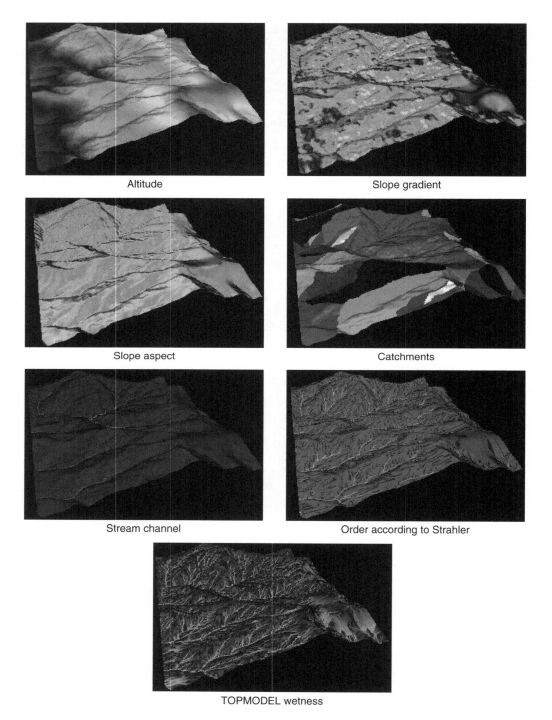

Altitude

Slope gradient

Slope aspect

Catchments

Stream channel

Order according to Strahler

TOPMODEL wetness

Figure 4.4   Topographic and hydrological factors derived from the Tambito DEM using the D8 algorithm for flow routing

the flowlines according to the summation of flowline confluences using the method of Strahler. TOPMODEL wetness is a compound topographic index commonly used in hydrology to account for spatial variations in moisture. It is compound because it combines upslope area, A (the propensity to receive water) with local slope, s (the propensity to shed water) to yield an overall index of wetness as $\ln(A/\tan(s))$.

4.2.3 The hydrologically significant properties of catchments

We have considered some of the complexities of catchment definition, now let us look a little at the hydrologically significant properties of catchments themselves. These include the following:

(a) The topological structure of their drainage network, which determines the lag time for arrival of rainfall to a point in the network and the temporal concentration of the streamflow hydrograph which results.

(b) Their geomorphological and pedological characteristics (and spatial variation) which determine the potential for infiltration and local storage over runoff and thus contribution to streamflow. The large-scale geomorphology will also control the topology of the drainage network and, in large catchments, this may be the most important control on a catchment's hydrology (Vörösmarty *et al.*, 2000).

(c) Their climatic characteristics (and spatial variation). The spatial distribution of temperature, radiation and rainfall, which are themselves highly correlated with altitude, will determine the spatial distribution of contributions from areas within a catchment.

(d) Their vegetation cover and land-use characteristics and its spatial variation. Chapter 3 on hillslope hydrology indicates the significance of vegetation for the hydrological balance of hillslopes and the partitioning of rainfall into local infiltration and infiltration excess, that is, runoff.

(e) The spatial distribution of their populations. Populations are the source of demand for water and their locations may determine the locations of local extractions and artificial storage of water from the channel network or from locally generated runoff (as in the Aljibes of North Africa and the Mediterranean[www]) and local groundwater sources. The location of populations will also determine the magnitude of local land-use change with corresponding impacts and the sources of point and nonpoint

agricultural, industrial and domestic pollution to the water courses.

Because of surface and subsurface lateral flows, hydrological catchments are highly connected such that a change in any one part of them can have implications for a number of other parts downstream. Furthermore, the spatial distribution of all of these factors relative to the direction of streamflow is particularly important since it determines the potential for cumulation or diminution of streamflow along the flow path. A series of source areas in line with the flow path will cumulate outflow along the flowline, whereas a mixture of source and sink areas will tend to diminute outflow along the flow path[www].

4.2.4 A brief review of catchment hydrological modelling

The use of models to better understand and to predict the behaviour of water in catchments has a long history, and, because models are rarely successful in application outside the catchments for which they were developed, there are many models to be found. The models range from simple 'black box' representations of input and output which are often successful in the prediction of runoff, from rainfall, through to more complex representations of some of the spatio-temporal complexity of catchments which are more capable of fostering a better understanding of the reasons for observed behaviour. Catchments themselves are superb simplifiers, converting a spatial complexity of patterns and processes into a relatively simple and well understood output – the hydrograph. The range of models available reflects the need to predict the outcomes of specific interventions and scenarios for change, the emphasis on prediction as well as scientific investigation (for the purposes of flood prediction and mitigation or water resource management) and the paucity of data available for larger catchments compared with smaller ones.

The types of catchment model available include physically based models, based solidly on an understanding of the physical processes, empirical models based on the patterns observed in data, and conceptual models which pay little attention to the physics of the processes but, rather, represent the catchment conceptually as, for example, a series of cascading stores for water and the fluxes between them. Models may be deterministic models in which a given set of inputs will always produce the same output, or stochastic models, which represent the variability of processes or events using probability

---

[www] indicates material covered in more detail in the website which accompanies this book: http://www.kcl.ac.uk/envmod

distributions and which thus attempt to handle some of the inherent uncertainty in modelling and in data. Models may be lumped at the catchment scale, meaning that data and modelling are aggregated at this scale, they may be lumped at the subcatchment scale (and thus semi-distributed at the catchment scale) or they may be fully distributed, that is lumped at the raster grid cell or TIN polygon scale.

Empirical models tend to be lumped, conceptual models tend to be semi-distributed and physically based models tend to be fully distributed. The increase in computing power and of available spatial data in the form of GIS datasets, especially DEMs, and remotely sensed imagery has vastly increased the potential for distributed modelling. At the catchment scale, to be based on physics, physically based models have to be distributed and so 'distributed' and 'physically based' often go together in catchment hydrological modelling. Moreover, many large catchments are ungauged and thus cannot provide the calibration data necessary for the development and parameterization of conceptual or empirical models. A driving force for the development of physically based models is their application in ungauged catchments, though we will see later that gaps in parameterization data and process knowledge create model uncertainty and thus the requirement for these models to be calibrated against gauging station data.

4.2.5 Physically based models

Since it was first 'blueprinted' by Freeze and Harlan in 1969, distributed physically based modelling has become very widespread, on the assumption that a spatially variable physical system is inherently more realistic than a lumped one. This is likely to be true but must be considered within the context of spatially distributed models being themselves often crude simplifications of any spatial variability that does exist in real catchments. Remote sensing has gone some way towards improving the observability of surface properties at the catchment scale but subsurface properties are still largely unobservable at any scale other than the point or line transect. Examples of current distributed, physically based models include the SHE model (Système Hydrologique Européen, see Abbott *et al.*, 1986) and the MIKE SHE and SHETRAN descendants of it (Bathurst *et al.*, 1995; Refsgaard and Storm, 1995), the IHDM model (Institute of Hydrology Distributed Model, e.g. Calver and Wood, 1995), the CSIRO TOPOG model (e.g. Vertessy *et al.*, 1993), Thales (Grayson *et al.*, 1992), and WEC-C (Croton and Barry, 2001), among others. Physically based models should be derived deductively from established

physical principles and produce results that are consistent with observations (Beven, 2002). In reality they are often one of these but rarely both.

According to Ward and Robinson (2000), SHE was developed jointly by the UK Institute of Hydrology (IH), the Danish Hydraulic Institute (DHI) and the Société Grenobloise d'Études et d'Applications Hydrauliques (SOGREAH). It was specifically designed to address the impact of human activities on catchment processes. This type of scenario analysis is very difficult to address with empirical or conceptual models but is the main focus of most physically based models. In SHE a finite difference (grid-based) approach is used in three dimensions with up to 30 horizontal layers. Surface and groundwater flow are two-dimensional while flow in the unsaturated zone is one-dimensional. The model has been widely applied and continuously updated since 1986.

While remote sensing provides some high quality spatial datasets for properties such as vegetation cover and topography, there are many hydrologically important properties of catchments, notably those of the soils and the surface that cannot be measured remotely and thus require intensive field campaigns which still cannot provide the kind of spatial coverage required to justify their distribution in models[www]. For example, most hydrological models are highly sensitive to the (saturated) hydraulic conductivity of soils (K_{sat}), see Davis *et al.* (1999) and Chappell *et al.* (1998), and this is notoriously difficult to measure for volumes greater than a few hundred cm^3, in part because of Baird's (bloody) macropores of Chapter 3. It is fair to say that there is a significant mismatch in the sophistication of our physically based models and the sophistication of the data collection technologies used to parameterize them. Moreover, as highlighted in Chapter 1, modelling is rather inexpensive compared with fieldwork and is also perhaps more glamorous, more comfortable and more suited to producing publications – see the discussion by Klemeš (1997: 48). So, the gap between the models and the data to fill them widens.

It is the imbalance between the model sophistication and the availability of data at appropriate scales (as well as our incomplete understanding and thus mathematization of the processes themselves), that means that even the most sophisticated models rarely perform well in a predictive capacity. Empirical models tend to be better predictors. Thus, a process of parameter calibration is often exercised on physically based distributed models to ensure agreement of predicted versus observed

---

[www] indicates material covered in more detail in the website which accompanies this book: http://www.kcl.ac.uk/envmod

runoff. This process, of course, compromises the physical realism of the model and thus its ability to *explain* as well as to predict. Since explanation is why physically based distributed models exist, this compromise is a serious one. If it were not for explanation then an equally predictive empirical model would always be the best model because of its parsimony. Some authors have argued that calibration can render physically based distributed models closer to over-parameterized conceptual, lumped models (Beven, 1989, 1996). Furthermore, physically based, distributed models – even the very widely used ones – have rarely been validated on variables other than the output variables. Very few studies have validated the internal state variables of these models in order to understand whether the models are producing validatable results for the correct physical (internal) reasons. In other words, have the lack of data, lack of detailed process understanding and over-calibration rendered physically based models into overly sophisticated conceptual models or do they indeed retain some useful explanatory power? Fewer researchers still concentrate on validation which is focused specifically on the purpose for which the model is intended, such as suitability for the analysis of the impact of climate or land use change (Ewen and Parkin, 1996). Furthermore, few studies focus on the evaluation of distributed behaviour as opposed to catchment-integrated outcomes such as runoff (Grayson *et al.*, 1995); the correct prediction of patterns may be a better measure of explanatory power than is success in the prediction of catchment outflows. In complex models, a number of different parameter sets can give rise to the correct outflow (the problem of equifinality). As a result, and given the uncertainty in most parameterization datasets, understanding whether your calibrated model has produced the right answers for the right physical reasons is rather difficult. Beven and Binley (1992) and a number of others since have used generalized likelihood uncertainty estimator (GLUE) approaches to look at the implications of parameter uncertainty on model outcomes.

It is these problems and others, amply spelled out by Rosso (1994), Beven (1989, 1996) and Refsgaard (1997), that are the reasons why physically based distributed models have tended to remain in the research domain and have had relatively little impact in the world of practical hydrology. The problems with building and using distributed, physically based models as per the Freeze and Harlan (1969) blueprint may be so great that an alternative blueprint is required (Reggiani *et al.*, 2000; Beven, 2002). Those problems associated with the application of Darcy's law for matrix flow (see Chapter 3) at a range of scales where it does not

apply or, if it does apply, we cannot test that it does, are particularly serious[www]. The alternative blueprint of Beven (2002) emphasized a more observation-based inductive approach over the more theoretical deductive approach of Freeze and Harlan (1969). See the data-based mechanistic approach outlined by Young *et al.* in Chapter 22 for an example. In this way the observations, and not the theory, determine which models are appropriate.

4.2.6 Conceptual and empirical models

Simpler models have been shown to give a good empirical fit to observed behaviour, though their very nature means that they must be calibrated to runoff records and thus cannot easily be used in ungauged catchments or transferred between catchments. IHACRES (Jakeman and Hornberger, 1993) is an example of a lumped parameter rainfall-runoff model with only five calibration parameters. It consists of two modules, one of which converts rainfall to rainfall excess and another which transforms rainfall excess to streamflow. A compromise between the lumped and the distributed approaches is to use a probability distribution approach which recognizes variability but says nothing of its spatial arrangement (e.g. Moore, 1985) or the subcatchment or flow-segment based semi-distributed approach.

TOPMODEL (Beven and Kirkby, 1979; Beven *et al.*, 1995) is a very widely used conceptual approach with some physical basis. It is a collection of concepts as much as a model (Beven *et al.*, 1995) and thus a wide range of versions have been implemented. It uses DEMs to recognize the importance of catchment topography in controlling the spatial pattern of stormflow source areas and has, more recently, been extended to include sediment, geochemical fluxes, evapotranspiration and attachment to land surface–atmosphere transfer models. Though TOPMODEL originates in the description of humid temperate environments, it has also been applied to a wide range of other environments. TOPMODEL has been widely applied on a variety of scales from small headwaters (Molicova *et al.*, 1997) to very large catchments where it is semi-distributed on a subcatchment basis (e.g. Band and Wood, 1988). TOPMODEL makes conceptual advances over many other simple approaches by taking into account the propensity for differences in catchment saturation in accordance with the compound topographic index and its effect on the generation of saturation overland flow.

[www] indicates material covered in more detail in the website which accompanies this book: http://www.kcl.ac.uk/envmod

These differences are represented statistically rather than physically. In TOPMODEL, rainfall occurs and water enters the soil along with water draining from other parts of the catchment causing the groundwater table to rise. When groundwater rises to the land surface, the area becomes saturated and saturation overland flow occurs. Water below the surface is also assumed to move downslope as throughflow since the hydraulic gradient of the saturated zone is approximated as the local topographic slope. Total outflow is throughflow plus saturation and infiltration-excess overland flow. More recent versions have added further sophistication to processes such as subsurface flow routing (Beven and Freer, 2001) and a spatially variable soil thickness (Saulnier et al., 1997).

The Stanford Watershed Model (SWM) is one of the earliest and best-known conceptual catchment models. The model has an hourly soil-moisture budget and storage and routing functions for the redistribution of water entering the channels to provide catchment scale runoff on a daily timestep (Crawford and Linsley, 1966; Viessman and Lewis, 1996). The model requires climate data and some 34 parameters describing the physical properties of the catchment. More recent models have given greater importance to the overarching control of the geomorphological properties of the catchment as exercised through the influence of the Horton–Strahler stream order configuration. Furthermore, it is envisaged that this type of model can be applied in ungauged catchments for the calculation of geomorphologically controlled unit hydrographs (Gupta et al., 1996; Schmidt et al., 2000; Yang et al., 2001), particularly for large catchments. The fractal scaling properties of river basins help significantly to simplify runoff modelling. Szilagyi and Parlange (1999) describe a semi-distributed conceptual model in which the catchment is conceptualized as a series of stores whose dimensions are derived from the Horton–Strahler stream order (see Figure 4.4). Overland flow fills these stores and water is routed between them. A separate groundwater model provides baseflow. The model requires the calibration of only seven parameters on the basis of a year-long rainfall-runoff record.

4.2.7 Recent developments

Recent trends in catchment modelling have seen the integration of catchment hydrology models with ecological and physiological ones (e.g. Hatton et al., 1992) and with erosion and sediment-transport models (De Roo, 1998; Ewen et al., 2000). Modelling scales have increased from regional through to continental and global (Vörösmarty et al., 2000) in part

to provide interfaces with general circulation models of the atmosphere for climate change (Stieglitz et al., 1997) and climate change impact (Bronstert et al., 2002) studies. These larger scales and this deeper integration have been in response to increasing pressures to make the models address some of the most serious environmental problems that face governments and citizens – accelerated land use and climate change. Since computer power has risen exponentially in the last decades, the models (and the modellers) have been able to keep up with these demands. However, the gap between capability and parameterizability (the C:P gap) continues to increase.

Recent years have also seen better integration of data resources within flexible and intuitive databases, particularly in GIS. More catchment models are designed to work with data held in common GIS formats or indeed to run within GIS software. This trend has both facilitated the spatialization of models and the potential for modelling over larger catchments and has also prevented the C:P gap from being even wider. The hydrological functionality of GIS and the links between common spatial hydrological models and GIS are reviewed by Ogden et al. (2001) while the advances in data availability and manipulation within a GIS context that have facilitated this integration of hydrology and GIS are reviewed in the companion paper by Garbrecht et al. (2001)[WWW]. De Roo (1998) outlines the state of the art in dynamic spatial modelling languages for the development and application of hydrological models but reminds us that all of this spatial detail does not necessarily mean better results.

In line with the greater emphasis on having hydrological models contribute to the solution of environmental problems and in line with the greater accessibility of personal computers, catchment hydrological models have also moved into the public domain. Decision-support tools, which incorporate hydrological models for the purposes of integrated catchment management (Walker and Johnson, 1996; Matens and Di Biase, 1996) or spatial land-use planning (Greiner, 1996), are now not uncommon[WWW]. These tools are often used for better understanding the unforeseen consequences of human intervention in the hydrological system (e.g. Engelen, 2000) or in larger systems of which hydrology is a small, but significant, part (Engelen, this volume). Though it is generally policy-makers who use these tools to support the decision-making process, there are even efforts to make catchment scale model results available direct to

---

[WWW] indicates material covered in more detail in the website which accompanies this book: http://www.kcl.ac.uk/envmod

the public in real time over the WWW for flood warning purposes (Al-Sabhan *et al.*, 2002)[WWW].

4.3 THE SIMPLICITY

4.3.1 Simplifying spatial complexity

Thus far we have concentrated on understanding the implications of spatial complexity for modelling needs at the catchment scale. We know that lumped models do not represent this complexity and that distributed models do, insofar as they can, but do not insofar as the required data is often not available. Semi-distributed (or semi-lumped!) approaches are a compromise between the two end points but say nothing of the interaction between neighbouring patches, which can be important in hydrological studies. Different aspects of the spatial variability of catchments have different magnitudes of impact on the hydrological system and different spatio-temporal scales of interaction with it. Soil and vegetation properties are likely to be more important for small catchments and short periods but their influence will become less as the size of the catchment or length of the period increases, at which point geomorphological and geomorphometric processes become more important. A nested hierarchy of catchment response units might be identified with climate, geomorphology, vegetation and soils having progressively greater variability and progressively smaller scales of influence (see Mulligan, 1996).

4.3.2 Simplifying temporal complexity

In addition to spatial complexity, catchment models must also handle the temporal complexity of hydrological processes and their interaction with each other. Catchments integrate rapid rate processes such as the partitioning of rainfall into runoff and infiltration or the routing of water through channels, with slower rate processes such as the trickle of groundwater recharge and the continuous draw of evapotranspiration. Hydrological models are sensitive to the timestep of simulation; large errors can ensue by aggregating evapotranspiration calculations (from hourly through day/night to daily) and by representing instantaneous rainfall intensities that can reach over 100 mm/hr as hourly averages, which are unlikely ever to do so (see Wainwright *et al.*, 1999, for an extended analysis). The temporal and spatial complexity of a catchment model must match the

spatio-temporal complexity of the processes, though the availability of data is usually the greater constraint. For simplicity, natural timesteps are always preferable to artificial ones (day and night separated at sunrise and sunset instead being aggregated into a whole day; rainfall-rate-based timesteps reflecting the real passage of rainfall intensities instead of hourly lumps of rainfall).

4.3.3 Simplifying process complexity

There are many areas of hydrology where our understanding of processes is basic but still sufficient to develop models. But there are still areas in which the complexity of hydrological processes is so great, or the information so little, that we do not understand the processes well enough to develop reliable models. This complexity of processes is separate from the issues related to spatial and temporal variation and the lack of data available to represent them as outlined above. Some of the areas in which there is still much progress to be made are outlined below:

(a) *The hydrology of sparse vegetation.* Though techniques for modelling the interception, drainage and evapotranspiration from forest canopies are now well established, there are still difficulties in understanding the role of canopy properties in determining the partitioning of water between the various fluxes. These difficulties are particularly clear for nonclosed canopies or sparse vegetation where the impact of surface roughness is less well known and the parameterization of evapotranspiration is much more difficult. Furthermore, the separation of vegetation into patches may change evapotranspiration loads in complex ways (see Veen *et al.*, 1996). Patchiness is not just important for evapotranspiration but also affects the generation and propagation of runoff and sediment, especially in arid and semi-arid environments (see Dunkerly and Brown, 1995)[WWW].

(b) *Subsurface quickflow mechanisms.* The parameterization of K_{sat} at scales greater than a few hundred cm^3 remains a major obstacle to progress in subsurface hydrological modelling, particularly given the importance of macropores in providing a mechanism for subsurface quickflow in many environments (Elsenbeer and Vertessy, 2000; Uchida *et al.*, 2001). Though our understanding of the mechanisms of runoff generation through Hortonian and saturation-excess mechanisms has improved considerably in recent years with the extensive incorporation of partial contributing area concepts

---

[WWW] indicates material covered in more detail in the website which accompanies this book: http://www.kcl.ac.uk/envmod

into hydrological models, our understanding of subsurface quickflow mechanisms through macropore pipe networks and shallow subsurface quickflow is much less advanced. This lack of progress is partly the result of the difficulty in measuring variations in the physical properties that control these processes at the catchment scale, and partly the result of the relatively recent recognition of subsurface quickflow as hydrologically important.

(c) *Hillslope-channel coupling.* Modelling catchments is about modelling the hydrology of hillslopes *and* of channels. Many of the processes that determine channelization are still poorly understood and the coupling of hillslopes to channels is an area in which new insights are being made but further research is required to help improve catchment models of streamflow, of the storm hydrograph and of flooding (see Chapter 5).

(d) *Nonrainfall precipitation.* There is still a relative dearth of modelling efforts focused upon the hydrology of nonrainfall inputs such as snow (Blöschl, 1999) and occult precipitation (Mulligan, 2003) at the catchment scale. These inputs are significant in many catchments but are more difficult to measure and to model than rainfall.

(e) *Tropical lowlands and tropical mountains.* The hydrological impacts of land-use and cover change in the tropics are much discussed in the literature but there are very few studies which apply modelling to better understand these systems (Mulligan, 2003) and fewer still that combine modelling with intensive field monitoring programmes (Chappell *et al.*, 1998). As a result, there is still much confusion about the implications of land use change in these environments and myths abound (Bruijnzeel, 1989; Calder, 1999; Mulligan, 2003).

(f) *Hydrological connectance.* A key characteristic of catchment scale studies is the lateral connectance between patches which results from surface and subsurface runoff. The properties of this connectance determine the manner in which patch scale output aggregate to catchment totals. Total catchment runoff is not a simple sum of the constituent patch-level runoff. Recent work by the author using a distributed hydrological model applied in tropical montane environments indicates the importance of hydrological connectance for catchment level outputs and the implications for the study of the hydrological impacts of land use change[www]. This study, which

is described in more detail on the accompanying website, looks at the implications of progressive deforestation on the catchment scale runoff and the sensitivity of runoff to forest loss. The study concludes that in the initial stages of deforestation (0–75% forest cover lost), the sensitivity of runoff to forest loss is significant but low compared with the situation beyond 75% loss. It is concluded that where patches of forest remain along the surface and subsurface flowlines, these absorb excess water generated on the deforested patches and thus extra runoff does not cumulate downstream. When so much forest is lost that flowlines are more or less deforested along their whole length, runoff generation accumulates down the flowlines and saturated wedges penetrate further upslope, thus providing a positive feedback for further runoff generation. In this way the sensitivity of catchment scale runoff to the deforestation of patches in this last 25% is very high, each patch contributing greatly to the continuity of deforested flow paths and thus to enhanced (flashy) streamflow. Furthermore, the location of the deforestation relative to the geometry of the flow network and the properties of the soil beneath and vegetation above it determine the exact outcome of forest loss. This example only goes to indicate the importance of lateral connectivity in driving the overall response of catchments and in integrating (in a relatively simple way) the complex spatial hydrology of them.

4.3.4 Concluding remarks

We have seen how catchments are defined by the flow networks that landscapes generate as a function of their topography. These flow networks can be represented within a geographical information system and can facilitate the modelling of hydrological processes at the catchment scale in a distributed manner such that the processes of precipitation, interception, evapotranspiration, infiltration and recharge can be calculated for individual cells or patches and routed between them towards the catchment outlet. We have reviewed the main types of model applied at the catchment scale and highlighted some of the best of these, some of which are discussed further and linked to from the website that accompanies this book. Further we have seen some of the complexities of modelling at the catchment scale and some of the obstacles to improved models. Finally we have looked at some of the simple but key controls on catchment behaviour that emerge from the connected interaction of spatial, temporal and process complexity at the catchment scale.

[www] indicates material covered in more detail in the website which accompanies this book: http://www.kcl.ac.uk/envmod

Catchment behaviour at scales from the magnificent Amazon through to the most modest drainage system is still poorly understood. Studies of catchment response to meteorological events, climate change and land use change is confounded by the interacting effects of spatial variability, spatio-temporal variation, process complexity and scaling, not to mention the scarcity and paucity of hydrological data at this scale. We are only just beginning to understand the simple outcomes that result from flow connectivity across these complex, spatially organized systems. Distributed modelling has far to go but distributed data has much further. While distributed modelling can help us understand the inner workings of catchments and the interactions within and between them, massive advances in data acquisition (rather than data integration) are required before they can reach their full worth in practical as well as academic enterprise.

REFERENCES

Abbott, M.B., Bathurst, J.C., Cunge, J.A., O'Connell, P.E. and Rasmussen, J. (1986) An introduction to the European Hydrological System – Système Hydrologique Européen, SHE. 2. Structure of a physically-based, distributed modelling system, *Journal of Hydrology* **87**, 61–77.

Al-Sabhan, W., Mulligan, M. and Blackburn, A. (2002) A real-time hydrological model for flood prediction using GIS and the WWW, *Computers, Environment and Urban Systems* **27** (1), 5–28.

Band, L.E. and Wood, L.F. (1988) Strategies for large-scale, distributed hydrologic simulation, *Applied Mathematics and Computation* **27** (1), 23–37.

Bathurst, J.C., Wicks, J.M. and O'Connell, P.E. (1995) The SHE/SHESED basin scale water flow and sediment transport modelling system, in V.P. Singh (ed.) *Computer Models of Watershed Hydrology*, Water Resources Publications, Highland Ranch, CO, 563–594.

Beven, K. (1989) Changing ideas in hydrology – the case of physically based models, *Journal of Hydrology* **105**, 157–172.

Beven, K. (1996) A discussion of distributed hydrological modelling, in M.B. Abbott and J.C. Refsgaard (eds) *Distributed Hydrological Modelling*, Kluwer Academic, Dordrecht, 255–278.

Beven, K. (2002) Towards an alternative blueprint for a physically based digitally simulated hydrologic response modelling system, *Hydrological Processes* **16**, 189–206.

Beven, K. and Binley, A. (1992) The future of distributed models: model calibration and uncertainty prediction, *Hydrological Processes* **6**, 279–298.

Beven, K. and Freer, J. (2001) A dynamic TOPMODEL, *Hydrological Processes* **15**, 1993–2011.

Beven, K.J. and Kirkby, M.J. (1979) A physically based variable contributing area model of basin hydrology, *Hydrological Sciences Bulletin* **24** (1), 43–69.

Beven, K.J., Lamb, R., Quinn, P.F., Romanowicz, R. and Freer, J. (1995) 'TOPMODEL' in V.P. Singh (ed.) *Computer Models of Watershed Hydrology*, Water Resources Publications, Highland Ranch, CO, 627–668.

Blöschl, G. (1999) Scaling issues in snow hydrology, *Hydrological Processes* **13**, 2149–2175.

Bronstert, A., Niehoff, D. and Burger, G. (2002) Effects of climate and land-use change on storm runoff generation: present knowledge and modelling capabilities, *Hydrological Processes* **16**, 509–529.

Bruijnzeel, L.A. (1989) (De)forestation and dry season flow in the humid tropics: a closer look, *Journal of Tropical Forest Science* **1**, 229–243.

Calder, I. (1999) *The Blue Revolution: Land Use and Integrated Water Resources Management*, Earth Scan, London, 208pp.

Calver, A. and Wood, W.L. (1995) The Institute of hydrology distributed model, in V.P. Singh (ed.) *Computer Models of Watershed Hydrology*, Water Resources Publications, Highland Ranch, CO, 595–626.

Chappell, N.A., Franks, S.W. and Larenus, J. (1998) Multiscale permeability estimation for a tropical catchment, *Hydrological Processes* **12**, 1507–1523.

Costa-Cabral, M.C. and Burges, S.J. (1994) Digital elevation model networks (DEMON): a model of flow over hillslopes for computation of contributing and dispersal areas, *Water Resources Research* **30**, 1681–1692.

Crawford, N.H. and Linsley, R.K. (1966) *Digital Simulation in Hydrology: Stanford Watershed Model IV*, Tech. Report No 39, Dept. Civil Engineering, Stanford University, Stanford, CA.

Croton, J.T. and Barry, D.A. (2001) WEC-C: a distributed, deterministic catchment model – theory, formulation and testing, *Environmental Modelling & Software* **16**, 583–599.

Davis, S.H., Vertessy, R.A. and Silberstein, R.P. (1999) The sensitivity of a catchment model to soil hydraulic properties obtained by using different measurement techniques, *Hydrological Processes* **13**, 677–688.

De Roo, A.P.J. (1998) Modelling runoff and sediment transport in catchments using GIS, *Hydrological Processes* **12**, 905–922.

Dietrich, W., Wilson, C., Montgomery, D. and Mckean, J. (1993) Analysis of erosion thresholds, channel networks and landscape morphology using a digital terrain model, *Journal of Geology* **101**, 259–278.

Dunkerly, D.L. and Brown, K.L. (1995) Runoff and runon areas in a patterned chenopod shrubland, arid Western New South Wales, Australia: characteristic and origin, *Journal of Arid Environments* **20**, 41–55.

Elsenbeer, H. and Vertessy, R.A. (2000) Stormflow generation and flowpath characteristics in an Amazonian rainforest catchment, *Hydrological Processes* **14**, 2367–2381.

Engelen, G. (2000) *MODULUS: A Spatial Modelling Tool for Integrated Environmental Decision Making*, (Final Project report to European Commission ENV4-CT97-0685), http://www.riks.nl/RiksGeo/proj_mod.htm

Ewen, J. and Parkin, G. (1996) Validation of catchment models for predicting land use and climate change impacts. 1 Method, *Journal of Hydrology* **175**, 583–594.

Ewen, J., Parkin, G. and O'Connell, P.E. (2000) SHETRAN: a coupled surface/subsurface modelling system for 3D water flow and sediment and solute transport in river basins, *ASCE Journal of Hydrologic Engineering* **5**, 250–258.

Fairfield, J. and Leymarie, P. (1991) Drainage networks from grid digital elevation models, *Water Resources Research* **27** (5), 709–717.

Freeze, R.A. and Harlan, R.L. (1969) Blueprint for a physically-based, digitally-simulated hydrologic response model, *Journal of Hydrology* **9**, 237–258.

Garbrecht, J., Ogden, F.L., DeBarry, P.A. and Maidment, D.R. (2001) GIS and distributed watershed models. I: data coverages and sources, *Journal of Hydrologic Engineering* **6** (6), 506–514.

Grayson, R.B., Blöschl, G. and Moore, I.D. (1995) Distributed parameter hydrological modelling using vector elevation data: THALES and TAPES-C, in V.P. Singh (ed.) *Computer Models of Watershed Hydrology*, Water Resources Publications, Highland Ranch, CO, 669–696.

Grayson, R.B., Moore, I.D. and McMahon, T.A. (1992) Physically based hydrological modelling. 1 Terrain based modelling for investigative purposes, *Water Resources Research* **28** (10), 2639–2658.

Greiner, R. (1996) SMAC: spatial optimisation model for analysing catchment management, *Environmental Software* **11** (1–3), 159–165.

Gupta, V.K., Castro, S.L. and Over, T.M. (1996) On scaling exponents of spatial peak flows from rainfall and river network geometry, *Journal of Hydrology* **187**, 81–104.

Hatton, T.J., Dawes, W.R. and Dunin, F.X. (1992) Simulation of hydroecological responses to elevated CO_2 at the catchment scale, *Australian Journal of Botany* **40**, 679–696.

Horn, B.K.P. (1981) *Hill Shading and the Reflectance Map*, Proceedings of the IEEE **69** (1), 14–47.

Jakeman, A.J. and Hornberger, G.M. (1993) How much complexity is warranted in a rainfall-runoff model? *Water Resources Research* **29**, 2637–2649.

Klemeš, V. (1997) Of carts and horses in hydrological modeling, *Journal of Hydrologic Engineering* **1** (4), 43–49.

Lea, N.L. (1992) An aspect-driven kinematic routing algorithm, in A.J. Parsons and A.D. Abrahams, *Overland Flow: Hydraulics and Erosion Mechanics*, UCL Press, London, 147–175.

Martens, D.M. and Di Biase, J.F. (1996) TCM-Manager: a PC-based total catchment management decision support system, *Environmental Software* **11** (1–3), 1–7.

Molicova, H., Grimaldi, M., Bonell, M. and Hubert, P. (1997) Using TOPMODEL towards identifying and modelling the hydrological patterns within a headwater, humid, tropical catchment, *Hydrological Processes* **11**, 1169–1196.

Moore, I., O'Loughlin, E. and Burch, G. (1988) A contour based topographic model for hydrological and ecological applications, *Earth Surface Processes and Landforms* **13**, 305–320.

Moore, I.D., Grayson, R.B. and Ladson, A.R. (1991) Digital terrain modelling: a review of hydrological, geomorphological, and biological applications, *Hydrological Processes* **5** (1), 3–30.

Moore, R.J. (1985) The probability-distributed principle and runoff production at point and basin scales, *Hydrological Science Journal* **30**, 273–297.

Mulligan, M. (1996) Modelling the complexity of landscape response to climatic variability in semi-arid environments, in M.G. Anderson and S.M. Brooks (eds) *Advances in Hillslope Processes*, John Wiley & Sons, Chichester, 1099–1149.

Mulligan, M. (2003) Modelling the hydrological response of tropical mountainous environments to land cover and land use change, in M. Ulli, H. Huber, K.M. Bugmann and M.A. Reasoner (eds) *Global Change and Mountain Regions: A State of Knowledge Overview*, Kluwer, Dordrecht.

O'Callaghan, J.F. and Mark, D.M. (1984) The extraction of drainage networks from digital elevation data, *Computer Vision, Graphics, and Image Processing* **28**, 323–344.

Ogden, F.L. Garbrecht, J., DeBarry, P.A. and Johnson, L.E. (2001) GIS and distributed watershed models. II: modules, interfaces, and models, *Journal of Hydrologic Engineering* **6** (6), 515–523.

O'Loughlin, E.M. (1986) Prediction of surface saturation zones in natural catchments by topographic analysis, *Water Resources Research* **22** (5), 794–804.

Palacios-Velez, O. and Cuevas-Renaud, B. (1986) Automated river-course, ridge and basin delineation from digital elevation data, *Journal of Hydrology* **86**, 299–314.

Quinn, P., Beven, K., Chevallier, P. and Planchon, O. (1991) The prediction of hillslope flow paths for distributed hydrological modelling using digital terrain models, *Hydrological Processes* **5**, 59–79.

Refsgaard, J.C. (1997) Parameterisation, calibration and validation of distributed hydrological models, *Journal of Hydrology* **198**, 69–97.

Refsgaard, J.C. and Storm, B. (1995) MIKE SHE, in V.P. Singh (ed.) *Computer Models of Watershed Hydrology*, Water Resources Publications, Highland Ranch, CO, 809–846.

Reggiani, P., Sivapalan, M. and Hassanizadeh, S.M. (2000) Conservation equations governing hillslope responses: exploring the physical basis of water balance, *Water Resources Research* **36**, 1845–1863.

Rosso, R. (1994) An introduction to spatially distributed modelling of basin response, in R. Rosso, A. Peano, I. Becchi and G.A. Bemporad (eds) *Advances in Distributed Hydrology*, Water Resources Publications, Highland Ranch, CO, 3–30.

Saulnier, G.M., Beven, K. and Obled, C. (1997) Including spatially variable effective soil depths in TOPMODEL, *Journal of Hydrology* **202**, 158–172.

Schmidt, J., Hennrich, K. and Dikau, R. (2000) Scales and similarities in runoff processes with respect to geomorphometry, *Hydrological Processes* **14**, 1963–1979.

Skidmore, A.K. (1989) A comparison of techniques for calculating gradient and aspect from a gridded digital elevation model, *International Journal of Geographical Information Systems* **3**, 323–334.

Stieglitz, M., Rind, D., Famiglietti, J.S. and Rosenzweig, C. (1997) An efficient approach to modeling the topographic control of surface hydrology for regional and global climate modeling, *Journal of Climate* **10**, 118–137.

Szilagyi, J. and Parlange, M.B. (1999) A geomorphology-based semi-distributed watershed model, *Advances in Water Resources* **23**, 177–187.

Tarboton, D.G. (1997) A new method for the determination of flow directions and upslope areas in grid digital elevation models, *Water Resources Research* **33**, 309–319.

Tarboton, D.G., Bras, R.L. and Rodriguez-Iturbe, I. (1992) A physical basis for drainage density, *Geomorphology* **5**, 59–76.

Uchida, T., Kosugi, K. and Mizuyama, T. (2001) Effects of pipeflow on hydrological process and its relation to land-slide: a review of pipeflow studies in forested headwater catchments, *Hydrological Processes* **15**, 2151–2174.

Van Deursen, W.P.A. (1995) *A geographical information systems and dynamic model: development and application of a prototype spatial modelling language*, PhD thesis, Utrecht Knag/Faculteit Ruimtelijke Wetenschappen Universiteit Utrecht, Netherlands.

Veen, A.W.L., Klaassen, W., Kruijt, B. and Hutjes, R.W.A. (1996) Forest edges and the soil–vegetation–atmosphere interaction at the landscape scale: the state of affairs, *Progress in Physical Geography* **20**, 292–310.

Vertessy, R.A., Hatton, T.J., O'Shaughnessy, P.J. and Jaya-suriya, M.D.A. (1993) Predicting water yield from a mountain ash forest using a terrain analysis based catchment model, *Journal of Hydrology* **150**, 665–700.

Viessman, W. and Lewis, G.L. (1996) *Introduction to Hydrology*, 4th edn, HarperCollins, London.

Vogt, J.V. Colombo, R. and Bertolo, F. (2002) Deriving drainage networks and catchment boundaries: a new methodology combining digital elevation data and environmental characteristics, *Geomorphology* **1285**, 1–18.

Vörösmarty, C.J., Fekete, B.M., Meybeck, M. and Lammers, R.B. (2000) Geomorphometric attributes of the global system of rivers at 30-minute spatial resolution, *Journal of Hydrology* **237**, 17–39.

Wainwright, J., Mulligan, M. and Thornes, J.B. (1999) Plants and water in drylands, in A. Baird and R. Wilby (eds) *Ecohydrology*, Routledge, London, 78–126.

Walker, D.H. and Johnson, A.K.L (1996) NRM Tools: a flexible decision support environment for integrated catchment management, *Environmental Software* **11** (1–3), 19–24.

Ward, R.C. and Robinson, M. (2000) *Principles of Hydrology*, McGraw-Hill, London.

Yang, D., Herath, S. and Musiake, K. (2001) Spatial resolution sensitivity of catchment geomorphologic properties and the effect on hydrological simulation, *Hydrological Processes* **15**, 2085–2099.

Zadroga, F. (1981) The hydrological importance of a montane cloud forest area of Costa Rica, in R. Lal and E.W. Russell (eds) *Tropical Agricultural Hydrology*, John Wiley & Sons, New York, 59–73.

Zevenbergen, L.W. and Thorne, C.R. (1986) Quantitative analysis of land surface topography, *Earth Surface Processes and Landforms* **12**, 47–56.

Zhang, W. and Montgomery, D.R. (1994) Digital elevation model grid size, landscape representation, and hydrologic simulations, *Water Resources Research* **30** (4), 1019–1028.

5

Modelling Fluvial Processes and Interactions

KATERINA MICHAELIDES AND JOHN WAINWRIGHT

5.1 INTRODUCTION

Since Bathurst (1988), it has been recognized that there is a growing need to address issues of complexity in the modelling of fluvial systems. Despite the ability of computers to solve numerical approximations to the equations of unsteady flow, models are still unable to estimate flows in real channels to a consistently high level of accuracy. Recent studies have highlighted the need to address interactions between elements in the fluvial system to provide a more complete understanding of processes. Coupled with the issues of data quality and representation of processes at different scales (see Chapters 2 and 19), we are still faced with challenges of how to make progress in modelling the fluvial system. There are a huge number of models in the literature, some of which are discussed below, some simple and some complex. As we will see, model complexity does not always guarantee success!

5.2 THE COMPLEXITY

5.2.1 Form and process complexity

A complex system is one that consists of interconnected or interwoven parts (Bar-Yam, 1997). The fluvial system is a complex system that consists not only of the river channel but also the surrounding hillslopes and floodplains together with their subsystems (e.g. soil subsurface, soil surface, vegetation, etc.). In order to understand the behaviour of the fluvial system, we need to understand not only how the individual parts behave in isolation but also how they interact to form the behaviour of the whole system (ibid.). Furthermore, if the individual components of the fluvial system are themselves complex systems, then it is harder to describe and understand the behaviour of the whole. What makes the fluvial system complex is the large number of interconnected, complex parts, the interactions and feedbacks between them over a range of time- and space-scales, the large spatio-temporal heterogeneity in the surface characteristics and the presence of chaotic (or random) processes in most hydrological systems (Rodríguez-Iturbe et al., 1989; Thornes, 1990; Phillips, 1992).

The morphology of natural river channels is determined by the interaction of the fluid flow with the erodible materials in the channel boundary (Figure 5.1; Knighton, 1998). This nonstatic property of river channels means that there is a complex interdependence between flow characteristics and the form of the channel. Flow velocity varies with resistance within the channel boundary while the shape of the channel varies according to discharge and channel sediment composition. Furthermore, flow velocity is highly variable through time and space and in turn affects patterns and thresholds of erosion of the channel boundaries. Erosion of the channel boundary also depends on the physical properties of the material that makes up the channel. In general, erosion will vary according to whether the channel is made up of cohesive or noncohesive sediment although again nonuniform grain-size distributions along and across the channel add to the further complexity of erosion patterns and feedbacks with flow characteristics. Fluvial interactions between the fluid flow and the channel boundary take place across a wide range of spatio-temporal scales, ranging from the movement of individual particles at the instantaneous time-scale to the movement of sediment through a drainage basin over the long term. Moreover, the fluvial system contains

Environmental Modelling: Finding Simplicity in Complexity. Edited by J. Wainwright and M. Mulligan
© 2004 John Wiley & Sons, Ltd ISBNs: 0-471-49617-0 (HB); 0-471-49618-9 (PB)

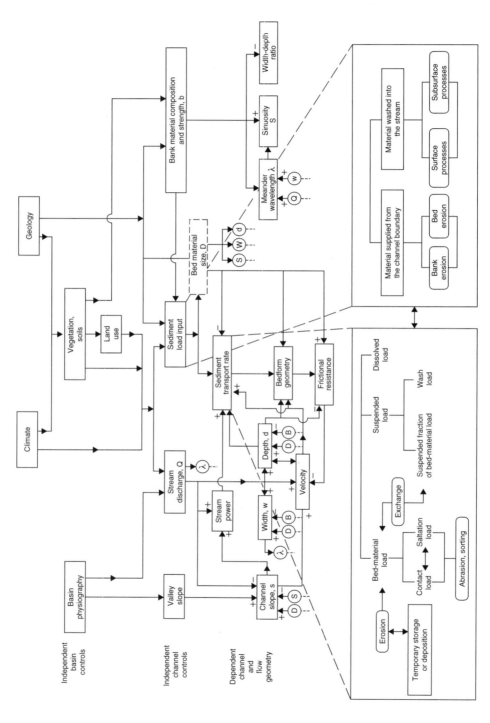

Figure 5.1 Complexity of the fluvial system (based on Knighton, 1998). + means direct, – means inverse link between subsystem elements; zoomed-in boxes illustrate the presence of complex interactions within each subsystem

a variety of landforms that can change through time as the system adjusts to climatic changes and human influences. As such, the fluvial system also has a complex history of adjustment of all the variables to externally imposed effects. Fluvial processes can be investigated on a range of time and space scales, from the drainage basin over geological time and the river reach, through to individual bed forms such as riffles and pools over instantaneous time scales. However, despite our ability to reduce process investigations to minute scales, individual elements of the fluvial system cannot be adequately viewed in isolation as there is an interaction between the hydrological, hydraulic, geological and geomorphological processes at all scales (Schumm, 1988).

Primarily, this interaction between and within scales of the fluvial system takes place across the hillslope, floodplain and channel elements of the catchment through what is termed 'hillslope-channel coupling' (Michaelides and Wainwright, 2002). The term 'hillslope-channel coupling' has been used sporadically in the literature since the late 1970s (e.g. Nortcliff and Thornes, 1984; Caine and Swanson, 1989; Harvey, 1992; Brunsden, 1993; Rice, 1994) to denote the degree of connectivity between the hillslope and channel in terms of the runoff- and sediment-delivery rates from the hillslopes to the channel and in terms of the fluvial activity imposed by the channel on the hillslope base. Coupling takes place at varying time and space scales.

In the long term, the coupling between hillslopes and channels determines drainage-network evolution since drainage patterns and hillslopes show a high degree of adjustment to each other (Kirkby, 1993a,b; Willgoose *et al.*, 1991a–d). The channel network provides a fundamental link between hillslopes and rivers in that it determines, largely through its density, the level of interaction between these two systems of water and sediment transfer (Knighton, 1998). Hillslope form is influenced by channels, which act as local base levels through lateral undercutting and vertical incision or aggradation (Kirkby, 1993a). In return, channels are influenced by hillslopes, which are a source of water and sediment to which they must respond. Kirkby (1993a) stresses the need to learn how the channel network evolves and coevolves with the hillslopes in the catchment. As hillslopes evolve, their hydrological response changes with both gradient and soil and vegetation development. As a result, there are consistent differences in hillslope response from different parts of a catchment, and these differences are related to the long-term evolution of the hillslope. In the long term, hillslope form is controlled by the way in which sediment-transport processes vary along the length of the hillslope and their interaction with channel processes acting at the slope base (Kirkby, 1993b). Dynamic interactions between hillslope- and channel-sediment transport therefore determine the mutual coevolution of channels and hillslopes as a result of varying rates of cut and fill. Steep slope-base gradients promote higher rates of hillslope erosion that may lead to higher inputs of sediment into the channel thereby causing aggradation at the slope base. Aggradation may, however, reduce the slope-base gradient, thereby also reducing hillslope-erosion rates, leading to a relative equilibrium between the hillslope and corresponding channel.

In the short term, hillslopes provide the input of water to the channels together with direct rainfall on the channels and floodplains themselves. Coupling is affected by various factors within the catchment (Michaelides and Wainwright, 2002) discussed further below.

The floodplain is an integral part of the fluvial system as its deposits affect channel form through the composition of the channel boundary, and therefore the type of material supplied to a stream (Knighton, 1998). Furthermore, floodplains are storage elements for sediment as it moves through a catchment, the capacity of which generally increases as they become wider with distance downstream. This effect has implications for the hillslope–channel relationship, which becomes more disjunct as the floodplain becomes wider, due to the buffering effects that a floodplain has both on the outputs from the hillslopes into the channel, and on the action of the channel at the hillslope base. In terms of water and sediment, their position means that they interact with both the hillslope and the channel, and therefore they can either buffer inputs from the hillslope or they can directly contribute significant inputs to the channel.

Work on temperate rivers by Archer (1989) and McCartney and Naden (1995) has shown that floodplain storage significantly reduces the magnitude of major floods and has a major impact on flood frequency. The reasons for this are that a floodplain acts as a weak form of lake or reservoir, providing an area of extra water storage during periods of high flow (McCartney and Naden, 1995). In addition, the dispersal of water from the main channel to the floodplain modifies the flow velocity because of the transfer of momentum between areas of different depths (Knight, 1989) and floodplain storage may change the shape of the hydrograph (McCartney and Naden, 1995). Above bankfull flow, increasing water storage on the floodplain and associated delays due to frictional resistance suppress flood growth downstream (Archer, 1989).

Below bankfull, as flood peak and volume increase, attenuation decreases and downstream flood growth becomes steeper.

5.2.2 System complexity

Are fluvial systems inherently difficult to model? At the smallest scales, flow features in most rivers are dominated by turbulence. Turbulence is the phenomenon by which flows exhibit apparently random motions as well as more structured bursts, whorls and eddies. Despite decades of intensive research, the causes of turbulence are still not well understood (e.g. Hunt, 1995, 1999; Lesieur, 1997) not least because the nonlinear character of the system at these scales causes the development of chaotic behaviour. Modelling at small scales requires a great deal of attention to the behaviour of turbulence, generally requiring simplification of the process to the level of the grid scale being used, not least because of computational power limitations (see Chapter 20 for a more detailed discussion). However, if turbulent bursts, where the local flow energy dramatically increases over a short period of time, are responsible for initiating sediment movement (e.g. Nearing, 1994) and this sediment movement causes a modification of the channel conditions, it is unlikely that a fully specified model of fluvial systems at this scale using such averaging is ever fully meaningful: 'The mechanics of sediment transport is so complex that it is extremely unlikely that a full understanding will ever be obtained. A universal sediment transport equation is not and may never be available' (Simons and Şentürk, 1977). In such cases, stochastic approaches may be more appropriate not only because of the process variability, but because of our inability to describe the system in sufficient detail in any practical application. To do so would require a grain-by-grain description of size and relative location. Stochastic models of sediment transport in gravel-bed channels have been relatively successful in overcoming this limitation (e.g. Naden, 1987; Malmaeus and Hassan, 2002).

Streamflows at larger scales of observation – for example, at the outflow of individual reaches or larger catchments also tend to exhibit chaotic behaviour. Porporato and Ridolfi (1997) found chaotic behaviour in streamflow from 28 US catchments ranging in scale from 31 km^2 to $35\,079 \text{ km}^2$, while Wang and Gan (1998) found chaotic and long-term persistent behaviour in their study of six Canadian rivers. Liu *et al.* (1998) and Sivakumar (2000) discuss the difficulties of evaluating whether such signals exist in detail. An important, ongoing research question is whether

this chaotic behaviour is an emergent property of the turbulence phenomena at smaller scales or whether it is related to interactions at larger spatial and temporal scales.

A tentative answer to this question can be arrived at from the opposite perspective. Channel networks and fluvial landscapes overall tend to exhibit fractal characteristics (Rodríguez-Iturbe and Rinaldo, 1996). At the finer scale, studies by Robert and Richards (1988) and Robert (1988, 1991) have shown that the elevations of a river bed exhibit fractal properties. Thus, it seems that fractal characteristics of fluvial systems occur over nine orders of magnitude. Studies looking at catchment development at large spatial and long temporal scales have demonstrated that fluvial systems can be described as self-organizing systems (De Boer, 1999). Self-organization implies the evolution of a system over a long period of time into a state where rapid change can occur at all spatial scales (Bak, 1997) so that a system can communicate 'information throughout its entire structure, [with] connections being distributed on all scales' (Rodríguez-Iturbe and Rinaldo, 1996: 356) implying fractal characteristics of river basins (see also Chapter 21 for a more detailed discussion). This form of model takes a simple process description and initial topography and attempts to produce realistic (i.e. complex) fluvial landscapes. Bak (1997) suggests that this form of self-organized criticality can explain the fractal characteristics of landscapes in a way that chaotic dynamics cannot. For this reason, we will present more detail on this approach in the case study section below.

In summary, fluvial systems (and the landscape features that interact with them) tend to demonstrate chaotic behaviour at all scales. Realistic models of fluvial systems should therefore attempt to characterize this behaviour in a reasonable way, whether it be by using simple or complex models. While carrying out the simplifications necessary for simulating for different purposes, it is important to maintain the elements that reproduce these characteristics.

5.3 FINDING THE SIMPLICITY

The increasing focus on small-scale processes and the complexity of fluvial processes has led to a reductionist approach that has tended to concentrate on individual fluvial elements in isolation from each other. Such an approach has tended to ignore important interactions in a system where the process complexity lies precisely in these interactions that are paramount in controlling the behaviour of the fluvial system.

For the purpose of understanding the complexity within fluvial systems, catchments are often subdivided into stores and process units (Baird, 1997), such as vegetation, the ground surface, the unsaturated zone of the soil, the saturated zone of the soil, groundwater and the stream channel, or spatial units (Baird, 1997), which could be discrete catchment areas, each with its own set of stores and processes. Commonly, modelling the fluvial system has also involved breaking down the system into its constituent components and modelling them separately. As a result, there have been many modelling attempts on hillslope processes, floodplain processes and river-channel processes, but all in isolation from each other. There have been efforts to incorporate pairs of fluvial elements in models (i.e. hillslope–floodplain, river–floodplain, hillslope–river), but rarely are all elements considered together in a dynamic way.

The following sections consider some approaches currently adopted in modelling interactions within the fluvial system.

5.3.1 Mathematical approaches

Mathematical approaches to the complexity of the fluvial system depend on two levels (Figure 5.2). First, the complexity of the equations used to represent the processes at work within the system will vary from relatively simple to highly complex representations, accounting for a wide range of interactions. Second, the topography and other parameters representing system behaviour in the specific environment can also vary from very simplistic to highly detailed approximations. These two levels of complexity interact and it is the role of the modeller to choose the appropriate intersection of process and environmental representation.

In Chapter 2, we saw how descriptions of flow processes can be very simple. At the simplest level, they may be represented by a form of regression equation that links rainfall to runoff. Neural network models are currently being developed (e.g. Campolo *et al.*, 1999) that follow a very similar empirical approach. Topographic and other parameter representation is often limited to a single variable, which may deal with the compound effects of a number of variables including topography within a basin. Spatial complexity may be dealt with by using a cascading model, where the inputs of one spatial component are derived from the outputs of the upstream components. The required complexity in this sort of approach has been evaluated by Perrin *et al.* (2001) who compared model structure for 429 catchments in France, the USA, Australia, the Ivory Coast and Brazil. Their sample covers catchments with annual average rainfall from 300 to 2300 mm, mean annual potential evapotranspiration of 630 to 2040 mm and mean annual streamflow of 0.2 to 2040 mm (equivalent depth over the catchment) and thus represents a wide range of environments. Using a split-sample procedure, they found that models with a greater number of parameters tended to do better during the calibration step, but during the testing step,

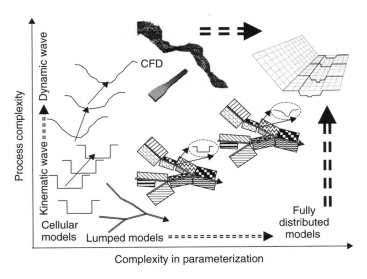

Figure 5.2 Schematic comparison of the levels of complexity used in modelling fluvial systems, comparing process and parameter representations

models with a smaller number of parameters (three compared to nine calibrated parameters) tended to do equally as well. Parsimony in parameterization is thus an advantage that can be derived from this approach to modelling streamflow. However, all such models tend to be limited by the lack of generality beyond the catchment for which they have been developed and deal poorly with nonlinear behaviour, particularly when extrapolation is required for understanding extreme events. Thus, although this sort of model is often useful for looking at the ongoing behaviour of a particular catchment, perhaps within the remit of real-time flood forecasting (e.g. Lardet and Obled, 1994), it is otherwise restricted, particularly when we are trying to use a model to understand the processes at work in more detail.

At the next level of mathematical complexity are the simplified routing schemes. The Muskingum routing method relates the change in storage to the inflow and outflow within a reach and the travel time required for water to flow through the reach. It is thus a simple storage form of the continuity equation. Cunge (1969) demonstrated that it is possible to derive this approach from the kinematic wave approximation to the St Venant equations that describe flow (see Chapter 2), and that the Muskingum–Cunge routing method is equivalent to an approximate solution of the diffusion-wave equation (see Singh, 1996, for a full derivation and worked example). The topographic information required to run such models can be very simple, being simply the length of the reach, slope of the channel bed, and cross-sectional area of the channel at the computation point. Developments of this approach have been used to account for far more variability in the channel system, including complex channel networks and more realistic shapes of the channel cross section (Garbrecht and Brunner, 1991). Attempts to make this type of approach more geomorphically realistic have been carried out using the concept of a geomorphic instantaneous unit hydrograph approach (Rodríguez-Iturbe and Valdez, 1979; Rodríguez-Iturbe, 1993; Snell and Sivapalan, 1994).

Interactions with the catchment system are a fundamental part of fluvial systems as outlined above and in Chapter 4. The KINEROS model (Smith *et al.*, 1995b; Woolhiser *et al.*, 1990) provides an example of the next level of complexity in that the models use a kinematic wave approximation to the routing of flow within channel networks but allow input from hillslope elements within the catchment. These elements are defined as a series of rectangular planes of fixed size and slope. The hydrographs produced from these kinematic planes (e.g. Smith and Woolhiser, 1971) are thus highly stylized.

KINEROS2 provides a further improvement in that it has a representation of floodplains (El-Shinnawy, 1993; Smith *et al.*, 1995a), albeit a highly simplified one that assumes for routing purposes that the combined width of the floodplains on both sides of the channel is only located on one side of the channel. The sensitivity of channel flows to inputs from the hillslope will therefore be overestimated on the bank where the floodplain has been removed, but under-estimated on the bank where the floodplain has been augmented.

The CASC2D model links a fully dynamic wave representation of the St Venant equations for channel flow in different channel cross-sections, with two-dimensional diffusive wave modelling of overland flow from hillslopes and more detailed, spatially distributed catchment accounting of rainfall, interception, retention and infiltration (Senarath *et al.*, 2000). This model requires a separate interface between GIS representations of datasets and their generation as model inputs. The sensitivity analysis of Senarath *et al.* demonstrated that outflows are most sensitive to three hillslope parameters – soil saturated hydraulic conductivity, overland flow roughness and overland flow retention depth – and then to the channel hydraulic roughness, demonstrating the importance of understanding the whole catchment behaviour in low-order catchments. Calibrations followed by model testing on the Goodwin Creek, Mississippi, USA, showed that larger rainfall events tend to be better simulated than smaller ones, which is a common outcome of this form of model, because of the lesser relative importance of errors in estimating infiltration rates. These forms of model provide a better approach to the understanding of the catchment processes, although it is important to remember that a model is only as good as the severity of the test imposed upon it. The results of Senarath *et al.* (2000), for example, have been evaluated at a number of locations within the channel system, which provides more confidence than a simple evaluation of the model at the catchment outlet. Grayson *et al.* (1992a,b) demonstrated using their THALES model of distributed flow within a catchment in Australia that it is possible to 'obtain the right results for the wrong reasons' by calibrating different infiltration models (saturated and Hortonian) to produce equivalent outputs at the catchment boundary. The dangers of calibrating such models can lead to overconfidence in their output, particularly when trying to extrapolate beyond the original calibration dataset so that the meaning of the process basis to the model can be lost (cf. Beven, 1996). Large amounts of time, effort and money are required to apply such an approach in terms of collecting appropriately detailed data for parameterization and testing of the model. Often, the final stage of checking whether calibrated parameters are realistic estimates is missing from the design of research projects that use this approach.

At the most extreme end of the level of mathematical complexity is the application of Computational Fluid Dynamics models (CFDs) to solve the full Navier–Stokes equations of fluid flow. CFDs will be dealt with in far more detail in Chapter 20, so we will only provide a brief overview here. Finite element and finite volume techniques are generally used in CFDs to solve the two- or three-dimensional flow fields in terms of their velocity, pressure and density characteristics, taking account of some simplification of the turbulence of the flow. Thus, they tend to be highly complex numerically and more limited in terms of the topography that can be accommodated. Most studies tend to concentrate on the detail of flow processes within a short reach, relating, for example, to channel confluences (e.g. Biron *et al.*, 2002). Furthermore, only very recently have approaches been developed that can accommodate interactions with the sediment-transport components of fluvial processes (e.g. Lane *et al.*, 2002). Thus, such models provide a great opportunity for enhancing our understanding of process operation at the small scale, but at the cost of significant effort in producing the model grids and data for model testing at appropriate spatial and temporal scales. In a recent example by Horritt and Bates (2001) two approaches of modelling floodplain inundation are compared. The first is a relatively simple raster-based approach with simple process representation and with channel flows being routed separately from the floodplain using a kinematic wave approximation called LISFLOOD-FP (Bates and De Roo, 2000). The second is a more complex finite-element hydraulic model that solves the full two-dimensional shallow-water equations called TELEMAC 2D. They found that the raster and finite-element models offer similar levels of performance when compared with SAR-derived shoreline data. They conclude that given the likely accuracy of the validation data, and the increasing complexity of the calibration process for models representing more complex processes, the simple raster-based model using a kinematic wave approximation over the channel is the simplest (and fastest) model to use and adequate for inundation prediction over this reach.

All of the above model descriptions deal with models that are deterministic. A great deal of effort has been expended since the 1970s to develop stochastic approaches that can reproduce the apparent variability in the flow process. Stochastic models can either incorporate random functions into the representation of the flow process (e.g. Naden, 1987) or in terms of the parameterization (e.g. Fisher *et al.*, 1997; Veihe and Quinton, 2000), often employing Monte Carlo simulation techniques and recently fuzzy logic approaches (e.g. Özelkan and Duckstein, 2001). Random fluctuations in flow patterns have been studied since the work of Hurst in the 1950s on the flood flows of the River Nile. This work was employed by Mandelbrot and Wallis (1969; Kirkby, 1987) to demonstrate the presence of deterministic chaos in fluvial systems. Deterministic chaos can produce the same apparently random results as random processes because of nonlinearities in the process behaviour. Baird (1997) suggested that deterministic chaos is a fundamental consideration in hydrological systems whose presence is hard to distinguish from simple stochastic behaviour. However, most models of fluvial flow are formulated in a way that precludes the development of chaotic behaviour. A reappraisal of the modelling approaches described above is required to make them compatible with observations exhibiting chaotic behaviour, and would be one explanation of the difficulty of producing consistently successful simulations.

5.3.2 Numerical modelling of fluvial interactions

5.3.2.1 Hillslope–channel interactions

Since the early work of Wooding (1965a,b, 1966) on one of the first explicit investigations into catchment hydrological response, there have been numerous models of hillslope–channel interactions. In his initial paper (Wooding, 1965a) developed a simple model for examining catchment hydrological response by describing hillslope and channel responses using the kinematic wave formulation. Wooding's initial analyses subsequently prompted a number of similar studies, albeit of increasing sophistication, which have provided insight into the problem of understanding how catchments of different sizes and composition respond to rainfall events (Calver *et al.*, 1972; Kirkby, 1976; Beven and Wood, 1993; Robinson *et al.*, 1995).

Calver *et al.* (1972) investigated the hydrological response of small catchments with a flow model that accounted specifically for hillslope flows, thus overcoming limitations associated with empirical basin-hydrograph models that have tended to ignore hillslope processes. They modelled the catchment by dividing it into zones of overland flow and channel flow in which the corresponding processes were represented. Their results suggest that the ratio between the flow travel times of the hillslope and channel is the most important factor affecting the catchment hydrograph, implying that the longest flow time will also have the most dominant effect on the overall flow. Therefore, in small catchments with low rainfall, the hillslope hydrograph will dominate, whereas for larger catchments and storms, the channel hydrograph will tend to become increasingly important on the catchment response.

More recently, Beven and Wood (1993) have developed a scale-independent model of catchment network response in order to investigate the effect of variable lateral inflows from hillslopes on the catchment hydrograph. They used hypothetical hillslope hydrographs to represent the variable inputs to the channel that were routed through the appropriate link in the channel network. Their results demonstrated how the predicted catchment hydrograph varies in relation both to the relative time scales of the hillslope and channel network responses, and to the degree of heterogeneity in runoff production. Lateral inflows from the hillslopes affected the catchment hydrograph, depending on the runoff production and network-routing processes. A similar study carried out by Robinson *et al.* (1995) examined the relative roles of hillslope processes, channel routing and network geomorphology in the hydrologic response of catchments. Their results illustrate, first, that small catchment response is governed primarily by the hillslope response; second, that large catchment response is governed primarily by network geomorphology; and, third, that nonlinearity in catchment response is just as great in large catchments as in small catchments. The nonlinearity exhibited in hillslope and channel responses occurs as a result of the spatial variability of runoff across hillslopes, and due to the delay between rainfall and the onset of flow (Calver *et al.*, 1972). The advantage of most hillslope–channel models is that they can be applied at varying scales from small catchment up to drainage basin scale in a relatively simple way due to the simple representations of the topography into channel and hillslope areas. However, these models do not account for the presence of a floodplain that has been shown to significantly affect both channel and hillslope flows.

5.3.2.2 Channel–floodplain interactions

There has also been an increasing surge in floodplain and channel–floodplain numerical modelling. These models have increased in numerical sophistication with time, and range from simple one-dimensional to complex three-dimensional models, mostly formulated for and applied to temperate catchments.

Wolff and Burges (1994) used a one-dimensional flood-routing model (DAMBRK) to investigate the combined influence of some channel and floodplain hydraulic and geometric properties on the frequency distribution of peak flood flow rates as flood waves propagate downstream. The channel and floodplain properties explored included channel and floodplain roughness, channel bed slope, channel width-to-depth ratio and floodplain width. The results emphasize the importance of floodplain and channel characteristics in flood routing and highlight the fact that channels and floodplains should be coupled when investigating the hydrological response of a catchment. Their results indicated that the variability in the flood-frequency distribution was reduced in the downstream direction due to the effects of the channel and floodplain geometry and hydraulic roughness, while the amount of reduction in peak flow rates and flood-frequency distribution variability depended on the effective storage of the reach which again is influenced both by geometry (wider floodplains gave greater attenuation), and hydraulic roughness.

Two-dimensional modelling of floodplains and channel–floodplain interactions has also been reported extensively in the literature. These models either solve a two-dimensional form of the simplified St. Venant equations with finite difference schemes (Cunge, 1975; Cunge *et al.*, 1980; Hromadka *et al.*, 1985; Nicholas and Walling, 1997), or solve the depth-integrated Reynolds equation for two-dimensional free-surface flow using the finite element technique (Gee *et al.*, 1990; Anderson and Bates, 1994; Bates *et al.*, 1997) with the benefit of being able to generate floodplain velocity vectors and inundation zones in addition to flow hydrographs. These floodplain models are generally concerned with predicting floodplain hydraulics, overbank sedimentation rates and patterns of inundation over extensive areas of floodplain. However, they require high resolution of spatial data and are thus difficult to parameterize, and can produce nonconservative results (see Chapter 1).

Recent advances in numerical techniques have also seen the development of three-dimensional models of fluid flow that have been used in the simulation of flow patterns in channel and floodplain areas (Falconer and Chen, 1996; Hervouet and Van Haren, 1996). Krishnappan and Lau (1986) developed a river-floodplain model based on the equations of continuity and three-dimensional turbulent momentum. It has the advantage of representing certain flow processes in a more advanced way, such as lateral momentum transfer, secondary circulation and other three-dimensional effects associated with river channel and floodplain flow interactions. However, three-dimensional models are even more complex and difficult to parameterize than the two-dimensional models and in addition tend to only be useful for modelling flows at small spatial and temporal scales.

5.3.2.3 Hillslope–floodplain–channel interactions

Many models attempt to account for the various fluvial elements by combining different models (e.g. Bates *et al.*, 1997; El-Hames and Richards, 1998; Stewart *et al.*, 1999) and using the output from one model as an input to another. Such an approach ignores important interactions in a setting where the process complexity lies precisely in the interactions and feedbacks that are paramount in controlling the behaviour of the fluvial system. The model COUP2D (Michaelides and Wainwright, 2002) is a two-dimensional, distributed catchment hydrological model of hillslope–channel coupling during a single rainfall event in a semi-arid river reach or small catchment. COUP2D combines a grid-based representation of the hillslope elements with a raster-based representation of the channel component. Hillslopes are represented by a two-dimensional DEM, which is used to route flow according to the kinematic wave approximation. The channel is represented as a one-dimensional series of cross-sections. Flow routing is again carried out along the channel using the kinematic wave approximation which was chosen for this model because it has been found to be particularly suitable for representing many steady uniform flow conditions for channel slopes greater than 1% (Moramarco and Singh, 2000) and for its relative simplicity and wide applicability to a variety of conditions (Moussa and Bocquillon, 1996). The model differs from the many existing catchment runoff models in that it accounts for a low-flow active channel and for changes in its width and location within the main channel belt with discharge. This particular function of the model enables the investigation of the effect that the low-flow channel may have on the coupling. The model differs from existing catchment hydrology models in that the hillslopes, floodplains and channel are represented in a dynamic and unified way. An extensive sensitivity analysis (Michaelides and Wainwright, 2002) has shown that under the appropriate conditions of hillslope and floodplain infiltration rates, roughness and hillslope-floodplain geometry, a topographically-decoupled catchment can produce more runoff than a topographically-coupled catchment. This result means that hillslope-channel coupling is determined by more than just the topographic linkages within the catchment and implies that the presence of a barrier between a hillslope and the channel can either attenuate or reinforce flows by altering the runoff response depending on the surface characteristics of both the barrier and the adjacent hillslope. The degree of coupling or decoupling has been shown to be influenced by the catchment topography, the catchment surface properties, the spatial variability of the surface properties and the rainfall intensity. However, we found that rainfall intensity and duration were actually more important than any other factor in affecting the hydrological response and the relative importance of other factors decreased as the rainfall intensity and duration increased. Thus, the effects of hillslope–channel coupling decreased as the magnitude of the rainfall event increased. Overall, the implications of this study are that it is important to represent the detail of catchment structure and its variability when trying to predict its hydrological response. The structure interacts in important ways with the surface characteristics, so that further field investigations into these interactions are necessary. Although topographic coupling was found to be significant, the interactions between rainfall variability and topographic and surface factors are far more important. Thus, accurate characterization of the spatial and temporal patterns in rainfall is fundamental in making accurate predictions. Although the model has produced some interesting findings about hillslope–floodplain–channel interactions and highlighted some potential threshold responses within the fluvial system, it has the disadvantage of requiring quite large amounts of topographic information for the characterization of the floodplain–channel cross-sections and thus it is impractical to apply for very large catchments or very long river reaches (see Case Study 1 below for further information).

5.3.3 Physical modelling of fluvial processes and interactions

Modelling of fluvial processes and interactions has not just been carried out using numerical techniques. Experiments are also being carried out in miniature, laboratory models that attempt to replicate a real fluvial system at a smaller scale. These laboratory models are called scaled physical models. Scaled physical models have been used quite widely throughout the past two centuries by scientists in various fields (Ashworth *et al.*, 1994). Their popularity lies in the fact that complex natural processes can be investigated in a controlled and simplified laboratory setting. Nowadays, laboratory-modelling experiments can be highly sophisticated and enable the investigation of many processes and interactions within the fluvial system. In order to use or construct a realistic physical model of the fluvial system, some sort of procedure must be applied which would correctly represent the characteristics of the system while reducing it in scale. This procedure is called 'scaling' and it is based on the concept that complete similarity between model and prototype is achieved when the model displays geometric, kinematic and dynamic similitude (Wallerstein *et al.*, 2001; Yalin, 1971). Geometric, kinematic and dynamic similitude imply that all homologous geometric ratios and particle-path lengths to travel-time ratios are equal, as well as that all homologous forces, work magnitudes and power rates are in the same proportion (Wallerstein *et al.*, 2001; Yalin, 1971). However, a true model would require that all the above ratios are equal, something that cannot be achieved when using the same fluid (i.e. water) in both prototype and model. Therefore, in order to reduce the scale while retaining similar characteristics as the prototype, one or more of the controlling variables must be relaxed (Ashworth *et al.*, 1994). The most common method of scaling in fluvial models is called Froude scaling and is based on the principle whereby the flow Reynolds (*Re*) number (a dimensionless measure of flow rate that indicates the conditions under which laminar or turbulent flow occur) is relaxed while correctly modelling the Froude (*Fr*) number (used to distinguish different flow states such as subcritical, critical and supercritical). Yalin (1971) provides a comprehensive and detailed description of the theory behind the construction of hydraulic models, while Wallerstein *et al.* (2001) and Ashworth *et al.* (1994) outline the main principles behind scaling in a concise manner.

5.3.3.1 Fluvial processes

In the field of fluvial hydrology and geomorphology, scaled physical models have mostly been used to investigate fluvial processes. Earlier examples include work by Leopold and Wolman (1957) and Schumm and Khan (1972) on the development of channel patterns and the effects of flow structure and bedload transport on channel geometry. More recently, physical modelling of fluvial processes has involved investigations into feedbacks between channel morphology and bedload transport in gravel-bed rivers (Ashworth and Ferguson, 1986; Ashmore, 1991; Hoey and Sutherland, 1991), mechanisms of anabranch avulsion within gravel-bed braided rivers (Leddy *et al.*, 1993), processes of fine-grained sediment deposition within braided rivers (Ashworth

et al., 1994) and the geomorphic and hydraulic impact of large woody debris elements within a river channel (Wallerstein *et al.*, 2001).

5.3.3.2 Channel–floodplain interactions

Laboratory modelling of channel–floodplain interactions has been ongoing since the 1950s and has been undertaken by scientists from a variety of disciplines ranging from engineering to geography. These experiments have aimed primarily at understanding the ways in which channels behave during floods, especially at the transition between bankfull and overbank stage, and determining the effect of a floodplain on flood flow.

Subsequently, these laboratory studies have involved experiments on a number of different factors that might affect either discharge or velocity distributions. A significant number of studies have been concerned with assessing the role of floodplains in the flood discharge of channels (Barishnikov and Ivanov, 1971; Wormleaton and Merrett, 1990; Chatila and Townsend, 1995; Rashid and Chaudhry, 1995). This work has generally shown that floodplains are important components of the channel whose effects on flood discharge should always be accounted for, because they act as storage features and can significantly attenuate the flood hydrograph by suddenly increasing the width and therefore reducing the velocity of flow. However, further experiments have concluded that it is not sufficient merely to represent the floodplain but that the interactions and exchanges of momentum between the channel and floodplain should also be accounted for (Sellin, 1964; Knight and Demetriou, 1983; Wormleaton *et al.*, 1982). Not including these interactions can lead to an overestimation of the total discharge in the main channel, particularly when the overbank region contributes a small proportion of the total cross-sectional area and when the depth of water on the floodplain is small compared with the bankfull depth of the channel (Sellin, 1964). Other experiments have investigated the effects of channel shape on the resistance in floodplain channels (Yen and Overton, 1973), while others have examined the three-dimensional mixing which occurs in meandering compound flows (Ervine *et al.*, 1994). Several studies have also combined laboratory and numerical modelling of flood hydraulics in compound channels for purposes of comparison and validation of the numerical models (Kohane and Westrich, 1994; Lyness and Myers, 1994; Wark and James, 1994).

Generally, these studies demonstrate the importance of floodplain and channel characteristics for the frequency distribution of peak flood flows, and of allowing for the interaction between the overbank flow and the channel flow. They also illustrate how great the effects of varying roughness and geometry can be on the shape of the hydrograph. However, as all these studies have been carried out in relation to temperate environments where the hillslope input is not significant and the floodplains are dominant in the landscape, they do not account for lateral inflows from hillslopes or for any hillslope–floodplain coupling. Ignoring a potentially important element in experiments like these may lead to an erroneous, or at least incomplete, picture of what is going on in the fluvial system. This means that conclusions regarding the importance of variables on the observed processes are, at best, only tentative as they may miss out important interactions.

5.3.3.3 Hillslope–floodplain–channel interactions

Recently, a unique physical model was constructed in order to investigate hillslope–floodplain–channel coupling. A tributary–junction recirculating channel flume was modified to include tiltable hillslope elements with overflow troughs for applying flow (Figure 5.3). The

aim of this work was to obtain data from a controlled setting in order to validate the numerical model Coup2d (Michaelides and Wainwright, 2002) as well as to observe the processes for a further understanding of coupling. By varying factors in a controlled and systematic way observations can be made about how the system adjusts to the change in one single factor. Factors varied include channel and hillslope angles, channel and hillslope discharges, channel and floodplain sediment size and the presence of a floodplain. In terms of using the physical model to validate the numerical model Coup2d, the results were encouraging, and indicated interesting thresholds of behaviour under varying conditions of channel discharge and hillslope–floodplain lateral inputs. In terms of understanding hillslope–channel coupling, the results indicated the importance of the inputs from lateral elements such as hillslopes and floodplains on channel flow and sediment movement, both in terms of total discharge and timing of discharge patterns throughout the event. The results also demonstrated the significance of varying hillslope, floodplain and channel characteristics (i.e. angle, infiltration rates, roughness,

Figure 5.3   Scaled physical model used to investigate hillslope–channel coupling processes

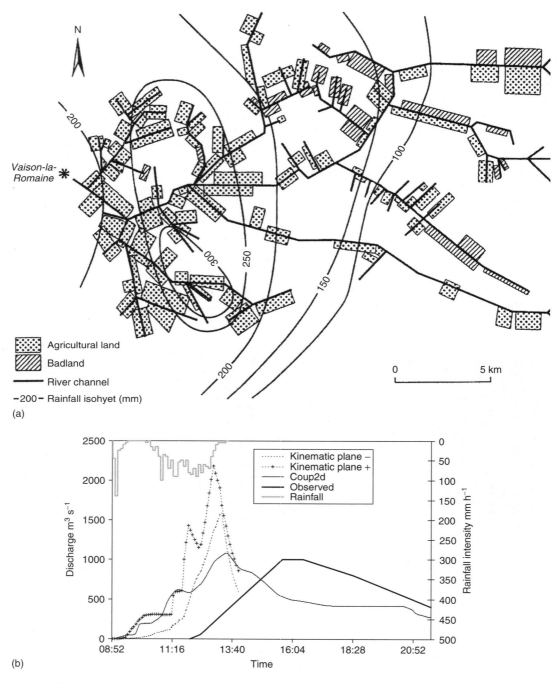

(a)

(b)

Figure 5.4   Modelling of the 22 September 1992 flood in the Ouvèze basin: (a) the discretization of the Ouvèze catchment using a kinematic plane approach; (b) model results; and (c) discretization using the Coup2d approach of Michaelides and Wainwright (2002); image to the left is the bridge at Vaison-la-Romaine which was overtopped in the event

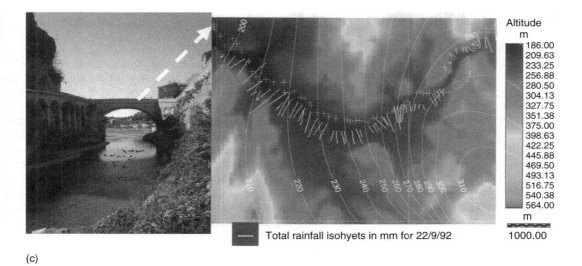

Total rainfall isohyets in mm for 22/9/92

(c)

Figure 5.4 (*continued*)

etc.) on the overall hydrological and hydraulic response of the river system. As with most physical models, our model of hillslope–floodplain–channel coupling suffers from the disadvantage that it can only be configurated to simulate a limited range of conditions and in discrete (rather than continuous) steps. Data collection is also prone to errors, as the instrumentation and replication of model runs under the exact same conditions can never be guaranteed due to the human input in the set-up of each run.

5.4 CASE STUDY 1: MODELLING AN EXTREME STORM EVENT

On 22 September 1992, a major storm event affected large areas of southern France. Up to 300 mm of rain fell in a period of five hours in the basin of the River Ouvèze, with recorded intensities of 190 mm h^{-1} over a six-minute duration at surface stations and 370 mm h^{-1} over a fifteen-minute duration from radar images (Benech *et al.*, 1993; Wainwright, 1996). Over 35 people were killed as flood levels rose to 15 m in the town of Vaison-la-Romaine, sweeping away a camp site located on a floodplain level 6 m above the current active channel. Based on field data from rainfall-simulation experiments, Wainwright (1996) modelled the flood event using a kinematic plane model for runoff from hillslope sections with flow routed through the channel system using a kinematic wave approximation. Active hillslope segments were defined based on land use, which suggested that only areas under cultivation and heavily

gullied badlands were sources of runoff. The discretization of the catchment using this approach is shown in Figure 5.4a. The results of this simulation suggested both an overprediction of the peak discharge and a flood wave that was significantly too early (Figure 5.4b).

To attempt to address these problems, a second model (COUP2D) was developed that uses a grid-based representation of the catchment topography together with a raster-based representation of the channel network (Michaelides and Wainwright, 2002). Flow routing is by two-dimensional kinematic wave on the hillslopes and floodplains (at low flows) and one-dimensional kinematic wave in the channel (and the compound channel–floodplain at high flows). The advantage of this approach is that it enables a better representation of the connectivity of the system, particularly in terms of the dynamics of the hillslope–floodplain–channel coupling. Catchment discretization using a digital elevation model (DEM) is shown in Figure 5.4c. This model was better able to reproduce the peak flow of the event and improves on the timing of the main flood wave, but still tends to produce a peak too early.

One possible explanation for the simulation of a flood wave that arrives too rapidly is in the simplifications that are still present in the catchment representation. Massive limestones are common throughout the Ouvèze catchment area, and thus karstic phenomena could be important in transferring water through the basin at a slower rate than the surface flows. Indeed, Lastennet and Mudry (1995) demonstrated that there was a significant element of flow from the karst system in this event.

This case study illustrates the use of modelling in improving the understanding of processes at work in the fluvial system. In one sense, both of our attempts at modelling the storm event of 22 September 1992 in the Ouvèze basin were unsuccessful, in that they simulate either too high or too rapid flows (or both). However, they do demonstrate the importance of linkages within the fluvial system and that dealing with the hillslope–floodplain–channel coupling is fundamental in process-based simulations of this type of system. There have been few attempts at this sort of modelling, and none that attempt to link subsurface flow pathways at this scale. There is seldom a linkage of research into karstic and surface pathways given the current research agenda within geomorphology and hydrology, which tends to focus on reductionist approaches.

5.5 CASE STUDY 2: HOLISTIC MODELLING OF THE FLUVIAL SYSTEM

An alternative strategy that attempts to introduce a holistic approach to the modelling of fluvial systems is that based on self-organized criticality (SOC) by Rodríguez-Iturbe and Rinaldo (1996). These authors attempt to generate river basins using a simple set of rules for their evolution (Figure 5.5a–f). An initial topography, which may be random or have a simple slope, is defined using a grid of cells (i.e. a DEM). For each cell, the drainage direction is defined by looking for the lowest immediate cell in the neighbourhood of the eight surrounding cells. The cumulative drainage area is then evaluated using the drainage direction of each cell to add catchment areas progressively in a downstream direction. The

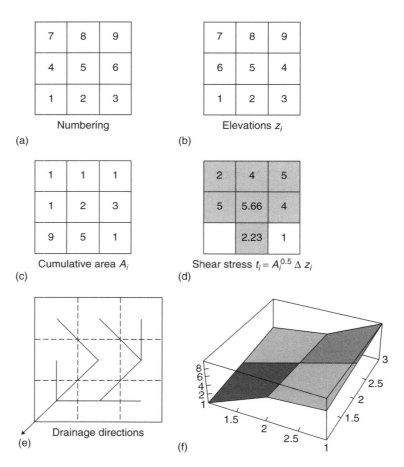

Figure 5.5 Cell-based models used to investigate the existence of self-organized criticality in fluvial networks: (a–f) rules used to generate channel networks; and (g) an example of simulated network
Source: Rodríguez-Iturbe and Rinaldo (1996)

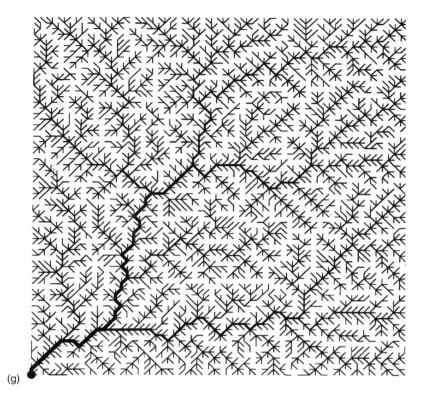

(g)

Figure 5.5 (*continued*)

value of the cumulative drainage area is used as a simple indicator of how much flow would be present in a particular cell. The network evolves by erosion at specific cells. Erosion occurs where the shear stress exceeds a threshold value (i.e. the shear strength of the sediment or surface). The shear stress is defined as a simple function of area (and therefore flow) and slope angle, in other words providing a simple representation of stream power, and the shear strength is assumed to be spatially constant. This approach can produce river networks that both appear realistic (Figure 5.5g), and follow network-scaling behaviour observed in actual networks (see also Caldarelli, 2001) and that is characteristic of SOC. Simulations that incorporate elements of climatic variability demonstrate that these networks can reproduce the characteristic complex responses of real fluvial systems when undergoing externally forced change (Schumm, 1977). Other examples of this sort of approach can be found in Willgoose *et al.* (1991a–d), Tucker *et al.* (2001) and Coulthard *et al.* (2002), which use a variety of approaches and levels of complexity in process and parameter representation.

An important question arises as to whether these approaches are really holistic in a meaningful way. There are two critical issues here. First, there is the question of how we evaluate whether the simulated networks are realistic or not. Kirchner (1993) has demonstrated that the fit of Horton's 'laws' to networks is statistically inevitable, rather than demonstrating the presence of geomorphic phenomena. Thus, we need to exercise caution in the interpretation of these results – even if they look sensible, do they actually mean anything in terms of understanding real-world fluvial systems? Second, there is a significant element of process reductionism at play in this approach. Although we know that many processes are involved in channel development, they have been reduced to one form here – a simple excess shear stress, reflecting when a channel can remove material from its bed. However, we know that rill and gully initiation (and thus channel development) is one of the most difficult unsolved problems of geomorphology (see discussion in Chapter 21). Furthermore, some channels develop by mass movements and subsurface flow processes (Bocco, 1991; Collison,

1996; Rinaldi and Casagli, 1999), so the simple shear-stress approximation is not appropriate in all cases. We have also seen how interactions within the entire system (surface and subsurface) are important in understanding landscape-forming events. Do these objections matter? The usual response is that the stream-power process essentially 'drowns out' the effects of the other processes. In the long term, it is therefore not necessary to understand them to evaluate the overall development of the network. However, the development of new branches to the channel system quite literally produces important bifurcations in the behaviour of the system, so it cannot be possible to ignore other ways in which these bifurcations can occur and still understand the development of fluvial systems. As with the smaller-scale examples noted above, we may have the right answer, but the reason for arriving at it is incorrect. Thus, although these holistic approaches are important, we believe that much progress is required before they can provide meaningful understandings of real-world fluvial systems.

5.6 THE RESEARCH FRONTIER

Models need to be developed that account for both the process and form complexity. This requirement does not simply mean the addition of increasingly complex equations and subsystems, but accounting better for different elements and their interactions within the system as a whole. There remains scope for using relatively simple models for a whole range of applications with success. Simple models may be more successful at identifying underlying patterns of nonlinear behaviour that it will never be possible to simulate using increasingly complex, linearized models. Also, a better understanding of how processes operate and may be linked across different spatial and temporal scales is a fundamental requirement.

More successful models will not be developed on their own. In most cases, there are insufficient data available to test models appropriately. We simply cannot say with any certainty in most cases that the model is wrong compared to our parameterization of it or uncertainties in the data used to test it. Without an interactive link between modelling and data collection, there will be no progress. It is not simply a case of collecting more data, but more a requirement to collect data of an appropriate quality for input and testing of our results. If spatial models are being developed, then data for parameterization and testing need to be spatially distributed as well. Otherwise, the tests we make will continue to be insufficient.

Interactions with other components seem to be a fundamental part of understanding the fluvial system (see also Chapter 4). Yet research into channel processes is often carried out independently of that into hillslopes (and vice versa of course). There is a need to take a holistic view in disciplinary terms as well as in relation to the system we are trying to understand.

ACKNOWLEDGEMENTS

The work on hillslope–floodplain–channel interactions discussed in this chapter was sponsored by NERC grant GR3/12067. We would also like to thank Jonathan Tooby for drawing Figure 5.1 and Diren-Paca and MétéoFrance for the provision of data used in the Ouvèze modelling case study.

REFERENCES

Anderson, M.G. and Bates, P.D. (1994) Evaluating data constraints on two-dimensional finite element models of floodplain flow, *Catena* **22**, 1–15.

Archer, D.R. (1989) Flood wave attenuation due to floodplain storage and effects on flood frequency, in K. Beven and P. Carling (eds) *Floods: Hydrological, Sedimentological and Geomorphological Implications*, John Wiley & Sons, Chichester, 37–46.

Ashmore, P.E. (1991) Channel morphology and bed load pulses in braided, gravel-bed streams, *Geografiska Annaler* **73A**, 37–52.

Ashworth, P.J., Best, J.L., Leddy, J.O. and Geehan, G.W. (1994) The physical modelling of braided rivers and deposition of fine-grained sediment, in M.J. Kirkby (ed.) *Process Models and Theoretical Geomorphology*, John Wiley & Sons, Chichester, 115–139.

Ashworth, P.J. and Ferguson, R.I. (1986) Interrelationships of channel processes, changes and sediments in a proglacial river, *Geografiska Annaler* **68**, 361–371.

Baird, A.J. (1997) Continuity in hydrological systems, in R.L. Wilby (ed.) *Contemporary Hydrology*, John Wiley & Sons, Chichester, 25–58.

Baird, A.J. (1999) Modelling, in A.J. Baird and R.L. Wilby (eds) *Eco-Hydrology: Plants and Water in Terrestrial and Aquatic Environments*, Routledge, London, 300–345.

Bak, P. (1997) *How Nature Works*, Oxford University Press, Oxford.

Barishnikov, N.B. and Ivanov, G.V. (1971) Role of floodplain in flood discharge of a river channel, *Proceedings of the 14th International Conference of the International Association for Hydraulic Research*, Paris, France, 141–144.

Bar-Yam, Y. (1997) *Dynamics of Complex Systems*, Addison-Wesley Longman, Reading, MA.

Bates, P.D., Anderson, M.G., Hervouet, J.-M. and Hawkes, J.C. (1997) Investigating the behaviour of two-dimensional finite element models of compound channel flow *Earth Surface Processes and Landforms* **22**, 3–17.

Bates, P.D. and De Roo, A.P.J. (2000) A simple raster based model for flood inundation simulation, *Journal of Hydrology* **236**, 54–77.

Bathurst, J.C. (1988) Flow processes and data provision for channel flow models, in M.G. Anderson (ed.) *Modelling Geomorphological Systems*, John Wiley & Sons, Chichester, 127–152.

Benech, B., Brunet, H., Jacq, V., Payen, M., Rivrain, J.-C. and Santurette, P. (1993) La catastrophe de Vaison-la-Romaine et les violentes précipitations de septembre 1992: aspects météorologiques, *La Météorologie* **8** (1), 72–90.

Beven, K. (1996) The limits of splitting: hydrology, *The Science of the Total Environment* **183**, 89–97.

Beven, K. and Wood, F.E. (1993) Flow routing and the hydrological response of channel networks, in K. Beven and M.J. Kirkby (eds) *Channel Network Hydrology*, John Wiley & Sons, Chichester, 99–128.

Biron, P.M., Richer, A., Kirkbride, A.D., Roy, A.G. and Han, S. (2002) Spatial patterns of water surface topography at a river confluence, *Earth Surface Processes and Landforms* **27**, 913–928.

Bocco, G. (1991) Gully erosion: processes and models, *Progress in Physical Geography* **15**, 392–406.

Brooks, R.J. and Tobias, A.M. (1996) Choosing the best model: level of detail, complexity, and model performance, *Mathematical and Computer Modelling* **24**, 1–14.

Brunsden, D. (1993) The persistance of landforms, *Zeitschrift für Geomorphologie*, Supplementband **93**, 13–28.

Caine, N. and Swanson, F.J. (1989) Geomorphic coupling of hillslope and channel systems in two small mountain basins, *Zeitschrift für Geomorphologie* **33** (2), 189–203.

Caldarelli, G. (2001) Cellular models for river networks, *Physical Review E* **63**, art. no. 021118 Part 1 FEB 2001.

Calver, A., Kirkby, M.J. and Weyman, D.R. (1972) Modelling hillslope and channel flows, in R.J. Chorley (ed.) *Spatial Analysis in Geomorphology*, Methuen, London, 197–218.

Campolo, M., Andreussi, P. and Soldati, A. (1999) River flood forecasting with a neural network model, *Water Resources Research* **35**, 1191–1197.

Chatila, J.G. and Townsend, R.D. (1995) Modelling floodplain conveyance in compound channel flows, *Canadian Journal of Civil Engineering* **22**, 660–667.

Chwif, L., Barretto, M.R.P. and Paul, R.J. (2000) On simulation model complexity, in J.A. Joines, R.R. Barton, K. Kang and P.A. Fishwick (eds) *Proceedings of the 2000 Winter Simulation Conference*, The Society for Computer Simulation, San Diego, CA, 449–455.

Collison, A.J.C. (1996) Unsaturated strength and preferential flow as controls on gully head development, in M.G. Anderson and S.M. Brooks (eds) *Advances in Hillslope Processes*, John Wiley & Sons, Chichester, 753–770.

Coulthard, T.J., Macklin, M.G. and Kirkby, M.J. (2002) A cellular model of Holocene upland river basin and alluvial fan evolution, *Earth Surface Processes and Landforms* **27**, 269–288.

Cunge, J. (1969) On the subject of a flood propagation computation method (Muskingum method), *Journal of Hydraulic Research* **7**, 205–230.

Cunge, J.A. (1975) Two-dimensional modelling of floodplains, in K. Mahmood and V. Yevjevich (eds) *Unsteady Flow in Open Channels*, Vol. II, Water Resources Publications, Fort Collins, CO, 705–762.

Cunge, J.A., Holly, F.M. and Verwey, A. (1980) *Practical Aspects of Computational River Hydraulics*, Pitman Publishing, London.

De Boer, D.H. (1999) Self-organization in fluvial landscapes: sediment dynamics as an emergent property, in *Proceedings of the Geocomputation 99 Conference*, http://www.geovista.psu.edu/sites/geocomp99/Gc99/074/gc_074.htm

El-Hames, A.S. and Richards, K.S. (1998) An integrated, physically based model for arid region flash flood prediction capable of simulating dynamic transmission loss, *Hydrological Processes* **12**, 1219–1232.

El-Shinnawy, I.A. (1993) Evaluation of transmission losses in ephemeral streams, Unpublished PhD thesis, Department of Civil Engineering, University of Arizona.

Ervine, D.A., Sellin, R.J. and Willetts, B.B. (1994) Large flow structures in meandering compound channels, in W.R. White and J. Watts (eds) *2nd International Conference on River Flood Hydraulics*, H.R. Wallingford Ltd. and John Wiley & Sons, Chichester, 459–469.

Falconer, R.A. and Chen, Y. (1996) Modelling sediment transport and water quality processes on tidal floodplains, in M.G. Anderson, D.E. Walling and P.D. Bates (eds) *Floodplain Processes*, John Wiley & Sons, Chichester, 361–398.

Fisher, P., Abrahart, R.J. and Herbinger, W. (1997) The sensitivity of two distributed non-point source pollution models to the spatial arrangement of the landscape, *Hydrological Processes* **11**, 241–252.

Garbrecht, J. and Brunner, G. (1991) Hydrologic channel-flow routing for compound sections, *Journal of Hydraulic Engineering* **177**, 629–642.

Gee, M.D., Anderson, M.G. and Baird, L. (1990) Large-scale floodplain modelling, *Earth Surface Processes and Landforms* **15**, 513–523.

Grayson, R.B., Moore, I.D. and McMahon, T.A. (1992a) Physically based hydrologic modelling. 1. A terrain-based model for investigative purposes, *Water Resources Research* **28**, 2639–2658.

Grayson, R.B., Moore, I.D. and McMahon, T.A. (1992b) Physically based hydrologic modelling. 2. Is the concept realistic?, *Water Resources Research* **28**, 2659–2666.

Harvey, A.M. (1992) Process interactions, temporal scales and the development of hillslope gully systems: Howgill Fells, Northwest England, *Geomorphology* **5**, 323–344.

Hervouet, J.-M. and Van Haren, L. (1996) Recent advances in numerical methods for fluid flow, in M.G. Anderson, D.E. Walling and P.D. Bates (eds) *Floodplain Processes*, John Wiley & Sons, Chichester, 183–214.

Hoey, T.B. and Sutherland, A.J. (1991) Channel morphology and bedload pulses in braided rivers: a laboratory study, *Earth Surface Processes and Landforms* **16**, 447–462.

Horritt, M.S. and Bates, P.D. (2001) Predicting floodplain inundation: raster-based modelling versus the finite element approach, *Hydrological Processes* **15**, 825–842.

Hromadka, T.V. II., Berenbrock, C.E., Freckleton, J.R. and Guymon, G.L. (1985) A two-dimensional dam-break flood plain model, *Advances in Water Resources* **8**, 7–14.

Hunt, J.C.R. (1995) Practical and fundamental developments in the computational modelling of fluid flows, *Journal of Mechanical Engineering Scientific Proceedings C* **209**, 297–314.

Hunt, J.C.R. (1999) Environmental forecasting and turbulence modelling, *Physica D* **133**, 270–295.

Jakeman, A.J. and Hornberger, G.M. (1993) How much complexity is warranted in a rainfall-runoff model? *Water Resources Research* **29**, 2637–2649.

Kirchner, J.W. (1993) Statistical inevitability of Horton's laws and the apparent randomness of stream channel networks, *Geology* **21**, 591–594.

Kirkby, M.J. (1976) Tests of the random network model and its application to basin hydrology, *Earth Surface Processes* **1**, 197–212.

Kirkby, M.J. (1987) The Hurst effect and its implications for extrapolating process rates, *Earth Surface Processes and Landforms* **12**, 57–67.

Kirkby, M.J. (1993a) Network hydrology and geomorphology, in K. Beven and M.J. Kirkby (eds) *Channel Network Hydrology*, John Wiley and Sons, Chichester, 1–11.

Kirkby, M.J. (1993b) Long term interactions between networks and hillslopes, in K. Beven and M.J. Kirkby (eds) *Channel Network Hydrology*, John Wiley & Sons, Chichester, 255–293.

Klemeš, V. (1983) Conceptualization and scale in hydrology, *Journal of Hydrology* **65**, 1–23.

Knight, D.W. (1989) Hydraulics of flood channels, in K. Beven and P. Carling (eds) *Floods: Hydrological, Sedimentological and Geomorphological Implications*, John Wiley & Sons, Chichester, 83–105.

Knight, D.W. and Demetriou, J.D. (1983) Flood plain and main channel flow interaction, *Journal of the Hydraulics Division, ASCE* **109** (8), 1073–1092.

Knighton, D. (1998) *Fluvial Forms and Processes: A New Perspective*. Arnold, London.

Kohane, R. and Westrich, B. (1994) Modelling of flood hydraulics in compound channels, in W.R. White and J. Watts (eds) *2nd International Conference on River Flood Hydraulics*, H.R. Wallingford Ltd and John Wiley & Sons, Chichester, 415–423.

Krishnappan, B.G. and Lau, Y.L. (1986) Turbulence modelling of flood plain flows, *Proceedings of the American Society of Civil Engineers, Journal of Hydraulic Engineering* **112** (4), 251–266.

Lane, S.N., Hardy, R.J., Elliott, L. and Ingham, D.B. (2002) High-resolution numerical modelling of three-dimensional flows over complex river bed topography, *Hydrological Processes* **16**, 2261–2272.

Lane, S.N. and Richards, K.S. (1997) Linking river channel form and process: time, space and causality revisited, *Earth Surface Processes and Landforms* **22**, 249–260.

Lardet, P. and Obled, C. (1994) Real-time flood forecasting using a stochastic rainfall generator, *Journal of Hydrology* **162**, 391–408.

Lastennet, R. and Mudry, J. (1995) Impact of an exceptional storm episode on the functioning of karst system – the case of the 22/9/92 storm at Vaison-la-Romaine (Vaucluse, France), *Comptes Rendus de l'Académie de Science Série II*, **320**, 953–959.

Leddy, J.O., Ashworth, P.J. and Best, J.L. (1993) Mechanisms of anabranch avulsion within gravel-bed braided rivers: observations from a scaled physical model, in J.L. Best and C.S. Bristow (eds) *Braided Rivers*, Geological Society Special Publication No. **75**, London, 119–127.

Leopold, L. and Wolman, M.G. (1957) River channel patterns: braided, meandering, and straight, *US Geological Survey Professional Paper* **262-B**.

Lesieur, M. (1997) *Turbulence*, 3rd edn. Kluwer Academic Publishers, Dordrecht.

Liu, Q., Islam, S., Rodríguez-Iturbe, I. and Le, Y. (1998) Phase-space analysis of daily streamflow: characterization and prediction, *Advances in Water Resources* **21**, 463–475.

Lyness, J.F. and Myers, W.R.C. (1994) Comparisons between measured and numerically modelled unsteady flow in a compound channel using different representations of friction slope, in W.R. White and J. Watts (eds) *2nd International Conference on River Flood Hydraulics*, H.R. Wallingford Ltd and John Wiley & Sons, Chichester, 383–391.

Malmaeus, J.M. and Hassan, M.A. (2002) Simulation of individual particle movement in a gravel streambed, *Earth Surface Processes and Landforms* **27**, 81–97.

Mandelbrot, B.B. and Wallis, J.R. (1969) Some long-run properties of geophysical records, *Water Resources Research* **5**, 321–340.

McCartney, M.P. and Naden, P.S. (1995) A semi-empirical investigation of the influence of floodplain storage on flood flow, *Journal of CIWEM* **9**, 236–246.

Michaelides, K. and Wainwright, J. (2002) Modelling the effects of hillslope–channel coupling on catchment hydrological response, *Earth Surface Processes and Landforms* **27**, 1441–1457.

Moramarco, T. and Singh, V.P. (2000) A practical method for analysis of river waves and for kinematic wave routing in natural channel networks, *Hydrological Processes* **14**, 51–62.

Moussa, R. and Bocquillon, C. (1996) Criteria for the choice of flood routing methods in natural channels, *Journal of Hydrology* **186**, 1–30.

Naden, P. (1987) Modelling gravel-bed topography from sediment transport, *Earth Surface Processes and Landforms* **12**, 353–367.

Nearing, M.A. (1994) Detachment of soil by flowing water under turbulent and laminar conditions, *Soil Science Society of America Journal* **58**, 1612–1614.

Nicholas, A.P. and Walling, D.E. (1997) Modelling flood hydraulics and overbank deposition on river floodplains, *Earth Surface Processes and Landforms* **22**, 59–77.

Nortcliff, S. and Thornes, J.B. (1984) Floodplain response of a small tropical stream, in T.P. Burt and D.E. Walling (eds) *Catchment Experiments in Fluvial Geomorphology*, GeoBooks, Norwich.

Özelkan, E.C. and Duckstein, L. (2001) Fuzzy conceptual rainfall-runoff models, *Journal of Hydrology* **253**, 41–68.

Perrin, C., Michel, C. and Andréassian, V. (2001) Does a large number of parameters enhance model performance? Comparative assessment of common catchment model structures on 429 catchments, *Journal of Hydrology* **242**, 275–301.

Phillips, J.D. (1992) Deterministic chaos in surface runoff, in A.J. Parsons and A.D. Abrahams (eds) *Overland Flow: Hydraulics and Erosion Mechanics*, UCL Press, London, 275–305.

Porporato, A. and Ridolfi, L. (1997) Nonlinear analysis of river flow time sequences, *Water Resources Research* **33**, 1353–1367.

Rashid, R.S.M.M. and Chaudhry, M.H. (1995) Flood routing in channels with floodplains, *Journal of Hydrology* **171** (1–2), 75–91.

Rice, S. (1994) Towards a model of changes in bed material texture at the drainage basin scale, in M.J. Kirkby (ed.) *Process Models and Theoretical Geomorphology*, John Wiley & Sons, Chichester, 159–172.

Rinaldi, M. and Casagli, N. (1999) Stability of streambanks formed in partially saturated soils and effects of negative pore water pressures: the Sieve River (Italy), *Geomorphology* **26**, 253–277.

Robert, A. (1988) Statistical properties of sediment bed profiles in alluvial channels, *Mathematical Geology* **20** (3), 205–225.

Robert, A. (1991) Fractal properties of simulated bed profiles in coarse-grained channels, *Mathematical Geology* **23** (3), 367–382.

Robert, A. and Richards, K.S. (1988) On the modelling of sand bedforms using the semivariogram, *Earth Surface Processes and Landforms* **13** (5), 459–473.

Robinson, J.S., Sivapalan, M. and Snell, J.D. (1995) On the relative roles of hillslope processes, channel routing, and network geomorphology in the hydrological response of natural catchments, *Water Resources Research* **31** (12), 3089–3101.

Rodríguez-Iturbe, I. (1993) The geomorphological unit hydrograph, in K. Beven and M.J. Kirkby (eds) *Channel Network Hydrology*, John Wiley & Sons, Chichester.

Rodríguez-Iturbe, I., de Power, B.F., Sharifi, M.B. and Georgakakos, K.P. (1989) Chaos in rainfall, *Water Resources Research* **25**, 1667–1675.

Rodríguez-Iturbe, I. and Rinaldo, A. (1996) *Fractal River Basins: Chance and Self-Organization*, Cambridge University Press, Cambridge.

Rodríguez-Iturbe, I. and Valdez, J.B. (1979) The geomorphologic structure of hydrology response, *Water Resources Research* **15**, 1409–1420.

Schumm, S.A. (1977) *The Fluvial System*, John Wiley & Sons, Chichester.

Schumm, S.A. (1988) Variability of the fluvial system in space and time, in T. Rosswall, R.G. Woodmansee and P.G. Risser (eds) *Scales and Global Change: Spatial and Temporal Variability in Biospheric and Geospheric Processes*, SCOPE Publication, No. 35, John Wiley & Sons, Chichester, 225–250.

Schumm, S.A. and Khan, H.R. (1972) Experimental study of channel patterns, *Geological Society of America Bulletin* **93**, 1755–1770.

Schumm, S.A. and Lichty, R.W. (1965) Time, space, and causality in geomorphology, *American Journal of Science* **263**, 110–119.

Sellin, R.H.J. (1964) A laboratory investigation into the interaction between the flow in the channel of a river and that over its flood plain, *La Houille Blanche* **7**, 793–801.

Senarath, S.U.S., Ogden, F., Downer, C.W. and Sharif, H.O. (2000) On the calibration and verification of two-dimensional, distributed, Hortonian, continuous watershed models, *Water Resources Research* **36**, 1495–1510.

Simons, D.B. and Şentürk, F. (1977) *Sediment Transport Technology*, Water Resources Publications, Littleton, CO.

Singh, V.P. (1996) *Kinematic Wave Modelling in Water Resources: Surface-Water Hydrology*, John Wiley & Sons, Chichester.

Sivakumar, B. (2000) Chaos theory in hydrology: important issues and interpretations, *Journal of Hydrology* **227**, 1–20.

Smith, R.E., Goodrich, D.C. and Quinton, J.N. (1995a) Dynamic, distributed simulation of watershed erosion: the KINEROS2 and EUROSEM models, *Journal of Soil and Water Conservation* **50** (5), 517–520.

Smith, R.E., Goodrich, D.C. and Woolhiser, D.A. (1995b) KINEROS, A KINematic runoff and EROSion model, in V.P. Singh (ed.) *Computer Models of Watershed Hydrology*, Water Resources Publication, Fort Collins, CO, 697–732.

Smith, R.E. and Woolhiser, D.A. (1971) Overland flow on an infiltrating surface, *Water Resources Research* **7**, 899–913.

Snell, J.D. and Sivapalan, M. (1994) On geomorphological dispersion in natural catchments and the geomorphological unit hydrograph, *Water Resources Research* **30**, 2311–2323.

Stewart, M.D., Bates, P.D., Anderson, M.G., Price, D.A. and Burt, T.P. (1999) Modelling floods in hydrologically complex lowland river reaches, *Journal of Hydrology* **223**, 85–106.

Thornes, J.B. (1990) Big rills have little rills..., *Nature* **345**, 764–765.

Tucker, G.E., Lancaster, S.T., Gasparini, N.M. and Bras, R.L. (2001) The channel-hillslope integrated landscape development (CHILD) model, in R.S. Harmon and W.W. Doe III (eds) *Landscape Erosion and Evolution Modelling*, Kluwer Academic/Plenum Publishers, Dordrecht, 349–388.

Veihe, A. and Quinton, J. (2000) Sensitivity analysis of EUROSEM using Monte Carlo simulation. I: hydrological, soil and vegetation parameters, *Hydrological Processes* **14**, 915–926.

Wainwright, J. (1996) Hillslope response to extreme storm events: the example of the Vaison-la-Romaine event, in M.G. Anderson and S.M. Brooks (eds) *Advances in Hillslope Processes*, John Wiley & Sons, Chichester, 997–1026.

Wallerstein, N.P., Alonso, C.V., Bennett, S.J. and Thorne, C.R. (2001) Distorted Froude-scaled flume analysis of large woody debris, *Earth Surface Processes and Landforms* **26**, 1265–1283.

Wang, Q. and Gan, T.Y. (1998) Biases of correlation dimension estimates of streamflow data in the Canadian prairies, *Water Resources Research* **34**, 2329–2339.

Wark, J.B. and James, C.S. (1994) An application of a new procedure for estimating discharges in meandering overbank flows to field data, in W.R. White and J. Watts (eds) *2nd International Conference on River Flood Hydraulics*, H.R. Wallingford Ltd and John Wiley & Sons, Chichester, 405–414.

Willgoose, G., Bras, R.L. and Rodríguez-Iturbe, I. (1991a) A coupled channel network growth and hillslope evolution model, 1: Theory, *Water Resources Research* **27**, 1671–1684.

Willgoose, G., Bras, R.L. and Rodríguez-Iturbe, I. (1991b) A coupled channel network growth and hillslope evolution model, 2: Nondimensionalization and applications, *Water Resources Research* **27**, 1685–1696.

Willgoose, G., Bras, R.L. and Rodríguez-Iturbe, I. (1991c) A physical explanation of an observed link area-slope relationship, *Water Resources Research* **27**, 1697–1702.

Willgoose, G., Bras, R.L. and Rodríguez-Iturbe, I. (1991d) Results from a new model of river basin evolution, *Earth Surface Processes and Landforms* **16**, 237–254.

Wolff, C.G. and Burges, S.J. (1994) An analysis of the influence of river channel properties on flood frequency, *Journal of Hydrology* **153** (1–4), 317–337.

Wooding, R.A. (1965a) A hydraulic model for the catchment-stream problem. I. Kinematic wave theory, *Journal of Hydrology* **3**, 254–267.

Wooding, R.A. (1965b) A hydraulic model for the catchment-stream problem. II. Numerical solutions, *Journal of Hydrology* **3**, 268–282.

Wooding, R.A. (1966) A hydraulic model for the catchment-stream problem. III. Comparisons with runoff observations, *Journal of Hydrology* **4**, 21–37.

Woolhiser, D.A., Smith, R.E. and Goodrich, D.C. (1990) *KINEROS: A Kinematic Runoff and Erosion Model: Documentation and User Manual*, USDA – Agricultural Research Service, ARS-77, Washington, DC.

Wormleaton, P.R. and Merrett, D.J. (1990) An improved method of calculation for steady uniform flow in prismatic main channel/flood plain sections, *Journal of Hydrological Research* **28** (2), 157–174.

Wormleaton, P.R., Allen, J. and Hadjipanos, P. (1982) Discharge assessment in compound flow, *Journal of the Hydraulics Division, ASCE* **182**, 975–994.

Yalin, M.S. (1971) *Theory of Hydraulic Models*, Macmillan, London.

Yen, C.L. and Overton, D.E. (1973) Shape effects on resistance in floodplain channels, *Journal of the Hydraulics Division, ASCE* **99**, 219–238.

6

Modelling the Ecology of Plants

COLIN P. OSBORNE

6.1 THE COMPLEXITY

Plant ecology encompasses an enormous range of spatial and temporal scales, covering several orders of magnitude. Its models span the microscopic scale of plankton populations (Krivtsov *et al.*, 1999) and the global dimensions of forest biogeography (Cox *et al.*, 2000), with timescales ranging from the geological, for leaf evolution in the earliest land plants, 400 million years ago (Beerling *et al.*, 2001), to the seconds needed for photosynthesis to respond to sunflecks (Gross *et al.*, 1991). Despite this diversity, all are united by a general purpose – to understand plant function, structure and dynamics in terms of the interactions among plants, animals, microbes, atmosphere and soil (Figure 6.1).

The interactions between plant function and the biotic and abiotic (Figure 6.1) environments are studied in plant physiological ecology, and commonly considered in terms of the flow of energy or biogeochemical (carbon, nitrogen, hydrological) cycles. Particular emphasis is placed on the physiological processes that exchange these essential resources with the environment. Carbon is acquired in the form of atmospheric or dissolved CO_2 by photosynthesis (Lawlor, 2000), and lost as a by-product of energy generation in respiration (Amthor, 2000). Nutrients such as nitrogen or phosphorus are acquired from the soil by the roots of land plants or absorbed directly from the surrounding water by algae, and lost by the shedding of leaves and ephemeral roots (Aerts and Chapin, 2000). Water is obtained by roots on land and lost by transpiration from leaves, and absorbed light in leaves provides the energy for chemical fixation of CO_2 by photosynthesis (Lawlor, 2000). Biotic interactions may be critical in mediating the acquisition and use of these resources. For example,

the chemical forms of soil nitrogen utilized by a plant depend critically on the type of association formed between roots and mycorrhizal fungi (Read, 1990).

The structure – i.e. architecture, shape and size – of a plant is intimately linked with its physiological function (Figure 6.1). For example, the number and size of leaves in a plant canopy are key factors in determining its total water consumption. Since excessive water use will lead to soil drought, plant wilting and leaf death, the size of a canopy is thus tightly coupled with the physiology of water transport and loss in transpiration (Woodward, 1987). Similar coupling between structure and function exists for stems/wood (Osborne and Beerling, 2002b) and root systems (Woodward and Osborne, 2000). These relationships between form and function offer the potential for generalizing across the enormous diversity of structure seen in the world's plant species. An accurate representation of this structure is vital in ecological models, because it is the feature of plants most readily measured by observation and used in model testing. For example, inferences from satellite remote sensing measurements allow assessment of canopy structure models on both large spatial and long temporal scales (Osborne and Woodward, 2001).

The dynamics of specific plant parts and individual plants within a community are key determinants of plant and vegetation structure. For example, the long-lived leaves of some conifers may develop into very deep and dense canopies compared with those of nearby deciduous hardwood trees that shed their leaves annually (Gower *et al.*, 1997). Plant structure is therefore governed by life history (Schulze, 1982), population dynamics (Smith and Shugart, 1996) and disturbance by climate, catastrophic events such as hurricanes and fires (Turner *et al.*, 1997), and grazing (Hester *et al.*,

Environmental Modelling: Finding Simplicity in Complexity. Edited by J. Wainwright and M. Mulligan
© 2004 John Wiley & Sons, Ltd   ISBNs: 0-471-49617-0 (HB); 0-471-49618-9 (PB)

Abiotic factors

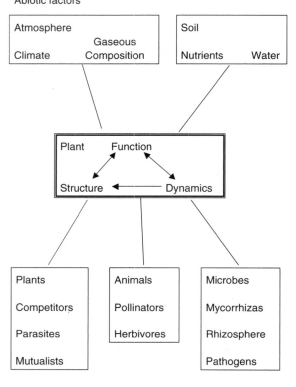

Biotic factors

Figure 6.1 Ecological interactions between an individual plant and other plants, animals, the atmosphere and soil. Abiotic (physical, or nonbiological) and biotic (biological) interactions are distinguished

2000), particularly at the scale of communities/stands of vegetation. Accounting for catastrophic events is problematic, given their massive disturbance effect but rarity and unpredictability.

Recent evidence also suggests a close link between dynamics and function in plants, principally in their leaves and roots. In order to remain intact and functional throughout their lifetime, long-lived leaves must invest heavily in a thick, tough structure and chemical defences to resist mechanical damage and herbivory (Coley, 1988). These modifications directly impair physiology (Niinemets, 1999) and have an indirect, energetic cost, leaving fewer plant resources for producing and maintaining the energetically expensive enzyme systems involved in photosynthesis and respiration (Reich *et al.*, 1992). These trade-offs lead to well-characterized, general correlations between increasing leaf lifespan,

rising mass per unit area, and decreasing rates of photosynthesis and respiration (Reich *et al.*, 1997, 1999). Similar reductions in physiological activity occur in individual leaves and ephemeral roots as they age, with important consequences for evolutionary adaptation to climate (Kikuzawa, 1991) and the soil environment (Eissenstat and Yanai, 1997).

The relationships between plant function, structure and dynamics are therefore complex and highly integrated. However, an emerging picture of this integration, and its interactions with climate, soils and other organisms, offers great promise for the ecological modelling of plants using general principles.

6.2 FINDING THE SIMPLICITY

6.2.1 A question of scale

Plant function forms the basis for models of plant physiological ecology, whereas vegetation structure and dynamics are captured by models of plant community and population ecology. A key difference between these approaches is the explicit consideration by community models of individual plants and their reproduction, in contrast to the tendency of physiological approaches to simulate homogeneous stands of vegetation. In fact, the two approaches must be married for reasonable simulations of plant ecology at large scales, with ecosystem structure and dynamics being governed principally by community dynamics, and plant function determined by physiological ecology (Beerling and Woodward, 2001; Smith *et al.*, 2001).

The modelling of plant ecology at large spatial scales (ecosystem to global) and over long time scales (decades to centuries) is an important area of current research, because plants are key regulators of energy, carbon and water exchange between the atmosphere, vegetation and soil. These functions have critical implications for climate and changes in atmospheric greenhouse gas concentrations (IPCC, 2001). However, research in ecology has traditionally focused on individual organisms, typically for periods of less than a decade, adopting a largely reductionist approach. Furthermore, experiments on the large-scale impacts of environmental change often cannot be conducted for moral and logistic reasons. The use of upscaling with mathematical models, the integration of information from small to larger scales, is therefore assuming increasing importance in research on the interactions between the biosphere and global change.

The processes of an ecological system and its structural organization both vary according to the scale at which they are observed. Accounting for the control of ecological processes by the appropriate negative and positive feedbacks is particularly important for upscaling, since they vary greatly in significance with scale. In the case of the structural organization of ecological systems, accounting for heterogeneity can be critical for accurate upscaling, especially if properties of the system show nonlinear responses to structural changes. For example, the clumping of leaves within a vegetation canopy, the clustering of individuals within a plant community and the patches of different vegetation types within a landscape all significantly modify the absorption of solar radiation by vegetation in comparison with uniform distributions of each (Asner *et al.*, 1998). This modification occurs because absorption with respect to leaf area is nonlinear. Good estimates of solar energy absorption are vital in models of vegetation function because this energy source is the driver of photosynthesis. Heterogeneity in the absorption of solar energy therefore has important consequences for both the upscaling of energy balance and photosynthesis, and for the downscaling of satellite remote sensing measurements. Scaling in models of plant ecology is discussed in greater depth by Jarvis (1995).

6.2.2 What are the alternatives for modelling plant function?

The remainder of this chapter focuses on the modelling of plant physiological ecology, however, many of the principles covered have a more general application. The reader is referred to Pacala (1997) for an introduction to the modelling of plant-community dynamics and to Silvertown and Lovett-Doust (1993) for information on the simulation of plant-population biology. Approaches to modelling plant physiological ecology may be grouped into three broad categories: mechanistic, teleonomic and empirical (Thornley and Johnson, 1990). Each differs in the extent to which it accounts for causal mechanisms, and is best viewed as a convenient way to describe the philosophy underpinning a particular model, rather than a rigid way of defining it.

The mechanistic approach is reductionist, seeking to build models based on plant processes occurring at lower spatial and temporal scales than the scale of interest (Loomis *et al.*, 1979). In the case of a whole plant or its organs (e.g. leaves and roots), this involves an explicit consideration of the physiological or biochemical pathways involved. A mechanistic model is more flexible in its usage than the other two approaches because, by considering the way in which an ecological system works, it permits limited extrapolation to new circumstances. In addition, by providing insight into the

operation of this system, it is a useful investigative tool for explaining ecological phenomena.

Many aspects of plant ecology cannot be described using mechanistic models because their operation is just too poorly understood. However, if the net outcome of a process is well described, a model can be devised to reproduce this end-point. Such a 'goal-directed', or teleonomic model empirically simulates the operation of a plant process and its relationship with external factors such as climate, without describing the underlying physiological or biochemical mechanisms, i.e. *how* this outcome is reached (Thornley and Johnson, 1990). For example, while the physiological mechanisms controlling the acquisition of carbon in photosynthesis are well characterized, those involved in the partitioning of this carbon between the growth of different plant organs remain poorly understood. Partitioning depends on the balance between carbon supply from photosynthesis, its demand in the growth of specific plant organs and its rate of transport through plant vascular systems (Farrar and Jones, 2000). However, models based on these principles remain largely theoretical and, while they are useful tools for interpreting and investigating plant physiology, remain impractical for most ecological applications (Minchin *et al.*, 1993; Bidel *et al.*, 2000).

By contrast, the relative proportions of plant carbon in each organ type (e.g. leaves, roots, stem) are broadly predictable within a particular environment and consistent between species (Friedlingstein *et al.*, 1999). These generalized relationships have long been utilized in teleonomic models of carbon partitioning, following a functional approach (Woodward and Osborne, 2000). Carbon is invested predominantly in organs that acquire the most limiting resource for plant growth so that, say, in a nutrient-limited ecosystem, root growth is favoured over stem and leaf growth. This leads to an increase in root:shoot ratio that is typical of nutrient-limited plants (Gleeson, 1993). A functional scheme of this type has recently met reasonable success in a global model for terrestrial vegetation, which simulates carbon partitioning relative to light, water and nitrogen availability (Friedlingstein *et al.*, 1999). Presumably this 'optimization' of plant resources mimics the evolutionary process at some level, because it represents the most efficient use of carbon, maximizing reserves for reproduction or the survival of extreme conditions, and thereby increasing fitness.

A purely empirical model takes no account of the processes underlying the aspect of plant ecology being simulated, although biological meaning may be attached to the mathematical formula used. For example, the zero intercept of a curve relating the abundance of a species to mean annual temperature (MAT) presumably reflects some critical threshold of plant cold tolerance. However, it is unlikely to be the MAT *per se* that limits abundance, but rather the minimum winter temperature, a feature of climate that correlates reasonably well with the mean. Although this results in a mathematical relationship between plant abundance and MAT, the two are not causally linked. In this case, a teleonomic model would recognize a linkage between minimum temperature and the tolerance or resistance of a plant to freezing, but would not account for the physiological mechanisms responsible (Woodward, 1987). In theory, a fully mechanistic model would simulate the freezing of plant tissues and the lethal effects of cellular desiccation and damage caused by ice crystals, but these processes are insufficiently understood for such an approach to be practical (ibid.).

As a descriptive tool for making broad generalizations, a purely empirical model may be both practical and useful, but gives no information beyond the data used in its construction (Thornley and Johnson, 1990). For example, the net primary productivity (NPP) or 'annual growth' of terrestrial biomes is broadly related to mean annual precipitation (MAP) and MAT (Lieth, 1975). These relationships can therefore be used to produce integrated estimates of today's global NPP, an approach that shows reasonable agreement with two independent assessments (Figure 6.2).

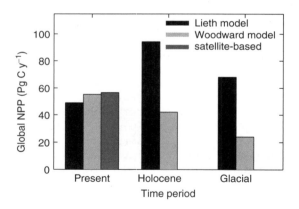

Figure 6.2 Estimates of global net primary productivity (NPP) for terrestrial vegetation during the present, mid-Holocene (6000 years before present, BP) and last Glacial (21 000 years BP) time periods. These were made using either an empirical model based on modern vegetation–climate relationships (Lieth, 1975), a mechanistic model of vegetation function (Beerling, 1999), or a method for inferring vegetation activity from satellite-based reflectance measurements (Field *et al.*, 1998)

However, without the explanatory detail of a mechanistic approach, an empirical model is unlikely to be reliable as a predictive or explanatory tool when conditions are encountered beyond those used to fit model relationships. The Lieth method suggests high values of global NPP during the last ice age and the subsequent warm mid-Holocene period compared with the present (see Figure 6.2). These changes would have been driven by the warmer or wetter past climate. By contrast, a sophisticated mechanistic model (Woodward *et al.*, 1995) estimates significantly reduced NPP during both the mid-Holocene and last ice age compared with the present (see Figure 6.2; Beerling, 1999). The cause of this marked qualitative difference is a failure of the empirical model to account for atmospheric CO_2 concentration, which was 45% (glacial) and 27% (Holocene) lower than at present. Experiments demonstrate a severe limitation of growth by these concentrations (Polley *et al.*, 1993), suggesting that estimates using the mechanistic model are more likely, since they account for the interacting influences of CO_2, temperature and precipitation (Woodward *et al.*, 1995). For most ecological purposes, some consideration of mechanisms is therefore advantageous, and the remainder of this section discusses the practical application of this approach.

6.2.3 Applying a mechanistic approach

The problems, practicalities and compromises involved in applying a mechanistic approach are well illustrated by reference to the example of leaf-nitrogen content. The concentration of nitrogen within plant foliage is a key component of many process-based models of plant ecology, because it has critical physiological implications at the leaf scale, but is closely coupled with processes at whole-plant and ecosystem scales (Osborne and Beerling, 2002a). At the leaf scale, nitrogen concentration is tightly linked with physiological function, because the proteins required for photosynthesis are rich in the element (Evans, 1989), and tissues high in protein are metabolically expensive to maintain (Penning de Vries, 1975). Nitrogen also has important effects on the consumption of leaves by herbivores, although the mechanisms at work remain unclear (Hartley and Jones, 1997). The availability of nitrogen is a primary limitation on plant growth in natural ecosystems because soil reservoirs of mineral nitrogen are highly mobile, and regulated by the biological mineralization of organic matter (Vitousek and Howarth, 1991). Foliar nitrogen concentration is therefore closely related to the limitation of growth rate at the whole-plant scale, and the leaves of plants growing in nitrogen-poor soils tend to have a lower nitrogen content than those of well-nourished plants (Gutschick, 1993). At the ecosystem scale, differences in nitrogen concentration between species reflect contrasting ecological strategies with, for example, evergreen plants on infertile or dry soils tending to have lower concentrations than their deciduous counterparts on more fertile or wetter sites (Aerts, 1995). The nitrogen content of a leaf has further implications after its abscission, being an important correlate of leaf litter decomposition rate (Aerts, 1997), and hence nitrogen cycling within the soil. An accurate estimate of leaf-nitrogen concentration is therefore vital for reasonable simulations of a whole suite of physiological, plant and ecosystem processes.

Model estimates of leaf-nitrogen concentration generally follow one of two approaches. Both are widely used in plant ecological modelling, with the aim of simplifying the vast range of plant form and function into a manageable and well-characterized scheme, while retaining a realistic representation of biological diversity (Woodward and Beerling, 1997):

1. A division of plants into 'functional types' or 'functional groups', using shared physiological and ecological responses to the biotic or abiotic environment (Smith *et al.*, 1997). Classification schemes are designed according to functional responses that are of particular interest and, for this reason, may differ significantly between models, even for the same biome (e.g. Pausas, 1999; Osborne *et al.*, 2000). Support for classifications based on climate comes from the observation that plants of taxonomically distant species growing in similar, but geographically separated, climates tend to evolve comparable suites of physical characteristics (Wolfe, 1985). This convergence suggests that evolution in each climate regime is constrained by the same basic limitations of plant physiological ecology. In the case of leaf-nitrogen concentration, functional groups or individual species may each be assigned a value based on direct measurements.

2. The definition of general rules that can be safely applied to a wide range of plants. In seeking this goal, a common approach is to attempt simulations of plant structure, function and dynamics in terms of the underlying, causal mechanisms (e.g. Woodward *et al.*, 1995; Osborne *et al.*, 2000). This method is powerful because, at a physiological or biochemical level, there may be a high degree of similarity between plants from disparate groups, suggesting minimal evolutionary radiation in some of these basic

properties (e.g. Larcher, 1994; Long *et al.*, 1996). For example, leaf-nitrogen concentration is correlated strongly with the rate of plant-nitrogen uptake, and the latter can be simulated using soil properties and temperature using a basic understanding of root associations with mycorrhizal fungi (Woodward *et al.*, 1995).

The mechanistic model of photosynthesis developed by Farquhar and co-workers (Farquhar *et al.*, 1980; von Caemmerer, 2000) is proven and widely used, and illustrates further some of the advantages and problems in adopting a process-based approach. Photosynthesis fixes gaseous CO_2 to form sugars, which are used to fuel plant growth and ultimately feed most organisms in every ecosystem. It is a complex, highly integrated physiological process, involving numerous interacting biochemical pathways (Baker, 1996). Despite this complexity, it can be simulated accurately by representing regulation in only a few key steps (Farquhar *et al.*, 1980). Scaling up from biochemical reactions to photosynthesis by the whole leaf turns out to be straightforward because the physiology of cells adjusts to the profile of absorbed light in the leaf (Farquhar, 1989). Further scaling of the model to whole plant canopies is more difficult, but possible given appropriate estimates of light absorption and physiological properties (de Pury and Farquhar, 1997). The Farquhar model has therefore proved useful at cellular, leaf and whole plant scales and, in the 20 years since its first publication, has been widely adopted by the plant ecology modelling community (Farquhar *et al.*, 2001). An important reason for its appeal is the highly conserved nature of many simulated processes, allowing the same equations and constant parameters to be used for a wide range of species – a useful property in ecological modelling. For example, catalysis by the primary enzyme responsible for fixing CO_2 into sugars is similar in diverse plant species (Parry *et al.*, 1989; Delgado *et al.*, 1995).

Farquhar's model has been widely applied at global and regional scales, where it forms the basis for several important models of the interactions between terrestrial biosphere and climate (Sellers *et al.*, 1996; Betts *et al.*, 1997; Cox *et al.*, 2000). The model performs adequately in the context of these applications, operating with a temporal resolution of 30–60 minutes. By contrast, during very short periods in a rapidly changing light regime as experienced on a forest floor, the dynamics of photosynthesis become critical, and the basic model requires refinement (Cheeseman and Lexa, 1996). For better reliability in these conditions, it must be modified using empirical functions and parameters with no direct biological significance (Gross *et al.*, 1991; Naumberg *et al.*, 2001), undermining its conceptual value as a mechanistic model.

To address the problem of dynamic photosynthesis, Laisk and co-workers have developed a significantly more complex model by considering the individual biochemical reactions concerned (Laisk *et al.*, 1997; Laisk and Oja, 1998). The Laisk model has proved a valuable investigative tool for photosynthesis research (Laisk *et al.*, 1997), but is of limited utility for whole-plant simulations because its high level of complexity means that many photosynthetic reactions must be excluded because they are poorly understood and not quantified experimentally (Cheeseman and Lexa, 1996). Mechanistic models therefore involve a trade-off between the more realistic simulation of a process that results from greater attention to detail, and the concurrent increase in limitations imposed by experimental evidence. In the case of dynamic photosynthesis, a modified version of the Farquhar model provides a practical solution (Gross *et al.*, 1991; Naumberg *et al.*, 2001). In summary therefore, the development of a mechanistic model in plant ecology always involves a compromise between complexity, reliability and practicality.

6.3 THE RESEARCH FRONTIER

One of the most important challenges currently facing modellers of plant ecology is the elucidation of generic relationships that can be applied universally throughout the plant kingdom. These must represent the physical, chemical or evolutionary constraints that operate on all plants, irrespective of their life form or phylogeny. Tomorrow's models are likely to incorporate more fully integrated generic relationships between plant structure, function and dynamics. These are already well quantified in empirical terms for leaves (Reich *et al.*, 1999), and on-going research looks set to clarify their evolutionary (Westoby *et al.*, 2000) and mechanistic basis (Niinemets, 1999, 2001). Leaf lifespan is closely related to physiological function and chemical composition in plants from vastly differing ecosystems, from tropical wet forest to Arctic tundra, and in contrasting phylogenetic groups (Reich *et al.*, 1992, 1997, 1999). Similar work linking the structure and function of roots with their dynamics is still in its infancy, but recent experimental and model results show the potential for a generic approach in modelling belowground processes (Eissenstat and Yanai, 1997; Pregitzer, 2002). Generic relationships offer great potential for scaling from individual plant organs to patch and regional models of plant ecology (Osborne and Beerling, 2002a).

Regional model simulations of the distribution and strength of terrestrial sinks for atmospheric CO_2 are becoming increasingly important for setting appropriate controls on greenhouse gas emissions (Amthor and Jacobs, 2000). This political significance has driven the parallel development of global vegetation models by research groups throughout the world (Cramer *et al.*, 2001; McGuire *et al.*, 2001). Each is underpinned by a different philosophy and methodology, meaning that simulations may differ significantly between models. However, policy-makers require definitive answers, and the testing of model accuracy by comparison with 'real-world' observations has therefore become an important area of activity. Testing models at appropriate regional scales is not an easy exercise, but the increasing usage of satellite observations of vegetation (Hicke *et al.*, 2002), comparisons with sensitive measurements of the seasonal and latitudinal changes in atmospheric CO_2 levels (Nemry *et al.*, 1999), and tower measurements of ecosystem CO_2 exchange (Baldocchi *et al.*, 2001) are helping greatly in this effort.

Current large-scale models of ecosystem function simplify biological diversity by grouping species on a functional basis, so that a simulated ecosystem typically comprises a small, but manageable, number of functional types. However, new work suggests that such simplification may cause errors, because species diversity *per se* may be an important determinant of ecosystem function. Experiments using artificial mixtures of species to manipulate biodiversity report correlations between species number and primary productivity (Hector *et al.*, 1999), and similar relationships with nutrient-use efficiency and stability. These findings are attributed to positive species interactions and niche complementarity, the fulfilment of different ecological functions and occupation of alternative ecosystem spaces by different species (ibid.). However, they remain controversial because of problems with the nonrandom selection of species and the 'sampling effect' in randomly selected species groups (Huston, 1997). The latter is the increasing probability in high biodiversity treatments of selecting species with extreme ecological properties such as high seed germination rate or stress tolerance (Huston, 1997). Further work is clearly needed to resolve these issues, but biological diversity may yet prove to be an important factor in future models of ecosystem function.

6.4 CASE STUDY

Mediterranean ecosystems are key hotspots of global biodiversity and play an important economic role in pastoral and agricultural systems (Blondel and Aronson,

1999). Continuing changes in the Mediterranean climate therefore have potentially huge consequences for human societies in the region (Lavorel *et al.*, 1998). Increases in atmospheric CO_2 and other greenhouse gases since the Industrial Revolution have caused regional warming of around 0.5°C and, more recently, a localized decline in precipitation (Kutiel *et al.*, 1996; Esteban-Parra *et al.*, 1998; Jones *et al.*, 1999). These atmospheric and climatic trends are set to continue well into the twenty-first century (Palutikof *et al.*, 1994), and have sharpened concern for the future welfare of Mediterranean ecosystems (Lavorel *et al.*, 1998). The recent development of models designed to investigate the ecology of vegetation in the region (Pausas, 1999; Osborne *et al.*, 2000) is an important first step in assessing the scale of this problem.

Osborne *et al.* (2000) developed a patch-scale model of Mediterranean vegetation to examine climatic responses of plants in the region, and evaluate the potential mediating role of rising atmospheric CO_2. This vegetation model was initially a component of a landscape model, used in assessing land degradation (Kirkby *et al.*, 1998), but was subsequently expanded for independent applications (Osborne *et al.*, 2000; Woodward and Osborne, 2000; Osborne and Woodward, 2001). Within the model, plant diversity was managed by grouping species into functional types based on their physiological and life history responses to drought. Some of these are illustrated in Figure 6.3. Seasonal drought is a characteristic feature of the Mediterranean climate that has exerted a critical control on the evolution of plant ecological strategies (Dallman,

(a)

(b)

(c)

Figure 6.3 Examples of functional types used in modelling and some key distinctions between them: (a) Evergreen sclerophyllous shrub (*Quercus coccifera* in Murcia, Spain) – retains drought-tolerant foliage during the summer, and has an extensive rooting system which maximizes access to soil water during this dry period; (b) Drought-deciduous shrub (*Anthyllis cytisoides* in Almeria, Spain) – sheds foliage at the beginning of summer, avoiding drought by remaining dormant until the autumn, when new shoot growth occurs; (c) Perennial tussock grass (*Stipa tenacissima* in Almeria, Spain) – has a much more limited capacity for the storage of energy reserves than shrubs, and a shallow, fibrous root system, but is extremely drought-tolerant

1998). Each model functional type was differentiated by a key set of attributes, principal distinctions being in leaf physiology, e.g. the sensitivity of stomata to soil moisture; phenology, i.e. timing of events in the life cycle relative to the seasons; morphology, e.g. rooting depth; and carbon allocation patterns. In other respects the functional types were identical, with model processes being based on the typical behaviour of terrestrial plants. So, for example, many canopy properties were related to leaf longevity (Reich *et al.*, 1997), and the energy and water balances of vegetation and soil were derived from physical principles (Shuttleworth and Gurney, 1989). Full details of the model may be found at http://www.shef.ac.uk/uni/academic/A-C/aps/medveg.pdf

A mechanistic approach was used where possible in this model, with particular emphasis being placed on plant interactions with climate. These included both direct canopy relationships with light, air temperature, humidity and CO_2 concentration, and indirect, drought-related interactions with soil-water availability (Figure 6.4). The level of detail and processes included were appropriate to the scale of large patches (Figure 6.5; $10^2-10^4\,m^2$), in this case defined as even-aged plant stands of roughly homogeneous species composition, i.e. areas with a uniform history of large-scale disturbance like fire (Gardner *et al.*, 1996). A mechanistic basis for these relationships allowed the model to be applied with confidence across a large geographical

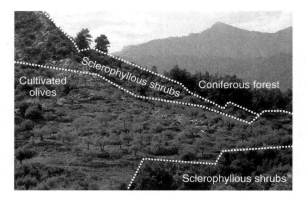

Figure 6.5 A Mediterranean landscape covered by patches of different vegetation types: cultivated olives, coniferous forest and evergreen sclerophyllous shrubland

region and long time sequence, after testing for a limited number of sites and shorter time period.

Model simulations of vegetation structure and function were tested in two key respects: first, their ability to capture the spatial variation throughout the Mediterranean region; second, their capacity to accurately represent both seasonal variation and decadal trends more typical of long-term climatic change. Since spatial variation within the Mediterranean is caused by climatic differences and phylogenetic distinctions between distant parts of the region (Dallman, 1998), the former test assessed both model responses to climate and the appropriateness of functional types in grouping species.

The model was initially applied to the sclerophyllous shrub functional type (Figure 6.3), including species such as *Quercus coccifera* (Kermes Oak), *Pistacia lentiscus* (Mastic Tree) and *Arbutus unedo* (Strawberry Tree). This group of species retains a leaf canopy and remains physiological active throughout the summer drought period, via conservative water use and continuing access to deep groundwater facilitated by an extensive root system (Archibold, 1995). Its leaves are long-lived, being toughened to resist mechanical damage (Turner, 1994). Model simulations for this vegetation type reproduced spatial variation in above-ground productivity and canopy size/density (Osborne *et al.*, 2000), key features of its function and structure, respectively. Seasonal variation in modelled physiological function also compared favourably with direct observations (ibid.).

Direct measurements of Mediterranean vegetation structure and function tend to be labour-intensive, and have not been made repeatedly at the same site over long time scales, however, each may be approximated

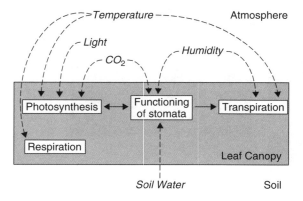

Figure 6.4 The key model relationships between atmosphere, soil and climate, with arrows indicating interactions and their direction. Photosynthesis takes up, and respiration releases, CO_2, together governing net canopy rates of CO_2 exchange. Stomata are tiny pores in the surface of leaves, whose aperture is regulated by changes in cell shape. Their functioning may limit CO_2 exchange and the loss of water from leaves by transpiration

using an indirect technique. First, canopy size and light absorption may be assessed with satellite remote sensing. Since appropriate satellite records are available from 1981, this allows canopy properties to be studied on a decadal time scale (Myneni *et al.*, 1997, 1998). Second, leaf physiological function can be investigated using its stable carbon isotope content. Herbarium specimens from the same geographical area, dating from the nineteenth and twentieth centuries, permit centennial-scale reconstructions of leaf physiological function using this technique (Peñuelas and Azcón-Bieto, 1992; Beerling and Woodward, 1995). Since reconstructed climatologies for the Mediterranean region now stretch from 1901 to the present day (New *et al.*, 2000), model simulations may be compared with these findings. The model's mechanistic basis means that the properties of leaf canopies measured by satellite can be explicitly simulated from first principles, allowing direct comparison between model and satellite data (Figure 6.6). Similarly, mechanistic modelling of leaf physiology allows stable carbon-isotope discrimination by key leaf processes to be simulated, and direct comparison between model and herbarium leaf isotope compositions (Figure 6.7). Both comparisons have revealed good agreement between the magnitude of, and trends in, observed and modelled values (Figures 6.6 and 6.7), supporting long-term simulations of canopy structure and function with twentieth-century global change. Because the model has a mechanistic basis, these trends can be interpreted in terms of the underlying control of vegetation properties by climate and atmospheric CO_2.

Model simulations for Mediterranean sclerophyllous shrubland have produced two key findings. First, that water availability, rather than temperature, exerts the dominant control on both canopy size and productivity in this vegetation type (Osborne and Woodward, 2001). This result provides the first confirmation and quantification of this control on a decadal time scale. Second, it suggests that rising concentrations of atmospheric CO_2 have significantly altered plant water-use, increasing carbon fixation in relation to water loss, and offsetting the negative impacts on plant growth of all but the harshest droughts (Osborne *et al.*, 2000). Under these severe droughts, CO_2 had no effect on simulated growth. Modelled effects occurred during the past 50 years, in response to a modest increase in CO_2 from 300 to 365 ppm, and demonstrate the extreme sensitivity of water-limited Mediterranean shrubland to concentrations of the gas.

Interestingly, since these model results were first published, similar interactions between CO_2 and drought have been observed experimentally for the desert shrub

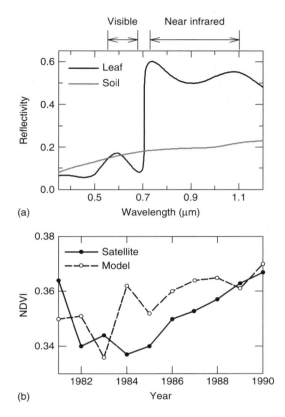

Figure 6.6 Model testing using a time series of satellite measurements. (a) Reflectivity by a typical leaf and soil at a range of wavelengths. By measuring radiation reflected in the visible and near infrared wavebands from the land surface, a satellite may detect the density of leaves covering a particular area. Soil and leaves reflect visible radiation approximately equally, but differ strongly in their reflectance of near infrared. The difference between measurements of these wavebands therefore indicates green plant cover, and is standardised to produce the normalized-difference vegetation index (NDVI). (b) Comparison of NDVI measured by US weather satellites over the Mediterranean region, and NDVI calculated using simulated canopy properties in the vegetation model (redrawn from Woodward & Osborne, 2000)
Source: Reproduced by permission of Blackwell Publishing

Larrea tridentata in Nevada. In a year when annual rainfall was typical of a moderate Mediterranean drought but wet for the desert site, shoot growth doubled with an increase in CO_2 from 365 to 550 ppm (Smith *et al.*, 2000). By contrast, growth was completely unresponsive to CO_2 under severe drought conditions, which occur only rarely in the Mediterranean (Osborne *et al.*,

are technically and ethically difficult with experiments, or simply impossible. Comparison of simulations by these models with large-scale field observations of vegetation provides both confidence in their results and insight into the drivers and mechanisms of change. Ongoing developments in vegetation modelling focus on the inter-relationships between plant structure, function and dynamics, and aim to produce widely applicable, generic models. Until such generic relationships are developed, the modelling of plant ecology using functional groups of species provides a practical way of simplifying biological complexity, while retaining a realistic description of functional diversity.

ACKNOWLEDGEMENTS

I thank David Beerling for helpful suggestions, discussion and financial support during the preparation of this manuscript, Ian Woodward for his comments, Josep Peñuelas for access to stable carbon-isotope data, and Glyn Woods for help in preparing the illustrations. Funding from the European Community Environment Programme, Mediterranean Desertification and Land Use (MEDALUS) Project supported the modelling work on Mediterranean vegetation.

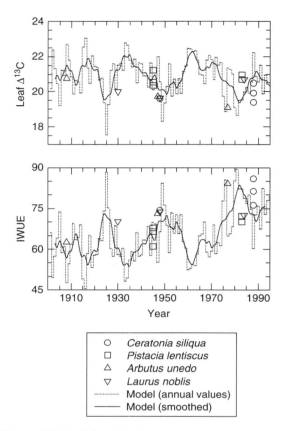

Figure 6.7 Leaf discrimination against the heavy isotope of carbon ($\Delta^{13}C$, top), and instantaneous leaf water-use efficiency (IWUE, bottom). A comparison is made between values derived from herbarium specimens of four sclerophyllous shrub species, collected at intervals during the twentieth century, and values calculated using the vegetation model. Stable carbon-isotope composition reflects the efficiency of water-use during leaf growth, following a well-characterized relationship (Farquhar *et al.*, 1989) *Source*: Reproduced from Osborne *et al.* (2000) by permission of Blackwell Science Ltd

2000), but were typical of the desert environment (Smith *et al.*, 2000). Thus experimental observations replicated modelled growth characteristics, providing further independent support for these simulations.

6.5 CONCLUSION

Mechanistic models provide a useful means of integrating our current knowledge of plant ecology, allowing investigations at large spatial and temporal scales that

REFERENCES

Aerts, R. (1995) The advantages of being evergreen, *Trends in Ecology and Evolution* **10**, 502–507.

Aerts, R. (1997) Nitrogen partitioning between resorption and decomposition pathways: a trade-off between nitrogen-use efficiency and litter decomposibility?, *Oikos* **80**, 603–606.

Aerts, R. and Chapin, F.S. (2000) The mineral nutrition of wild plants revisited: a re-evaluation of processes and patterns, *Advances in Ecological Research* **30**, 1–67.

Amthor, J.S. (2000) The McCree-de Wit-Penning de Vries-Thornley respiration paradigms: 30 years later, *Annals of Botany* **86**, 1–20.

Amthor, J.S. and Jacobs, G.K. (2000) Carbon sequestration in terrestrial ecosystems: goals, processes, and limits, *Eos, Transactions of the American Geophysical Union* **81** (suppl.), F245.

Archibold, O.W. (1995) *Ecology of World Vegetation*, Chapman & Hall, London.

Asner, G.P., Wessman, C.A. and Archer, S. (1998) Scale dependence of absorption of photosynthetically active radiation in terrestrial ecosystems, *Ecological Applications* **8**, 1003–1021.

Baker, N.R. (ed.) (1996) *Photosynthesis and the Environment*, Kluwer Academic Publishers, Dordrecht.

Baldocchi, D., Falge, E., Gu, L., Olson, R., Hollinger, D., Running, S., Anthoni, P., Bernhofer, C., Davis, K., Evans, R., Fuentes, J., Goldstein, A., Katul, G., Law, B., Lee, X.,

Malhi, Y., Meyers, T., Munger, W., Oechel, W., Paw, K.T., Pilegaard, K., Schmid, H.P., Valentini, R., Verma, S., Vesala, T., Wilson, K. and Wofsy, S. (2001) FLUXNET: a new tool to study the temporal and spatial variability of ecosystem-scale carbon dioxide, water vapor, and energy flux densities, *Bulletin of the American Meteorological Society* **82**, 2415–2434.

Beerling, D.J. (1999) New estimates of carbon transfer to terrestrial ecosystems between the last glacial maximum and the Holocene, *Terra Nova* **11**, 162–167.

Beerling, D.J., Osborne, C.P. and Chaloner, W.G. (2001) Evolution of leaf form in land plants linked to atmospheric CO_2 decline in the Late Palaeozoic era, *Nature* **410**, 352–354.

Beerling, D.J. and Woodward, F.I. (1995) Leaf stable carbon isotope composition records increased water-use efficiency of C_3 plants in response to atmospheric CO_2 enrichment, *Functional Ecology* **9**, 394–401.

Beerling, D.J. and Woodward, F.I. (2001) *Vegetation and the Terrestrial Carbon Cycle*, Cambridge University Press, Cambridge.

Betts, R.A., Cox, P.M., Lee, S.E. and Woodward, F.I. (1997) Contrasting physiological and structural vegetation feedbacks in climate change simulations, *Nature* **387**, 796–799.

Bidel, L.P.R., Pages, L., Riviere, L.M., Pelloux, G. and Lorendeau, J.Y. (2000) MassFlowDyn I: a carbon transport and partitioning model for root system architecture, *Annals of Botany* **85**, 869–886.

Blondel, J. and Aronson, J. (1999) *Biology and Wildlife of the Mediterranean Region*, Oxford University Press, Oxford.

Cheeseman, J.M. and Lexa, M. (1996) Gas exchange: models and measurements, in N.R. Baker (ed.) *Photosynthesis and the Environment*, Kluwer Academic Publishers, Dordrecht, 223–240.

Coley, P.D. (1988) Effects of plant growth rate and leaf lifetime on the amount and type of anti-herbivore defence, *Oecologia* **74**, 531–536.

Cox, P.M., Betts, R.A., Jones, C.D., Spall, S.A., Totterdell, I.J. (2000) Acceleration of global warming due to carbon-cycle feedbacks in coupled climate model, *Nature* **408**, 184–187.

Cramer, W., Bondeau, A., Woodward, F.I., Prentice, I.C., Betts, R.A., Brovkin, V., Cox, P.M., Fisher, V., Foley, J.A., Friend, A.D., Kucharik, C., Lomas, M.R., Ramankutty, N., Sitch, S., Smith, B., White, A. and Young-Molling, C. (2001) Global response of terrestrial ecosystem structure and function to CO_2 and climate change: results from six dynamic global vegetation models *Global Change Biology* **7**, 357–373.

Dallman, P.R. (1998) *Plant Life in the World's Mediterranean Climates*, Oxford University Press, Oxford.

Delgado, E., Medrano, H., Keys, A.J. and Parry, M.A.J. (1995) Species variation in Rubisco specificity factor, *Journal of Experimental Botany* **46**, 1775–1777.

de Pury, D.G.G. and Farquhar, G.D. (1997) Simple scaling of photosynthesis from leaves to canopies without the errors of big-leaf models, *Plant Cell and Environment* **20**, 537–557.

Eissenstat, D.M. and Yanai, R.D. (1997) The ecology of root lifespan, *Advances in Ecological Research* **27**, 1–60.

Esteban-Parra, M.J., Rodrigo, F.S. and Castro-Diez, Y. (1998) Spatial and temporal patterns of precipitation in Spain for the period 1880–1992, *International Journal of Climatology* **18**, 1557–1574.

Evans, J.R. (1989) Photosynthesis and nitrogen relationships in leaves of C_3 plants, *Oecologia* **78**, 9–19.

Farquhar, G.D. (1989) Models of integrated photosynthesis of cells and leaves, *Philosophical Transactions of the Royal Society of London, Series B* **323**, 357–367.

Farquhar, G.D., von Caemmerer, S. and Berry, J.A. (1980) A biochemical model of photosynthetic CO_2 assimilation in leaves of C_3 species, *Planta* **149**, 78–90.

Farquhar, G.D., von Caemmerer, S. and Berry, J.A. (2001) Models of photosynthesis, *Plant Physiology* **125**, 42–45.

Farrar, J.F. and Jones, D.L. (2000) The control of carbon acquisition by roots, *New Phytologist* **147**, 43–53.

Field, C.B., Behrenfield, M.J., Randerson, J.T. and Falkowski, P. (1998) Primary production of the biosphere: integrating terrestrial and oceanic components, *Science* **281**, 237–240.

Friedlingstein, P., Joel, G., Field, C.B. and Fung, I.Y. (1999) Toward an allocation scheme for global terrestrial carbon models, *Global Change Biology* **5**, 755–770.

Gardner, R.H., Hargrove, W.W., Turner, M.G. and Romme, W.H. (1996) Climate change, disturbances and landscape dynamics, in B. Walker and W. Steffan (eds) *Global Change and Terrestrial Ecosystems*, Cambridge University Press, Cambridge, 149–172.

Gleeson, S.K. (1993) Optimization of tissue nitrogen and root-shoot allocation, *Annals of Botany* **71**, 23–31.

Gower, S.T., Vogel, J.G., Norman, J.M., Kucharik, C.J., Steele, S.J. and Stow, T.K. (1997) Carbon distribution and above-ground net primary production in aspen, jack pine, and black spruce stands in Saskatchewan and Manitoba, Canada, *Journal of Geophysical Research* **102**, D24, 29029–29041.

Gross, L.J., Kirschbaum, M.U.F. and Pearcy, R.W. (1991) A dynamic model of photosynthesis in varying light taking account of stomatal conductance, C3-cycle intermediates, photorespiration and Rubisco activation, *Plant, Cell and Environment* **14**, 881–893.

Gutschick, V.P. (1993) Nutrient-limited growth rates: roles of nutrient-use efficiency and of adaptations to increase uptake rate, *Journal of Experimental Botany* **44**, 41–51.

Hartley, S.E. and Jones, C.G. (1997) Plant chemistry and herbivory, or why the world is green, in M.J. Crawley (ed.) *Plant Ecology*, Blackwell Science, Oxford, 284–324.

Hector, A., Schmid, B., Beierkuhnlein, C., Caldeira, M.C., Diemer, M., Dimitrakopoulos, P.G., Finn, J.A., Freitas, H., Giller, P.S., Good, J., Harris, R., Hogberg, P., Huss-Danell, K., Joshi, J., Jumpponen, A., Korner, C., Leadley, P.W., Loreau, M., Minns, A., Mulder, C.P.H., O'Donovan, G., Otway, S.J., Pereira, J.S., Prinz, A., Read, D.J., Scherer-Lorenzen, M., Schulze, E.D., Siamantziouras, A.S.D., Spehn, E.M., Terry, A.C., Troumbis, A.Y., Woodward, F.I., Yachi, S. and Lawton, J.H. (1999) Plant diversity and productivity experiments in European grasslands, *Science* **286**, 1123–1127.

Hester, A.J., Edenius, L., Buttenschon, R.M. and Kuiters, A.T. (2000) Interactions between forests and herbivores: the role of controlled grazing experiments, *Forestry* **73**, 381–391.

Hicke, J.A., Asner, G.P., Randerson, J.T., Tucker, C., Los, S., Birdsey, R., Jenkins, J.C., Field, C. and Holland, E. (2002) Satellite-derived increases in net primary productivity across North America, 1982–1998, *Geophysical Research Letters* **29**, 10.1029/2001GL013578.

Huston, M.A. (1997) Hidden treatments in ecological experiments: re-evaluating the ecosystem function of biodiversity, *Oecologia* **110**, 449–460.

IPCC (2001) *Climate Change 2001: The Scientific Basis. Contribution of Working Group I to the Third Assessment Report of the Intergovernmental Panel on Climate Change*, J.T. Houghton, Y. Ding, D.J. Griggs, M. Noguer, P.J. van der Linden, X. Dai, K. Maskell and C.A. Johnson (eds) Cambridge University Press, Cambridge.

Jarvis, P.G. (1995) Scaling processes and problems, *Plant, Cell and Environment* **18**, 1079–1089.

Jones, P.D., New, M., Parker, D.E., Martin, S. and Rigor, I.G. (1999) Surface air temperature and its changes over the past 150 years, *Reviews of Geophysics* **37**, 173–199.

Kikuzawa, K. (1991) The basis for variation in leaf longevity of plants, *Vegetatio* **121**, 89–100.

Kirkby, M.J., Abrahart, R., McMahon, M.D., Shao, J. and Thornes, J.B. (1998) MEDALUS soil erosion models for global change, *Geomorphology* **24**, 35–49.

Krivtsov, V., Bellinger, E.G. and Sigee, D.C. (1999) Modelling of elemental associations in *Anabaena*, *Hydrobiologia* **414**, 77–83.

Kutiel, H., Maheras, P. and Guika, S. (1996) Circulation and extreme rainfall conditions in the Eastern Mediterranean during the last century, *International Journal of Climatology* **16**, 73–92.

Laisk, A. and Oja, V. (1998) *Dynamics of Leaf Photosynthesis: Rapid-Response Measurements and their Interpretations*, CSIRO Publishing, Melbourne.

Laisk, A., Oja, V., Rasulov, B., Eichelmann, H. and Sumberg, A. (1997) Quantum yields and rate constants of photochemical and nonphotochemical excitation quenching: experiment and model, *Plant Physiology* **115**, 803–815.

Larcher, W. (1994) Photosynthesis as a tool for indicating temperature stress events, in E.D. Schultze and M.M. Caldwell (eds) *Ecophysiology of Photosynthesis*, Springer-Verlag, Berlin, 261–277.

Lavorel, S., Canadell, J., Rambal, S. and Terradas, J. (1998) Mediterranean ecosystems: research priorities on global change effects, *Global Ecology and Biogeography Letters* **7**, 157–166.

Lawlor, D.J. (2000) *Photosynthesis*, Bios Scientific Publishers, Oxford.

Lieth, H. (1975) Modeling the primary productivity of the world, in H. Lieth and R. Whittaker (eds) *Primary Productivity of the Biosphere*, Springer-Verlag, New York.

Long, S.P., Postl, W.F. and Bolhár-Nordenkampf, H.R. (1996) Quantum yields for uptake of carbon dioxide in C_3 vascular plants of contrasting habitats and taxonomic groupings, *Planta* **189**, 226–234.

Loomis, R.S., Rabbinge, R. and Ng, E. (1979) Explanatory models in crop physiology, *Annual Review of Plant Physiology* **30**, 339–367.

McGuire, A.D., Sitch, S., Clein, J.S., Dargaville, R., Esser, G., Foley, J., Heimann, M., Joos, F., Kaplan, J., Kicklighter, D.W., Meier, R.A., Melillo, J.M., Moore, B., Prentice, I.C., Ramankutty, N., Reichenau, T., Schloss, A., Tian, H., Williams, L.J. and Wittenberg, U. (2001) Carbon balance of the terrestrial biosphere in the twentieth century: analyses of CO_2, climate and land use effects with four process-based ecosystem models, *Global Biogeochemical Cycles* **15**, 183–206.

Minchin, P.E.H., Thorpe, M.R. and Farrar, J.F. (1993) A simple mechanistic model of phloem transport which explains sink priority, *Journal of Experimental Botany* **44**, 947–955.

Myneni, R.B., Keeling, C.D., Tucker, C.J., Asrar, G. and Nemani, R.R. (1997) Increased plant growth in the northern high latitudes from 1981 to 1991, *Nature* **386**, 698–702.

Myneni, R.B., Tucker, C.J., Asrar, G. and Keeling, C.D. (1998) Interannual variations in satellite-sensed vegetation index data from 1981–1991, *Journal of Geophysical Research* **103**, 6145–6160.

Naumberg, E., Ellsworth, D.S. and Katul, G.G. (2001) Modeling dynamic understory photosynthesis of contrasting species in ambient and elevated carbon dioxide, *Oecologia* **126**, 487–499.

Nemry, B., François, L., Gérard, J.C., Bondeau, A., Heimann, M. and the participants of the Potsdam NPP model intercomparison (1999) Comparing global models of terrestrial net primary productivity (NPP): analysis of the seasonal atmospheric CO_2 signal, *Global Change Biology* **5** (suppl. 1), 65–76.

New, M., Hulme, M. and Jones, P. (2000) Representing twentieth-century space–time climate variability. Part II: Development of 1901–1996 monthly grids of terrestrial surface climate, *Journal of Climate* **13**, 2217–2238.

Niinemets, U. (1999) Components of leaf mass per area – thickness and density – alter photosynthetic capacity in reverse directions in woody plants, *New Phytologist* **144**, 35–47.

Niinemets, U. (2001) Global-scale climatic controls of leaf dry mass per unit area, density, and thickness in trees and shrubs, *Ecology* **82**, 453–469.

Osborne, C.P. and Beerling, D.J. (2002a) A process-based model of conifer forest structure and function with special emphasis on leaf lifespan, *Global Biogeochemical Cycles* **16** (4), doi: 10.1029/2001GB001467, 1097.

Osborne, C.P. and Beerling, D.J. (2002b) Sensitivity of tree growth to a high CO_2 environment – consequences for interpreting the characteristics of fossil woods from ancient 'greenhouse' worlds, *Palaeogeography, Palaeoclimatology, Palaeoecology* **182**, 15–29.

Osborne, C.P., Mitchell, P.L., Sheehy, J.E. and Woodward, F.I. (2000) Modelling the recent historical impacts of atmospheric CO_2 and climate change on Mediterranean vegetation, *Global Change Biology* **6**, 445–458.

Osborne, C.P. and Woodward, F.I. (2001) Biological mechanisms underlying recent increases in the NDVI of

Mediterranean shrublands, *International Journal of Remote Sensing* **22**, 1895–1907.

Pacala, S.W. (1997) Dynamics of plant communities, in M.J Crawley (ed.) *Plant Ecology*. Blackwell Science, Oxford, 532–555.

Palutikof, J.P., Goodess, C.M. and Guo, X. (1994) Climate change, potential evapotranspiration and moisture availability in the Mediterranean Basin, *International Journal of Climatology* **14**, 853–869.

Parry, M.A.J., Keys, A.J. and Gutteridge, S. (1989) Variation in the specificity factor of C_3 higher plant Rubiscos determined by the total consumption of Ribulose-P_2, *Journal of Experimental Botany* **40**, 317–320.

Pausas, J.G. (1999) Mediterranean vegetation dynamics: modelling problems and functional types, *Plant Ecology* **140**, 27–39.

Penning de Vries, F.W.T. (1975) The cost of maintenance processes in plant cells, *Annals of Botany* **39**, 77–92.

Peñuelas, J. and Azcón-Bieto, J. (1992) Changes in leaf $\Delta^{13}C$ of herbarium plant species during the last 3 centuries of CO_2 increase, *Plant, Cell and Environment* **15**, 485–489.

Polley, H.W., Johnson, H.B., Marino, B.D. and Mayeux, H.S. (1993) Increase in C_3 plant water-use efficiency and biomass over Glacial to present CO_2 concentrations, *Nature* **361**, 61–64.

Pregitzer, K.S. (2002) Fine roots of trees – a new perspective, *New Phytologist* **154**, 267–273.

Read, D.J. (1990) Mycorrhizas in ecosystems, *Experientia* **47**, 376–391.

Reich, P.B., Walters, M.B. and Ellsworth, D.S. (1992) Leaf life-span in relation to leaf, plant, and stand characteristics among diverse ecosystems, *Ecological Monographs* **62**, 365–392.

Reich, P.B., Walters, M.B. and Ellsworth, D.S. (1997) From tropics to tundra: global convergence in plant functioning, *Proceedings of the National Academy of Sciences USA* **94**, 13730–13734.

Reich, P.R., Ellsworth, D.S., Walters, M.B., Vose, J.M., Gresham, C., Volin, J.C. and Bowman, W.D. (1999) Generality of leaf trait relationships: a test across six biomes, *Ecology* **80**, 1955–1969.

Schulze, E.-D. (1982) Plant life forms and their carbon, water and nutrient relations, in O.L. Lange, P.S. Nobel, C.B. Osmond and H. Ziegler (eds) *Plant Physiological Ecology II: Water Relations and Carbon Assimilation*, Springer-Verlag, Berlin and Heidelberg, 615–676.

Sellers, P.J., Bounoua, L., Collatz, G.J., Randall, D.A., Dazlich, D.A., Los, S.O., Berry, J.A., Fung, I., Tucker, C.J., Field, C.B. and Jensen, T.G. (1996) Comparison of radiative and physiological effects of doubled atmospheric CO_2 on climate, *Science* **271**, 1402–1406.

Shuttleworth, W.J. and Gurney, R.J. (1989) The theoretical relationship between foliage temperature and canopy resistance in sparse crops, *Quarterly Journal of the Royal Meteorological Society* **116**, 497–519.

Silvertown, J.W. and Lovett-Doust, J. (1993) *Introduction to Plant Population Biology*, Blackwell Scientific Publications, Oxford.

Smith, B., Prentice, I.C. and Sykes, M.T. (2001) Representation of vegetation dynamics in the modelling of terrestrial ecosystems: comparing two contrasting approaches within European climate space, *Global Ecology and Biogeography* **10**, 621–637.

Smith, S.D., Huxman, T.E., Zitzer, S.F., Charlet, T.N., Housman, D.C., Coleman, J.S., Fenstermaker, L.K., Seemann, J.R. and Nowak, R.S. (2000) Elevated CO_2 increases productivity and invasive species success in an arid ecosystem, *Nature* **408**, 79–82.

Smith, T.M. and Shugart, H.H. (1996) The application of patch models in global change research, in B. Walker and W. Steffan (eds) *Global Change and Terrestrial Ecosystems*, Cambridge University Press, Cambridge, 127–148.

Smith, T.M., Shugart, H.H. and Woodward, F.I. (eds) (1997) *Plant Functional Types*, Cambridge University Press, Cambridge.

Thornley, J.H.M. and Johnson, I.R. (1990) *Plant and Crop Modelling: A Mathematical Approach to Plant and Crop Physiology*, Clarendon Press, Oxford.

Turner, I.M. (1994) Sclerophylly: primarily protective?, *Functional Ecology* **8**, 669–675.

Turner, M.G., Dale, V.H. and Everham, E.H. (1997) Fires, hurricanes and volcanoes: comparing large disturbances, *Bioscience* **47**, 758–768.

Vitousek, P.M. and Howarth, R.W. (1991) Nitrogen limitation on land and in sea. How can it occur?, *Biogeochemistry* **13**, 87–115.

von Caemmerer, S. (2000) *Biochemical Models of Leaf Photosynthesis*, CSIRO Publishing, Melbourne.

Westoby, M., Warton, D. and Reich, P.B. (2000) The time value of leaf area, *The American Naturalist* **155**, 649–656.

Wolfe, J.A. (1985) Distribution of major vegetation types during the Tertiary, *American Geophysical Union Monographs* **32**, 357–375.

Woodward, F.I. (1987) *Climate and Plant Distribution*, Cambridge University Press, Cambridge.

Woodward, F.I. and Beerling, D.J. (1997) The dynamics of vegetation change: health warnings for equilibrium 'dodo' models, *Global Ecology and Biogeography Letters* **6**, 413–418.

Woodward, F.I. and Osborne, C.P. (2000) The representation of root processes in models addressing the responses of vegetation to global change, *New Phytologist* **147**, 223–232.

Woodward, F.I., Smith, T.M. and Emanuel, W.R. (1995) A global land primary productivity and phytogeography model, *Global Biogeochemical Cycles* **9**, 471–490.

7

Spatial Population Models for Animals

GEORGE L.W. PERRY AND NICK R. BOND

7.1 THE COMPLEXITY: INTRODUCTION

Ecology has come to play an increasingly important, sometimes central, role in natural resource management. The problems to be addressed are diverse, but common are two particular factors that present ecologists with a particularly strong challenge: complexity and scale. The inherent complexity of ecosystems has long been recognized by ecologists (Bradbury *et al.*, 1996; Shugart, 1998). Ecosystems consist of large numbers of species, each cycling through patterns of births and deaths as individual organisms interact with one anther via processes such as competition and predation. At the same time, these biotic processes (births, deaths and interactions) are themselves affected by the physical environment; factors such as resource supply rates (nutrients, water, habitat, etc.) as well as punctuated (often unpredictable) 'disturbance' events such as floods, fires and storms, which can induce sudden and dramatic changes to the biota. These interactions occur over different spatial and temporal scales depending on the mobility and longevity of the organisms involved (Wiens and Milne, 1989; MacNally, 1999), but played out in real landscapes, such as in forests and oceans, these scales can be large. Likewise, management decisions are often based on outcomes across large spatial scales. Consequently, in areas of applied research and management, ecologists are (quite reasonably) now being asked to address large-scale questions directly.

The tradition, however, has been largely to sidestep the problems of complexity and scale by conducting relatively simple experiments (often with fewer species) at small spatio-temporal scales, in which just one or a few factors are allowed to vary at the one time (Eberhardt and Thomas, 1991). It is generally argued

that results from such experiments can then be scaled up (or generalized) to provide a picture of what will happen in 'real', much larger and more complex, systems (Beck, 1997; Thrush *et al.*, 1997). This philosophy of experimentation developed under the belief that manipulative experiments provide significantly stronger inference than most other types of evidence that one might acquire (Underwood, 1990; Weiner, 1995; Beyers, 1998). In adopting an experimental approach, the difficulties (both practical and ethical) of manipulating and observing whole ecosystems, and hence the need to upscale, are quite obvious, although similar logistical difficulties can also apply to nonmanipulative studies (Eberhardt and Thomas, 1991). The question of scaling up has stimulated considerable debate among ecologists (e.g. see Rastetter *et al.*, 1992), but the emerging view acknowledges the pitfalls that beset inappropriately scaled research (Schneider, 1994; Mac Nally, 1999). Yet, while an holistic ecosystem-level focus makes an attractive goal for experiments, the sheer logistics and expense of conducting even single trials (ignoring issues such as replication and the availability of logical controls) will often prove too difficult (Turner *et al.*, 1995). In addition, the decision to manipulate at the ecosystem level typically comes only once ideas about important processes at work within the system have become reasonably well established, typically from conducting much smaller-scale experiments.

The point of the foregoing discussion is to highlight the difficulties that complexity and scale impose on our ability to understand the dynamics of ecosystems using direct methods of observation and experimentation, particularly at large spatial scales. As we discuss in this chapter, ecological modelling is increasingly being used to answer questions at these large spatial scales because

Environmental Modelling: Finding Simplicity in Complexity. Edited by J. Wainwright and M. Mulligan
© 2004 John Wiley & Sons, Ltd ISBNs: 0-471-49617-0 (HB); 0-471-49618-9 (PB)

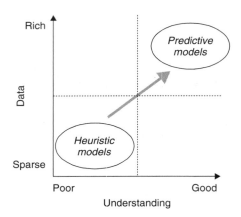

Figure 7.1 Position of ecological models in a data-under-standing conceptual space. Often ecological modelling at larger scales occurs with limited data and mechanistic understanding – this colours the way model outputs are (best) used (after Starfield and Beloch, 1986)

of the relative ease with which logistical problems can be overcome. Equally, modelling has aided our understanding of ecological processes at quite small spatial scales (e.g. With and Crist, 1996). Modelling has thus become a valuable strategy for increasing our ecological understanding at both small and large spatial scales (Jackson *et al.*, 2000). Modelling now plays several key roles in the progression and consolidation of knowledge in ecology, from cases where we are relatively knowledge- and data-poor (in which heuristic models help guide development), to cases where our knowledge and data are rich, permitting the creation of complex predictive models (Figure 7.1; Peters, 1991; Jackson *et al.*, 2000). In keeping with the title of this book, our focus in this chapter is on the role that heuristic models can play in tackling any particular ecological problem, and we are of the opinion that such models play a vital role in forwarding our current understanding of how ecosystems work. We begin by examining the historical development of ecological modelling, as well as more recent approaches (focusing on spatially explicit models) drawing examples from the 'animal ecology' literature. Nevertheless, many of our general points are relevant to all types of spatial (ecological) modelling.

7.2 FINDING THE SIMPLICITY: THOUGHTS ON MODELLING SPATIAL ECOLOGICAL SYSTEMS

7.2.1 Space, spatial heterogeneity and ecology

Mathematical modelling has a long and rich history in ecology (Kingsland, 1995). Models such as the logistic

and the Lotka-Volterra have played a key role in the development of ecological theory and have been termed 'classical ecological models'. These classical models are characterized by low dimensionality, a focus on finding system equilibria and an interest in temporal, rather than spatial, patterns. Typically, classical equilibrium ecological models have assumed that ecological interactions take place in a spatially homogeneous environment. The reasons behind the adoption of this view are many, but are largely centred on a desire to keep theoretical and experimental studies relatively simple (Wiens, 1997). The classical approach to modelling ecological systems ignores space by assuming, at least implicitly, that every individual is equally accessible to every other individual. The resulting model takes the form of a series of difference or differential equations for the mean abundance of the various model elements (e.g. individuals in a population, different species etc.). Because every individual is assumed to see the average or mean-field, this approach is termed the 'mean-field' solution (Hastings, 1994). Although several seminal early papers (notably Watt, 1947; Huffaker, 1958) addressed the role of space in ecological dynamics, it has only been in the past 15 to 20 years that space has been more routinely considered by most ecologists. Recently, theoretical ecologists have begun to consider how an explicit representation of space may alter the dynamics of nonspatial (or spatially implicit) ecological models. Over the past 15 to 20 years as spatial heterogeneity has been explicitly incorporated into many ecological field studies and models, it has become obvious that it is important in many situations. For example, species diversity within an area may be related to habitat heterogeneity (Tilman, 1994), the dynamics of populations and predator–prey interactions may be more stable or persistent in patchy environments (Kareiva, 1990; Hassell *et al.*, 1994), the spread of contagious disturbances such as fire or pathogen outbreaks is altered by patchiness (Turner and Romme, 1994; Li and Apps, 1996; Logan *et al.*, 1998) and dispersal and recruitment dynamics are affected by the patch structure of the environment (Pulliam, 1988; Pulliam and Danielson, 1991; Wiens *et al.*, 1993). Interestingly the development of spatial ecology has varied greatly between the various ecological subdisciplines; Perry (2002) discusses the development of 'spatial ecology' in more detail.

7.2.2 Three approaches to spatially explicit ecological modelling

Ecological models fall into two broad domains: analytical models and simulation models (Horn *et al.*, 1989).

Analytical models may be regarded as mathematical models that are potentially tractable or solvable in a closed form (e.g. differential systems, Markov models, etc.). Such models tend to be abstract and not concerned with the dynamics of a particular ecosystem. Conversely, simulation models tend to incorporate more biological detail, often including explicit nonlinearities, are usually system-specific, and are also much more constrained to the 'real' region of parameter space. However, this increased detail carries with it a loss of mathematical tractability and generality. We will describe both types of model approach before using a simulation model to highlight specific issues associated with spatial ecological modelling.

Hastings (1994) considered there to be three broad 'frameworks' for the introduction of space into ecological models. The first of these is based on reaction-diffusion systems, which may be either discrete or continuous in both time and space. A second approach is to consider changes in the occupation of different patches over time; such models may be termed patch-occupancy or metapopulation models (Hanski and Simberloff, 1997). A third, and spatially explicit, approach is to use cellular automata or grid-based models; these may be individual-based or more aggregated.

7.2.2.1 Reaction-diffusion systems

Reaction-diffusion systems model the growth and interaction of populations in the absence of dispersal (the 'reaction' component) and then add a representation of dispersal processes to the system (the 'diffusion' component, see Kareiva, 1990; Hastings, 1994). They are of the general form (for one species in one-dimension) in continuous space–time:

$$\frac{\partial N}{\partial t} = \frac{\partial}{\partial x}\left(D(x)\frac{\partial N}{\partial x}\right) + Nf(N, x) \qquad (7.1)$$

where $N(x, t)$ is the density function for the population size at x at t, $D(x)$ is the diffusion coefficient (dispersal rate), and the function $f(N, x)$ is the per capita rate of increase of the population at position x.

A discrete space analogue of the continuous form (Equation 7.1) is given by:

$$\frac{dN_i}{\partial t} = \sum_{j=1,n} D_{ij}(N_j - N_i) + Nf_i(N) \qquad (7.2)$$

where $N_i(t)$ is the population size at position i at time t, $D_{ij} = D_{ji}$ is the rate of exchange between patches i and j, and the function $f_i(N)$ is the per capita increase of the population at position i.

A widely used example of a continuous form of a reaction-diffusion model is the Fisher equation, which has been frequently used to model invasion processes (e.g. see Holmes *et al.*, 1994):

$$\frac{\partial N_{(x,t)}}{\partial t} = D\frac{\partial^2 N_{(x,t)}}{\partial x^2} + rN_{(x,t)}\left(1 - \frac{N_{(x,t)}}{K}\right) \qquad (7.3)$$

where r = the rate of population increase, and K is the carrying capacity of the environment.

In this example the reaction component is the familiar logistic equation. The population will spread in a wave at a rate dependent upon the parameters r and D. Note that the diffusion term will equal zero if the environment is either entirely uniform or the spatio-temporal variation in N is strictly linear (Tilman *et al.*, 1997). An advantage of this approach is that space can be considered to be continuous. Reaction-diffusion systems are mathematically tractable and provide a rich framework in which the implication of changes in dispersal dynamics and population growth and interactions may be analysed. The major assumptions made by such models are that (a) local population dynamics are deterministic, and (b) although individuals move randomly throughout their lifetime, at the population level movement is deterministic.

Andow *et al.* (1990) used the example of muskrats (*Ondatra zibethicus*) spreading from Prague after their release there in 1905 to highlight some of the strengths of this approach. Based on data for location and timing of first sightings after release, a contour map of population range over time could be created and the real rate of spread could be compared with the theoretical mean squared displacement of the wave front. Predicted rate of spread was between 6 and 32 km a^{-1}; observed spread ranged between 1 and 25 km a^{-1}. Thus, at least in this case, the diffusion model offers clear and testable hypotheses about the movement of organisms. Further, the examples described by Andow *et al.* (1990) show a good relationship between prediction and observed spread; variation in these measurements is most likely due to landscape heterogeneity.

7.2.2.2 Metapopulation and patch-occupancy models

The use of the 'metapopulation' as a framework for spatial population modelling has become popular in the past 10 to 15 years (Hanski, 1998). A metapopulation refers to a group of spatially separated subpopulations that exchange individuals infrequently (two to three generations). Metapopulation ecology is concerned with the dynamic consequences of migration among local populations and the conditions of regional populations of species with unstable local populations. The essence of the idea is that despite repeated local extinctions, a metapopulation may persist indefinitely so long as there is sufficient recolonization of empty patches. Importantly, this conception of populations is increasingly relevant to natural landscapes, which are becoming highly fragmented, forcing once contiguous populations into a metapopulation structure.

Levins (1969) described the first metapopulation model. Levins's model occurs in the context of an implicit landscape in which sites may either be occupied or empty. It may be conceptualized in two ways. First, as a metapopulation model with each site in the grid representing the occurrence of individual populations that may become locally extinct (see Hanski, 1991), or, second, each site may be viewed as the occurrence of individual organisms in a single population with each site in the grid being the size required by one adult individual of the species of interest (see Tilman *et al.*, 1997). The model is formulated as:

$$\frac{dp}{dt} = cp(1 - p) - mp \qquad (7.4)$$

The change in site occupancy over time (dp/dt) is a function of the rate of propagule production (cp), the proportion of empty or available sites ($1 - p$), and the mortality rate (mp). The population will persist if $c > m$, and when $c > m$, p approaches a globally stable equilibrium (p^*) at $1 - m/c$ (see Figure 7.2).

Two key assumptions of Levins's patch model are that propagule dispersal occurs randomly and that seeds are available globally. To examine the implications of

this assumption Tilman *et al.* (1997) built a stochastic cellular automata representation of the model in which dispersal occurs locally within a set neighbourhood of dimension d. Although the cellular automata representation is not mathematically tractable, it is directly comparable to Levins's model and allows us to explore the implications of the explicit consideration of space. The incorporation of space into the model has two important results. First, sites occupied by individuals are aggregated and not randomly distributed in space; this is an outcome of localized dispersal occurring around occupied sites (a result more thoroughly described by Durrett and Levin, 1994b, 1994a). That individuals become clustered is also significant because it shows that spatial patterning, in this case aggregation, may arise solely from local interaction and movement in a spatially uniform environment, and not necessarily as a result of any spatial heterogeneity. A second effect of the explicit consideration of space in the cellular automata model is that the mean level of occupation (p) is lower than predicted by the deterministic model. This outcome is a result of propagules falling on previously occupied sites and being lost; this does not occur in Levins's original model. Comparison of spatial and nonspatial models is of growing interest as a way of distinguishing the specific effects of space and spatial pattern on ecological processes (see Dieckman *et al.*, 2000 and references therein).

The metapopulation concept also has been applied to specific species in real systems. For example, the metapopulation has been used as a framework to explore

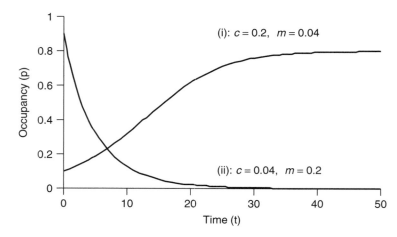

Figure 7.2   Dynamics of the patch occupancy model (Equation 7.4) with (i) $m = 0.04$ and $c = 0.2$ ($c > m$ so $p^* > 0$ [$p^* = 0.8$]) and (ii) $m = 0.2$ and $c = 0.04$ ($m > c$ so $p^* = 0$). Note that the trajectory of the system follows that of the familiar logistic curve

the population dynamics of various arboreal marsupials in south-eastern Australia (e.g. Leadbeater's Possum *Gymnobelideus leadbeateri*, Mountain Brushtail Possum *Trichosurus caninus*, and the Greater Glider *Petauroides volans*). Possingham *et al.* (1994) coupled spatial habitat information in a GIS database, relationships between habitat attributes and quality, and the dynamics of those habitat attributes, to assess long-term metapopulation viability of the Greater Glider in old-growth *Eucalyptus* forest, south-eastern Australia. They found that patch size is the key factor in metapopulation persistence, with large patches having the longest median persistence time for the populations. Other similar studies have shown the potential importance of inter-patch dispersal in metapopulation persistence, and the high extinction risk of small and isolated subpopulations (Lindenmayer and Lacy, 1995a,b). McCarthy and Lindenmayer (1999) used metapopulation models, modified to include disturbance (fire), to assist in reserve design for arboreal marsupials in old-growth forests, and explored the influence of the spatial configuration of reserves on extinction risk of greater glider populations. Similar to the study of Possingham *et al.* (1994), their model suggests that large, contiguous areas of old-growth forest were important for population persistence. The importance of large old-growth forest patches identified by these models is in direct conflict with the current use of these forests for timber production, and in this context the modelling has served a particularly important role in recognizing the need to modify the current management strategy.

7.2.2.3 Individual-based simulation models

Individual-based models (IBMs) have become widely used over the past 10 to 15 years (Huston *et al.*, 1988; Judson, 1994; Grimm, 1999). Unlike the frameworks presented above, in IBMs individual organisms are represented as the basic entity, thus, they may be used to explore how aggregate system properties emerge from interactions between these fundamental units. In essence they conceptualize the dynamics of the system of interest as a result of individual interactions. There are a number of reasons why IBMs are useful tools: (a) in evolutionary theory the individual is used as the basic unit on which natural selection acts; (b) no two biological organisms are ever identical (even when they have identical genes) because they will differ in age, size, condition, behavioural characteristics, etc.; (c) ecological populations are often very small and subject to stochasticity and so it is unrealistic to model them as continuous state variables; and (d) the behaviour of most organisms is spatially complex. IBMs may be

grid-based (e.g. Bond *et al.*, 2000) or represent space in a more continuous way by assigning individuals *x,y* positions on some plane (e.g. Deutschmann *et al.*, 1997). IBMs are being increasingly used to aid in the conservation of threatened species (e.g. Wiegand *et al.*, 1999; Mortia and Yokota, 2002); this approach is valuable where models are needed to evaluate case-specific conditions (Dunning *et al.*, 1995).

The earliest ecological IBMs were developed for plants with an emphasis on economic applications (e.g. stand growth and production). In the early 1970s the JABOWA model of hardwood forest dynamics was developed by Botkin and co-workers (Botkin *et al.*, 1972b; Botkin *et al.*, 1972a). Since then JABOWA-type models ('gap models') have been used in a wide range of systems with some success (in terms of their ability to reproduce, if not always explain, temporal patterns observed in the 'real' world). Typically they involve simulation of stand dynamics on a small area (typically less than 1 ha); all individuals above a certain size threshold are followed on the plot with birth, death, recruitment and growth processes modelled for all. While the earliest gap models focused on horizontal heterogeneity (diffusion of light through the canopy to the forest floor, etc.) recent incarnations, such as SORTIE (Deutschmann *et al.*, 1997), have been spatially explicit and include detailed mechanistic representations of the processes of interest.

A large number of individual-based models have also been developed for animals across a diverse range of taxonomic groups (e.g. fish, arthropods, invertebrates, mammals and birds). A recent trend in the development of these models has been their integration with GIS to look at resource and organism dynamics in real landscapes (e.g. Liu *et al.*, 1995). An example of this approach is provided by the BACH-MAP model developed by Pulliam *et al.* (1992) to explore the fate of small populations of Bachman's Sparrow (*Aimophila aestivalis*) in southern US woodlands. The BACHMAP model has been extended to create a model (ECOLE-CON, see Liu *et al.*, 1994) that links the original ecological model with an economic model. In ECOLECON, forest economic indices are calculated to determine the economic costs of managing the species. Integration of ecological models with, for example, economic models is likely to become more frequent in the future (Peterson, 2000; Liu, 2001). Individual-based models of animal population dynamics are also becoming widely used as components of population viability assessment for rare or threatened species (e.g. Wiegand *et al.*, 1999; Mortia and Yokota, 2002).

7.2.3 Top-down or bottom-up?

From the description above it may appear that IBMs represent an (the?) ideal approach or framework for modelling spatial population dynamics. However, criticisms have been levelled at the approach, primarily along the lines of the generality, or otherwise, of such models. While IBMs do offer great flexibility model formulation, the cost of this flexibility is a loss of tractability and generality. For example, IBMs are not usually amenable to exploration with analytical means and instead Monte Carlo-type methods need to be applied to them. Furthermore, it has proved difficult to compare results between systems using IBMs as differences may well be due to model structure rather than the systems themselves (see Smith and Shugart, 1996). The issues here hinge on why a specific model is being designed and how it is applied. Model building may take either a bottom-up approach, where conceptualization starts at the lowest level, with system dynamics arising from interactions between the entities at these levels are explored, or a top-down approach which usually involves development and application of a general framework (e.g. reaction-diffusion models) to different systems (see Figure 7.3). IBMs inherently take a bottom-up approach.

Grimm (1999) believed that ecological models may be built either as pragmatic (i.e. specific purpose with limited reference to a theoretical framework) or paradigmatic (i.e. with specific reference to some theoretical framework and toolkit). A review of animal-based IBMs suggests that the majority fall into the pragmatic camp. As a result, Grimm argued the individual modeller might learn a great deal from the development of an IBM but ecology as a whole may not. He argued that the individual-based approach is a bottom-up approach (Figure 7.3) that starts with the 'parts' of a system and then tries to understand how its properties emerge from the interaction among these parts. However, bottom-up approaches alone will never lead to theories at the systems level. State variable or top-down approaches are needed to provide an integrated view (Grimm, 1999).

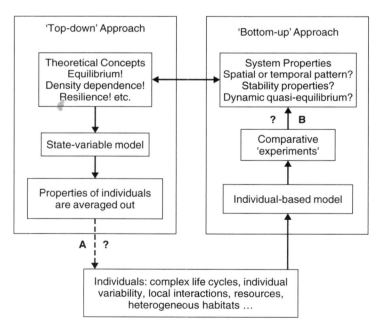

Figure 7.3 Schematic overview of how top-down and bottom-up approaches could complement each other and hence help develop ecological theory. Testing models developed from a top-down perspective (via Path A) is difficult. However, using a bottom-up view, theories emerge (Path B) as the models are used to perform comparative experiments with the goal of improving understanding of the phenomena of interest. The role of the top-down approach thus becomes one of identifying relevant questions; see Grimm (1999) for a fuller account.
Reprinted from *Ecological Modelling*, **115**, Grimm, Ten years of individual-based modelling in ecology: what have we learned and what could we learn in the future? 129–148, © (1999), with permission from Elsevier Science

7.3 THE RESEARCH FRONTIER: MARRYING THEORY AND PRACTICE

The past two decades have seen advances in the development of theoretical models, which have examined a growing number of questions of fundamental importance to ecology. Due in part to the speed with which models can be developed (one of their inherent virtues), experimental and empirical testing of much of this theory now lags well behind the modelling 'frontier' (Steinberg and Kareiva, 1997). While the need to link theory and practice better is well recognized (e.g. Kareiva, 1989; Steinberg and Kareiva, 1997), the problem stands that many theoretical developments remain intractable from a practical standpoint. In part, this stems from the difficulties of designing and replicating large-scale field experiments capable of putting model predictions to the test (e.g. see Kareiva *et al.*, 1996 in the context of genetically modified organisms [GMOs]). Equally, empiricists have been slow to embrace the benefits that modelling can provide when combined with empirical research. One possible avenue to marry models and empirical/experimental ecology at large spatial scales, and to apply the outcomes of this approach, is through the idea of 'adaptive management' (Walters, 1986, 1993). Under an adaptive management framework, model predictions and empirical observations influence, in an iterative sequence, management decisions and in turn enhance model development. In this framework, successive management decisions represent the experimental treatments of a traditional experimental framework, although thus far, failure to properly monitor the outcomes of particular management decisions has limited the effectiveness of the strategy as a whole (e.g. see Downs and Kondolf, 2002, in the context of river restoration). Clearly, though, models developed and applied for management must make predictions that managers need and can use, and as much as possible these predictions should be validated (Mac Nally, 2002). There is also a need to think more clearly about issues of parameterization and model error. Surrogate measures for difficult-to-obtain data could be developed, for example, a key parameter in many animal IBMs is dispersal and dispersal-mortality – parameters that are very difficult to measure but often ones to which models are very sensitive (e.g. Kareiva *et al.*, 1997; Ruckleshaus *et al.*, 1997). In our eyes, overcoming the divide between empirical and theoretical developments stands at the forefront of current research needs. A number of possibilities exist. One is for theory to take a more pattern-orientated approach (i.e. to consider specific 'real-world' patterns Wiener, 1995; Grimm *et al.*, 1996), while at the same time adopting an experimental approach to model analysis (Grimm, 1999). Perhaps the greatest successes will come where ecological modelling is applied to specific research needs in which empirical methods (at least at present) limit our ability to study particular phenomena. Such a problem is explored in detail below. It is also perhaps worth considering that in some situations quantitative prediction is too much to ask of even the best-designed experiment or model. In such cases models may be best used in an heuristic sense, for example, in aiding the design and implementation of environmental monitoring programmes (e.g. Kareiva *et al.*, 1996; Urban, 2000).

Finally, scaling of ecological models remains a pressing issue. Increasingly ecologists are being asked regional and global-scale questions yet much ecological research is conducted at much smaller scales. The problem of scale has been a critical impediment to incorporating important fine-scale processes and information into coarser-scale ecosystem and landscape models. Our knowledge of fine-scale physiological and ecological processes comes from a variety of measurements, ranging from forest plot inventories to remote sensing, made at spatial resolutions considerably smaller than the large scale at which global ecosystem models are defined. The development of methods to link information at these two disparate scales is a critical challenge. Thus, there is still a need to develop methods of up- and downscaling ecological data (Rastetter *et al.*, 1992) and there have been some developments in that area through meta-modelling (statistical abstraction of fine-scale models to coarser scales, see Urban *et al.*, 1999) and other methods (e.g. Moorcroft *et al.*, 2001).

7.4 CASE STUDY: DISPERSAL DYNAMICS IN STREAM ECOSYSTEMS

Our aim here is to illustrate the way in which a relatively simple spatial model can produce quite complex and unexpected outcomes when applied to a relatively basic, yet experimentally 'challenging', ecological phenomenon.

7.4.1 The problem

The beds of streams and rivers typically harbour a rich and abundant assemblage of invertebrates, dominated by insects, crustaceans and gastropods. These organisms fill a wide range of ecological roles (predators, grazers, detritivores, etc.), and many exhibit complex life cycles, spending only part of their life in the stream (Allan, 1995). Common to most, however, is their tendency (while in the stream) to spend most of their time in direct contact with the bed; hence acquiring the label of 'benthic' fauna. Yet, despite their strong streambed relation, the tendency to periodically leave the streambed and 'drift' downstream in the water current (eventually to return to the streambed) is a behavioural characteristic common to many species, and a distinctive feature of most benthic stream communities (Brittain and Eikekland, 1988; Allan, 1995). While the behavioural basis of this drift phenomena has long been debated (Waters, 1972), its importance in the dispersal of benthic fauna and the colonization of downstream areas is now quite well understood (Downes and Keough, 1998). To ecologists interested in drift, factors influencing drift distances have come to the fore as an important theme – particularly how drift might affect upstream–downstream population linkages and the recovery rates of benthic assemblages following disturbance (Lancaster *et al.*, 1996; Moser and Minshall, 1996).

There has thus been great interest in determining the drift distances of benthic fauna in real streams (e.g. Elliott, 1971, 2002; Lancaster *et al.*, 1996). Three initial observations have an important bearing on the development of such models. First, most of these animals are weak swimmers relative to the current speeds they experience when they drift downstream (Fonseca, 1999). This means that drift can be modelled as passive particles (Fonseca, 1999). Second, the flow environment in streams is often extremely complex when examined at small spatial scales, but this complexity in flow can be factored out when examined at larger spatial scales (Carling, 1992). Finally, attempting to track the drift and settlement of individual animals is a complex task that has met with limited success (Downes and Keough, 1998).

As a mean-field problem (i.e. at the larger scale), the field of drift-distance models developed quickly, describing drift distance as a simple exponential decay function in which drift distance is governed by height above the bed at which animals enter the drift (itself an interesting issue!) and average stream velocity (e.g. Ciborowski, 1983; Reynolds *et al.*, 1990). These models were, in places, quite successful at describing actual observed drift distances, but discrepancies in drift distances between sites with similar mean-field velocities were also common (Lancaster *et al.*, 1996).

Generalizing from both theoretical and empirical studies in other systems, we hypothesized that spatial variation in localized flow environments could influence dispersal even where mean-field conditions are similar. Central to this argument is the assumption that drifting animals settle in complex flow environments when they

enter local areas of near-zero flow (so called dead-water zones [DWZ]) behind small obstacles and in backwaters along the stream margins (Lancaster and Hildrew, 1993). Thus while average velocities will be influenced only by the abundance of such DWZ within a stream reach, differences in the spatial distribution of DWZ may have independent effects on average drift distances (Bond *et al.*, 2000).

7.4.2 The model

Because of our interest in understanding how the spatial pattern of DWZ (settlement sites) might affect drift distances of individual organisms, we adopted a cellular automata or grid-based model structure that predicts the fate of each individual. We used a spatial lattice to represent the stream reaches, which provided a simple control on the proportion and spatial pattern of obstacles (which created downstream DWZ) within the land or streamscape. Individual organisms were introduced into this landscape and allowed to move (drift) according to a set of predefined behavioural rules. These rules varied depending on the position of individuals relative to DWZ and obstacles, and were designed to encapsulate in a simple fashion the ways in which turbulent flow occurs around obstacles in real streams.

Complete details of the model can be found in Bond *et al.* (2000). Essentially, streams were represented as a two-dimensional lattice (depth was ignored) 30 cells wide and 1000 cells long. Cells within the lattice took one of three states: flowing water, obstacles or DWZ. DWZ were always located directly below obstacles and were also always of equal size. Obstacles represented rocks and other obstructions observed in streams; their size (and that of associated DWZ) were taken from field surveys of several streams. Organisms moved through the landscape according to a biased (downstream) random walk, with a set of simple probabilistic behavioural rules determining their movement patterns around obstacles, and their likelihood of entering DWZ, where settlement occurred (ending drift). These simplified rules provided a practical way of overcoming the otherwise intractable problem of modelling complex flow dynamics (see Carling, 1992).

7.4.3 The question

Here we discuss two central questions addressed by the model. First, how do the proportion and spatial patchiness of obstacles in a stream reach affect the mean distance travelled by drifting animals? We fixed the proportion of obstacles at either 5% or 15%, and at each

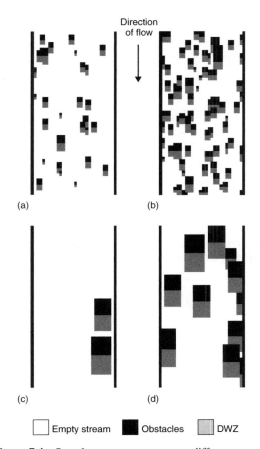

Direction of flow

(a)   (b)

(c)   (d)

☐ Empty stream   ■ Obstacles   ▨ DWZ

Figure 7.4   Sample streamscapes at two different proportions of obstacles and with two different obstacle size distributions. (a) 5% obstacles; size (mean ± SD) 2 ± 1 cells; (b) 15% obstacles, size 2 ± 1 cells; (c) 5% obstacles, size 6 ± 1 cells; (d) 15% obstacles, size 6 ± 1 cells
Source: Reprinted from N.R. Bond *et al.* (2000) Dispersal of organisms in a patchy stream environment under different settlement scenarios, *Journal of Animal Ecology* **69**, 608–620, Copyright Blackwell Publishing

of these proportions we varied the size (mean ± SD) of individual DWZs from 2 ± 1 to 6 ± 1 square cells (Figure 7.4). We hypothesized that both of these factors (quantity and patchiness of DWZ) would influence the distance that individual organisms drift. We tested this idea by measuring the downstream distance travelled by 1000 organisms dispersing through each of 250 random landscapes constructed under each of the four scenarios of DWZ quantity and patchiness. Second, we examined the central assumption of our model and many of the field studies examining drift (e.g. Lancaster *et al.*, 1996), which is that settlement is restricted to DWZ. We did

this by running the four scenarios above (proportion and patchiness of DWZ), but allowing settlement to occur with a low probability anywhere in the landscape, and then again with settlement restricted only to DWZ.

7.4.4 Results

When settlement was random, average downstream drift distances decreased by roughly 50% when the proportion of DWZ was increased from 5% to 15%, but spatial patchiness had little effect. This picture was complicated by restricting settlement to DWZ. Increasing the proportion of DWZ again halved drift distances, but at each proportion, the average drift distance was roughly doubled when there were fewer, but larger obstacles (Figure 7.5). The spread of drift distances also increased greatly with fewer (larger) obstacles, particularly at lower (5%) obstacle densities, and again when settlement occurred only in DWZ (Figure 7.6). Thus, even with these simple movement rules, spatial patterning of settlement sites appears to have a strong bearing on the potential drift distances of our model organisms. According to the model, the effects of spatial patchiness are contingent on the realities of where organisms actually settle in streams, and whether this is restricted to particular areas such as DWZ. Testing this assumption is a critical challenge. Nevertheless, the results of this modelling provide a simple explanation for discrepancies in previous empirical studies (e.g. Lancaster *et al.*, 1996).

Together, these two results – the apparent importance of space, and the importance of where organisms actually do settle – have provided a significant development in our understanding of drift in streams. At a more general level, the model demonstrates the importance of spatial pattern in stream systems, a system in which spatially explicit models have been slow to take hold (Wiens, 2002). This example demonstrates our earlier points on the potential for modelling and empirical studies to contribute synergistically toward a better understanding of ecological systems.

7.5 CONCLUSION

Ecologists are increasingly making use of (spatial) models. This is a relatively recent development as for much of its history ecology has downplayed space by focusing on time. Three general approaches have been taken to spatially explicit modelling of population dynamics: (a) reaction-diffusion systems; (b) metapopulation and patch-occupancy model; and (c) individual-based models. These approaches have their own strengths and weaknesses and represent, in some ways, a tractability–realism gradient. Indeed, finding an appropriate

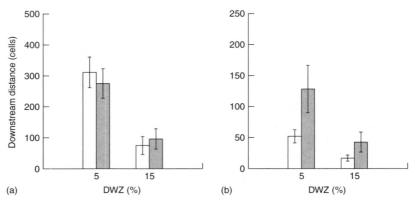

Figure 7.5   Mean downstream distance (±1 SD) travelled in each landscape type with (a) random settlement and (b) settlement in DWZ only. Mean obstacle sizes are 2 × 2 cells (white columns) and 6 × 6 cells (hatched columns)
Source: Reprinted from N.R. Bond *et al.* (2000) Dispersal of organisms in a patchy stream environment under different settlement scenarios, *Journal of Animal Ecology* **69**, 608–620, Copyright Blackwell Publishing

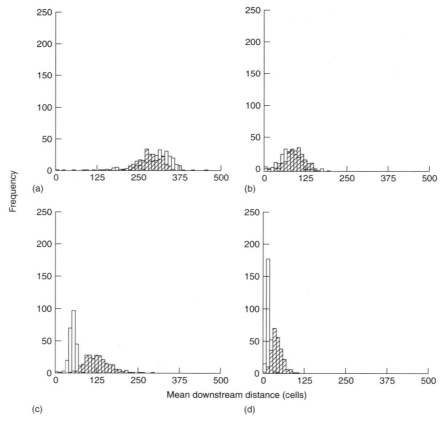

Figure 7.6   Frequency distributions of average downstream distance travelled per landscape, for each landscape type. (a) 5% obstacles, random settlement; (b) 15% obstacles, random settlement; (c) 5% obstacles, settlement in DWZ only; (d) 15% obstacles. Mean obstacle sizes are 2 × 2 cells (white columns) and 6 × 6 cells (hatched columns)
Source: Reprinted from N.R. Bond *et al.* (2000) Dispersal of organisms in a patchy stream environment under different settlement scenarios, *Journal of Animal Ecology* **69**, 608–620, Copyright Blackwell Publishing

balance between tractability and realism (complexity) remains a major challenge to ecological modellers.

A spatial model of drift by invertebrate larvae in streams was presented to illustrate some of these points. The drift of invertebrates in streams has been the focus of a considerable amount of research yet is empirically challenging (e.g. the difficulty of tracking individual larvae, etc.). Using a relatively simple spatial model we show how such a model can produce quite complex and unexpected outcomes when applied to a relatively basic phenomenon. It has also provided a possible solution to some puzzling discrepancies in previous empirical studies. The recent interest in spatial patterns and processes in ecology has resulted in the development of a wealth of spatial ecological theory. However, empirical testing of this theory has not proceeded as quickly, in part, because of inherent problems in experimental design and analysis at extended spatial scales, but also because some of this spatial theory has not been framed in an obviously testable manner.

ACKNOWLEDGEMENTS

We would like to thank Ralph Mac Nally for his useful comments on an earlier draft of this chapter.

REFERENCES

Allan, J.D. (1995) *Stream Ecology: Structure and Function of Running Waters*, Chapman & Hall, London.

Andow, D.A., Kareiva, P.M., Levin, S.A. and Okubo, A. (1990) Spread of invading organisms, *Landscape Ecology* **4**, 177–188.

Beck, M.W. (1997) Inference and generality in ecology: current problems and an experimental solution, *Oikos* **78**, 265–273.

Beyers, D.W. (1998) Causal inference in environmental impact studies, *Journal of the North American Benthological Society* **17**, 367–373.

Bond, N.R., Perry, G.L.W. and Downes, B.J. (2000) Dispersal of organisms in a patchy stream environment under different settlement scenarios, *Journal of Animal Ecology* **69**, 608–620.

Botkin, D.B., Janak, J.F. and Wallis, J.R. (1972a) Rationale, limitations and assumptions of a northeast forest growth simulator, *IBM Journal of Research and Development* **16**, 101–116.

Botkin, D.B., Janak, J.F. and Wallis, J.R. (1972b) Some ecological consequences of a computer model of forest growth, *Journal of Ecology* **60**, 849–872.

Bradbury, R.H., van der Laan, J.D. and Green, D.G. (1996) The idea of complexity in ecology, *Senckenbergiana Maritima* **27**, 89–96.

Brittain, J.E. and Eikekland, T.J. (1988) Invertebrate drift – a review, *Hydrobiologia* **166**, 77–93.

Carling, P.A. (1992) In-stream hydraulics and sediment transport, in P. Calow and G.E. Petts (eds) *The Rivers Handbook*, Blackwell, Oxford, 101–125.

Ciborowski, J.J.H. (1983) Downstream and lateral transport of nymphs of two mayfly species (Ephemeroptera), *Canadian Journal of Fisheries and Aquatic Sciences* **40**, 2025–2029.

Deutschmann, D.H., Levin, S.A., Devine, C. and Buttle, L.A. (1997) Scaling from trees to forests: analysis of a complex simulation model, *Science (Online)*: http://www.sciencemag.org/feature/data/deutschman/index.htm

Dieckman, U., Law, R. and Metz, J.A.J. (eds) (2000) *The Geometry of Ecological Interactions: Simplifying Spatial Complexity*, Cambridge, Cambridge University Press.

Downes, B.J. and Keough, M.J. (1998) Scaling of colonization processes in streams: parallels and lessons from marine hard substrata, *Australian Journal of Ecology* **23**, 8–26.

Downs, P.W. and Kondolf, G.M. (2002) Post-project appraisals in adaptive management of river channel restoration, *Environmental Management* **29**, 477–496.

Dunning, J.B.J., Stewart, D.J., Noon, B.R., Root, T.L., Lamberson, R.H. and Stevens, E.E. (1995) Spatially explicit population models: current forms and future uses, *Ecological Applications* **5**, 2–11.

Durrett, R. and Levin, S.A. (1994a) The importance of being discrete (and spatial), *Theoretical Population Biology* **46**, 363–394.

Durrett, R. and Levin, S.A. (1994b) Stochastic spatial models: a user's guide to ecological applications, *Philosophical Transactions of the Royal Society of London, Series B* **343**, 329–350.

Eberhardt, L.L. and Thomas, J.M. (1991) Designing environmental field studies, *Ecological Monographs* **61**, 53–73.

Elliott, J.M. (1971) The distances travelled by drifting invertebrates in a Lake District stream, *Oecologia* **6**, 350–379.

Elliott, J.M. (2002) Time spent in the drift by downstream-dispersing invertebrates in a Lake District stream, *Freshwater Biology* **47**, 97–106.

Fonseca, D.M. (1999) Fluid-mediated dispersal in streams: models of settlement from the drift, *Oecologia* **121**, 212–223.

Grimm, V. (1999) Ten years of individual-based modelling in ecology: what have we learned and what could we learn in the future?, *Ecological Modelling* **115**, 129–148.

Grimm, V., Frank, K., Jeltsch, F., Brandl, R., Uchmanski, J. and Wissel, C. (1996) Pattern-oriented modelling in population ecology, *Science of the Total Environment* **183**, 151–166.

Hanski, I.A. (1991) Single-species metapopulation dynamics: concepts, models and observations, *Biological Journal of the Linnean Society* **42**, 17–38.

Hanski, I. (1998) Metapopulation dynamics, *Nature* **396**, 41–49.

Hanski, I.A. and Simberloff, D. (1997) The metapopulation approach, its history, conceptual domain and application to conservation, in I.A. Hanski and M.E. Gilpin (eds) *Metapopulation Biology: Ecology, Genetics and Evolution*, Academic Press, San Diego, 5–26.

Hassell, M.P., Comins, N.N. and May, R.M. (1994) Species-coexistence and self-organising spatial dynamics, *Nature* **370**, 290–292.

Hastings, A. (1994) Conservation and spatial structure: theoretical approaches, in S.A. Levin (ed.) *Frontiers in Mathematical Biology*, Springer-Verlag, Berlin, 494–503.

Holmes, E.E., Lewis, M.A., Banks, J.E. and Veit, R.R. (1994) Partial differential equations in ecology: spatial interactions and population dynamics, *Ecology* **75**, 17–29.

Horn, H.S., Shugart, H.H. and Urban, D.L. (1989) Simulators as models of forest dynamics, in J. Roughgarden, R.M. May, and S.A. Levin (eds) *Perspectives in Ecological Theory*, Princeton University Press, Princeton, NJ, 256–267.

Huffaker, C.B. (1958) Experimental studies on predation: dispersion factors and predator–prey oscillation, *Hilgardia* **27**, 343–383.

Huston, M.A., DeAngelis, D.L. and Post, W. (1988) New computer models unify ecological theory, *BioScience* **38**, 682–691.

Jackson, L.J., Trebitz, A.S. and Cottingham, K.L. (2000) An introduction to the practice of ecological modelling, *BioScience* **50**, 694–706.

Judson, O.P. (1994) The rise of the individual-based model in ecology, *Trends in Ecology and Evolution* **9**, 9–14.

Kareiva, P. (1989) Renewing the dialogue between theory and experiments in population ecology, in J. Roughgarden, R.M. May and S.A. Levin (eds) *Perspectives in Ecological Theory*, Princeton University Press, Princeton, NJ, 69–88.

Kareiva, P. (1990) Population dynamics in spatially complex environments: theory and data, *Philosophical Transactions of the Royal Society of London, Series B* **330**, 175–190.

Kareiva, P., Parker, I.M. and Pascual, M. (1996) Can we use experiments and models in predicting the invasiveness of genetically engineered organisms? *Ecology* **77**, 1670–1675.

Kareiva, P.M., Skelly, D.K. and Ruckelshaus, M. (1997) Reevaluating the use of models to predict the consequences of habitat loss and fragmentation, in S.T.A. Pickett, R.S. Ostfeld, H. Schchak and G.E. Likens (eds) *Enhancing the Ecological Basis of Conservation: Heterogeneity, Ecosystem Function and Biodiversity*, Chapman & Hall, New York, 156–167.

Kingsland, S. (1995) *Modeling Nature: Episodes in the History of Population Ecology*, 2nd edn, University of Chicago Press, Chicago.

Lancaster, J. and Hildrew, A.G. (1993) Characterizing in-stream flow refugia, *Canadian Journal of Fisheries and Aquatic Science* **50**, 1663–1675.

Lancaster, J., Hildrew, A.G. and Gjerlov, C. (1996) Invertebrate drift and longitudinal transport processes in streams, *Canadian Journal of Fisheries and Aquatic Science* **53**, 572–582.

Levins, R. (1969) Some demographic and genetic consequences of environmental heterogeneity for biological control, *Bulletin of the Entomological Society of America* **15**, 237–240.

Li, C. and Apps, M.J. (1996) Effects of contagious disturbance on forest temporal dynamics, *Ecological Modelling* **87**, 143–151.

Lindenmayer, D.B. and Lacy, R.C. (1995a) Metapopulation viability of arboreal marsupials in fragmented old-growth forests: comparison among species, *Ecological Applications* **5**, 183–199.

Lindenmayer, D.B. and Lacy, R.C. (1995b) Metapopulation viability of Leadbeater's possum, *Gymnobelideus leadbeateri*, in fragmented old-growth forests, *Ecological Applications* **5**, 164–182.

Liu, J. (2001) Integrating ecology with human demography, behavior, and socioeconomics: needs and approaches, *Ecological Modelling* **140**, 1–8.

Liu, J., Cubbage, F. and Pulliam, H.R. (1994) Ecological and economic effects of forest structure and rotation lengths: simulation studies using ECOLECON, *Ecological Economics* **10**, 249–265.

Liu, J., Dunning, J.B.J. and Pulliam, H.R. (1995) Potential effects of a forest management plan on Bachman's sparrows (*Aimophila aestivalis*): linking a spatially explicit model with GIS, *Conservation Biology* **9**, 62–75.

Logan, J.E., White, P., Bentz, B.J. and Powell, J.A. (1998) Model analysis of spatial patterns in mountain pine beetle outbreaks, *Theoretical Population Biology* **53**, 236–255.

Mac Nally, R. (1999) Dealing with scale in ecology, in J.A. Wiens and M.R. Moss (eds) *Issues in Landscape Ecology*, International Association for Landscape Ecology, Guelph, ON, Canada.

Mac Nally, R. (2002) Improving inference in ecological research: issues of scope, scale and model validation, *Comments on Theoretical Biology* **7**, 235–254.

McCarthy, M.A. and Lindenmayer, D.B. (1999) Incorporating metapopulation dynamics of greater gliders into reserve design in disturbed landscapes, *Ecology* **80**, 651–667.

Moorcroft, P.R., Hurtt, G.C. and Pacala, S.W. (2001) A method for scaling vegetation dynamics: the ecosystem demography model (ED), *Ecological Monographs* **71**, 577–586.

Mortia, K. and Yokota, A. (2002) Population viability of stream-resident salmonids after habitat fragmentation: a case study with white-spotted charr (*Salevnius leucomaenis*) by an individual based model, *Ecological Modelling* **155**, 85–94.

Moser, D.C. and Minshall, G.W. (1996) Effects of localized disturbance on macroinvertebrate community structure in relation to mode of colonization and season, *American Midland Naturalist* **135**, 92–101.

Perry, G.L.W. (2002) Landscapes, space and equilibrium: some recent shifts, *Progress in Physical Geography* **26**, 339–359.

Peters, R.H. (1991) *A Critique for Ecology*, Cambridge University Press, Cambridge.

Peterson, G. (2000) Political ecology and ecological resilience: an integration of human and ecological dynamics, *Ecological Economics* **35**, 323–336.

Possingham, H.P., Lindenmayer, D.B., Norton, T.W. and Davies, I. (1994) Metapopulation viability analysis of the greater glider (Petauroides volans) in a wood production area, *Biological Conservation* **70**, 227–236.

Pulliam, H.R. (1988) Sources, sinks and population regulation, *American Naturalist* **132**, 652–661.

Pulliam, H.R. and Danielson, B.J. (1991) Sources, sinks and habitat selection: a landscape perspective on population dynamics, *American Naturalist* **137** (suppl): S50–S66.

Pulliam, H.R., Dunning, J.B.J. and Liu, J. (1992) Population dynamics in complex landscapes: a case study, *Ecological Applications* **2**, 165–177.

Rastetter, E.B., King, A.W., Cosby, B.J., Hornberger, G.M., O'Neill, R.V. and Hobbie, J.E. (1992) Aggregating fine-scale ecological knowledge to model coarser-scale attributes of ecosystems, *Ecological Applications* **2**, 55–70.

Reynolds, C.S., White, M.L., Clarke, R.T. and Marker, A.F. (1990) Suspension and settlement of particles in flowing water: comparison of the effects of varying water depth and velocity in circulating channels, *Freshwater Biology* **24**, 23–34.

Ruckleshaus, M., Hartway, C. and Kareiva, P.M. (1997) Assessing the data requirements of spatially explicit dispersal models, *Conservation Biology* **11**, 1298–1306.

Schneider, D.C. (1994) *Quantitative Ecology: Spatial and Temporal Scaling*, Academic Press. San Diego.

Shugart, H.H. (1998) *Terrestrial Ecosystems in Changing Environments*, Cambridge University Press, Cambridge.

Smith, T.M. and Shugart, H.H. (1996) The application of patch models to global change research, in B.H. Walker and W.L. Steffen (eds) *Global Change and Terrestrial Ecosystems*, Cambridge University Press, Cambridge, 127–148.

Starfield, A.M. and Beloch, A.L. (1986). *Building Models for Conservation and Wildlife Management*, Macmillan, New York.

Steinberg, E.K. and Kareiva, P.M. (1997) Challenges and opportunities for empirical evaluation of 'spatial theory', in D.A. Tilman and P.M. Kareiva (eds) *Spatial Ecology: The Role of Space in Population Dynamics and Interspecific Interactions*, Princeton University Press, Princeton, NJ, 318–333.

Thrush, S.F., Schneider, D.C., Legendre, P., Whitlatch, R.B., Dayton, P.K., Hewitt, J.E., Hines, A.H., Cummings, V.J., Lawrie, S.M., Grant, J., Pridmore, R.D. and Turner, S.J. (1997) Scaling-up from experiments to complex ecological systems: where to next?, *Journal of Experimental Marine Biology and Ecology* **216**, 243–254.

Tilman, D.A. (1994) Competition and biodiversity in spatially structured habitats, *Ecology* **75**, 2–16.

Tilman, D.A., Lehman, C.L. and Kareiva, P.M. (1997) Population dynamics in spatial habitats, in D.A. Tilman and P.M. Kareiva (eds) *Spatial Ecology: The Role of Space in Population Dynamics and Interspecific Interactions*, Princeton University Press, Princeton, NJ, 3–21.

Turner, M.G., Arthaud, G.J., Engstrom, R.T., Hejl, S.J., Liu, J., Loeb, S. and McKelvey, K. (1995) Usefulness of spatially explicit population models in land management, *Ecological Applications* **5**, 12–16.

Turner, M.G. and Romme, W.H. (1994) Landscape dynamics in crown fire ecosystems, *Landscape Ecology* **9**, 59–77.

Underwood, A.J. (1990) Experiments in ecology and management: their logics, functions and interpretations, *Australian Journal of Ecology* **15**, 365–389.

Urban, D.L. (2000) Using model analysis to design monitoring programs for landscape management and impact assessment, *Ecological Applications* **10**, 1820–1832.

Urban, D.L., Acevedo, M.F. and Garman, S.L. (1999) Scaling fine-scale processes to large-scale patterns using models derived from models: meta-models, in D.J. Mladenoff and W.L. Baker (eds) *Spatial Modeling of Forest Landscapes: Approaches and Applications*, Cambridge University Press, Cambridge, 125–163.

Walters, C.J. (1986) *Adaptive Management of Renewable Resources*, Macmillan Publishing Co., New York.

Walters, C.J. (1993) Dynamic models and large scale field experiments in environmental impact assessment and management, *Australian Journal of Ecology* **18**, 53–61.

Waters, T.F. (1972) The drift of stream insects, *Annual Review of Entomology* **17**, 253–272.

Watt, A.S. (1947) Pattern and process in the plant community, *Journal of Ecology* **35**, 1–22.

Weiner, J. (1995) On the practice of ecology, *Journal of Ecology* **83**, 153–158.

Wiegand, T., Naves, J., Stephan, T. and Fernandez, A. (1999) Assessing the risk of extinction for the Brown Bear (Ursus arctos) in the Cordillera Cantabrica, Spain, *Ecological Applications* **68**, 539–570.

Wiener, J. (1995) On the practice of ecology, *Journal of Ecology* **83**, 153–158.

Wiens, J.A. (1997) The emerging role of patchiness in conservation biology, in S.T.A. Pickett, R.S. Ostfeld, H. Schchak and G.E. Likens (eds) *Enhancing the Ecological Basis of Conservation: Heterogeneity, Ecosystem Function and Biodiversity*, Chapman & Hall, New York, 93–108.

Wiens, J.A. (2002) Riverine landscapes: taking landscape ecology into the water, *Freshwater Biology* **47**, 501–515.

Wiens, J.A. and Milne, B.T. (1989) Scaling of 'landscapes' in landscape ecology, or, landscape ecology from a beetle's perspective, *Landscape Ecology* **3**, 87–96.

Wiens, J.A., Stenseth, N.C., van Horne, B. and Ims, R.A. (1993) Ecological mechanisms and landscape ecology, *Oikos* **66**, 369–380.

With, K.A. and Crist, T.O. (1996) Translating across scales: simulating species distributions as the aggregate response of individuals to heterogeneity, *Ecological Modelling* **93**, 125–137.

8

Ecosystem Modelling: Vegetation and Disturbance

STEFANO MAZZOLENI, FRANCISCO REGO, FRANCESCO GIANNINO AND COLIN LEGG

8.1 THE SYSTEM COMPLEXITY: EFFECTS OF DISTURBANCE ON VEGETATION DYNAMICS

Vegetation is continuously changing in time and space. The changes may be autogenic, driven by the properties of the component plant species, or allogenic, driven by external factors. Allogenic change may be a result of gradual changes in the environment, but more often results from disturbance events, and the way vegetation responds will be very dependent upon the disturbance history (Miles, 1979).

The term disturbance is used in ecology to refer to the partial or complete reduction of biomass due to some external factors and does not include natural mortality or decay processes. Typical causes of disturbance are grazing, cutting, fire and frost. Disturbance can also be defined as an event that causes a significant change from the normal pattern in an ecological system (Forman and Godron, 1986), but the problem with this relative definition of disturbance stems from the difficulty of defining the 'normal' range of conditions for an ecosystem (White and Harrod, 1997). This is why an absolute definition requiring measures of real change is more appropriate for a mechanistic modelling approach. This was the choice taken by Waring and Running (1998), for example, who defined disturbance in a forest as any factor that brings about a significant reduction in the overstory leaf area index for a period of more than one year. Grubb (1985) reviewed the concept of disturbance and distinguished vegetation resilience 'in situ' and 'by migration' (Grubb and Hopkins, 1986), referring to the

regeneration capacity of surviving individuals in the disturbed area and recolonization from neighbouring areas respectively.

In the simplest case of one type of disturbance on a homogeneous site the disturbance regime can be generally described by its frequency and its intensity. However, if intensity is associated with biomass accumulated, as in the case of fire, there is an inverse relationship between intensity and frequency. These considerations were very well established for fire regimes in North America (see, for instance, Heinselman, 1978) and in Europe (e.g. Angelstam, 1997). Therefore the simplest way to describe a fire regime is to determine the average time between disturbances (years) or conversely, the mean frequency of the disturbance (expressed in $year^{-1}$).

This chapter focuses only on fire and grazing. Henderson and Keith (2002) performed a canonical correspondence analysis on the sources of variation in shrub composition of eucalypt forests which showed that fire and grazing accounted for more variation than the combination of all other environmental and spatial variables. Although they are perhaps the two most important sources of disturbance, they differ very much in character; the former causes an abrupt discontinuity in the system while the latter is characterized by a gradual and continuous process of selective biomass removal.

8.1.1 Fire

The immediate effects of fire are the removal of the majority of above-ground biomass and the pulse of

---

Environmental Modelling: Finding Simplicity in Complexity. Edited by J. Wainwright and M. Mulligan
© 2004 John Wiley & Sons, Ltd ISBNs: 0-471-49617-0 (HB); 0-471-49618-9 (PB)

heat through the upper soil horizons. Some plants are killed, but those that survive and can resprout are changed physiologically because of the imbalance in the root:shoot ratio. The exposure of the modified soil horizons results in a new post-fire microclimate and changes in hydrology. However, the resilience of vegetation to fire has been reported for several ecosystems. For example, *Quercus coccifera* garrigue and *Ulex parviflorus* and *Rosmarinus officinalis* shrublands regained their total cover in less than five years after burning (Ferran and Vallejo, 1998). Similarly, few changes following fire were reported for oak scrub in Florida uplands where species composition and structure largely recovered in less than two years (Abrahamson and Abrahamson, 1996). In general, after fire, the most obvious change is that of the composition and relative abundance of herbaceous species. As succession proceeds and the canopy closes, the early flush of herbaceous vegetation is either restricted to small canopy openings or remains dormant in the soil in the form of seeds waiting for the next fire to come (Arianoutsou, 1998). Similar results were found in California (Hanes, 1971), in Australia (Specht *et al.*, 1958), in Italy (Mazzoleni and Pizzolongo, 1990) and in Portugal (Espírito-Santo *et al.*, 1993).

Two broad groups of communities can be distinguished related to the reproductive strategy of the dominant plant species and their comparative rates of plant cover regenerating after fire:

- Plant communities dominated by resprouters with deep root systems (e.g. species of *Quercus, Pistacia, Arbutus, Rhamnus, Phyllirea, Erica, Daphne* and some *Ulex*) that recover very rapidly after fire from organs which have been preserved below ground.
- Plant communities dominated by species regenerating after fire from seeds (e.g. species of *Cistus, Rosmarinus officinalis* and *Ulex parviflorus*). The response of these obligate seeders to fire in terms of projected plant cover is much slower than that of the resprouters.

In general, tree survival can vary according to the intensity of the fire, the height of the crown and bark thickness (Ryan and Reinhardt, 1988). Other considerations such as fire intensity, fire season and water availability at the time of the fire and during the immediate post-fire period can be of great importance in determining rate of recovery for seeders and, to a lesser extent, for sprouters (Moreno and Cruz, 2000). Also the shade tolerance of species and light conditions after fire can be fundamental in determining the post-fire recovery of vegetation.

In most cases, however, plants are top-killed by fire; plant succession in fire-adapted systems, as, for example, in most Mediterranean communities, consists largely of the recovery of the species present before fire through their respective life-cycle and growth processes. This process is called autosuccession (Hanes, 1971) in which the burned stand, although initially appearing to be different from the pre-burned one, retains its floristic identity in time.

8.1.2 Grazing

Grazing effects on vegetation have been studied in many ecosystems. It is clear that herbivores cause changes in species richness (Basset, 1980; Persson, 1984; Bakker *et al.*, 1983; Wimbush and Costin, 1979).

An important parameter of grazing disturbance is the specificity of grazers to species, to age and size classes and to community structure. However, there are large differences in selectivity between animals. Large animals are generally less selective in their diet than are smaller animals. In general, animals select the smaller or younger plants within a species, they select only some species within a plant community and select only some plant community types within a landscape (Dumont, 1997; Legg *et al.*, 1998).

Grazing animals exert controls on the rates of several important processes in ecosystems. The significance of this fact can be easily understood from the conclusion by Wolfe and Berg (1988) that selective browsing of young deciduous trees by ungulates has effectively eliminated many deciduous hardwood species from European forests, while conifers usually escape damage.

Many similar studies can be found in the literature for other herbivores. The spatial distribution of wild rabbit (*Oryctolagus cuniculus*) in dry conditions, for example, was found to be correlated with water and forage availability, but also with the existence of a protective cover of shrubs or oak trees. At the same time consumption of acorns and oak seedlings by rabbits can affect natural oak regeneration (Martins, 2001).

8.1.3 Fire and grazing interactions

It has been suggested that where multiple disturbances occur simultaneously they generally have nonadditive effects on ecosystems (White and Pickett, 1985). These effects may be either synergistic (greater than additive) or antagonistic (less than additive).

Fire and grazing may have antagonistic effects over relatively short time scales. Grazing reduces plant growth and removes available fuel, thus reducing the intensity of fire, and their combined effect is less than expected. A good example of these relationships is presented by Turner (1985) who found that the impacts of multiple disturbances (clipping, grazing and fire) on vegetation were not as severe as expected based on the disturbances applied singly. It is also known that woody species that evolved to recover rapidly following disturbance have a greater ability to recover from browsing than those species that have slower growth rates (Bryant *et al.*, 1991).

Early work by Braun-Blanquet described different successional pathways following changes in land use with different regimes of burning and grazing. Naveh (1974) described the regrowth of Mediterranean plants and recognized that the presence of heavy grazing pressure could affect the regeneration processes. Interactions between fire and herbivory have been described in Mediterranean environments for dwarf shrub communities of *Calicotome villosa* and *Sarcopoterium spinosum*; the dominance of the latter species is reduced with the exclusion of goat grazing (Henkin *et al.*, 1999). Similarly, *Calluna* heathland dynamics have been reported to show huge differences with or without sheep grazing (Ellenberg, 1988). Miles (1985) clearly showed how the successional transitions between different vegetation types could be changed and redirected by different combinations of grazing and fire frequency (Figure 8.1).

Quinn (1986) produced a detailed review of the interactions between mammalian herbivory and post-fire resilience in Mediterranean ecosystems. Clear evidence has been reported on how the selective feeding of post-fire regrowth by grazing animals affects the interspecific plant competition (Leigh and Holgate, 1979; Hesp *et al.*, 1983; Mills, 1983). After a fire, herbivores can either delay the recovery processes or change the successional outcome (Quinn, 1986). Grazing animals have a preference for young palatable vegetation, so they tend to congregate on recently burned areas (Cates and Orians, 1975). However, the grazing pressure on any one patch of vegetation depends heavily on the other vegetation types available to the grazing animals within the vicinity. It is clear that the spatial habitat patterns in the landscape and the relative location and size of burned areas will affect the distribution of herbivores (Quinn, 1986; Oom *et al.*, 2002).

8.2 MODEL SIMPLIFICATION: SIMULATION OF PLANT GROWTH UNDER GRAZING AND AFTER FIRE

The representation of such complex systems in computer models requires considerable simplification. Models must have clear objectives and these, together with the availability of data, will determine the key processes that should be included and the level of aggregation that will be possible without unacceptable loss of precision. Models with different objectives may therefore have very different structures. These have ranged in the past from very simple probability-based models of transitions between vegetation states as Markovian processes through to detailed spatially explicit individual-based models where each individual plant is represented by a physiological growth model.

Markov models are the simplest form of succession model. Markov models can be applied where a patch of vegetation can be classified into one of a limited number of possible states and vegetation changes by transitions from one state to another. If the probability of transition to another particular state depends only on the current state, then the process is Markovian. These models have been used successfully by several authors to describe succession (e.g. Horn, 1975, for forest trees; Gimingham *et al.*, 1981, for heathland; and Rego *et al.*, 1993, for Mediterranean garrigue). However, as has frequently been pointed out, vegetation is not Markovian (Usher, 1992). More complex models can be developed (e.g. Malanson, 1984, for coastal sage scrubs dominated by *Artemisia californica*, *Salvia leucophylla* and *S. mellifera*) but these models are perhaps better used as descriptions of past change, or as simple null-models for detecting nonrandom processes (e.g. Hobbs and Legg, 1984; Lippe *et al.*, 1985).

A simple development of Markov models is to replace the transition probabilities with rules that express the circumstances in which a transition between two different states will occur. In fact, many descriptive successional models use qualitative information and have been simply expressed in verbal, diagrammatic or tabular form (e.g. Braun-Blanquet, 1954; Walter, 1973; Nahal, 1981). These qualitative rule-based models can also be implemented as computer models that predict directions of change, though predicting rates of change may require a more quantitative approach (McIntosh *et al.*, 2002; Oxley, 2000).

A different approach has been to disaggregate vegetation to the species level and develop rules that express the ecological properties and behaviour of the component species. Noble and Slatyer (1980) identified the 'vital attributes' that relate to recurrent disturbance by fire. The effects of different fire regimes could be predicted from knowledge of the life-history characteristics of the species and the mechanism of surviving or recolonizing an area after fire. A rule-based model of *Banksia* species, taking into account interspecific competition for space, provided some evidence that species coexistence

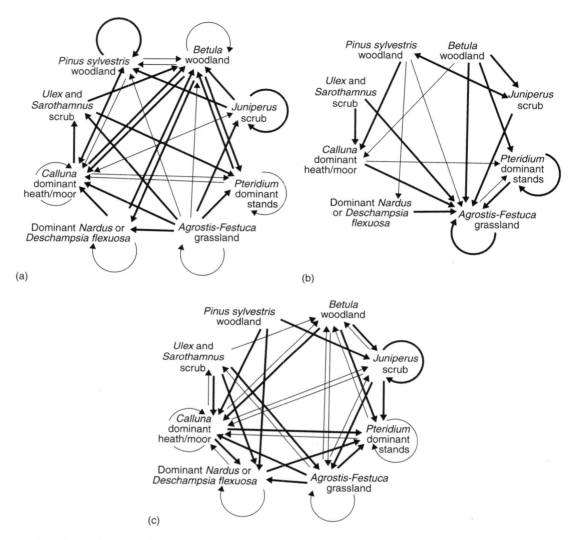

Figure 8.1   Successional transitions in northeast Scotland under (a) low grazing pressure and no fire; (b) high grazing pressure and frequent fires; and (c) intermediate grazing pressure and infrequent burning. Arrow thickness refers to frequency of transition
Source: Miles (1985)

depended on both fire regime and topographic gradients (Groeneveld *et al.*, 2002).

Historically, models of forest dynamics have been developed either at the stand level, with individual trees grown deterministically using an annual time-step difference equation, and by simulation calculating the fate of each individual, or at the level of some integrated unit, such as an age cohort. This approach was initially developed in JABOWA by Botkin *et al.* (1972) and applied to different forest ecosystems (e.g. FORET by Shugart

and West, 1977; BRIND by Shugart and Noble, 1981). However, in spite of their widespread use, this family of individual-based gap-phase models provided limited explanations for the simulated successional response and did not account for spatial processes.

Spatial structure modelling has largely been derived from individual-based ideology, taking into account nearest-neighbour interactions (Ford, 1975). These interactions could be a result of shading and/or redistribution of the soil and water resources among the trees.

Hybrid models that combine the initial individual-based models with process models were a further step forward (Mladenoff and Baker, 1999), now taking into consideration the whole forest stand-soil system including the role of ground vegetation in the redistribution of resources for growth (Chertov *et al.*, 1999).

Since the earliest days of systems ecology (Odum, 1983) modelling of forest ecosystems as whole entities has been seen as another important required step. As a consequence, biogeochemical process models were developed to simulate vertical fluxes of carbon, water and nitrogen between ecosystem components based on fundamental physical and ecophysiological relationships. As in FOREST-BGC, individual trees are generally not identified, and the model assumes horizontal homogeneity (Running and Coughlan, 1988; Running and Gower, 1991; Keane *et al.*, 1996).

The consideration of landscapes as heterogeneous land areas composed of clusters of interacting ecosystems initiated landscape ecology as a science (Forman and Godron, 1986). Regional and landscape models are now considered as indispensable for forestry and environmental planning and management but, in spite of some attempts (Waring and Running, 1998), it seems that this field of forest simulation has been developed less intensively than stand models of various types (Chertov *et al.*, 1999).

Recent modelling applications have shown how both patch distribution in the landscape and plant dispersal capability can affect the patterns of competition and dynamics of the vegetation (Higgins and Cain, 2002). Other simulation work has also shown the importance of the presence of 'stepping stones', or islands of vegetation, in the survival and distribution of species with short-range dispersal ability (Söndgerath and Schröder, 2002). Both growth rate and dispersal range influenced the outcome of a spatially explicit population model subjected to patch destruction and recovery dynamics (Johst *et al.*, 2002).

There have been several different approaches to modelling the use of plant resources by herbivores. Foraging theory has been used to make many predictions on foraging patterns of herbivores (e.g. Charnov, 1976). It was initially proposed that ungulates tend to maximize digestible energy intake (van Soest, 1982), but other authors claim that food selection cannot be based simply on the 'energetic optimization' theory, since a very few chemical constituents, which are usually volatile and unstable, play a major role in plant defence against browsing animals (Bryant *et al.*, 1991).

It is, however, generally accepted that the quality of browse is as important, or more important, than the amount available, and it is known that quality changes with plant species, plant age and environment conditions, making predictions more difficult. Nevertheless, there are some successful examples where, for example, it has been found that the spatial distribution of female red deer (*Cervus elaphus*) could be modelled by an interaction of vegetation biomass and quality (Fonseca, 1998); this is as expected from the dependence of reproductive success on their ability to acquire resources (Clutton-Brock and Harvey, 1978). Models of resource selection have been used extensively in predicting the spatial distribution of animals (Manly *et al.*, 1993). It is also important to recall that under drought conditions the spatial distribution of ungulates is often strongly constrained by the availability of drinking water (Western, 1975).

When simulating the long-term effects of the combination of fire and grazing in heterogeneous mixed forests of pine and oak, it was concluded that grazing may encourage the development of pine stands by selective foraging on oaks (Turner and Bratton, 1987). These pine stands probably burn more readily than oak forests, thus enhancing the spread of fire and reducing potential oak recovery. Thus, over the long term, grazing may interact synergistically with fire (effects greater than additive) resulting in the reduction of the forest matrix and the spread of fire-resilient shrubs (Turner and Bratton, 1987). Fire-effects models and models of resource selection by herbivores can be applied to predict the combined effects of the two factors.

One of the biggest difficulties associated with developing spatially explicit models is the volume of data and computational effort required to scale up from individual-level models to landscapes. It is clear that different ecological processes operate at different temporal and spatial scales (Levin, 1992; Legg *et al.*, 1998) and these can only be incorporated into ecological models through a hierarchical system with disturbances and processes represented at the correct scale (Forman and Godron, 1986; Chertov *et al.*, 1999).

One approach to this problem is described in Legg *et al.* (1998) where models of vegetation response to herbivory are constructed at three different spatial and temporal scales: individual level, community level and landscape level. The dispersal of animals across the landscape is determined by vegetation and topographic characteristics at the landscape level. The selection by herbivores of particular plant species, the effects of dunging and trampling of soils and seed dispersal by the animals are modelled community-level processes. The response of plants to damage through allocation of resources to roots, shoots and reproduction are individual-level processes. These three spatial scales can be modelled separately in a hierarchical system with each level modelling only a sample of the area represented at the level above, but a sufficient sample to capture the behaviour of that level of the whole system.

8.3 NEW DEVELOPMENTS IN ECOSYSTEM MODELLING

In general, simple population models (see Renshaw, 1991, for a review) have described the competition between species with few state variables, either without or with poor explicit representation of spatial interactions, whereas individual-based models of ecological populations (e.g. De Angelis and Gross, 1992) have been facing the complexity of competitive spatial interactions among neighbours. More recently, the increasing interest in spatial modelling has been associated with integrated applications of Geographic Information Systems (GIS) and with the development of specific object-oriented methodologies (Sequeira *et al.*, 1991; Muetzelfeldt and Massheder, 2003).

The understanding of systems as complex as vegetation and ecosystems requires integrated analysis at different scales. Engelen *et al.* (1995) recognized how GIS, beside their great capability of handling spatial information, have intrinsic limitations for representing dynamic processes. Thus, these authors proposed a modelling system based on two interacting components for micro and macro-level processes. They reported as an example an integrated simulation of socio-economic and environmental models applied at different scales (ibid.). This work was a very good proof of concept, but the system was not fully accessible to users since the change of input maps and model functions was difficult.

More recently, Boumans *et al.* (2001) pointed out the need for the 'formulation of general unit models for simulation of temporal processes' at the landscape level, in contrast to specific models for particular habitat types. They provided an example of the application of a nonspatial ecological model within elements of a river landscape.

An interesting example of the coupling of a biological model based on differential equations with a spatial cellular automata model has been presented recently by Aassine and El Jaï (2002), who underlined how this integrated approach is consistent with a natural way of describing different processes. We fully agree with their conclusion and believe that their approach needs to be implemented with an integration of generic models as pointed out by Boumans *et al.* (2001).

Another aspect to be considered is that traditional model implementation has been by conventional programming languages, such as Fortran, Basic and C, whereas recent trends in the scientific literature show an increasing application of generic modelling environments. These software tools (see Costanza and Gottlieb, 1998, and Costanza and Voinov, 2001) allow the construction of models by graphical interfaces that do not require knowledge of conventional programming languages because the executable program is automatically compiled by the system. Among these, STELLA (www.hps-inc.com) has a wide application in the field of ecology and agriculture modelling (e.g. Liddel, 2001; Pan and Raynal, 1995; Chiavaura-Mususa *et al.*, 2000). SIMULINK (www.mathworks.com) is a supporting tool for MATLAB package and is characterized by strong computational capacity and access to mathematical libraries. Other available packages are POWERSIM (www.powersim.com) and ModelMaker (www.cherwell.com).

SIMILE (www.simulistics.com) is a recent addition to the family of the system dynamic modelling environments and presents very interesting enhanced capabilities well suited for ecological modelling work (Muetzelfeldt and Massheder, 2003).

A new software tool named SIMARC (www.ecoap. unina.it/software.htm) was recently developed to interface models created by SIMILE with ArcView (ESRI Inc.) GIS environment. The tool enables users to link a model to input data from the GIS database. This essentially amounts to running a model in each polygon element of the ArcView map with inputs from layers of the relative shapefile and creating new GIS layers from any selected model output variable. Considering the intrinsic power of SIMILE for ecological modelling, its combination with the spatial analysis power of a GIS has great potential for many applications.

The system described in the following modelling exercise illustrates a step in the direction outlined above. We made use of two modelling environments: the SIMILE system and a raster-based spatial modelling system called LANDLORD (Mazzoleni and Legg, 2001). The former system has been used to model plant community-level processes, whereas the latter system handles spatial processes at the landscape level. An additional software tool named LSD (Landlord Simile Dynamic link) allows the dynamic integration of models produced by the two systems. The advantage of using a new dedicated spatial tool instead of an established GIS package lies in its enhanced modelling capabilities, i.e. the possibility of making use of temporal simulations at both local and spatial scales in a highly integrated way. The graphical interfaces of the integrated system are quite simple and user-friendly and allow the flexible creation of different modelling workspaces of increasing complexity in a transparent and modular way.

The future of advanced ecological modelling will be based on the possibility of implementation on new computational scenarios, such as enhanced power by grid computing and parallelization techniques. Moreover, the use of remote access to modelling workstations will make feasible the development of much more complex ecological models and simulation applications.

8.4 INTERACTIONS OF FIRE AND GRAZING ON PLANT COMPETITION: FIELD EXPERIMENT AND MODELLING APPLICATIONS

8.4.1 A case study

An experiment to study the interactions of fire and cutting with grazing in Mediterranean shrubland was established in the Tapada Nacional de Mafra, a forested hunting reserve 40 km north of Lisbon in Portugal. The main shrub species in the area are *Erica scoparia* and *E. lusitanica*, *Ulex jussiaei* and *Rubus ulmifolius*. The total area of the Tapada is about 900 hectares intensively used by around 500 fallow deer (*Dama dama*), 60 red deer (*Cervus elaphus*), and a similar number of wild boar (*Sus scrofa*).

Only part of the experiment is reported in this chapter, i.e. a fire treatment with four replicates plots of 5 m × 20 m subdivided into two subplots (grazed and protected). The experiment was established in December 1997, and vegetation sampling was carried out from January 1998 using the line-intercept method first proposed by Canfield (1941) for cover measurements made at each plant intercepted. Average densities (individual stems per m^2) were 2.1, 17.4 and 3.2 for *Ulex*, *Erica* and *Rubus* respectively.

In order to evaluate the effects of herbivory and its interactions with fire, the post-treatment recovery of major plant species was regressed on time for the different treatment combinations.

The results are shown in Figure 8.2 which highlights some interesting interactions. *Rubus ulmifolius* is known to have high palatability because of the high protein content of its leaves. Moreover, herbivores like the very young twigs and leaves produced by vegetative resprouting after fire. This is clearly reflected by the total suppression of this species under the grazing treatment. A similar trend, though to a lesser extent, is shown by *Ulex jussiaei*, whereas *Erica* spp. showed the opposite behaviour by gaining a competition advantage in the

grazed areas, thus reflecting the dislike of animals for this species. These results fully agree with the theoretical view of selective foraging behaviour affecting post-fire successional pathways (Quinn, 1986).

8.4.2 A model exercise

In order to reflect the post-fire regrowth patterns observed in the experimental study, a simple model was defined with plant species being characterized by different growth rates, competition sensitivities and palatability values. No differences between species were set in terms of post-fire resprouting capacity. The model resembles in some aspects Tilman's (1988, 1994) competition models, but it presents a more flexible behaviour without a predefined dominance hierarchy between species. In our model, the growth rate of all species is affected (with different sensitivities) by resource level and by the available space, whereas mortality is increased by grazing. Biomass growth is a function of growth rate and density of growing stems. This latter function makes the model sensitive to the initial composition and it does not show asymptotic convergence to equilibrium values as in the Tilman approach.

The SIMILE model structure is shown in Figure 8.3, the mathematical model functions are listed below:

$$\frac{dUlex}{dt} = MaxGrowthRateUlex$$
$$* (1 - AllCover) * Uw * Uc * Density$$
$$- (Ud + GrazingEffect) * Ulex^2 \quad (8.1)$$

$$\frac{dErica}{dt} = MaxGrowthRateErica$$
$$* (1 - AllCover) * Ew * Ec * Density$$
$$- (Ed + GrazingEffect) * Erica^2 \quad (8.2)$$

$$\frac{dRubus}{dt} = MaxGrowthRateRubus$$
$$* (1 - AllCover) * Rw * Rc * Density$$
$$- (Rd + GrazingEffect) * Rubus^2 \quad (8.3)$$

and Table 8.1 reports the description of model elements and inputs.

The exercise below starts with an application of the SIMILE model at the same spatial and temporal scale as the real study case reported above. The application is then extended both in time and space in order to show the system capability to upscale local (pixel) functional dynamics and to integrate them with other landscape-level processes.

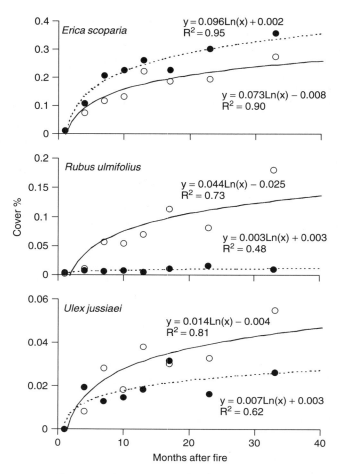

Figure 8.2   Post-fire recovery according to grazing treatment of main shrub species of Mediterranean shrubland, Mafra, Portugal. Data represent averages of field-plot measurements. Dotted lines and closed symbols represent grazed plots

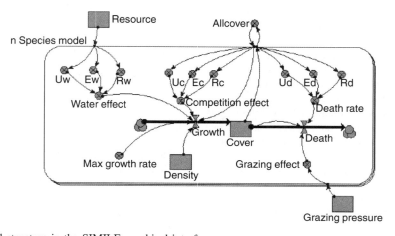

Figure 8.3   Model structure in the SIMILE graphical interface

Table 8.1   Model elements and inputs

Name	Meaning	Units
Erica	Erica cover	%
Rubus	Rubus cover	%
Ulex	Ulex cover	%
AllCover	Sum of cover of all species	%
MaxGrowthRateSpecie	Maximum growth rate (for all species)	
Death Rate (Ud, Ed, Rd)	Death rate (for all species)	
Water effect (Uw, Ew, Rw)	Growth resource function (for each species)	[0,1]
Competition effect (Uc, Ec, Rc)	Growth competition sensitivity (for each species)	[0,1]
Resource	Resource	[0,1]
Density	Number of initial individuals for each species	0–20
Grazing pressure	Animal presence	0,1
Grazing effect	Response to grazing pressure (for each species)	[0,1]

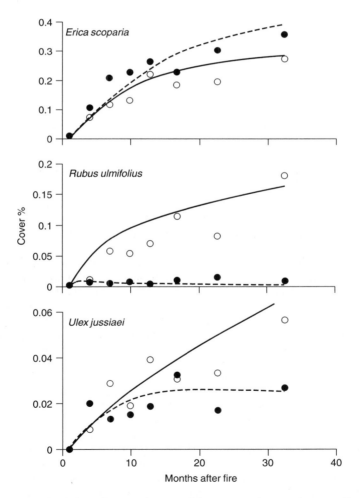

Figure 8.4   Results of model simulation of cover changes of three competing species according to grazing treatment. Data are the same as reported in Figure 8.2. Dashed lines and closed symbols represent grazed plots

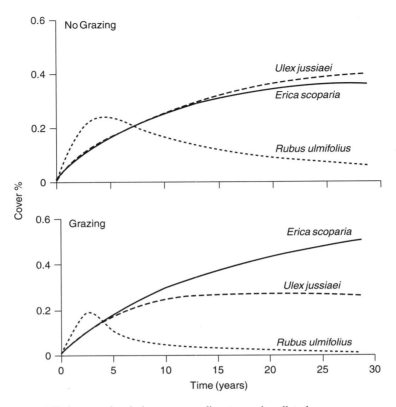

Figure 8.5 Long-term modelled successional changes according to grazing disturbance

First, the changes of relative cover in the three species were modelled under the same post-fire conditions (plant densities and grazing treatments) as in the real study case. The good fit between the regressions based on real data (Figure 8.2) and the outputs of the process-based model (Figure 8.4) are evident.

Second, another simulation of plant regrowth after fire was done with equal numbers of individuals for the different species; this removed the competitive advantage due to the higher initial density of *Erica* species. Moreover, the model run was extended to 30 years in order to investigate longer-term successional behaviour (Figure 8.5). In this case, without grazing, *Rubus* shows an initial dominance followed by a decline due to its sensitivity to competition, whereas *Erica* and *Ulex* can shift their final dominance in relation to the grazing pressure.

A schematic representation of the vegetation model linked to pixels of the spatial raster system is shown in Figure 8.6a. Results of a simple exercise simulation show the emergence of spatial patterns of species dominance being related to both the grazing pressure

and initial density conditions (Figure 8.7). Such patterns simply reflect the isolated model behaviour being distributed in space with changing conditions.

The complexity of the exercise can be increased (Figure 8.6b) by adding a fire disturbance with removal of biomass in burned pixels and by the use of a dynamic model to simulate the distribution of animals in the landscape and hence grazing pressure (Legg *et al.*, 1998). The simulation shows (Figure 8.8) a concentration of herbivores in the burned area, which reflects the higher grazing value for the early successional-stage vegetation. However, as grazing removes the highly palatable *Rubus*, and to a lesser extent *Ulex*, the animals spread again around the landscape with a more even distribution. The distribution of animals, of course, creates feedback to the growth of plants and the process of vegetation dynamics.

A final exercise included in the workspace a seed-dispersal model (Figure 8.6c) (Heathfield *et al.*, 2001). While in the previous examples all species were considered to be present in all pixels from the start of the simulation, in this case patchy distributions were considered

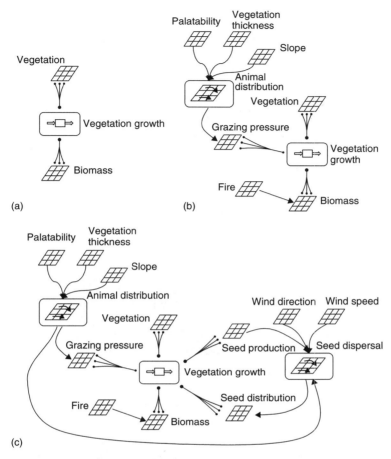

Figure 8.6 Schematic representation of modelling workspaces: (a) simple link between SIMILE n-species model and the raster system; (b) integration of animal distribution model with pixel level vegetation model; and (c) integration of two spatial processes (animal distribution and seed dispersal) with pixel level vegetation model

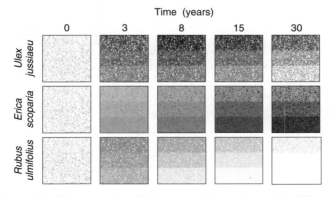

Figure 8.7 Results of 30-year simulation exercise of three competing species under different grazing pressures. Three rectangular areas were set to represent grazing exclosure (TOP) and low and high grazing pressure (MIDDLE and BOTTOM areas respectively)

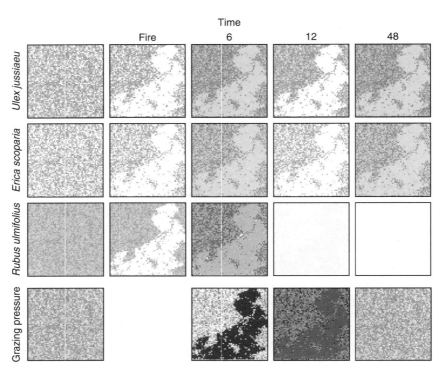

Figure 8.8   Results of 48-year model run of simulated post-fire recovery of three competing species with dynamic changes of grazing pressure

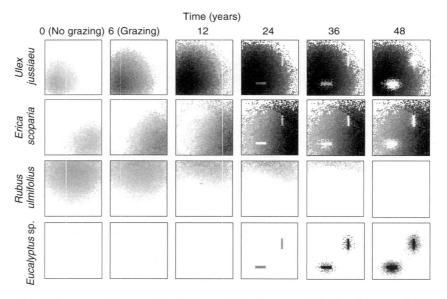

Figure 8.9   Results of 60-year model run of simulated vegetation dynamics including plant growth (pixel process) and seed dispersal (spatial process). At start, only three competing species are present, whereas the fourth is introduced as two plantations after 35 years. Dispersal from planted plots alter the dominance patterns of the natural vegetation in the surrounding area

with new establishment by colonization from neighbouring cells occurring during the modelling exercise.

The results are shown in Figure 8.9 and demonstrate an expanding area of dominance by *Ulex* and to a lesser extent by *Erica*, whereas *Rubus* is strongly reduced by competition with the other species.

8.5 CONCLUSION

The modelling exercise reported in this chapter shows the power of existing modelling tools to recover complex interactions of ecological processes, such as the combined effects of grazing and fire on patterns of plant competition.

Other modelling workspaces can be built which include other integrated objects and spatial models. In particular, the addition of changing resource levels (e.g. water uptake and infiltration processes) can be added to look at their effect on the spatial pattern of plant communities (Rietkerk *et al.*, 2002).

These new modelling environments can be considered revolutionary in their flexibility and power to represent complex systems and to integrate across scales. The traditional gap between 'field' ecologists and mathematically-minded or computer-programming modellers seems to have been closed by the newly available modelling technologies. It is now time to build up banks of models to be used as elementary building blocks of larger-scale dynamic systems. This should enable the development of truly process-based models that are more robust, with greater generality, and with behaviour that better resembles that in the real world.

ACKNOWLEDGEMENTS

The development of the modelling system used in the simulation exercise has been largely supported by the EU-DG.XII ModMED research projects (ENV4-CT97-0680).

We thank Antonello Migliozzi for support to input data and graphical preparation.

REFERENCES

Aassine, S. and El Jaï, M.C. (2002) Vegetation dynamics modelling: a method for coupling local and space dynamics, *Ecological Modelling* **154**, 237–249.

Abrahamson, W.G. and Abrahamson, C.R. (1996) Effects of fire on long-unburned Florida uplands, *Journal of Vegetation Science* **7**, 564–574.

Angelstam, P. (1997) Landscape analysis as a tool for the scientific management of biodiversity, *Ecological Bulletins* **46**, 140–170.

Arianoutsou, M. (1998) Aspects of demography in post-fire mediterranean plant communities of Greece, in P.W. Rundel, G. Montenegro and F.M. Jaksic (eds) *Landscape Disturbance and Biodiversity in Mediterranean-Type Ecosystems*, Springer-Verlag, Berlin, 273–295.

Bakker, J.P., de Bie, S., Dallinga, J.H., Tjaden, P. and de Vries, Y. (1983) Sheep-grazing as a management tool for heathland conservation and regeneration in the Netherlands, *Journal of Applied Ecology* **20**, 541–560.

Bassett, P.A. (1980) Some effects of grazing on vegetation dynamics in the Camargue, France, *Vegetatio* **43**, 173–184.

Botkin, D.B., Janak, J.F. and Wallis, J.R. (1972) Some ecological consequences of a computer model of forest growth, *Journal of Ecology* **60**, 849–872.

Boumans, R.M., Villa, F., Costanza, R., Voinov, A., Voinov, H. and Maxwell, T. (2001) Non-spatial calibrations of a general unit model for ecosystem simulations, *Ecological Modelling* **146**, 17–32.

Braun-Blanquet, J. (1954) *Pflanzensoziologie*, 2nd edn, Springer, Vienna.

Bryant, J.P., Provenza, F.D., Pastor, J., Reichardt, P.B., Clause, T.P. and du Toit, J.T. (1991) Interactions between woody plants and browsing animals mediated by secondary metabolites, *Annual Review of Ecology and Systematics* **22**, 431–446.

Canfield, R.H. (1941) Application of the line interception method in sampling range vegetation, *Journal of Forestry* **39**, 388–394.

Cates, R. and Orians, G. (1975) Successional status and the palatability of plants to generalized herbivores, *Ecology* **56**, 410–418.

Charnov, E.L. (1976) Optimal foraging: the marginal value theorem, *Theoretical Population Biology* **9**, 129–136.

Chertov, O.G., Komarov, A.S. and Karev, G.P. (1999) *Modern Approaches in Forest Ecosystem Modelling*, European Forest Institute Research Report 8, Koninklijke Brill NV, Leiden.

Chiavaura-Mususa, C., Campbell, B. and Kenyon, W. (2000) The value of mature trees in arable fields in the smallholder sector, Zimbabwe, *Ecological Economics* **33**, 395–400.

Clutton-Brock, T.H. and Harvey, P.H. (1978) Mammals, resources and reproductive strategies, *Nature* **273**, 191–195.

Costanza, R. and Gottilieb, S. (1998) Modelling ecological and economic systems with STELLA: Part II, *Ecological Modelling* **112**, 81–84.

Costanza, R. and Voinov, A. (2001) Modelling ecological and economic systems with STELLA: Part III, *Ecological Modelling* **143**, 1–7.

De Angelis, D.L. and Gross, L.J. (1992) *Individual-based Models and Approaches in Ecology*, Chapman & Hall, New York.

Dumont, B. (1997) Diet preferences of herbivores at pasture, *Annales de Zootechnie* **46** (2), 105–116.

Ellenberg, H. (1988) *Vegetation Ecology of Central Europe*, Cambridge University Press, Cambridge.

Engelen, G., White, R., Uljee, I. and Drazan, P. (1995) Using cellular automata for integrated modelling of socio-environmental systems, *Environmental Monitoring and Assessment* **34**, 203–214.

Espírito-Santo, M.D., Rego, F. and Costa, J.C. (1993) Vegetation dynamics in the Serra dos Candeeiros (central Portugal), in L. Trabaud and R. Prodon (eds) *Fire in Mediterranean Ecosystems*, Ecosystem Research Report no. 5, Commission of the European Communities, Brussels, 29–46.

Ferran, A. and Vallejo, V.R. (1998) Long-term plant regeneration after wildfires in Mediterranean ecosystems of NE Spain, in L. Trabaud (ed.) *Fire Management and Landscape Ecology*, International Association of Wildland Fire, Washington, DC, 155–166.

Fonseca, M.M. (1998) Plasticity of mating behaviour in red deer (*Cervus elaphus*) in a Mediterranean environment, PhD dissertation, University of London. London.

Ford, E.D. (1975) Competition and stand structure in some even-aged plant monocultures, *Journal of Ecology* **63**, 311–333.

Forman, R.T.T. and Godron, M. (1986) *Landscape Ecology*, John Wiley & Sons, New York.

Gimingham, C.H., Hobbs, R.J. and Mallik, A.U. (1981) Community dynamics in relation to management of heathland vegetation in Scotland, *Vegetatio* **46**, 149–155.

Groeneveld, J., Enright, N.J., Lamont, B.B. and Wissel, C. (2002) A spatial model of coexistence among three *Banksia* species along a topographic gradient in fire-prone shrublands, *Journal of Ecology* **90**, 762–774.

Grubb, P.J. (1985) Plant populations in relation to habitat. Disturbance and competition: problem of generalization, in J. White and W. Beeftink (eds) *The Population Structure of Vegetation. Handbook of Vegetation Science*, vol. 3, Harper and Row, New York, 83–97.

Grubb, P.J. and Hopkins, A.J.M. (1986) Resilience at the level of the plant community, in B. Dell, A.J.M. Hopkins and B.B. Lamont (eds) *Resilience in Mediterranean-Type Ecosystems. Task for Vegetation Science 16*, Dr. W. Junk Publishers, Dordrecht, 113–128.

Hanes, T.L. (1971) Succession after fire in the chaparral of southern California, *Ecological Monographs* **41**, 27–52.

Heathfield, D., Arianoutsou, M., Georghiou, K., Thanos, C. and Loureiro, C. (2001) Seed dispersal, in S. Mazzoleni and C.J. Legg (eds) *ModMED: Modelling Mediterranean Ecosystem Dynamics, Final Report*, ModMED III Project, EU-DGXII Environment (IV) Research Programme, Contract ENV4-CT97-0680, Brussels.

Heinselman, M.L. (1978) Fire intensity and frequency as factors in the distribution and structure of northern ecosystems, in Fire Regimes and Ecosystems Properties: Proceedings of the Conference, USDA – Forest Service. General Technical Report WO-26, Washington, DC, 7–57.

Henderson, M.K. and Keith, A.D. (2002) Correlation of burning and grazing indicators with composition of woody understorey flora of dells in a temperate eucalypt forest, *Australian Ecology* **27**, 121–131.

Henkin, Z., Seligam, G.S., Noy-Meir, I. and Kafkafi, U. (1999) Secondary succession after fire in a Mediterranean dwarf-shrub community, *Journal of Vegetation Science* **10**, 503–514.

Hesp, P., Wells, M., Ward, B. and Riches, J. (1983) *Land Resource Survey of Rottnest Island*, Bulletin No. 4086, Soil and Conservation Service Branch, Division of Resource Management, Department of Agriculture, Western Australia, Perth.

Higgins, S. and Cain, L.M. (2002) Spatially realistic plant metapopulation models and the colonization-competition trade-off, *Journal of Ecology* **90**, 616–626.

Hobbs, R.J. and Legg, C.J. (1984) Markov models and initial floristic composition in heathland vegetation dynamics, *Vegetatio* **56**, 31–43.

Horn, H.S. (1975) Markovian properties of forest succession, in M.L. Cody and J.M. Diamond (eds) *Ecology and Evolution of Communities*, Harvard University Press, Cambridge MA, 196–211.

Johst, K., Brandl, R. and Eber, S. (2002) Metapopulation persistence in dynamic landscapes: the role of dispersal distance, *Oikos* **98**, 263–270.

Keane, R.E., Morgan, P. and Running, S.W. (1996) *FIRE-BGC A Mechanistic Ecological Process Model for Simulating Fire Succession on Coniferous Forest Landscapes of the Northern Rocky Mountains*, USDA Forestry Service Research Paper, INT-RP-484, Washington, DC.

Legg, C., Papanastasis, V., Heathfield, D., Arianoutsou, M., Kelly, A., Muetzelfeldt, R. and Mazzoleni, S. (1998) Modelling the impact of grazing on vegetation in the Mediterranean: the approach of the ModMED Project, in V.P. Papanastasis and D. Peter (eds) *Ecological Basis of Livestock Grazing in Mediterranean Ecosystems*, EC Science Research and Development, Brussels, 189–199.

Leigh, J. and Holgate, W. (1979) The response of the understorey of forest and woodlands of the Southern Tablelands to grazing and browsing, *Australian Journal of Ecology* **4**, 23–43.

Levin, S.A. (1992) The problem of pattern and scale in ecology, *Ecology* **73**, 1943–1967.

Liddel, M. (2001) A simple space competition model using stochastic and episodic disturbance, *Ecological Modelling* **143**, 33–41.

Lippe, E., de Smidt, J.T. and Glenn-Lewin, D.C. (1985) Markov models and succession: a test from heathland in the Netherlands, *Journal of Ecology* **73**, 775–791.

Malanson, G.P. (1984) Linked Leslie matrices for the simulation of succession, *Ecological Modelling* **21**, 13–20.

Manly, B.F., McDonald, L.L. and Thomas, D.L. (1993) *Resource Selection by Animals: Statistical Design and Analysis for Field Studies*, Chapman & Hall, London.

Martins, H. (2001) Aspectos da ecologia e da gestão cinegética do coelho bravo (*Oryctolagus cuniculus* L.) em montado, PhD thesis, Instituto Superior de Agronomia, UTL, Lisbon.

Mazzoleni, S. and Legg, C.J. (eds) (2001) *ModMED: Modelling Mediterranean Ecosystem Dynamics: Final Report*, ModMED III Project, EU-DGXII Environment (IV) Research Programme, Contract ENV4-CT97-0680, Brussels.

Mazzoleni, S. and Pizzolongo, P. (1990) Post-fire regeneration patterns in Mediterranean shrubs in the Campania region, southern Italy, in J.G. Goldammer and M.J. Jenkins (eds) *Fire in Ecosystem Dynamics: Mediterranean and Northern Perspectives*, SPB Academic, The Hague, 43–52.

McIntosh, B.S., Muetzelfeldt, R.I., Legg, C.J., Mazzoleni, S. and Csontos, P. (2003) Reasoning with direction and rate of cange in vegetation state transition modelling, *Environmental Modelling and Software* (in press).

Miles, J. (1979) *Vegetation Dynamics*. Chapman & Hall, London.

Miles, J. (1985) The pedogenic effects of different species and vegetation types and the implications of succession, *Journal of Soil Science* **36**, 571–584.

Mills, J. (1983) Herbivory and seedling establishment in post-fire southern California chapparal, *Oecologia* **60**, 267–270.

Mladenoff, D.J. and Baker, W.L. (eds) (1999) *Spatial Modelling of Forest Landscape Change: Approaches and Applications*, Cambridge University Press, Cambridge.

Moreno, J.M. and Cruz, A. (2000) La respuesta de las plantas al fuego, in R. Velez (ed.) *La Defensa contra Incendios Forestales: Fundamentos y Experiencias*, McGraw-Hill, Madrid, 413–436.

Muetzelfeldt, R. and Massheder, J. (2003) The Simile visual modelling environment, *European Journal of Agronomy* **18**, 345–358.

Nahal, I. (1981) The Mediterranean climate from a biological viewpoint, in F. Di Castri, D.W. Goodall and R.L. Specht (eds) *Mediterranean-Type Shrublands: Ecosystems of the World 11*, Elsevier, Amsterdam, 63–86.

Naveh, Z. (1974) Effect of fire in the Mediterranean region, in T.T. Kozlowski and C.E. Ahlgren (eds) *Fire and Ecosystems*, Academic Press, New York, 401–432.

Noble, I.R. and Slatyer, R.O. (1980) The use of vital attributes to predict successional changes in plant communities subject to recurrent disturbance, *Vegetatio* **43**, 5–21.

Odum, H.T. (1983) *Systems Ecology: An Introduction*, John Wiley & Sons, New York.

Oom, S.P., Hester, A.J., Elston, D.A. and Legg, C.J. (2002) Spatial interaction models from human geography to plant-herbivore interactions, *Oikos* **98**, 65–74.

Oxley, T., McIntosh, B.S., Mulligan, M. and de Roode, R. (2000) Adaptation, integration and application of an integrative modelling framework in MODULUS, in G. Engelen (ed.) *MODULUS: A Spatial Modelling Tool for Integrated Environmental Decision Making: Final Report*, The Modulus Project, EU-DGXII Environment, Brussels.

Pan, Y. and Raynal, D.J. (1995) Decomposing tree annual volume increments and constructing a system dynamic model of tree growth, *Ecological Modelling* **82**, 299–312.

Persson, S. (1984) Vegetation development after the exclusion of grazing cattle in a meadow area in the south of Sweden, *Vegetation* **55**, 65–92.

Quinn, R.D. (1986) Mammalian herbivory and resilience in mediterranean-climate ecosystems, in B. Dell, A.J.M. Hopkins and B.B. Lamont (eds) *Resilience in Mediterranean-Type Ecosystems: Tasks for Vegetation Science 16*, Dr. W. Junk Publishers, Dordrecht, 113–128.

Rego, F., Pereira, J. and Trabaud, L. (1993) Modelling community dynamics of a *Quercus coccifera* L. garrigue in relation to fire using Markov chains, *Ecological Modelling* **66**, 251–260.

Renshaw, E. (1991) *Modelling Biological Populations in Space and Time*, Cambridge University Press, Cambridge.

Rietkerk, M., Boerlijst, M.C., van Langevelde, F., HilleRis-Lambers, R., van de Koppel, J., Kumar, L., Prins, H. and de Roos, A.M. (2002) Self-organization of vegetation in arid ecosystems, *The American Naturalist*, **160** (4), 524–530.

Running, S.W. and Coughlan, J.C. (1988) A general model of forest ecosystem processes for regional applications. I. Hydrologic balance, canopy gas exchange and primary production processes, *Ecological Modelling* **42**, 125–154.

Running, S.W. and Gower, S.T. (1991) FOREST-BGC, a general model of forest ecosystem processes for regional applications. II. Dynamic carbon allocation and nitrogen budgets, *Tree Physiology* **9**, 147–160.

Ryan, K.C. and Reinhardt, E.D. (1988) Predicting postfire mortality of seven western conifers, *Canadian Journal of Forest Research* **18**, 1291–1297.

Sequeira, R.A., Sharpe, P.J.H., Stone, N.D., El-Zik, K.M. and Makela, M.E. (1991) Object-oriented simulation: plant growth and discrete organ-to-organ interaction, *Ecological Modelling* **58**, 25–54.

Shugart, H.H. and Noble, I.R. (1981) A computer model of succession and fire response of the high altitude Eucalyptus forest of the Brindabella Range, Australian Capital Territory, *Australian Journal of Ecology* **6**, 149–164.

Shugart, H.H. and West, D.C. (1977) Development of an Appalachian deciduous forest succession model and its application to assessment of the impact of the chestnut blight, *Journal of Environmental Management* **5**, 161–179.

Söndgerath, D. and Schröder, B. (2002) Population dynamics and habitat connectivity affecting the spatial spread of population – a simulation study, *Landscape Ecology* **17**, 57–70.

Specht, R.L., Rayson, P. and Jackman, M.E. (1958) Dark Island heath (Ninety-Mile Plain, South Australia). VI. Pyric succession: changes in composition, coverage, dry weight and mineral nutrient status, *Australian Journal of Botany* **6**, 193–202.

Tilman, D. (1988) *Plant Strategies and the Dynamics and Structure of Plant Communities*, Princeton University Press, Princeton, NJ.

Tilman, D. (1994) Competition and biodiversity in spatially structured habitats, *Ecology* **75**, 2–16.

Turner, M. (1985) Ecological effects of multiple perturbations on a Georgia salt marsh, Doctoral dissertation, University of Georgia, Athens, GA.

Turner, M. and Bratton, S. (1987) Fire, grazing and the landscape heterogeneity of a Georgia Barrier Island, in M. Turner (ed.) *Landscape heterogeneity and disturbance, Ecological Studies 64*, Springer-Verlag, Berlin, 85–101.

Usher, M.B. (1992) Statistical models of succession, in D.C. Glenn-Lewin, R.K. Peet and T.T. Veblen (eds) *Plant Succession: Theory and Prediction*, Chapman & Hall, London, 215–248.

Van Soest, P.J. (1982) *Nutritional Ecology of the Ruminant*, O. and B. Books, Inc, Corvallis, OR.

Walter, H. (1973) *Vegetation of the Earth and Ecological Systems of the Geobiosphere*, Springer-Verlag, New York.

Waring, R.H. and Running, S.W. (1998) *Forest Ecosystems: Analysis at Multiple Scales*, 2nd edn, Academic Press, San Diego.

Western, D. (1975) Water availability and its influence on the structure and dynamics of a savannah large mammal community, *East African Wildlife Journal* 13, 265–268.

White, P.S. and Harrod, J. (1997) Disturbance and diversity in a landscape context, in J.A. Bissonette (ed.) *Wildlife and Landscape Ecology: Effects of Pattern and Scale*, Springer-Verlag, New York, 128–159.

White, P.S. and Pickett, S.T.A. (1985) Natural disturbance and patch dynamics: an introduction, in S.T.A. Pickett and P.S. White (eds) *The Ecology of Natural Disturbance and Patch Dynamics*, Academic Press, New York, 3–13.

Wimbush, D.J. and Costin, A.B. (1979) Trends in vegetation at Kosciusko. I Grazing trials in the sub-alpine zone, 1957–1971, *Australian Journal of Botany* 27, 741–787.

Wolfe, M.L. and Berg, F.C. (1988) Deer and forestry in Germany – half a century after Aldo Leopold, *Journal of Forestry* 86, 25–28.

9

Erosion and Sediment Transport

JOHN N. QUINTON

9.1 THE COMPLEXITY

Erosion and sediment transport appears to be a relatively straightforward topic for modelling. Rain falls and impacts on the soil surface, soil particles are dislodged and may be transported by rain splash or overland flow. Overland flow can also detach material and transport it. This conceptualization of erosion processes on an inter-rill area (Figure 9.1) was used by Meyer and Wischmeier (1969) in their mathematical model of the process of soil erosion and is still found at the heart of many process-based soil-erosion models today. Two key processes are identified: detachment and transport. Meyer and Wischmeier modelled splash detachment as a function of rainfall intensity, an approach also adopted in ANSWERS, KINEROS2, and GUEST (Morgan and Quinton, 2001). The transport capacity of the rainfall was described as a function of soil, slope and rainfall intensity. Flow detachment was modelled as a function of discharge slope and soil erodibility and the transport capacity of the flow as a function of slope, discharge and a soil factor. For splash and flow detachment, the sediment detached is compared with a value for transport capacity; if the value is below the transport capacity figure, the material is transported, above it is deposited. Meyer and Wischmeier recognized that their model was not comprehensive and highlighted the need for further components to be added to it.

Similar conceptual flowcharts can be found for a number of current generation erosion models, such as EUROSEM (Morgan et al., 1998) and WEPP (Flanagan and Nearing, 1995; Laflen, et al. 1997). However, unlike the model of Meyer and Wischmeier, the current generation of erosion models has tried to model the controls on these processes in more detail. Figure 9.2

illustrates some of the other factors that have to be taken into account when calculating detachment and transport. The inclusion of submodels to describe these processes makes the representation more complex. Many of the processes included are known to be important, but are only poorly described mathematically, with expressions only tested using data from a handful of situations. For example, EUROSEM contains an expression linking soil cohesion to the resistance of the soil to erosion by flowing water based on only a few measurements made in the laboratory.

Figure 9.2 is probably reasonably complete for describing erosion on an inter-rill area. However, erosion does not take place only on inter-rill areas. It is a process that often links inter-rill areas with transient or permanent channel networks, the position of which may vary in time and space, with some parts of the hill slope contributing sediment and water and others contributing none at all. The processes within rills and gullies, such as bank collapse and headwall retreat, are poorly understood and are almost universally ignored in erosion models, yet may contribute more than 50% of the sediment from a gully system (Blong, 1985). Where rills are included in models, such as EUROSEM and WEPP, sidewall and head processes are not simulated and the models erode the channels by assuming that material is removed equally from the sidewalls and bed.

This discussion focuses on processes that take place over timescales of minutes or hours. However, for many situations, such as global change studies (Williams et al., 1996), we may wish to understand erosion and sedimentation over years, decades or centuries. To do so requires us to understand changes in the controls over erosion that may take place over differing timescales, for example, the consolidation of soil after tillage and

Environmental Modelling: Finding Simplicity in Complexity. Edited by J. Wainwright and M. Mulligan
© 2004 John Wiley & Sons, Ltd ISBNs: 0-471-49617-0 (HB); 0-471-49618-9 (PB)

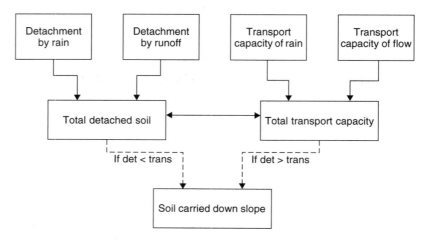

Figure 9.1   Conceptual model of inter-rill processes
Source: After Meyer and Wischmeier (1969)

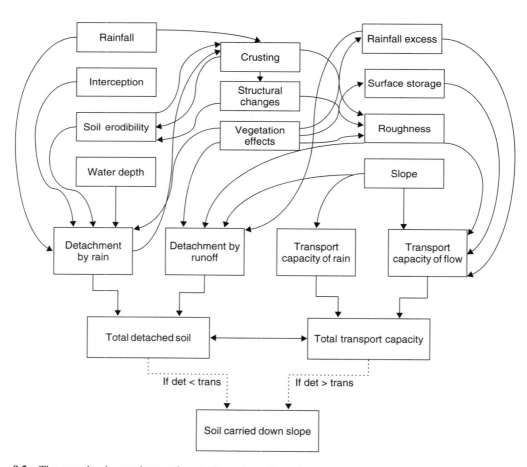

Figure 9.2   The complex interactions and controls on the soil erosion process

its effects on soil erodibility (Nearing *et al.*, 1988). Although water-erosion processes take place over fairly short timescales, the factors which control them may change over periods lasting several minutes, such as surface sealing, to several centuries for changes in topography or rates of pedogenesis. This spatial and temporal dimension to the erosion problem makes describing erosion processes even more complex.

Not only are the processes variable in time but also in space. In a study of 36 erosion plots all treated identically, Wendt (1986) found that erosion rates varied between 18 and 91%. This may be a response to the variability of parameters that control the generation of runoff and erosion. The spatial variability of hydraulic parameters, such as hydraulic conductivity, is well documented (Beckett and Webster, 1971; Warrick and Nilesen, 1980). Other parameters such as soil cohesion (Quinton, 1994) also exhibit considerable variation in the field.

So we find ourselves in a situation where many of the current generation of erosion models are getting more complex, with more processes being incorporated into them. They also suffer from considerable uncertainty (Quinton, 1997; Brazier *et al.*, 2000) associated with parameterization difficulties, and are generally poor at reproducing measured data (see, for example, Favis-Mortlock, 1998), which is highly variable (Wendt *et al.*, 1986; Nearing *et al.*, 1999). This chapter sets out to answer the question: does the level of uncertainty associated with both erosion model predictions and erosion model measurement mean that we can utilize simpler approaches to erosion prediction to those which are currently in vogue?

9.2 FINDING THE SIMPLICITY

As Kant (1885) states, 'hypotheses always remain hypotheses i.e. suppositions to the full reality of what we can never obtain'. In other words, even the most complex model of erosion and sedimentation processes can only still be a representation of reality, so how complex do they need to be?

As with most environmental modelling there are broadly two groups of people interested in building models: scientists and managers (Bunnel, 1989). Scientists usually do most of the model building, but tend to focus on representing all the processes which they believe control a particular system, while managers are more concerned about getting an accurate answer from the model so that they can make a decision. The fact that scientists have been driving the construction of erosion and sediment transport models perhaps explains the trend towards ever more complex models in recent years. While, on the face of it, this drive to ever more complex models has resulted in a better representation of the erosion system, it does have its problems.

9.2.1 Parameter uncertainty

Determining the inputs for soil-erosion models is difficult. Input parameters vary in time and space, and some are difficult to measure. Spatial variability poses difficulties for the modeller when trying to parameterize models. The spatial variability of many input parameters of commonly used erosion models is known to be high. For example, saturated hydraulic conductivity measurements may have coefficients of variation of 150% for a single soil-map unit (Warrick and Nilesen, 1980). Even though many erosion models are distributed, allowing different parameter sets to be used for different parts of the catchment, the model will impose a grid or element upon the landscape. Within each grid square or element, the model inputs may vary considerably, but they will be represented in the model as a single value. Beven (1989) gives the example of a variation in hydraulic conductivity causing runoff in one part of an element, where it is low, and infiltrating it in another area. If an average value has been used, it may be possible that the model erroneously predicts small amounts of runoff. In the case of EUROSEM, many of the parameters were identified as sensitive during the sensitivity analysis of the model (Veihe and Quinton, 2000; Veihe *et al.*, 2000), for example, saturated hydraulic conductivity and hydraulic roughness, and therefore have a great influence over the model outputs.

Temporal variation causes more uncertainty in the parameterization of erosion models. Here the problem is to identify the initial conditions for the model at the beginning of the simulation. The characteristics of the soil and vegetation are not static: their properties change through time owing to biological, chemical and physical processes. In the case of event-based models, even where regular measurement programmes to collect input data (e.g. soil and vegetation characteristics) have been undertaken, it is only by chance that a storm occurs directly after measurements have taken place. This problem can, for some of the inputs, be remedied by using automated sampling equipment, for example, tensiometers linked into a data-logger can give information on soil-water potential. However, at present the number of input parameters that can be measured in this way is limited.

Rather than measure the inputs, a different approach is to use a continuous simulation model to simulate the change in the key model inputs prior to an event. This is the approach that has been adopted for MIKE SHE SEM (Styczen and Nielsen, 1989) and WEPP (Flanagan and Nearing, 1995). This approach also has its limitations: describing the dynamics of soil structure, soil water and the vegetation is not straightforward, resulting in further uncertainty surrounding the values of input parameters.

Uncertainty surrounding the value of input-parameter values will have a limited impact on model output if the model is not sensitive to changes in the value of the parameter in question. Sensitivity analysis of the EUROSEM model by Quinton (1994), Veihe (2000) and Veihe *et al.* (2000) showed that many of EUROSEM's parameters did not invoke a sensitive response from the model. A similar pattern can be observed for WEPP in the work of Brazier *et al.* (2000). However, both models are sensitive to a number of input parameters and when this is combined with uncertainty over parameter values, there will be uncertainty in the model output results.

To investigate the impact of parameter uncertainty, Quinton (1997) applied the EUROSEM model to 0.1-ha erosion plots from the Woburn Erosion Reference Experiment (Catt *et al.*, 1994; Quinton *et al.*, 2001). He ran the EUROSEM model 525 times using different parameter sets. The parameter sets were constructed by taking four parameters to which the model was sensitive and constructing probability distributions for them based on the results of field sampling. Thus, all parameter sets had the same parameter values, except for the four sensitive parameters. All parameter sets could be assumed to be an equally likely representation of the system. The results showed that uncertainty bands on the output were large. The mean width of the 95% confidence interval for peak sediment discharge was $62.5 \, \text{kg} \, \text{min}^{-1}$, and for soil loss 217 kg. Using the same source of observed data, Brazier *et al.* (2000) assessed the uncertainty surrounding predictions of annual soil loss from the WEPP model and found uncertainty bands of up to $10 \, \text{t} \, \text{a}^{-1}$.

9.2.2 What are the natural levels of uncertainty?

Given the levels of uncertainty reported in the previous section, two things come to mind: is the natural world as uncertain as the virtual one described in erosion models? If so, can we use this uncertainty to make the case that we should describe the erosion system with simpler models than those that are currently in vogue?

Work carried out by Wendt *et al.* (1986) using 40 erosion plots, all with the same treatment, in Kingdom City,

Minnesota, found that coefficients of variation for the 25 storms ranged from 18% to 91%. The larger erosion events showed less variability. Very little of the variability could be attributed to measured plot properties, and the plots did not perform in the same manner relative to each other in subsequent events. Nearing *et al.* (1999) also considered the performance of replicates in erosion experiments. They took 2061 replicated storm events from 13 sites in the USA and found, like Wendt *et al.*, that the measured coefficient variation in the soil loss declined with event magnitude. It would appear that measured erosion rates will be subject to high degrees of variation. If this is the case and as Nearing *et al.* (1999) state, the study of Wendt *et al.* strongly implies that given our current ability to measure plot characteristics in the field, we will not be able to improve model-simulation results for these data, i.e. we cannot measure and therefore parameterize our models to reflect the differences between replicated plots. If this is the case, then is the addition of more process descriptions and their associated parameters simply a waste of time?

9.2.3 Do simple models predict erosion any better than more complex ones?

So far this chapter has demonstrated that predictions from erosion models suffer from considerable uncertainty and that there is considerable variation between similar sites exposed to the same rainfall. Do these high levels of uncertainty mean that even those models which offer an apparently more complete description of the erosion process perform no better than simple models?

Recently there have been a number of attempts to compare erosion models. Favis-Mortlock (1997) drew together the results from models simulating annual soil loss for a field site in the USA. The models ranged from simple models, such as CSEP (Kirkby and Cox, 1995) and those based on the USLE (Wischmeier and Smith, 1978) such as EPIC (Williams *et al.*, 1984) and GLEAMS (Leonard *et al.*, 1987), to the more process-based, and complex WEPP. Favis-Mortlock's results show no significant difference between the performance of the models. Morgan and Nearing (2000) compare the performance of the USLE, RUSLE and WEPP using 1700 plot years of data from 208 natural runoff plots (Figure 9.3). Once again it becomes clear that the more complex model, in this case WEPP, performs no better than the empirical, five parameter USLE (see Chapter 16).

It might be argued that the examples given above are based on models with quite different structures, and are therefore unfair. However, recent work by Morgan

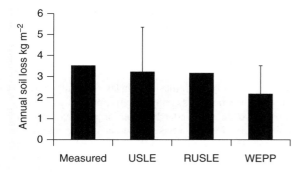

Figure 9.3 Comparison of the USLE, RUSLE and WEPP with measured data representing 1700 plot years of data from 208 natural runoff plots (average annual soil loss $3.51 \, \text{kg} \, \text{m}^{-2}$)
Source: Morgan and Nearing (2000)

(2001) allows us to consider the impact of increasing the number of parameters within a model, on that model's ability to simulate data from a wide variety of conditions. Morgan takes the Morgan–Morgan–Finney erosion model (Morgan *et al.*, 1984) and modifies it to deal with flow-erosion processes – the previous version dealt solely with inter-rill erosion. He then applied it to erosion data from 67 sites around the world, the same data set that he originally used to test the previous version of the model. The information in Morgan's

paper can be plotted (Figure 9.4), and illustrates that there is little difference in the performance of the two models. In fact, a comparison of the Nash and Sutcliffe (1970) coefficient of efficiency suggests that the older version with fewer parameters does marginally better at simulating mean annual soil loss across all the sites.

So it appears that more parameters do not produce better simulation results; that is not to say however, that the inclusion of further process descriptions in a model is not a worthy activity. From a scientific point of view, the inclusion of a flow detachment module in the MMF model makes the model a better representation of reality, but it does not make it better able to predict erosion rates.

There appear to be few comparisons between spatially complete data sets and incomplete data sets. One of the few is that of Loague (1990) who studied the effect that 157 extra infiltration measurements had on model simulations and found that there was not an 'overwhelming improvement in model performance'.

9.3 FINDING SIMPLICITY

In a situation where variability in observed sediment transport and difficulties in parameterization make the use of complex models unnecessary or simply not feasible, it might be expected that there would be efforts made to develop simpler approaches to erosion prediction.

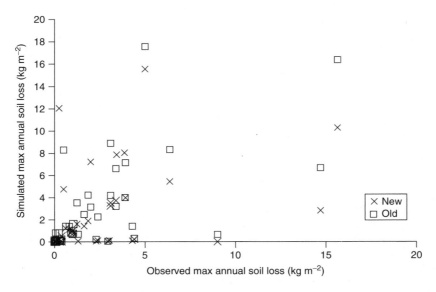

Figure 9.4 Comparison of simulated maximum annual soil loss against observed maximum annual soil loss for the revised (new) MMF model with the original (old) version of the model
Source: Morgan (2000)

To date, the response of the modelling community to this problem has been to try to mask the complexity. Both the EUROSEM and WEPP developers have developed user-friendly interfaces with database support to supply parameter values. As yet, there appear to be few attempts to develop models with a reduced number of processes or number of parameters. Where there is activity in producing simpler models is where there is a need to make regional-scale assessments of erosion rates or erosion risks. Kirkby *et al.* (2000) present a model for estimating erosion risk as a soil loss in tonne ha^{-1}. The model captures the key driver influencing erosion rates: topography. Soil erodibility, topography, vegetation and climate and data requirements appear low, with eight database layers and two tables needed to provide information for the model. However, a number of the parameters required by the model need to be derived from these data using transfer functions. The initial application of the model of Kirkby *et al.* was to the whole of France. However, as yet, they have conducted no validation of their approach.

So it seems that at present the search for simplicity is being driven forward by the need to make erosion estimates at ever increasing scales. However, erosion prediction is still dominated by the use of one model – the USLE. It provides the user with a simple model which can be executed using nothing more sophisticated than a pocket calculator, yet in model comparison studies predicts annual erosion as well as any model. This may not reflect any superiority in terms of the model, but simply the fact that a five-parameter empirical model can do as well as a multi-parameter model in capturing the extremely variable response of a hillside system.

9.4 THE RESEARCH FRONTIER

The combination of our inability to measure model inputs and the high levels of variation in the very things we are trying to predict suggest that increasing the number of processes represented in soil-erosion models and therefore the number of inputs required will not lead to better predictions of erosion and sedimentation. We need simpler ways to simulate erosion and sedimentation. Rule-based approaches, such as those described by Favis-Mortlock *et al.* (2000; see also Chapter 21), provide one possibility, although they still require large amounts of data. Perhaps it is more necessary to embrace the uncertainty in model simulation results and the uncertainty in data sets with which models are compared and use it as an opportunity to simplify our approaches to erosion prediction. The extra process descriptions and the extra parameters may make interesting science but they do not produce better predictions.

9.5 CASE STUDY

This case study tests the hypothesis that only a small number of parameters control the response of an erosion model and the remainder become largely inconsequential. To do this the revised Morgan–Morgan–Finney model (MMF) (Morgan, 2001) is selected. The MMF model is used because it is relatively simple, yet physically based, and can be programmed into a spreadsheet quickly. It is expected that other more complex models such as EUROSEM and WEPP will exhibit similar characteristics if they were to be tested in a similar way.

Full details of the model can be found in Morgan (2001) and the operating equations can be found in Table 9.1. A copy of the spreadsheet containing the program can be downloaded from the book's website (http://www.kcl.ac.uk/envmod). The MMF model separates erosion into the water and sediment phases. The energy available for soil detachment is derived from the rainfall and the volume of water determines the transport capacity. The model uses Meyer and Wischmeier's (1969) scheme in which sediment production is compared with the amount of sediment that the water is able to hold and the lower value taken as the sediment transport rate. The model operates on an annual timestep.

The model was applied to an erosion plot from the Woburn Erosion Reference Experiment (Catt *et al.*, 1994; Quinton, 1994; Quinton *et al.*, 2001). The soils at the site range in texture from loamy sand to sandy loam and correspond to the Cottenham and Lowlands series defined by Clayden and Hollis (1984) and classified as Lamellic Ustipsamment and Udic Haplustept respectively (Soil Survey Staff, 1999). The experiment was established after harvest in 1988 and the first crop of potatoes was planted in the following spring. This crop was followed by two winter cereals, then sugar beet and two more years of winter cereals – a crop rotation common in the area. The plot was cultivated up and down with residues removed before mouldboard ploughing. The plot measured approximately 25 m by 35 m (894 m^2) and was isolated from the rest of the slope by a low earth bank. Soil and water flowing off each plot was channelled to a collecting trough and through a pipe to two 2000-litre tanks where it was stored until sampled. The amount of runoff and soil loss from each plot was determined as soon after each runoff event as practically possible and usually within 48 hours. Annual soil losses were calculated for each agricultural year (crop and following fallow combined).

Table 9.1 Operating equations of the MMF model

No.	Description	Operating equation	Parameter d
9.1	Effective rainfall (mm)	$R_e = RA$	R = rainfall (mm) A rainfall reaching soil surface
9.2	Leaf drainage (mm)	$L_d = R_e\, C_c$	C_c = Canopy cover
9.3	Direct throughfall (mm)	$D_t = R_e - L_d$	
9.4	Kinetic energy of direct throughfall (J m^{-2})	$K_{et} = D_t(11.9 + \log I)$	I = typical rainfall intensity value for erosive rain (mm hr^{-1})
9.5	Kinetic energy of leaf drainage (J m^{-2})	$K_{el} = L_d(15.8\, P_h^{0.5}) - 5.87$	P_h is the plant canopy height (m)
9.6	Total Kinetic energy (J m^{-2})	$K_e = K_{et} + K_{el}$	
9.7	Annual runoff (mm) Soil moisture capacity (mm)	$Q = R e^{\frac{-R_c}{R_0}}$ $R_c = 1000\, M_c\, B_d\, H_d\, (E_t/E_o)$	R_o = mean rain per day (mm) M_c = soil moisture content at field capacity (% w/w) B_d = Bulk density of the soil (Mg m^{-3}) H_d = Effective hydraulic depth (m) E_t/E_o = the ratio of actual to potential evapotranspiration.
9.8	Soil particle detachment by raindrop impact (kg m^{-2})	$F = 0.001\, K\, K_e$	K the erodibility of the soil (g J^{-1})
9.9	Soil particle detachment by runoff (kg m^{-3})	$H = Z\, Q^{1.5} \sin S\, (1 - G_c)\, 10^{-3}$	S = slope steepness G_c = Ground cover
9.10	Resistance of the soil	$Z = \dfrac{1}{0.5\sigma}$	σ = soil cohesion (kPa)
9.11	Transport capacity of runoff	$T_C = C\, Q^2 \sin 10^{-3}$	C = the product of the C and P factors of the Universal Soil Loss Equation

Source: Morgan (2000)

Rather than apply the model in a deterministic fashion, probability distributions were assigned to each of the parameters (Table 9.2). These were either determined from field measurement, the literature or from best estimates. To facilitate this a commercial risk-analysis package – Crystal Ball[1] – was used. This package allows distributions to be assigned to single spreadsheet cells and makes Monte Carlo analysis possible within a spreadsheet environment. To avoid improbable combinations of parameters, where parameters depended upon one another, they were identified as either positively or negatively correlated prior to the simulation. The sampling of these parameter distributions is then constrained to produce the desired correlation.

Monte Carlo analysis was performed by randomly sampling from each distribution and executing the model; 1000 model runs were performed. To investigate the role of model parameters to the overall output, a sensitivity analysis was first performed. This was conducted by computing the Spearman rank correlation coefficient between every parameter and the simulated annual erosion. Three further simulations were then

run, with increasing numbers of parameters having their values frozen (Table 9.2), following their influence on calculated erosion, i.e. the parameters to which the model was least sensitive to were held constant first. For the last simulations only four of the parameters were allowed to vary.

The results of this exercise are presented in Figure 9.5. A log scale is used in an attempt to separate the distributions. No significant difference can be found between the four probability distributions simulated, demonstrating that the four most sensitive parameters are controlling the shape of the distribution. Changing some of the parameters which impact on the erosion equations can shift the position of the distribution on the y-axis. For example, if the slope steepness is increased from 9 to 14% the distribution can be shifted up (Figure 9.5).

This distribution can be compared with the observed data from the Woburn Erosion Reference Experiment (Figure 9.5). The observed data sit below the simulated distributions. However, since the observation period was only ten years, it could be argued that the extreme

Table 9.2 Parameter values and distributions used in the simulations and the run in which the parameter was held constant

Parameter		Mean	SD	Frozen value	Run number 1	2	3	4
Soil erodibility	g J⁻¹	0.8	0.08	0.77			X	X
Slope steepness	o	9	1	7.5	X	X	X	X
Cohesion	kPa	3.5	0.6	3.49	X		X	X
Ground cover		0.2	0.02	0.2			X	X
C factor		0.2	0.02	0.18			X	X
P factor							X	X
Mean annual rainfall	Mm	644.5	113.4					
Number of rain days		40	14.8					
Soil moisture at field capacity	%w/w	0.45	0.03	0.42				X
Bulk density	Mg m⁻³	1.2	0.12	1.17				X
Effective hydrological depth	m	0.08	0.02					
Ratio of actual to potential evaporation		0.8	0.08					
Interception ratio		0.7	0.07	0.62				X
Canopy cover		0.5	0.1	0.83				X
Typical value of rainfall intensity for erosive storms	mm hr⁻¹	10.0	5.0	10.37			X	X
Plant canopy height	m	0.7	0.07	0.71	X	X	X	X

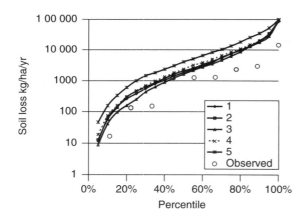

Figure 9.5 Cumulative distribution of total soil loss for plot seven of the Woburn Erosion Experiment (o) and as simulated by the MMF model with (1) all parameters enabled; (2) five parameters frozen; (3) ten parameters frozen; (4) 14 parameters frozen; and (5) with 14 parameters frozen, but slope increased from 9° to 14°

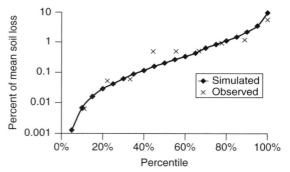

Figure 9.6 Comparison of the normalized distribution of soil loss for plot seven of the Woburn Erosion Reference Experiment with that simulated by the MMF model

values at the upper end of the distribution were not encountered. If this were the case, then the distribution could be shifted to the left, making the fit to the simulated data much closer. Normalizing the distribution

using the mean soil loss can test this idea to a certain extent. In Figure 9.6 the soil loss, expressed as a percentage of the mean, is compared for parameter set one and the observed data. The fit is extremely close. We should not be surprised to see the closer fit between the observed and simulated distributions than plots of observed and simulated points, such as for Figure 9.4, as comparing distributions of outputs in this way we remove the direct link between individual observed and simulated data points. In fact, if the data presented in Figure 9.4 are compared in the same way, the observed and simulated distributions also look very similar. However, by comparing distributions we do get a sense of how well erosion models are able to simulate longer-term means and the variability surrounding them, and this may be more useful for models producing annual predictions, given the likely variation in observed annual soil loss for any one site.

This example shows that the distribution of model outputs from the MMF model is controlled largely by four parameters: mean annual rainfall, the number of rain days, the effective hydrological depth and the ratio of actual to potential evapotranspiration. The other parameters influence the position of the distribution but not its shape.

The question therefore is, could a model, such as the MMF, be simplified by removing or modifying equations to reduce its complexity? If the results of the Monte Carlo analysis and the equations presented in Table 9.1 are considered, some reduction in complexity would seem possible. If leaf drainage is ignored, due to its limited influence over the final model simulations, then Equations 9.5 and 9.6 and their associated parameters can be removed. Equation 9.10 could also be simplified to combine the two resistance terms Z and Gc. If these changes are made, and the simulations are run again, the distribution is remarkably similar to those with all equations and parameters. The removal of the leaf drainage equations do reduce the complexity of the model, but it is arguable whether the other changes make much difference. A resistance term is still required by Equation 9.10, which then becomes an empirical parameter requiring calibration, whereas the two parameters currently included in Equation 9.10 are measurable in the field.

This approach demonstrates that even in a model such as MMF, which has a low number of input parameters and operating equations, there is still some, if limited, scope for simplification.

NOTE

[1] At the time of writing an evaluation copy of Crystal Ball could be downloaded from www.decisioneering. com

REFERENCES

Beckett, P.H.T. and Webster, R. (1971) Soil variability: a review, *Soils and Fertilisers* **34**, 1–15.

Beven, K. (1989) Changing ideas in hydrology – the case of physically-based models, *Journal of Hydrology* **105**, 157–172.

Blong, R.J. (1985) Gully sidewall development in New South Wales, Australia, in S.A. El-Swaify, W.C. Moldenhauer and A. Lo (eds) *Soil Erosion and Conservation*, Soil Conservation Society of America, Ames, IO, 575–584.

Brazier, R.E., Beven, K.J., Freer, J. and Rowan, J.S. (2000) Equifinality and uncertainty in physically based soil erosion models: application of the GLUE methodology to WEPP – The Water Erosion Prediction Project – for sites in the UK and USA, *Earth Surface Processes and Landforms* **25**, 835–845.

Bunnel, F.R. (1989) *Alchemy and Uncertainty: What Good Are Models?* USDA Forest Service, Pacific Northwest Research Station, Portland, OR.

Catt, J.A., Quinton, J.N., Rickson, R.J. and Styles, P.D.R. (1994) Nutrient losses and crop yields in the Woburn Erosion Reference Experiment, in R.J. Rickson (ed.) *Conserving Soil Resources: European Perspectives*, CAB International, Wallingford, 94–104.

Clayden, B. and Hollis, J.M. (1984) *Criteria for Differentiating Soil Series*, Soil Survey Technical Monograph No. 17, Harpenden, UK.

Favis-Mortlock, D.T. (1997) Validation of field scale soil erosion models using common data sets, in J. Boardman and D.T. Favis-Mortlock (eds) *Modeling Soil Erosion by Water*, NATO ASI series, Springer-Verlag, Berlin, 89–127.

Favis-Mortlock, D.T. (1998) A self-organizing dynamic systems approach to the simulation of rill initiation and development on hillslopes, *Computers in Geoscience* **24**, 353–372.

Favis-Mortlock, D.T., Boardman, J., Parsons, A.J. and Lascelles, B. (2000) Emergence and erosion: a model for rill initiation and development, *Hydrological Processes* **14**, 2173–2205.

Flanagan, D.C. and Nearing, M.A. (1995) *USDA-Water Erosion Prediction Project: Hillslope Profile and Watershed Model Documentation*, NSERL Report No. 10, USDA-ARS National Soil Erosion Research Laboratory, West Lafayette, IN 47097–1196.

Kant, I. (1885) *An Introduction to Logic* (trans. T.K. Abbott), Longmans Green and Company, London.

Kirkby, M.J. and Cox, N.J. (1995) A climatic index for soil-erosion potential (CSEP) including seasonal and vegetation factors, *Catena* **25** (1–4), 333–352.

Kirkby, M.J., Le Bissonais, Y., Coulthard, T.J., Daroussin, J. and McMahon, M.D. (2000) The development of land quality indicators for soil degradation by water erosion, *Agriculture, Ecosystems and Environment* **81**, 125–135.

Laflen, J.M., Elliot, W.J., Flanagan, D.C., Meyer, C.R. and Nearing, M.A. (1997) WEPP – Predicting water erosion using a process-based model, *Journal of Soil and Water Conservation* **52**, 96–102.

Leonard, R.A., Knisel, W.G. and Still, D.A. (1987) GLEAMS – Groundwater loading effects of agricultural management systems, *Transactions of the ASAE* **30**, 1403–1418.

Loague, K.M. (1990) R-5 revisited. 2, Reevaluation of a quasi-physically-based rainfall-runoff model with supplemental information, *Water Resources Research* **26**, 973–987.

Meyer, L.D. and Wischmeier, W.H. (1969) Mathematical simulation of the process of soil erosion by water, *Transactions of the ASAE* **12**, 754–758.

Morgan, R.P.C. (1994) A predictive model for the assessment of soil erosion risk, *Journal of Agricultural Engineering Research* **30**, 245–253.

Morgan, R.P.C. (2001) A simple approach to soil loss prediction: a revised Morgan–Morgan–Finney model, *Catena* **44**, 305–322.

Morgan, R.P.C. and Quinton, J.N. (2001) Erosion modelling, in W.W. Doe and R.S. Harmon (eds) *Landscape Erosion and Evolution Modelling*, Kluwer Academic Press, New York.

Morgan, R.P.C. and Newing, M.A. (2002) Soil erosion models: present and future, in J.L. Rubio, R.P.C. Morgan, S. Asins and V. Andreu (eds) *Man and Soil at the Third Millennium*, Geoforma Ediciones, Logroño, 187–205.

Morgan, R.P.C., Quinton, J.N., Smith, R.E., Govers, G., Poesen, J.W.A., Auerswald, K., Chisci, G., Torri, D. and Styczen, M.E. (1998) The European soil erosion model (EUROSEM): a process-based approach for predicting sediment transport from fields and small catchments, *Earth Surface Processes and Landforms* **23**, 527–544.

Nash, J.E. and Sutcliffe, J.V. (1970) River flow forecasting through conceptual models 1. A discussion of principles, *Journal of Hydrology* **10**, 282–90.

Nearing, M.A., West, L.T. and Brown, L.C. (1988) A consolidation model for estimating changes in rill erodibility, *Transactions of the ASAE* **31**, 696–700.

Nearing, M.A., Govers, G. and Norton, L.D. (1999) Variability in soil erosion data from replicated plots, *Soil Science Society of America Journal* **63**, 1829–1835.

Quinton, J.N. (1997) Reducing predictive uncertainty in model simulations: a comparison of two methods using the European Soil Erosion Model, *Catena* **30**, 101–117.

Quinton, J.N. (1994) The validation of physically based erosion models – with particular reference to EUROSEM. Unpublished PhD thesis, Cranfield University, Silsoe, Bedford, UK.

Quinton, J.N., Catt, J.A. and Hess, T.M. (2001) The selective removal of phosphorus from soil: is event size important? *Journal of Environmental Quality* **30**, 538–545.

Soil Survey Staff (1999) *Soil Taxonomy: A Basic System of Soil Classification for Making and Interpreting Soil Surveys*, 2nd edn, United States Department of Agriculture, Natural Resources Conservation Service, Washington, DC.

Styczen, M. and Nielson, S.A. (1989) A view of soil erosion theory, process research and model building: possible interactions and future developments, *Quaderni di Science de Suolo* **2**, 27–45.

Veihe, A. and Quinton, J.N. (2000) Sensitivity analysis of EUROSEM using Monte Carlo simulation: I hydrologic, soil and vegetation parameters, *Hydrological Processes* **14** (5), 915–926.

Veihe, A., Quinton, J.N. and Poesen, J.A. (2000) Sensitivity analysis of EUROSEM using Monte Carlo simulation: II the effects of rills and rock fragments, *Hydrological Processes* **14**, 927–939.

Warrick, A.W. and Nilesen, D.R. (1980) Spatial variability of soil physical properties in the field, in D. Hillel (ed.) *Applications of Soil Physics*, Academic Press, New York and London, 319–344.

Wendt, R.C., Alberts, E.E. and Hjelmfelt, A.T. (1986) Variability of runoff and soil loss from fallow experimental plots, *Soil Science Society of America Journal* **50**, 730–736.

Williams, J., Nearing, M., Nicks, A., Skidmore, E., King, K. and Savabi, R. (1996) Using soil erosion models for global change studies, *Journal of Soil and Water Conservation* **51**, 381–385.

Williams, J.R., Jones, C.A. and Dyke, P.T. (1984) A modeling approach to determining the relationship between erosion and productivity, *Transactions of the ASAE* **27**, 129–144.

Wischmeier, W.H. and Smith, D.D. (1978) *Predicting Water Erosion Losses: A Guide to Conservation Planning*, US Department of Agriculture Handbook No. 537, Washington, DC.

10

Modelling Slope Instability

ANDREW COLLISON AND JAMES GRIFFITHS

10.1 THE COMPLEXITY

Slope instability encompasses an enormous temporal and spatial range of processes, from soil creep in which individual particles move a few centimetres over hundreds of years, to catastrophic landslides where entire mountainsides run several kilometres in a few minutes. What binds this continuum of processes together is the underlying system of movement based predominantly on the gravitational force acting on the slope material, rather than stresses imposed by an external source.

Slope instability occurs when a mass of soil or rock experiences a greater downslope driving force than the force binding it into the slope. The balance between these forces can be expressed as a Factor of Safety (F_S) where:

$$F_S = \frac{\text{Resisting Forces}}{\text{Driving Forces}} \qquad (10.1)$$

When $F_S = 1$, the slope exists at the limit of equilibrium. Below this limit, slope instability will occur. When seeking to model slope instability, we must establish the nature of the resisting and driving forces for the particular processes and materials involved, and represent them in a spatially and temporally relevant manner.

This task is complicated by the fact that slope instability encompasses a wide range of physical processes, with ongoing debate about fundamentally new potential mechanisms. The failure mechanism can range from simple shear along a planar surface (e.g. translational landslide), through visco-plastic deformation (e.g. mudflow) to granular flow (e.g. debris avalanche). In each case the most significant driving forces will be different, and the resistance will be based on different physical properties of the material involved (e.g. interparticle friction and cohesion, viscosity or rock joint strength). Thus, before we can model an unstable hillslope, we must first determine the type of failure that has occurred, or is likely to occur. In the case of existing or historical slope failures, there are numerous classification schemes. Examples include the schemes by Varnes (1978), Hansen (1984) and Hutchinson (1988). However, use of a morphological classification scheme is no substitute for detailed field investigations, and in many cases even intensive study may not conclusively show the precise failure mechanism (see, for example, the debate over the mechanism in the 1963 Vaiont landslide; Muller, 1964; Trollope, 1981). This situation is even more complex when modelling first-time failures, as potentially several failure mechanisms may have to be simulated to see which is most likely.

In general, the driving force F_d can be considered to be a function of the weight of the material above a potential failure surface and the slope gradient:

$$F_d = z\gamma \sin \beta \qquad (10.2)$$

where:
z = thickness of material (m)
γ = unit weight of material (kN m^{-3})
β = slope angle (degrees).

However, while these relatively static (in human timescales) properties help determine the long-term susceptibility of a slope to instability, more dynamic processes act on the slope to trigger movement. These include acceleration due to seismic waves (see Murphy, 1995), slope loading due to precipitation (see Brooks and Richards, 1994; Capecchi and Focardi, 1988; Polemio and Sadao, 1999), and the effects of

Environmental Modelling: Finding Simplicity in Complexity. Edited by J. Wainwright and M. Mulligan
© 2004 John Wiley & Sons, Ltd   ISBNs: 0-471-49617-0 (HB); 0-471-49618-9 (PB)

vegetation (Greenway, 1987). While most of the effort in the geotechnical and geophysical community has been focused on the identification and representation of new process mechanisms, most of the recent modelling effort in environmental research has focused on ways of realistically incorporating these external triggering processes into landslide simulations. A major problem in modelling slope instability is deciding which external trigger processes to incorporate, and with what degree of complexity.

The issues surrounding resisting forces are equally complex. Resistance is a function of the inertia of the slope body (a function of mass and gradient) and its material strength. Depending on the materials involved, and the process mechanisms, different material strength models may be relevant.

In the simplest cases slope instability involves shearing along a failure plane. The strength along this plane is defined by the Terzaghi−Coulomb equation:

$$\tau_f = c' + (\sigma - u_w)\tan\phi' \qquad (10.3)$$

where:

τ_f = shear stress at failure (kPa)
c' = effective cohesion (kPa)
σ = normal stress (kPa)
u_w = pore-water pressure (kPa)
ϕ' = effective angle of friction (degrees).

While cohesion and friction vary little in human timescales, pore-water pressure is very dynamic, and on hillslopes composed of soil rather than rock, accounting for water movement is likely to be a prerequisite in order to successfully simulate slope instability. In recent years the role of *negative* pore-water pressure (matric suction) has also been recognized as an important control on instability, following the work of Fredlund (1987a, 1987b) who quantified the effect of suction on soil strength.

In rock hillslopes, a different set of factors needs to be taken into account, including the role of rock joint strength and the orientation of bedding planes relative to topography. The tensile or compressive strength of the rock may also have to be taken into account, rather than shear strength. More complex models of behaviour involving plastic or granular flow may also be required to simulate movement once instability has occurred in rock slopes.

In addition to the fundamental complexity caused by process uncertainty and differences in material behaviour, there is complexity due to parameter uncertainty. Parameter uncertainty in landslide models has

three sources: the inherent spatial or temporal variability of the property in question, problems in measuring the property accurately, and uncertainty as to the location and nature of the failure surface. While the first two sources are common to many environmental processes, the latter is a particular problem with landslides, where the most important boundary surface to be modelled may be several metres or more below ground level.

Slope instability therefore encompasses a highly complex set of processes, landforms and materials. There are many ways to tackle that complexity, ranging from better methods of data collection to more powerful modelling approaches. The most important sources of uncertainty are arguably fundamental disagreement over process mechanisms for several types of landslide, and the need to incorporate appropriate environmental triggering mechanisms.

10.2 FINDING THE SIMPLICITY

Models of slope instability vary from simple, static formulae that can be solved by hand, through to highly complex, physically based, distributed models that require high-performance computers. The oldest group of models simulates stability as a static balance of forces (limit equilibrium methods) while more recent models are rheological and dynamic.

10.2.1 One-dimensional models

The simplest model of slope instability is the infinite slope model (Figure 10.1). This one-dimensional model assumes a translational landslide with the shear surface parallel to the slope surface, and ignores all upslope, downslope and lateral boundaries. The only boundary simulated is the basal shear surface, which is assumed

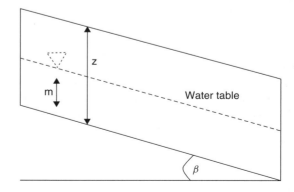

Figure 10.1 The infinite slope stability model

to be homogenous in angle, material properties and pore-water pressure. The factor of safety is calculated thus:

$$F_S = \frac{c' + (\gamma - m\gamma_w)z\cos^2\beta\tan\phi'}{\gamma\,z\sin\beta\cos\beta} \quad (10.4)$$

where:
γ_w = unit weight of water ($9.81\,\text{kN m}^{-3}$)
m = water table as a proportion of soil thickness z (dimensionless).

Though the infinite slope model is very simplistic, it has two great strengths for modellers. First, it reduces the system to its simplest, but most important, physical components, and is thus a good 'first pass' method of assessing relative slope stability for a situation where few data exist. Second, it can easily be executed within a GIS, making it a potentially powerful investigative tool. Slope angle can be derived from a DEM, as can relative water table using the TOPMODEL index (see Beven and Kirkby, 1979). Soil depth and strength can be derived or inferred from a soil or geology map. Running the infinite slope model in this way provides a very good method of identifying areas of potential instability over a wide scale, and a pseudo-three dimensional model of stability (Burroughs *et al.* 1986). The infinite slope model is also well suited to analysis of predominately frictional soils (soils where strength is largely due to a high friction angle ϕ, and little or no cohesion c') which tend to form shallow translational failures along the bedrock interface or a stratigraphic boundary such as an iron pan (Brooks *et al.*, 1995). The infinite slope model's biggest failing is arguably its inability to assess rotational failures (which are common in most sediments where strength is largely due to cohesion), and to represent slopes with variable gradients and nonhomogeneous mechanical properties.

10.2.2 Two-dimensional models

A number of two-dimensional models have been developed to achieve these tasks. All take a two-dimensional slice through a hillslope and simulate the driving and resisting forces acting on the shear surface (or potential shear surfaces). They thus capture the upslope and downslope boundaries that the infinite slope model ignores, but continue to exclude the lateral boundaries.

The simplest two-dimensional models represent the shear surface as an arc, and divide the material above the surface into slices. The factor of safety is calculated for each slice (with any toe slices where the shear surface is orientated 'uphill' acting against the downslope driving force), the sum of the forces being used to calculate the overall safety factor. Numerous numerical schemes exist to account for the inter-slice forces that occur, and this is the main source of differences between the models. Analysis can either be by force equilibrium, or as a moment equilibrium about a radius. The 'standard' two-dimensional limit equilibrium slope stability model is the Simplified Bishop Method (Bishop, 1955), in which individual slices are assessed for factor of safety accordingly (Figure 10.2):

$$F_S = \sum_{i=1}^{i=n} \frac{\left\{[c'b + (W - ub)\tan\phi'] \cdot \left[1 \Big/ \left[\cos\alpha\left(1 + \frac{\tan\alpha\tan\phi'}{F_S}\right)\right]\right]\right\}}{\sum_{i=1}^{i=n}[W\sin\alpha]} \quad (10.5)$$

where:
b = width of slice (m)
W = weight of slice (kN m^{-3})
u = pore water pressure (kPa)
α = angle between radius of slippage and vertical line through slice (degrees).

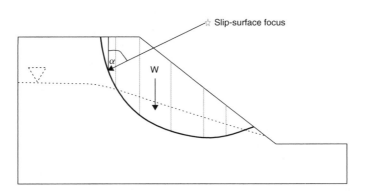

Figure 10.2 The rotational slip stability model

Since F_S appears on both sides of the equation, the solution must be iterative, with the term being varied until both sides of the equation balance (see Chapter 1). Where the potential shear surface location is not known, the analysis is performed using a matrix of potential slip surface foci, and a set of shear surfaces of different radius, so that the surface with the lowest factor of safety can be located. An automated procedure for carrying this search out is described by Wilkinson *et al.* (2000). More complex adaptations of these models (for example, the Morgenstern–Price method) can employ irregularly shaped slip surfaces, and can therefore be used to simulate translational and complex-form landslides as well as true rotational slides. Since both the shear surface and the pore pressure conditions can be physically distributed, two-dimensional models are well suited to linkage with hydrological models that predict pore pressure (see, for example, Anderson and Burt, 1990; Anderson and Brooks, 1996; Geo-Slope International, 1992).

Two-dimensional rotational slip models such as the Simplified Bishop Method are the most common types of analysis used in detailed slope-stability simulations. They have the advantage of capturing most of the complexity of 'real' landslides, simulating the upper and lower boundaries as well as the base, and encompassing variability in slope profile, shear surface depth, material properties and pore-water pressures, while avoiding the complexity of three-dimensional solutions. The rotational shape is suitable for modelling most cohesive-soil landslides, and can be adapted for application to failures on river banks and gully heads (Collison, 2001). As with the infinite slope stability model, rotational models can be incorporated into GISs to allow automated searches to be performed, though this is much rarer and more complex (see, for example, Wilkinson *et al.*, 2000). However, a limitation of these models is that the assumption of a purely circular slip surface, combined with failure by exceeding the Coulomb strength criteria, places *a priori* constraints on the geometry, and so the calculated factor of safety, of the slope. These limitations have led Griffiths and Lane (1999) to argue that limit equilibrium models of slope instability are becoming redundant, and that the widespread availability of relatively powerful computers means that the finite element method of slope stability analysis should become the new standard method for most applications.

The finite element method (FEM) provides a completely different approach to assessing hillslope instability. The FEM breaks the slope down into a number of elements, each having internally homogeneous properties. Numerous elements (typically hundreds to thousands for a hillslope) may be defined, usually with a relatively small number (typically one to five) of different material types, depending on the problem and the particular model. Each material is defined according to one of a number of mechanical models that describe the relationship between stress and strain. For soils this is most likely to be an elasto-plastic (St Venant) or viscoelastic-plastic model. Where the FEM is used to simulate rock-slope instability, a Kelvin or Burger model may be more appropriate (Selby, 1993). Within a FEM simulation the distribution of stresses is calculated, taking into account the unit weight of the material acting on each computational node, the pore-water pressure and the material strength at that point. Stresses can act in any direction (unlike limit equilibrium analysis which only considers stress normal to the assumed shear surface), and where local strength is exceeded failure and deformation will occur. Similarly, FEM models also permit the extent of the movement to be calculated, provided that the final yield strength of the material is not exceeded, enabling processes such as creep and bulging to be modelled.

Slope instability modelling by FEM is considered to be the most physically realistic approach at present, since stresses are modelled in all directions and failure is 'allowed' to occur 'naturally' rather than along a prescribed shear surface. However, FEM models are more complex to set up and run than limit equilibrium models, and require more user input. While the geotechnical engineering community has begun to shift to FEM modelling, the geographical community is only just beginning to use this approach.

10.2.3 Three-dimensional models

In reality, landslides have three-dimensional forms, and on deep-seated rotational landslides the shear surface curves in both the x and y planes. Two-dimensional models tend to under-estimate the factor of safety of such slides by about 10% when used to simulate the central line. While this is often acceptable or even desirable for engineers who wish to have a conservative safety margin, researchers may wish to investigate a landslide in three dimensions. Computationally this is very expensive; as a rule of thumb, computational time increases by an order of magnitude every time an additional dimension is added to a model. In addition, it is debatable whether the additional numerical accuracy is relevant, given the problems of parameter uncertainty involved in three-dimensional modelling because of the difficulties of making detailed, spatially distributed measurements.

The simplest method of representing a three-dimensional landslide is to perform a series of two-dimensional analyses along cross-sections through the slope, from the centreline to the margins. The factor of safety values resulting from these analyses are then summed to produce a pseudo-three-dimensional value. An extension of this approach is to carry out three-dimensional analysis where the slices of the Simplified Bishop Method are subdivided into columns, allowing the full geometry of the slope to be captured (see, for example, Lam and Fredlund, 1993). In this model the limit equilibrium is calculated for each column, and assumptions made concerning inter-column forces analogous to those carried out on the slices of a two-dimensional model. The factor of safety is calculated for each column and then summed for the slope as a whole.

Arguably the most physically realistic method currently available to model slope instability for non-flowing materials is to use a three-dimensional finite element model. This approach has been taken by Cai *et al.* (1998) in their simulation of the effects of installing drains into a hillslope. Their approach combined a three-dimensional saturated–unsaturated hydrology model with a three-dimensional finite element elasto-plastic shear strength model, and probably represents the current state-of-the-art both in terms of triggering mechanism and instability assessment for this type of landslide.

10.2.4 Slope instability by flow

The models outlined above have been developed to predict the stability of a coherent mass of slope material. However, many failures (for example, debris and mud flows, and rock avalanches) occur by visco-plastic flow processes rather than by shearing (e.g. Brunsden, 1979; Selby, 1993). These failures are often modelled as Bingham processes, in which a solid plug of material moves, surrounded by a viscous layer which is in contact with the external shear surface. Movement occurs when the plastic yield strength (τ_o) of the material is exceeded, and then proceeds at a rate determined by the force acting and the viscosity of the material. The strength of a visco-plastic material can be expressed as follows:

$$\tau = \tau_o + \eta \frac{dv}{dy} \qquad (10.6)$$

where:

τ = shear stress (kPa)
τ_o = yield stress (kPa)
η = viscosity ($N\,s\,m^{-2}$)

dv = flow rate ($m\,s^{-1}$)
dy = change in depth (m).

This modelling approach has been successfully used to simulate several landslides, but parameterization remains a problem, particularly for models that incorporate kinematic viscosity (i.e. viscosity is a function of rate of shear, which in turn controls velocity in a positive feedback loop). An alternative approach which has been used to model 'dry' debris flows and rock avalanches is to simulate granular flow processes using Bagnold's model or other approaches to account for momentum transfers between particles.

10.2.5 Modelling external triggers

In addition to modelling the failure process, we usually need to simulate the triggering mechanism. This requirement is especially common where the field conditions at the time of failure are unknown, for example, when simulating the consequences of environmental change on a hillslope. The most common trigger is pore-water pressure, and schemes exist to model this with widely varying degrees of complexity.

The simplest models assume a fixed water table and calculate the resulting pore-water pressure. In engineering applications this is often a 'worst-case scenario' where water is assumed to saturate the complete slope profile. In the case of a saturated slope with seepage parallel with the surface the pore pressure u is:

$$u = \gamma_w z \cos^2 \beta \qquad (10.7)$$

where:

z = thickness of the slope, or the slice (in 2-d analysis) (m)
β = slope angle (degrees).

However, for most environmental research applications this worst-case assumption is too crude. Greater realism can be obtained by using a hydrology model that generates pore-water pressures and passes them to the stability model. Detailed coverage of complexity in hydrological models is covered in Chapter 3, and this chapter will only provide a brief overview.

The choice of hydrology model is clearly closely related to the type of stability model, in terms of dimensionality (one, two or three), and the spatial and temporal resolution that is required. Where slope instability is being simulated using the infinite slope model, a simple 'tank' model may be appropriate. Tank models represent the soil hydrology as a small number of stores (typically unsaturated zone, saturated zone and groundwater) with movement between stores based on empirical power functions or simple transfer equations such as Darcy's law. The resulting water table height is then used in the safety factor calculation. This approach has also been successfully used to simulate pore pressure and resulting motion for a visco-plastic mudflow model (see Obrien *et al.*, 1993; Laigle and Coussot, 1997). For more detailed but shorter-term predictions, a one-dimensional finite difference hydrology model can be used to explore the effects of slope stability in translational slides with differentiated soil profiles (see, for example, Brooks and Collison, 1996).

Two-dimensional stability analyses require two-dimensional hydrology models. A well-known example is the Combined Hydrology and Stability Model (CHASM) developed by Anderson and Howes (1985). This uses a two-dimensional finite difference hydrology model to generate pore-water pressure and suction for a Simplified Bishop Method stability analysis. Applications have included assessing hydrological controls on mass failure in the tropics (Anderson *et al.*, 1988) and the mechanical and hydrological effects of vegetation (Collison *et al.*, 1995; Collison and Anderson, 1996). A similar framework is provided by the GeoSlope model SEEP/W coupled with the limit equilibrium model SLOPE/W, which uses a finite element hydrology model in a similar way (GeoSlope International, 1994).

Combined three-dimensional modelling of landslide hydrology is rare, though as computer power increases it should become more common. The work of Cai *et al.* (1998) referred to above is the only published study known to the authors so far. Pseudo-three-dimensional modelling is, however, common, using simple 'tank' hydrology models within a GIS. Examples include the work of Montgomery and Dietrich (1994), Dietrich *et al.* (1995) and Montgomery *et al.* (2000), who looked at spatial occurrence of landslides in forested catchments using a linked tank model and infinite slope stability model to assess the effects of timber removal. Hydrological triggers can also be incorporated statistically into empirical models of slope stability; for example, the work of Crozier (1999) on landslides in New Zealand where a rainfall threshold was determined based on antecedent condition and current rainfall.

Incorporating hydrological models presupposes that the triggering rainfall event is known. An extension of this method is to link General Circulation Models (GCMs) to hydrology models, to permit prediction of slope instability as a result of climate change scenarios. This approach has been used by Buma and Dehn (1998),

Dehn and Buma (1998) and Dehn *et al.* (2000) for the French Alps, and by Collison *et al.* (2000) for southeast England. Since this type of modelling involves long timescales (tens of years), using physically based distributed hydrology models is not feasible, and both examples referred to above apply the GCM output to simple 'tank' hydrology models to drive slope instability models.

10.2.6 Earthquakes

In addition to hydrological processes, earthquakes are another major trigger of landslides. Tectonic acceleration can easily be incorporated into limit equilibrium methods as a lateral (out of slope) acceleration term, allowing researchers to calculate the force required to cause slope instability, or the ability of a slope to withstand a given magnitude of earthquake. For wide-area investigations of seismic impact, simple slope stability models can be set up within a GIS and fixed, or where known, spatially variable, tectonic shocks applied (Miles and Ho, 1999). This approach has the advantage of permitting the investigator to vary acceleration spatially in accordance with ground conditions. More complex modelling is required to incorporate seismic factors other than simple acceleration. Holzer *et al.* (1988, 1989, 1999) have recorded excess pore-water pressures and liquefaction during earthquakes, and Murphy (1995) has shown analytically that these pressures may be required to explain slope failures during the 1908 earthquake at Messina, Italy. However, as with hydrological modelling, the main source of difficulty is not in incorporating the trigger into the landslide model, but rather in assessing the magnitude of the trigger effect at a local site.

10.3 THE RESEARCH FRONTIER

Modelling slope instability more successfully requires progress in four research areas: process understanding, process representation in models, modelling of environmental triggers, and data acquisition. For many types of slope instability, particularly in large rock landslides, there is still disagreement over the underlying processes that are responsible. For example, there is disagreement over whether long runout landslides are made possible by pore-water pressure effects or by granular rolling and momentum conservation within the moving mass. Likewise, the growing interest in small-scale landslide modelling has exposed a need for more information on soil-strength properties at very low normal loads, where soil mechanics models such as the Coulomb envelope do not fully apply. Clearly, these issues need to be resolved before realistic modelling can be performed.

There is also considerable variability in how processes are represented within models. While most new model *development* is focusing on rheological models, most model *applications* are still using the older limit equilibrium models. Though the limit equilibrium models still have a lot to offer for spatial investigations and large-scale research, detailed site modelling is likely to move towards finite element deformation modelling.

An alternative to searching for ever-greater degrees of process realism is the cellular automata approach, which has been used for spatial studies of predicted landslide occurrence as an alternative to the GIS methods (e.g. Di Gregorio *et al.*, 1999; Clerici and Perego, 2000; Avolio *et al.*, 2001). Early applications seem promising and this may prove to be the most effective method of predicting landslide occurrence where detailed data are unavailable, though its ability to shed light on the underlying processes is debatable.

For environmental scientists the most important research frontier is probably in the determination, representation and assessment of environmental triggers for slope instability. A huge area of research lies in predicting the impact of global climate change on landslide activity rates. Early research suggests that these impacts will vary hugely even within Europe, and much work is needed to assess the global pattern. Allied to this is the continuing need for research on the effects of land use change on slope instability, both as a research area in its own right and to assess possible remedies for slopes that appear to be vulnerable to climate change. In the more immediate timescale, the landslides in Honduras associated with Hurricane Mitch in 1999 highlight the associated need for wide-area landslide models that can be linked to real-time meteorological forecasts to provide hazard warnings for threatened communities.

Finally, slope instability research suffers greatly from the difficulty in obtaining data for model parameterization. In particular, it is difficult to measure strength properties for potential failure surfaces in statistically significant numbers where drilling or augering to great depth is required. *In situ* data collection methods such as the Iowa borehole shear test device offer some advantages here, and research is needed to develop other more rapid methods of collecting strength data. Research by Allison (1988, 1990) suggests that rock-strength properties can be obtained indirectly, and it is possible that similar progress might be made for granular materials such as soil.

10.4 CASE STUDY

10.4.1 The problem

The lower Greensand escarpment, the most landslide-prone geological formation in Britain, passes through Kent, SE England, and forms an abandoned sea cliff near Hythe. The escarpment is occupied by numerous dormant landslides that reactivate during periods of higher effective precipitation. The research question asked was, 'Given the projected change in rainfall and evapotranspiration predicted as a consequence of climate change, what will the effect be on landslide activity along the escarpment?' A secondary question was, 'If landslide activity does increase, can it be mitigated with a change in vegetation cover?'

10.4.2 The modelling approach

Since the question required long-term modelling, a physically based, distributed model was impractical. A one-dimensional 'tank' model was therefore used in conjunction with the infinite slope stability model to predict water table and factor of safety respectively. The potentially unstable soil profile was represented by just three layers: root zone, colluvium and underlying impermeable layer (see Figure 10.3). Total daily water content of each layer was calculated using the following mass-balance equations:

$$\text{Layer 1}: W_{t+1} = W_t + [r(1 - bp)]$$
$$- ET_a - d1 \qquad (10.8a)$$

$$\text{Layer 2}: W_{t+1} = W_t + [r(bp)] + d1 - d2 \quad (10.8b)$$

$$\text{Layer 3}: W_{t+1} = W_t + d2 - d3 \qquad (10.8c)$$

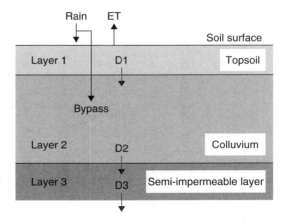

Figure 10.3   Schematic illustration of 1D tank model

where:

W_t = initial water content (mm)
W_{t+1} = resulting water content (mm)
r = net rainfall (mm)
ET_a = actual ET (mm)
bp = bypass coefficient
d1 = drainage from layer 1 (mm)
d2 = drainage from layer 2 (mm)
d3 = deep percolation (mm).

The rate of moisture drainage between layers was calculated as a function of hydraulic conductivity and soil moisture using the method described by van Genuchten (1980). A bypass-flow coefficient (bp) was also employed to represent the fraction of incident rainfall that passes directly from the surface into the lower soil layers via macropores created by surface desiccation and tension cracks. Evapotranspiration from the soil surface was calculated as a function of temperature and vegetation type.

10.4.3 Validation

The model was calibrated against a one-year period of observed water-table height, by varying values of saturated hydraulic conductivity used to calculate moisture drainage from each soil layer. These values were inevitably higher than field values in order to compensate for the lack of a lateral drainage component within the model. The determined values were validated against a second period of similar duration (see Figure 10.4). As the model was calibrated to predict critical water table threshold exceedance (i.e. water-table height at

which slope becomes unstable [−0.65 m]), the accuracy of water table predictions below −2 m was less good.

Long-term simulation of slope instability was made for a 30-year period of rainfall and temperature data. It can be seen from Figure 10.5 that the simulated water-table peaks correlated well with periods of known landsliding as described from local sources (Collison *et al.*, 1998). This information helped to confirm the existence of a critical water-table threshold at which slope stability tended to occur at the site (−0.65 m for a soil depth of 3.5 m).

The model was also used to predict water-table height under different vegetation types over a two-year period (Figure 10.6). It can be seen that predictions for grass-covered slopes, and recently felled woodland slopes, produced the most dynamic water table response (especially in winter). By contrast, the predictions for deciduous- and coniferous-covered slopes tended to respond more slowly, reflecting higher soil-moisture deficits and greater canopy interception and evapotranspiration losses. While the model correctly predicted the period of slope instability that occurred under the currently grass-covered slope in December 1998 (water table ≈ −0.65 m), model runs parameterized for other vegetation types suggested that more stable conditions would prevail under coniferous, deciduous or arable cover.

In order to assess the impact of climate change on the stability of the site, the hydrological model was run for 90 years of 'forecast' rainfall and temperature data (1990–2019; 2020–49; 2050–79), and for 30 years of 'hindcast' data (for the period 1960–90). Projected monthly climate data was first spatially downscaled from the UK Hadley Centre's second-generation, coupled ocean-atmosphere General Circulation Model (HadCM2

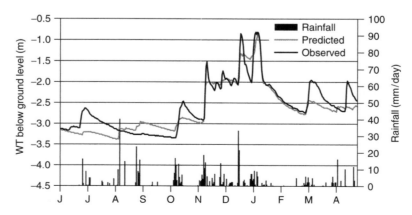

Figure 10.4   Comparison of observed and predicted water table for the period between June to April 1997 ($r^2 = 0.936$; Standard Error = 0.191)

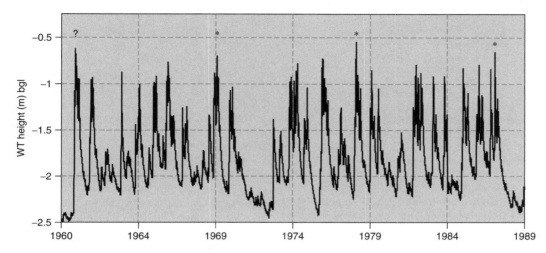

Figure 10.5   Water-table variation predicted for observed rainfall and average temperature data for the period 1960–89
Note: *Periods of known instability from local records and personal communication with landowners
Source: Collison *et al.* (1998)

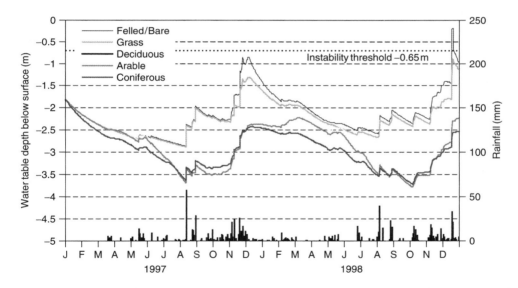

Figure 10.6   Predicted effects of vegetation-cover scenarios on water-table depth for observed temperature and rainfall data from the Roughs (1997–98)

GS) (see Buma and Dehn, 1998; Dehn and Buma, 1998). A Monte Carlo simulation technique was then used to generate daily data rainfall and temperature series.

10.4.4 Results

Comparison of model predictions showed a decrease of −0.024 m ± 0.0034 in the mean water-table for the

period 1990–2019, relative to the 1960–89 control period (see Table 10.1). This equates to a 17% decrease in the chance of actual slope failure. This decrease was surprising inasmuch as mean annual rainfall for this period actually increased, albeit due to greater summer as opposed to winter rainfall. By contrast, the subsequent 2020–49 period saw a slight increase in predicted mean water table (+0.01 m ± 0.005). This was more expected

Table 10.1 Descriptive statistics for water-table height and factor of safety for each modelled period

Period	Water-table height (m)		Factor of Safety	
	mean	s.d.	mean	s.d.
1960–89	−2.500 ± 0.001	0.252	1.265 ± 0.0002	0.043
1990–2019	−2.524 ± 0.002	0.284	1.268 ± 0.0003	0.041
2020–49	−2.515 ± 0.002	0.286	1.267 ± 0.0003	0.041
2050–79	−2.546 ± 0.001	0.283	1.217 ± 0.0001	0.041

and in response to a significant increase in mean winter rainfall. The probability of the water table rising to above −1 m increases by 90% in this period (relative to the control period).

Figure 10.7 illustrates predicted water table height exceedance frequency relative to the 1960–89 period. It can be seen that a relative decrease in water table height is predicted for each of the future scenarios, thus indicating that slope instability at the site will become less frequent as expected changes in climate take place. This trend is largely due to the greater evapotranspiration that will result from increased mean annual temperatures. One exception to this pattern will be the increased risk of slope instability for the period 2020–49 due to a predicted significant increase in winter rainfall.

Figure 10.8 illustrates water table exceedance frequencies predicted for each vegetation type for the period 1960–89. While the probability of slope instability under uniform grass cover was predicted to be 0.00026, it was found to be just 0.00018 when the model was parameterized for a coniferous canopy cover. Deciduous and arable vegetation cover, which exhibit greater summer canopy, both exhibited a probability of 0.0009. By contrast, when the model was parameterized to represent a lack of vegetation cover (recently felled woodland), a 0.01 risk of instability was predicted for the same period.

This modelling approach suggests that climate change is unlikely to have a significant effect on the frequency of landslides at this location. Small changes in the distribution of water-table heights predicted will result in decrease of slope instability, due to projected increases in mean annual temperatures. While some periods of higher landslide probability were predicted within each 30-year period assessed, it is thought that this potential hazard could be much reduced through the growth of different vegetation cover in the prone areas. Uncertainty within the results produced was found to be relatively high at each stage of the modelling chain, thus making transfer of the predicted relative change in landslide probability to actual values difficult. Similarly, the results produced are highly dependent on assumptions made within the stochastic weather generator. The tank model that was used however, has subsequently been applied to two other field sites and was found to perform equally well for different soil and vegetation types, and within different climate regimes (see Corominas *et al.*, 1998; Griffiths, 2001).

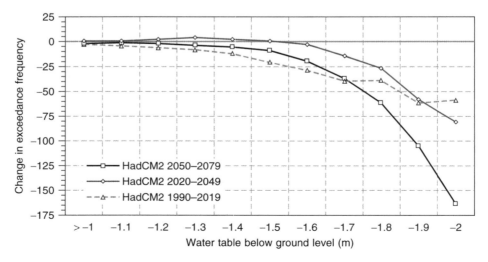

Figure 10.7 Change in frequency of water-table exceedance predicted for HadCM2 1990–2019, 2020–2049 and 2050–2079 scenarios, relative to HadCM2 1960–1989

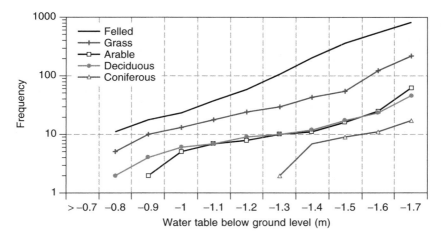

Figure 10.8 Water-table exceedance frequencies (−0.7 to −1.7 m bgl), for different vegetation-cover scenarios at the Roughs, derived for HadCM2 monthly rainfall and temperature data for the period 1960–89

10.5 CONCLUSION

Like most environmental phenomena, slope instability encompasses a complex set of processes, of which a small subset are usually capable of explaining most of the observed pattern of events. In many applied situations the benefit of modelling greater process complexity is offset by the punitive costs of data collection and by the uncertainty attached to the associated data, resulting in the application of very simple models driven primarily by slope and basic soil or rock properties. In research and highly complex engineering applications, however, the devil lies in the detail: insights into slope failure processes in the unsaturated materials, where vegetation is present, in earthquake-prone areas and where failure involves flow processes require much greater levels of model complexity in order to produce meaningful analyses and interpretations. Similarly, while it is relatively easy to predict where a landslide will start, it is very difficult to predict where it will stop, and models of landslide movement require high levels of rheological process information and data.

Two diverging trends are emerging in landslide modelling. There is incorporation of greater process complexity into models, made possible by better understanding and by new field parameterization techniques, especially remote sensing techniques that reveal information on subsurface material properties, at the site scale. By comparison there is a trend to use simpler models such as cellular automata to reveal spatial patterns of landslide activity at the landscape scale. Reconciliation of these trends may be achieved when remote sensing techniques enable detailed data to be gathered at sufficiently high spatial scale, though this appears some way off at present.

REFERENCES

Allison, R.J. (1988) A non-destructive method of determining rock strength, *Earth Surface Processes* **13** (8), 729–736.

Allison, R.J. (1990) Developments in a non-destructive method of determining rock strength, *Earth Surface Processes* **15** (6), 571–577.

Anderson, M.G. and Brooks, S.M. (eds) (1996) *Advances in Hillslope Processes*, Vol. 1, John Wiley & Sons, Chichester.

Anderson, M.G. and Burt, T.P. (eds) (1990) *Process Studies in Hillslope Hydrology*, John Wiley & Sons, Chichester.

Anderson, M.G. and Howes, S. (1985) Development and application of a combined soil water-slope stability model, *Quarterly Journal of Engineering Geology* **18**, 225–236.

Anderson, M.G., Kemp, M. and Lloyd, D.M. (1988) Applications of soil water finite difference models to slope stability problems, in G. Bonnard (ed.) *Fifth International Landslide Symposium*, Lausanne, Balkema, Rotterdam, 525–530.

Avolio, M.V., Ambrosio, D.D., Di Gregorio, S., Iovine, G., Lupiano, V., Rongo, R. and Spataro, W. (2001) Simulating different complexity landslides with cellular automata methods, *Geological Research Abstracts* (*CD*) Vol. 3, *General Assembly of the EGS, GRA3*, 8746, Issn: 1029-7006.

Beven, K.J. and Kirkby, M.J. (1979) A physically based, variable contributing area model of basin hydrology, *Hydrological Sciences Bulletin* **24** (1), 43–69.

Bishop, A.W. (1955) The use of the slip circle in stability analysis of earth slopes, *Geotechnic* **5**, 7–17.

Brooks, S.M. and Richards, K.S. (1994) The significance of rainstorm variations to shallow translational hillslope failure, *Earth Surface Processes and Landforms* **19**, 85–94.

Brooks, S.M. and Collison, A.J.C. (1996) The significance of soil profile differentiation to hydrological response and slope instability: a modelling approach, in M.G. Anderson and S.M. Brooks (eds) *Advances in Hillslope Processes*, John Wiley & Sons, Chichester, 471–486.

Brooks, S.M., Anderson, M.G. and Crabtree, K. (1995) The significance of frangipans to early-Holocene slope failure – application of physically-based modelling, *Holocene* **5** (3), 293–303.

Brunsden, D. (1979) Mass movement, in C. Embleton and J. Thornes (eds) *Process in Geomorphology*, Butler and Tanner, London, 130–186.

Buma, J. and Dehn, M. (1998) A method for predicting the impact of climate change on slope stability, *Engineering Geology* **35** (2–3), 190–196.

Burroughs, E.R., Hammond, C.J. and Booth, G.D. (1986) Relative stability estimation for potential debris avalanche sites using field data, in *Proceedings of the International Symposium on Erosion, Debris Flow and Disaster Prevention*, Erosion Control Engineering Society, Tsukuba, Japan, 335–339.

Cai, F., Ugai, K., Wakai, A. and Li, Q. (1998) Effects of horizontal drains on slope stability under rainfall by three-dimensional finite element analysis, *Computers and Geotechnics* **23** (4), 255–275.

Capecchi, F. and Focardi, P. (1988) Rainfall and landslides: research into a critical precipitation coefficient in an area of Italy, in C. Bonnard (ed.) *Landslides: Proceedings of the Fifth International Symposium on Landslides, 10–15 July 1988*, Lausanne, Balkema, Rotterdam, Vol. 2, 1131–1136.

Clerici, A. and Perego, S. (2000) Simulation of the Parma River blockage by the Corniglio landslide (Northern Italy), *Geomorphology* **33**, 1–23.

Collison, A.J.C. (2001) The cycle of instability: stress release and fissure flow as controls on gully head retreat, *Hydrological Processes* **15** (1), 3–12.

Collison, A.J.C. and Anderson, M.G. (1996) Using a combined slope hydrology/stability model to identify suitable conditions for landslide prevention by vegetation in the humid tropics, *Earth Surface Processes and Landforms* **21**, 737–747.

Collison, A.J.C., Anderson, M.G. and Lloyd, D.M. (1995) Impact of vegetation on slope stability in a humid tropical environment – a modelling approach, *Proceedings of the Institute of Civil Engineers – Water, Maritime and Energy* **112** (2), 168–175.

Collison, A.J.C., Wade, S. and Griffiths, J.A. (1998) UK national report in J. Corominas, J. Moya, A. Ledesma, J. Gili, A. Lloret and J. Rius (eds) *New Technologies for Landslide Hazard Assessment and Management in Europe: Final Report*, CEC Environment Programme (Contract ENV-CT966-0248), EU, Brussels.

Collison, A.J.C., Wade, S., Griffiths, J. and Dehn, M. (2000) Modelling the impact of predicted climate change on landslide frequency and magnitude in SE England, *Engineering Geology* **55**, 205–218.

Corominas, J., Moya, J., Ledesma, A., Gili, J.A., Lloret, A. and Rius, J. (1998) *New Technologies for Landslide Hazard Assessment and Management in Europe: Final Report*, CEC Environment Programme (Contract ENV-CT966-0248), EU, Brussels.

Crozier, M.J. (1999) Prediction of rainfall triggered landslides: a test of the antecedent water status model, *Earth Surface Processes and Landforms* **24** (9), 763–779.

Dehn, M. and Buma, J. (1998) Modelling future landslide activity based on general circulation models, *Geomorphology* **30**, 175–187.

Dehn, M., Burger, G., Buma, J. and Gasparetto, P. (2000) Impact of climate change on slope stability using expanded downscaling, *Engineering Geology* **55**, 193–204.

Dietrich, W.E., Reiss, R., Hsu, M. and Montgomery, D.R. (1995) A process-based model for colluvial soil depth and shallow landsliding using digital elevation data, *Hydrological Processes* **9**, 383–400.

Di Gregorio, S., Rongo, R., Siciliano, C., Sorriso-Valvo, M. and Spataro, W. (1999) Mount Ontake landslide simulation by the cellular automata model SCIDDICA-3, *Physics and Chemistry of the Earth Part A – Solid Earth Geodesy* **24**, 131–137.

Fredlund, D.G. (1987a) The shear strength of unsaturated soil and its relationship with slope stability problems in Hong Kong, *Hong Kong Engineer* **8**, 57–9.

Fredlund, D.G. (1987b) Slope stability analysis incorporating the effect of soil suction, in M.G. Anderson and K.S. Richards (eds) *Slope Stability*, John Wiley & Sons, Chichester, 113–144.

GeoSlope International (1992) *SEEP/W User's Guide*, Alta, Calgary.

GeoSlope International (1994) *SLOPE/W User's Guide*, Alta, Calgary.

Greenway, D.R. (1987) Vegetation and slope stability, in M.G. Anderson and K.S. Richards (eds) *Slope Stability*, John Wiley & Sons, Chichester, 187–230.

Griffiths, D.V. and Lane, P.A. (1999) Slope stability analysis by finite elements, *Geotechnique* **49** (3), 387–403.

Griffiths, J.A. (2001) Modelling the sensitivity and response of shallow landslides in SE England and SE Spain, unpublished PhD thesis, University of London.

Hansen, M.J. (1984) Strategies for classification of landslides, in D. Brunsden and D.B. Prior (eds) *Slope Instability*, John Wiley & Sons, Chichester, 363–418.

Holzer, T.L., Bennett, M.J., Ponti, D.J. and Tinsley, J.C. *et al.* (1999) Liquefaction and soil failure during 1994 Northridge earthquake, *Journal of Geotechnical and Geoenvironmental Engineering* **125** (6), 438–452.

Holzer, T.L., Bennett, M.J., Youd, T.L. and Chen, A.T.F. *et al.* (1988) Parkfield, California, liquefaction prediction, *Bulletin of the Seismological Society of America* **78** (1), 385–389.

Holzer, T.L., Youd, T.L. and Hanks, T.C. (1989) Dynamics of liquefaction during the 1987 Superstition Hills, California, earthquake, *Science* **244** (4900), 56–59.

Hutchinson, J.N. (1988) Morphological and geotechnical parameters of landslides in relation to geology and hydrogeology, in C. Bonnard (ed.) *Landslides, Proceedings of the 5th International Symposium on Landslides*, Balkema, Rotterdam, Vol. 1, 3–35.

Laigle, D. and Coussot, P. (1997) Numerical modeling of mudflows, *Journal of Hydraulic Engineering – ASCE* **123** (7), 617–623.

Lam, L. and Fredlund, D.G. (1993) A general limit equilibrium model for 3-dimensional slope stability analysis, *Canadian Geotechnical Journal* **30** (6), 905–919.

Miles, S.B. and Ho, C.L. (1999) Applications and issues of GIS as tool for civil engineering modeling, *Journal of Computing in Civil Engineering* **13** (3), 144–152.

Montgomery, D.R. and Dietrich, W.E. (1994) A physically based model for the topographic control on shallow landsliding, *Water Resources Research* **30** (4), 1153–1171.

Montgomery, D.R., Schmidt, K.M., Greenberg, H.M. and Dietrich, W.E. *et al.* (2000) Forest clearing and regional landsliding, *Geology* **28** (4), 311–314.

Muller, L. (1964) The rock slide in the Vaijont Valley, *Rock Mechanics and Engineering Geology* **2**, 148–228.

Murphy, W. (1995) The geomorphological controls on seismically triggered landslides during the 1908 Staits of Messina earthquake, Southern Italy, *Quarterly Journal of Engineering Geology* **28**, 61–74.

Obrien, J.S., Julien, P.Y. and Fullerton, W.T. (1993) 2-Dimensional water flood and mudflow simulation, *Journal of Hydraulic Engineering – ASCE* **119** (2), 244–261.

Polemio, M. and Sadao, F. (1999) The role of rainfall in the landslide hazard: the case of the Avigliano urban area (Southern Apennines, Italy), *Engineering Geology* **53** (3/4), 297–309.

Selby, M.J. (1993) *Hillslope Materials and Processes*, Oxford University Press, Oxford.

Trollope, D.H. (1981) The Vaiont slide failure, *Rock Mechanics* **13**, 71–88.

van Genuchten, M. Th. (1980) A closed-form equation for predicting the hydraulic conductivity of unsaturated soils, *Soil Science Society of America* **48**, 892–898.

Varnes, D.J. (1978) Slope movement types and processes, in R.L. Schuster and R.J. Krizek (eds) *Landslides: Analysis and Control*, National Academy of Sciences, Transport Research Board – Special Report 176, Washington, DC, 11–33.

Wilkinson, P.L., Brooks, S.M. and Anderson, M.G. (2000) Design and application of an automated non-circular slip surface search within a combined hydrology and stability model (CHASM), *Hydrological Processes* **14** (11–12), 2003–2017.

11

Finding Simplicity in Complexity in Biogeochemical Modelling

HÖRDUR V. HARALDSSON AND HARALD U. SVERDRUP

11.1 INTRODUCTION TO MODELS

Finding simplicity in complexity is the driving force behind scientific modelling processes. What is regarded as an achievement within research is the ability to test a hypothesis on any given problem by creating simple models that can explain a complex reality. Simplification is a process that is initiated by the desire to capture the essence of a complex problem. The simplification can be formed either objectively or subjectively. But total objectivity in research is a mere illusion, and modellers often find themselves slipping into the practice of over-complexity, or being locked into certain routines or subjective opinions.

A model is a simplified representation of some aspect observed in the real world. A model is any consequence or interpretation taken from a set of observations or experience. Many problems in natural systems are so complex, nonlinear and multi-dimensional that they require a nonlinear approach. Traditionally, simplification is seldom dealt with in a nonlinear fashion. Linear correlation between different independent components has instead been used. This requires complex explanations and reduces our understanding of the fundamental dynamics behind complex problems. The understanding is then not the focus of the study but the constructed model itself. The original purpose, understanding, is lost through explanations and comments such as:

> *... the biological system is determined by unknown forces so we cannot understand it ... There are thousands of factors affecting ... It cannot be observed, but it is very important for ... Well, it is always different in the real world, you know, so it is no use trying to explain it ...*

Models that require such explanations lack transparency of their principles and procedures, and are hard to communicate. Validating models requires insight and an understanding of the processes of how the essential parts of the model are constructed.

Models are important in research, not because they produce results in their own right, but because they allow complex and nonlinear systems to be investigated and data from such systems to be interpreted. With models, the interaction of several simultaneous processes in a single experiment can be studied. Basically all models serve one or both of two purposes:

- testing the synthesized understanding of a system, based on mathematical representation of its subsystems and the proposed coupling of subsystems;
- predicting what will happen in the future, based on the ability to explain how and why things have worked in the past.

All models can be classified into three different stages (Levenspiel, 1980). These stages are dependent on the analytical and predictive power of the model:

Environmental Modelling: Finding Simplicity in Complexity. Edited by J. Wainwright and M. Mulligan
© 2004 John Wiley & Sons, Ltd ISBNs: 0-471-49617-0 (HB); 0-471-49618-9 (PB)

Stage 1		Stage 2		Stage 3
Qualitative description	→	Direct quantitative description in terms of observable conditions	→	Differential rate based on underlying physics and processing

The stage 1 model is typical of classifications into categories where a certain occurrence is predictable based on the present conditions. Geological mapping is typical of stage 1 models; rocks and minerals occur according to geographical distributions. Such models have very limited predictive power.

Stage 2 models are based on 'case-by-case' predictive power, and they must be recalibrated on new data each time the initial and boundary conditions change. Plotting a pH over time in an acidified lake is an example of a stage 2 model. These type 2 models are limited by cases, and the properties cannot be transferred to another case.

Stage 3 models involve changes through time and use the differential approach first used in physics, and later in all natural science. They relate how changes at every point in time are related to the state of the system at that time. Stage 3 models introduce the mechanism of change which depends on state variables for the system and is generally valid. The state of the system is characterized by conditions in terms of order, spatial distribution, concentration and adaptation capabilities. A stage 3 model is generally valid and applicable when it can be parameterized properly and the coefficients estimated, but it is in differential form, and requires mathematical manipulation when used. This used to pose difficulties, but with modern computers and user-friendly modelling software, such problems are much reduced.

In modelling, it is important to distinguish between 'good models' and 'bad models', and 'good performance' and 'bad performance'. The definition of a 'good model' is when everything inside it is visible, inspectable and testable. It can be communicated effortlessly to others. A 'bad model' is a model that does not meet these standards, where parts are hidden, undefined or concealed and it cannot be inspected or tested; these are often labelled black box models. Intuitive models are 'bad' because they do not explain what they do. Often statistical models produced from automated statistical packages are bad models, as it remains totally unclear to the user what the implication of the package-proposed

model is, how the relation was established and, finally, what on earth it is good for. Models do have different performances depending on the model developing process and definitions. A model must work with inputs that can be defined and determined, and it must yield outputs that can be observed. A model can perform poorly, but still adhere to the principles of good models. Many of our models will start like this when we develop them. With a good model we can analyse its performance in order to change the model iteratively and improve its performance. Bad models may perform well, but since they can neither be tested nor inspected, there is no way to determine whether this is pure chance or something substantial, and as parts are not visible, there is not much we can do to improve them, hence, the term 'bad model'. A 'bad model' does not allow a learning process and it fails to communicate its principles. A good model is one that adheres to the following rules:

- The model must be transparent. It must be possible to inspect and understand the rules and principles the model is using.
- It must be possible to test the model. It must work on inputs that can be defined and determined, and it must yield outputs that can be observed.

'Goodness' or 'badness' of a model has nothing to do with the adequacy of the principles inside the model. If the model is 'good', then we can verify or falsify the performance of the model with a specific principle incorporated. If the model is 'bad', then we cannot verify or falsify the performance of the model with a specific principle incorporated. The model can be a mental understanding of a mechanism, system, pattern or principle, and it can be substantiated as an equation or a set of equations or rules. If the principles and rules are numerous, then it is practical to let a computer program keep track of all connections and the accounting of numbers.

11.2 DARE TO SIMPLIFY

All models are mental projections of our understanding of processes and feedbacks of systems in the real world. The general approach is that models are as good as the system upon which they are based. Models should be designed to answer specific questions and only incorporate the necessary details that are required to provide an answer. Collecting massive amounts of data and information ahead of the modelling procedure is costly and does not necessarily generate understanding of the problem. More often, it adds to the confusion. Modelling starts with problem definition and simplification of the causalities. It means raising the observation to a higher level in order to extract clear causal links and driving forces from the problem. Focus should be on what is essential for the model and what is not needed. One of the common pitfalls is to assume that models need to be complex and data-hungry. The performance of the model need not be perfect, it only needs to be good enough to answer the relevant questions – better than good enough is extra work with no purpose. Thus, it is always relevant to reflect: what was the objective of the model application in the first place?

- A simple model must make complex assumptions. A simple model is easy to use, and the input data can be obtained at relatively low cost. Because of the simplicity, the applicability may be limited, and we will have problems with addressing the effect of the assumptions.
- A complex model can make simple assumptions. The model will have better general applicability and less restriction on use. But it will require more input data and be relatively more expensive to use. Further increases in model complexity may remove assumptions and consider more feedbacks, but higher demands are made on input data.

The total complexity of a system in modelling is divided between the assumptions and the model itself. For every question there is an optimal complexity, and great care must be exercised to evaluate this aspect. Failing to do so will result in loss of control over uncertainties. And it is important to realize that we cannot get rid of the complexity in a system, but only decide if it goes into the model or into assumptions. Claims to any other effect can safely be laughed at. All models must fulfil some minimum requirements. They must be able to describe events at single sites based on real data. If a model cannot describe single sites and their past history, then it has no credibility in future predictions.

A model of causalities is a system and all systems are defined by their boundaries, internal structure and internal quantities. In order for us to understand a system properly, we need to understand how systems behave and what their properties are. Systems are usually confined by certain inflow and outflow of physical matter or energy. When we create mental models, we do not intend to capture the whole reality in one model. Such models are as complex as reality itself. What we want to do is to map part of the reality in such a way that it gives us a basic understanding of a complex problem. The level of details needed to explain and analyse a problem is dependent on the type of answer that is desired. The number of components depends on the level of detail when the observation takes place. When creating a model it is necessary to have a holistic perspective on the causal relations in the problem and understand the basic driving forces to hand. The following example uses a *causal loop diagram* (CLD) to demonstrate the phosphorus cycle in eutrophic lakes (Figure 11.1). The CLD concept (Richardson and Pugh, 1981) is a systematic way of thinking in causes and effects where variables either change in the same direction (indicated by a 'plus') or change in the opposite direction (indicated by a 'minus').

It is evident that model complexity depends directly on the question asked, as illustrated in Figures 11.1 and 11.2. In the case of the phosphorus cycle, the complex model did not enhance the understanding of the overall behaviour but increased the uncertainty that was involved by increasing the number of observation levels in the system. The simple model used a less complicated observation level but used a more sophisticated explanation of assumption to the causal links.

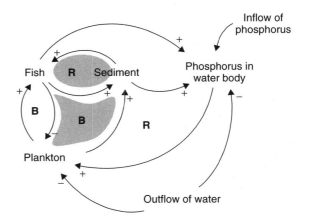

Figure 11.1 A simple conceptual model of the phosphorus cycle in lakes that is powerful enough to illustrate and predict the phosphorous dynamics between the key elements: phytoplankton, fish and sediment. The different fish species have been simplified to one 'fish'. All plankton and zooplankton have been simplified to one general 'plankton'. The shaded area illustrates the core-driving loop that runs the cycle. If lake-water transparency is the issue, then this will be a fully sufficient model

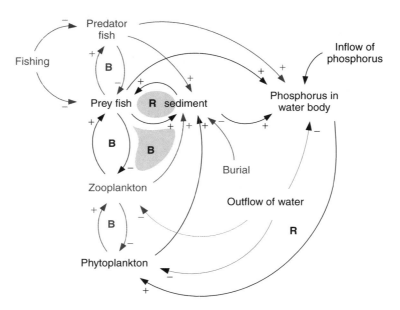

Figure 11.2   A complex model of the phosphorous cycle, which is based on the simple model but includes components that add details to the driving loop for eutrophication. This does not necessarily enhance the performance of the model with respect to the water transparency issue; the simple model will be good enough. If the issue is fish population dynamics, then the added details will be required

11.3 SORTING

Building models is a long process especially if the model is designed to answer many questions. The general rule of thumb is that a specific question requires one model. If a problem involves many questions, then equally many models may be required to address them and then the 'model' becomes a cluster of models. Nevertheless, constructing a model with many components can be costly (Figure 11.3). The art is to construct a model that is robust, answers the desired questions and is simple. The performance need not be perfect, it should only be sufficient, and nothing more. Such models can both save time and money, and be useful building blocks for further model developments.

The process starts by sorting the causalities in the problem in relation to their importance to answer the specific question. Then they are sorted according to their contribution to performance. Causalities with obvious driving forces explicitly related to the problem have the highest performance. Then the sorting process continues downwards. If several components are identified as outlining the main driving forces in the problem, these components may describe sufficiently the dynamics in the problem. Subsequently, the system boundaries can be drawn. A problem can never be fully explained due to

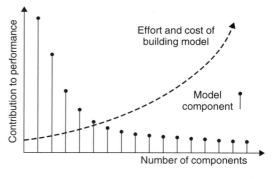

Figure 11.3   The number of components and their influence are in proportion with the cost and effort of building a model

uncertainties in the structure and components. Usually the model performance will level out (Figure 11.4) and decrease from a maximum level (Figure 11.5) even if further causes are added. This is because adding a cause involves adding uncertainty, and at some time the accumulated uncertainties will overtake the contribution to performance (see Chapter 1).

Causalities always involve some amount of uncertainty. Adding further causalities to the model will

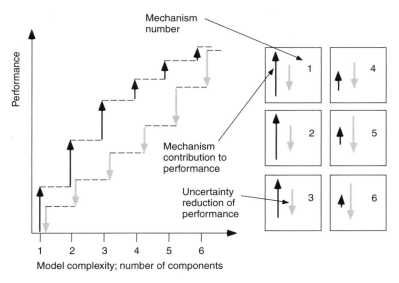

Figure 11.4   Adding a cause contributes to the overall performance of the model, but its contribution is not necessarily symmetrical to other causalities in the model

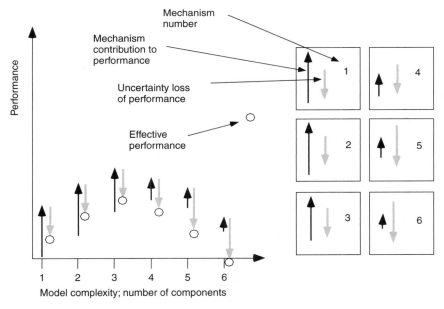

Figure 11.5   Adding a cause involves a certain amount of uncertainty, which can overshoot the contribution to the model performance by adding the cause

increase the necessary details that the model will have to incorporate, but at the cost of increased uncertainty. Increased uncertainty will lower the overall performance up to the level where uncertainty from added causalities outweighs added performance. This point of no improvement is characterized as the point where more complexity costs more in extra inputs and added inaccuracies than it achieves in improved performance. In Figure 11.5, the highest model performance is achieved with three causalities, but further addition involves

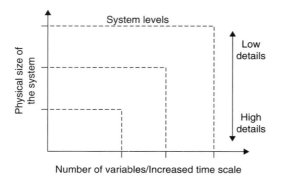

Figure 11.6 Focus of models depends on the observation level. Global-scaled models can require many variables and high time integration but must address lower detail levels, compared to local models

higher uncertainty, not a better contribution to performance.

All models have this type of peak performance in relation to number of causes, which can be several or many hundred. By adding further components, a complex model may end up being poorer in performance than a simple one. It may be useful for impressing one's colleagues with how large and complex a model you can make, but for predictions, it may be worse than a blind guess, and in reality completely useless.

Models are based on subsystems themselves and their components on subsystems, etc. It is important to identify the system level that we operate on. Greater numbers of components mean more interaction between them and it becomes important to observe the transparency between the components. A model can have components that are on different system levels depending on the level of details. Large models on the global scale can be based on numbers that are derived locally. Such models need simplification and balance between number of components and the observation level (Figure 11.6).

We should select the level needed to understand the interrelationship among our selected variables – the level we want to influence. After defining that level, detailed knowledge of the underlying components is not needed. That knowledge falls into the complex assumption made during the sorting process. Driving a car, for instance, does not require detailed knowledge of the construction of the engine; the vehicle can be operated without such knowledge.

11.4 THE BASIC PATH

The process from specifying the question and building the model starts by gathering information and listing all the variables we feel are appropriate for the system. Then we try to categorize the critical and the indicator variables. It gives us much more clarity when we start to assemble the causal links. This is highlighted in two points by Dörner (1996):

- We need to know what other variables those that we want to influence depend on. We need to understand, in other words, how the causal relationships among the variables in a system work together in that system.
- We need to know how the individual components of a system fit into a hierarchy of broad and narrow concepts. This can help us fill in, by analogy, those parts of a structure unfamiliar to us.

After performing basic sorting, previously unrecognized interactions between variables may require alteration or revision of the hypothesis. Some light may have been shed on the larger complexes outside the defined system boundaries, in which elements of the defined system are embedded. This process is iterative and may require several revisions before the model is fixed according to desired standards.

11.5 THE PROCESS

Generalization is often the key to understanding complex systems. Modelling is without exception based on some sort of recipe. Whatever approach is used, all methods focus on answering a specific question. We propose that the basic process should start by defining and confining system boundaries according to the specific question, or questions requiring an answer. The whole process should be considered iterative (Figure 11.7)

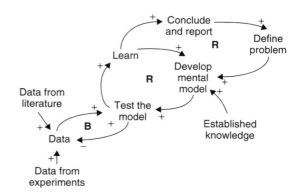

Figure 11.7 The learning loop is an iterative process where the steps from building a mental model and producing results can always be improved

where experience from previous evaluation is used to redefine the problem. The learning experience gained and the conclusions from the testing will help us re-evaluate our problem and our mental model.

Modelling never starts with data, contrary to what many believe. Just collecting 'even more' data without a clear model does not create greater understanding nor clarity, it merely adds to general confusion. Not everybody would agree, but the track record is just too consistent to accept any objections. All research starts with a definition of a problem and through that an understanding of that problem. This is 'the model', the mental image of how the problem is to be understood. Thus, no models are 'wrong'; if something does not work, the cause is to be found in faulty understanding. After a model has been created, we can test and challenge it. Then a specific demand for data will arise. Not all data is needed, only what is relevant to our problem. The rest must be sorted. From the test we will learn and redevelop our understanding, our 'model'. Model development requires iteration in such a cycle several times and will continue as long as there is sufficient data from experiments or literature to support

the testing of the model. In that way the communication of the model is effective, both to the user and the developer. Furthermore, it enables us to communicate the success and problems encountered.

11.6 BIOGEOCHEMICAL MODELS

We define biogeochemical models as models that describe the connection from biology through chemistry, to geology and back. Geology (the solid phase) is only visible to biology through its expression in chemistry. The only way that geology (solids) affects chemistry is through chemical weathering and ion exchange. Ion exchange is always in the form of reversible reactions, whereas chemical weathering reactions are almost always irreversible and slow. Decomposition of organic matter, however, we regard as part of the biotic cycle. For all biogeochemical models where the connection between biological and geological aspects is significant, the representation of the weathering process will be a vulnerable point for overall model performance (Figure 11.8).

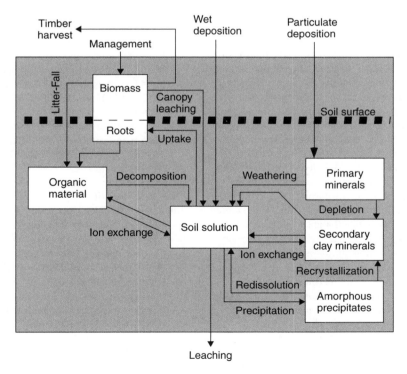

Figure 11.8 A box-arrow diagram for the process system in soil, typical for many biogeochemical models. Biogeochemical models range from describing a small part of this diagram to having it as a small subcompartment

There are many biogeochemical codes available for biogeochemical calculations in ecosystems. Some of these are easy to obtain and use, while others exist only on paper and in scientific articles. If the purpose of your modelling is to predict, assess or design, then using an existing model is often an efficient way to get the work done. If the issue is to explain your research, investigate your own understanding and develop it, it is recommended that you build your own model rather than using an existing one. If you use an existing model, do not start to use it before you have thoroughly understood the principles and assumptions it applies and preferably when you agree with how it is made. We can define three types of model:

1. models based on process-oriented kinetics;
2. models based on equilibrium principles;
3. empirical models.

Here we will be mainly concerned with the first two types. The traditional soil-chemistry models, originally developed for issues in agriculture and groundwater geochemistry, were based on equilibrium processes. This basis had historical reasons: before the advent of modern computers, systems of differential equations were unsolvable for any practical use. Still the traditional models rely heavily on large amounts of data for calibration. If this requirement is fulfilled, they will produce useful back-casting and can be used for making extrapolative predictions. These models quickly run into problems without large amounts of data to calibrate on, and this is caused by a specific shortcoming that must be pointed out. The traditional geochemical models use equilibrium formulations for processes which are valid for soluble salts, carbonate-dominated systems and other reversible reaction systems. But such models are not formally valid for kinetically controlled irreversible systems. The so-called 'thermodynamic equilibrium databases' for soil and geochemical silicate reactions referred to are not 'thermodynamic' at all, they are steady state balance coefficients that have been misinterpreted as equilibria. Thus, such models may be useful for simulations or design work after careful calibration, but do not expect them to explain processes. Without the calibration, these models lack the ability to generate *a priori* predictions.

In modelling the focus is often on the output instead of robustness and performance, which makes the 'cult of success' very strong. The 'cult of success' implies that the calculated output is expected to pass through all observation points. If not, the user of a model might soon hear 'The model is wrong ...' or 'Well, it is obvious that the model does not work ...'. On the other hand, if it is

obvious that the model was calibrated by adjusting one or several parameters, critics will soon remark, 'You can probably make the thing fit anything ...'. Resist this! Resist the temptation to overcalibrate! Ignore such derogative comments, they are never made from insight! The fit need not be perfect, the line need not go through all data points. Performance should only be sufficient to give an adequate answer! It is from the lack of perfect fit that learning is gained.

We may classify existing models into categories, depending on which scale they are applied to as well as the degree of integration of multiple processes over those scales. A list, not claiming to be comprehensive, would include examples as follows: Single problem models: Sverdrup and de Vries, 1994; Cosby *et al.*, 1985; de Vries *et al.*, 1989; Wright *et al.*, 1998; Botkin *et al.*, 1972; Kimmins *et al.*, 1999; Kram *et al.*, 1999; Chen, 1993; Crote *et al.*, 1997; Crote and Erhard 1999; Sverdrup and Stjernquist, 2002; Jansson, 1991; Sverdrup and Warfvinge, 1995; Warfvinge *et al.*, 1998; Kros, 2002; Berge and Jakobsen, 1998. Integrated multiple system models: Parton *et al.*, 1987; Sykes, 1996; Sverdrup and Stjernquist, 2002; Mohren *et al.*, 1993; Kram *et al.*, 1999. Higher hierarchy models: Alcamo *et al.*, 1990; den Elzen, 1994; Gough *et al.*, 1994:

- Single problem models
  — Acidification models (SAFE, MAGIC, SMART)
  — Eutrophication models (MERLIN, MAGIC-WAND, Vollenweider eutrophication model)
  — Forest models (JABOWA, FORECAST, TREE-GRO, FOREST-BGC/PnET, NuChem, FOR-SANA)
  — Simple cycle models (carbon, nitrogen, mercury, magnesium, etc.) (GOUDRIAN, DECOMP, SOIL, SOIL-N, COUP, PROFILE)
  — Atmosphere models (EMEP)
  — Groundwater substance transport models magic (FREEQUE).
- Integrated multiple system models
  — Nutrient cycle – population models
  — Nutrient cycle – tropic cascade models
  — Climate change – carbon cycle – vegetation models (CENTURY, FORSKA, BIOME)
  — Climate change – acidification models (FOR-SAFE)
  — Forest – management – nutrient cycle models (FORGRO, FORSAFE, PnET).
- Higher hierarchy models
  — Decision – management-biogeochemical models (RAINS, IMAGE, CASM)
  — Neural network models for complex societal systems (Adaptive Learning Networks).

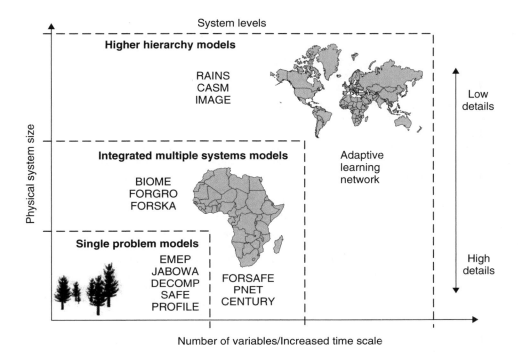

Figure 11.9 Biogeochemical models can be categorized depending on which scale they are applied to as well as the degree of integration of multiple processes over those scales

In the higher hierarchy models, the biogeochemistry only forms a subsystem of the whole model system, and the focus of the question may be totally outside the biogeochemical domain, even when the biogeochemical modules form essential parts of the interactive system (see Figure 11.9).

Most biogeochemical models do not exist as computer codes, nor do they have any fancy names or acronyms attached. They exist as mental models and are mostly only recorded in language on paper. Many computer models exist only to produce a few published articles, only to disappear when the originator gets a new job. The mental models will survive, but the code is lost.

Very few of these models are operable in reality or useful to other people than those that built them. Very few of the models have been user-adapted, and generally no support apparatus will be in existence. This problem is related to the fact that the answers are not driven by the models, but rather that the questions determine which model is required, and every new adaptation of the question will demand modification of the model or even a completely new model. Just taking a model, getting its answers and then searching for the appropriate question that might fit the answer often becomes a backward way of working. It will only work well if the questions are highly standardized.

Some useful models for bulk soil chemistry have come from acidification research and we will use these as examples. These all had the requirement that they should be regionally applicable, testable, have *a priori* capacity and not need calibration on essential parameters. These requirements excluded all of the earlier existing models for soil chemistry. Of course, there are as many biogeochemical models as there are questions to ask or problems to study. We will not produce a comprehensive list which would fast become redundant. Acidification soil and surface water models have been built for research and learning purposes, and from these the 'survivor' models that came to be used in the critical loads assessment were derived. They are survivors because they do more than exist in research laboratories, they are publicly available, and they can be used by anyone without excessive training, they work on data that can be found and they give useful results. They share many properties and differ mainly in focus. The steady state models for acidity are:

1. *SMB* is the general European default model for terrestrial ecosystems, which all countries are supposed to use. It is a mass balance, one-box formulation, involving one mass balance for ANC, coupled to a mass balance for base cations (Sverdrup and de Vries, 1994).

2. *F-factor models* are semi-empirical mass balance models developed for aquatic systems, applicable to acidity only. All the different models philosophically go back to the original ideas of Henriksen (the Henriksen Model). They make complex assumptions, but are easy to use on existing data (Brakke *et al.*, 1990).

3. *PROFILE* was developed for forest soils in Sweden, and it is a fully integrated process-oriented model. It focuses on the kinetics of chemical weathering, nutrient uptake to vegetation and cycling, nitrogen transformation reactions and solute flow in the soil horizon (Sverdrup and Warfvinge, 1995). It calculates soil chemistry layer by layer: it is a multi-layer model. It differs from the two other models by calculating the chemical weathering *a priori*, without any calibration, from physical and geological soil properties, and not using the weathering rate as a calibration parameter. PROFILE is the only existing model which will predict the weathering rate under field conditions from soil mineralogy and geophysics.

There are also a number of 'survivor' models for dynamic assessments for acidity. Dynamic models are important for interpreting the critical loads, the time aspect of effects and the devolution of effects after emission reductions. The dynamic models are no substitute for the steady state models, and do not do a better job just because they are more complex; often the complexity is offset by additional uncertainty brought by the increased complexity. Calculating critical loads with dynamic models are just a more complex and expensive way to do something that could be done at less cost and sometimes with better accuracy with models like SMB or PROFILE. All these 'survivor' models are available and easy to use:

1. *MAGIC* was originally developed in the United States (Cosby *et al.*, 1985), but it has found more use in Europe after the discontinuation of America's acid-rain research, and now the focus is on surface water. The core processes are alkalinity mass balances, sulphate adsorption and cation exchange. The minimum time resolution of the model is one year, equal to the numerical integration step in the model.

It is a one-layer soil model. The model is calibrated by adjusting weathering, initial base saturation, selectivity coefficients, gibbsite coefficients and present base saturation and present stream chemistry is used as optimizing targets (Cosby *et al.*, 1985). On surface waters, the model is acceptably well confined.

2. *SMART* is a more recent model (de Vries *et al.*, 1989), and its focus is on average major soil chemistry. The core processes are alkalinity mass balance, sulphate adsorption and cation exchange. The minimum time resolution of the model is one year, equal to the numerical integration step in the model. The model is calibrated by adjusting two parameters: the weathering rate and the initial base saturation (de Vries *et al.*, 1989); present base saturation and soil chemistry are used as optimizing targets. It is a one-layer soil model. The model is reasonably well confined on soils and surface waters.

3. *SAFE* was developed for forest soils in Sweden, and focuses on chemical weathering, simple nutrient cycling and development of cation exchange with time (Warfvinge *et al.*, 1998). It is the dynamic version of PROFILE. It calculates soil chemistry layer by layer. It is a multi-layer model. It differs from the two other models by calculating the chemical weathering from physical and geological soil properties, and not by using the weathering rate as a calibration parameter. The core processes are weathering and cation exchange. This tends to cause difficulties with application to catchments and lakes when the average soil depth for the whole watershed is needed. The model is calibrated by adjusting initial base saturation with present base saturation as the optimizing target. The model is uniquely defined for soils and acceptably well defined for surface waters.

In summary, the MAGIC and SMART models are easier to apply than SAFE because of the smaller amount of input data and the greater freedom in calibration. They are cheaper to use, but lead to simpler results. SAFE is significantly better constrained by its stricter calibration on fewer parameters, its higher demand on input data and its higher stratigraphic resolution. It is more expensive to use, but provides more elaborate results. Which is best depends on the purpose of the user.

There are several model proposals for nitrogen assessments, but these should be classified as tentative at best. Existing models operate with excessive calibration, the models are actually calibrated in such a way that inputs become output directly. Once the demand is made for no excessive calibration, there is virtually no model available. For calculation of critical loads for nitrogen, a SMB-N is used. It is a simple mass balance, where each term is estimated by largely empirical methods (uptake, immobilization, permitted runoff). SOIL-N, MERLIN and MAGIC-WAND are available and operable, but inputs leave so much to the user to predefine that we can safely say that a lot more work is required before we can trust our decisions to them.

The most important property of the models that survive is that all have observable parameters and input data are strongly simplified, and can be simplified further. These models also have in common that they were developed by strong but small groups of researchers who had long-term financing over more than five years and a strong drive to apply the models in practical life.

The models that have not survived in practical use have also certain properties in common. Several models never became regionally applicable, which was often caused by the lack of simplification. If a model is allowed to contain too many 'pet processes', it will be hampered by many unnecessary parts which still require input data and computation time. Inclusion of many processes does not necessarily improve performance. Beyond a certain point (see Figure 11.5), the model performance quickly deteriorates as more processes and parts are added. Too often political prestige or private, short-sighted ambition have prevented the necessary simplification of these models. Some models have process descriptions of such a nature that too many of the parameters of the model have no physical significance that can be determined by measurement, and the parameters are not observable. This effectively precludes generalization and transfer to regional use. A long list of such very impressive, but rather useless, models can be made.

In Figure 11.10 the 'critical load model' model is shown. However, not only is the specific computer code chosen, but in reality all the information and interpretations that actually take place before the computer code can come into play. The numbers for deposition are not objective data, they are measured concentrations in a plastic cup of rainwater that somebody collected under a tree or maybe beside it. The deposition value is the result of one of those undefined models we often forget about mentioning as a model. This can be repeated for all the other values we use to create our input data files. The computer codes like 'MAGIC' or 'PROFILE' are just one of the components of the 'critical load model'. Uncertainty arises at many places in this scheme, and the computer codes are not necessarily where most of the uncertainty is

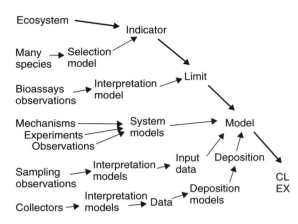

Figure 11.10 The 'critical load model' model is not only the specific computer code chosen, but in reality all the information reinterpretations that actually take place before the computer code can come into play. The computer codes like 'MAGIC' or 'PROFILE' are just one of the components of the 'critical load model'

generated. The experience from the European critical load mapping programme is that use and modification of these 'nonmodels' consumed 75% of work time, and that running MAGIC or SAFE only occupied 15% of the time.

11.7 CONCLUSION

The question asked or the issue investigated define the model to be used for any biogeochemical problem or issue. The chosen model defines the data needed to create the learning process required to produce the answer. No modelling starts by assembling 'all' data, and just adding more does not give more clarity, but less. If asked for a value judgement, no biogeochemical model is 'best'. The best model is the model that answers the question asked with the necessary amount of accuracy with the smallest cost or effort. This will vary depending on the question asked. Thus we want to make some definite statements:

- All models should start with problem definition and explanation of system boundaries.
- Understanding of a system implies a mental model. There are no models without a mental model, and the quality of the mental model is what defines the quality. There is no such statement that 'the model is wrong!': if something is wrong, then it applies to your mental model, your own understanding!

- Simplification is necessary in order to sort variables and get a basic understanding of the system functionalities.
- Data collected without purpose is not research, it is 'redundant information with potential for use', until a problem is found that defines a use for a specified part of it.
- Basic principles and driving forces need to be identified in order to determine effectively the required level of detail in the model building.
- Make sure that the calibration of the model is robust, in order to avoid telling the model what to give you (the input becomes the output).
- A transparent and testable model is required for effective communication of the model to others. Simpler models are easier to communicate than complex ones.

In terms of making research a learning process, the authors encourage researchers actively to build their own model using user-friendly software such as STELLA, POWERSIM, MATLAB, MAPLE, etc.

REFERENCES

Alcamo, J., Shaw, R. and Hordijk, L. (1990) *The RAINS Model of Acidification: Science and Strategies in Europe*, Kluwer Academic Publishers, Dordrecht.

Berge, E. and Jakobsen, H.A. (1998) A regional scale multilayer model for the calculation of long-term transport and deposition of air pollution in Europe, *Tellus* **50B**, 205–223.

Botkin, D.B., Janak, J.F. and Wallis, J.R. (1972) Some ecological consequences of a computer model of forest growth, *Journal of Ecology* **60**, 849–872.

Brakke, D.F., Henriksen, A. and Norton, S.A. (1990) A variable F-factor to explain changes in base cation concentrations as a function of strong acid deposition, *Verhandlungen Internationale Vereinigung für Theoretische und Angewandte Limnologie* **24**, 146–149.

Chen, C.J. (1993) The Response of Plants to Interacting Stresses, PGSM version 1.3 model documentation.

Cosby, B.J., Wright, R.F., Hornberger, G.M. and Galloway, J.N. (1985) Modelling the effects of acidic deposition, assessment of a lumped-parameter model of soil water and stream water chemistry, *Water Resources Research* **21**, 51–63.

Crote, R. and Erhard, M. (1999) Simulation of tree and stand development under different environmental conditions with a physiologically based model, *Forest Ecology and Management* **120**, 59–76.

Crote, R., Erhard, M. and Suckow, F. (1997) *Evaluation of the Physiologically-Based Forest Growth Model FORSANA 22*, Potsdam Institute for Climate Impact Research, Potsdam.

den Elzen, M. (1994) *Global Environmental Change and Integrated Modeling Approach*, International Books, Utrecht.

de Vries, W., Posch, M. and Kämäri, J. (1989) Simulation of the long-term soil response to acid deposition in various buffer ranges, *Water, Air and Soil Pollution* **48**, 349–390.

Dörner, D. (1996) *The Logic of Failure: Recognizing and Avoiding Error in Complex Situations*, Perseus Books, Cambridge, MA.

Gough, C.A., Bailey, P.D., Biewald, B., Kuylenstierna, J.C.I. and Chadwick, M.J. (1994) Environmentally targeted objectives for reducing acidification in Europe, *Energy Policy* **22**, 1055–1066.

Jansson, P.E. (1991) *SOIL Water and Heat Model: Technical Description*, Soil Science Department, University of Agricultural Sciences, Uppsala.

Kimmins, J.P., Mailly, D. and Seely, B. (1999) Modeling forest ecosystem net primary production: the hybrid simulation approach used in FORECAST, *Ecological Modeling* **122**, 195–224.

Kram, P., Santore, R.C., Driscoll, C.T., Aber, J.D. and Hruska, J. (1999) Application of the forest-soil-water model (PnET-BGC/CHESS) to the Lysina catchments, Czech Republic, *Ecological Modelling* **120**, 9–30.

Kros, H. (2002) Evaluation of biogeochemical modeling at local and regional scale, PhD thesis, Alterra Green World Institute, Wageningen, Netherlands.

Levenspiel, O. (1980) The coming of age of chemical reaction engineering, *Chemical Engineering Science* **35**, 1821–1839.

Mohren, G.M.J., Bartelink, H.H., Jorristma, I.T.M. and Kramer, K. (1993) A process-based growth model (FORGRO) for analysis of forest dynamics in relation to environmental factors, in M.E.A. Brockmeyer, W. Vos and H. Koep (eds) *European Forest Reserves*, Workshop, 6–8 May 1992, PUDOC, Wageningen, Netherlands, 273–280.

Parton, W.J., Schimel, D.S., Cole, C.V. and Ojima, D.S. (1987) Analysis of factors controlling soil organic levels of grasslands in the Great Plains, *Soil Science Society of America Journal* **51**, 1173–1179.

Richardson, P.G. and Pugh, A.L. (1981) *Introduction to System Dynamics Modelling with DYNAMO*, Productivity Press, New York.

Rotmans, J., van Asselt, M.B.A., de Bruin, A.J., den Elzen, M.G.J., de Greef, J., Hilderink, H., Hoekstra, A.Y., Janssen, M.A., Koster., H.W., Martens, W.J.M., Niessen, L.W. and de Vries, H.J.M. (1994) *Global Change and Sustainable Development: A Modelling Perspective for the Next Decade*, National Institute of Public Health and Environmental Protection (RIVM), Bilthoven.

Sykes, M.T., Prentice, I.C. and Cramer, W. (1996) A bioclimatic model for the potential distributions of north European tree species under present and future climates, *Journal of Biogeography* **23**, 203–233.

Sverdrup, H. and Stjernquist, I. (2002) *Developing Principles for Sustainable Forestry*, Kluwer Academic Publishers, Dordrecht.

Sverdrup, H. and de Vries, W. (1994) Calculating critical loads for acidity with the simple mass balance method, *Water, Air and Soil Pollution* **72**, 143–162.

Sverdrup, H. and Warfvinge, P. (1995) Estimating field weathering rates using laboratory kinetics, in A.F. White and

S.L. Brantley (eds) *Chemical Weathering Rates of Silicate*, Mineralogical Society of America, Washington, DC.

Warfvinge, P., Sverdrup, H. and Wickman, T. (1998). Estimating the weathering rate at Gårdsjön using different methods, in H. Hultberg and R. Skeffington (eds) *Experimental Reversal of Acid Rain Effects; The Gårdsjön Roof Project*, John Wiley & Sons, Chichester, 231–250.

Wright, R.F., Beier, C. and Cosby, B.J. (1998) Effects of nitrogen deposition and climate change on nitrogen runoff at Norwegian boreal forest catchments: the MERLIN model applied to Risdalsheia (RAIN and CLIMEX projects), *Hydrology and Earth System Sciences* **2**, 399–414.

12

Modelling Human Decision-Making

JOHN WAINWRIGHT AND MARK MULLIGAN

12.1 INTRODUCTION

This book is a complex system. It is made up of a series of interconnecting parts with inputs from a large number of external sources – the experiences of the contributing authors. These experiences combine to provide a whole that can function in a number of ways – as a manual, a reference, and a source of ideas. The way in which it is put together provides one means of using (reading) it, but as noted in the Introduction, you needn't have found this chapter about half-way through, because the best way to read the book is not necessarily from beginning to end. But then, who always chooses to read the Introduction? When we first decided it would be a good idea to put together a book about environmental modelling, we thought it would be best to benefit from the insights of specialists in particular areas rather than write everything ourselves. We also thought this approach would make the whole process quicker (something our publisher is beginning to doubt!), which is an advantage given the pressure on time and resources in the relatively under-funded world of UK academia. But consider what this means – we now have 32 different contributors, all with their different decisions to make about what to include, how to do so, and when to do it (even modellers occasionally prefer going to the cinema or a concert or to watch a football match than to sit in front of the computer for a few hours more). Being in the middle of the book, we cannot fit page numbers to the remainder until it is complete (and in fact are awaiting one or two preceding chapters so we can't even get this far!). So the whole thing is strongly interdependent in its structure and output. Why did we ever decide to do it in the first place? The answer may lie in a mix of prestige, expectation and satisfaction (certainly not for the money!). For any contributor working under the watchful eye of the Research Assessment Exercise in UK universities for which edited volumes score very little in comparison with journal articles and research monographs, the decision is something of an irrational one, and not necessarily one for which our heads of department will thank us. So is human behaviour rational? In this case obviously not. How can we hope to model something that can appear to be almost randomly irrational? What's more, why did any of the contributors decide to take up modelling in the first place rather than drive trains or become firefighters? What made you pick the book from the shelf of the library or bookshop (or have your mouse inexplicably drawn to it on the Internet)?

How this chapter got to be here and how you got to be there reading it is therefore a function of a long series of decisions whose ultimate explanation is, quite literally, lost in the mists of time. If we cannot explain these decisions, how can we expect to model the huge number of interacting human decisions that interact with and affect the environment at scales from local to global? At this point, should we just throw up our hands in dismay and wander off (or spend endless hours – and pages – engaged in discourse about the undecidability of decision-making . . .)?

Environmental Modelling: Finding Simplicity in Complexity. Edited by J. Wainwright and M. Mulligan
© 2004 John Wiley & Sons, Ltd ISBNs: 0-471-49617-0 (HB); 0-471-49618-9 (PB)

If you are still with us, then hopefully you will agree that some attempt to model human decision-making is necessary – especially if the models we come up with help us understand why we make decisions (both rational and irrational) and how the process works overall. Of course, having elements relating to human decisions is fundamental for most models, especially those in the applied field, as we will see in the next section. (Indeed, all models relate to human decisions by the very nature of how they are built and implemented. After a long day of trying to remove a 'totally obvious' error from our model code, we might wish that this human element were not the case! Developments in genetic algorithms [e.g. Mitchell, 1996], that is, models that essentially build themselves, have yet to reach environmental models in a significant way.)

Human activities within the environment continue to increase. (We will leave aside the debate about whether humans are part of the environment or 'nature' – see Phillips and Mighall, 2000, for discussions.) Hooke (2000) has demonstrated that human activity has now overtaken rivers as the principal transport agent of sediment. Human-induced climate change and land use are a fundamental part of this increase in impact, and the modelling of human behaviour is a central element of attempts to assess the nature and extent of change (IPCC, 2001). Indeed, the interest in environmental modelling (and part of the rationale for this book) at the present time is largely spurred on by the potential threats of climate change and its two-way relationship with land use. So, increasingly, all types of environmental modelling require consideration of at least some aspect of human behaviour. Why is this human behaviour so complex compared with other animals, plants and abiotic aspects of the environment?

12.2 THE HUMAN MIND

The structure of the human brain is complex. Despite being the most remarkable object we know, it resembles a bowl of cold porridge, at least according to one of the founding fathers of computation, Alan Turing (cited in Penrose, 1989). At the top is the deeply wrinkled, grey cerebral cortex, divided into two almost symmetrical parts and sitting above the ancestral mammalian brain or cerebellum (Figure 12.1; Carter, 1998). The limbic system is located beneath the cerebral cortex and communicates with the rest of the body via the brain stem, but is also responsible for most of our behaviour directly related to survival and the formation of memories. The cerebral cortex is itself divided into five main areas, each of which seem to serve specific functions. For

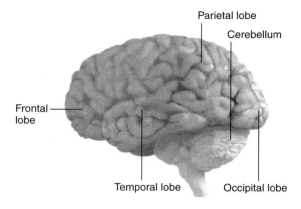

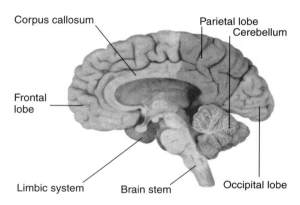

Figure 12.1   General structure of the human brain
Source: Carter (1998)

example, the parietal lobes deal with movement among other things, while the frontal lobes deal with conceptual and planning issues as well as emotions. One-tenth of the cells in the brain are the 10^{12} neurons – responsible for brain activity (see also Chapter 1) – and connected by long, thin nerve cells of two types. Axons carry information away from a neuron, while dendrites carry information towards it. The location where an axon joins a dendrite is known as a synapse, and information is passed chemically along the nerve by successively changing the balance of sodium then potassium ions across the cell membrane (see Penrose, 1989, for a simple explanation). There may be several thousand such connections from a single neuron to others. The nerves are sheathed in myelin, which permits this system to operate at very rapid speeds. As well as these specific connections, there are also general chemical connections within the brain, as a result of cells that produce specific substances called neurotransmitters, which themselves have specific effects on brain activity and hence behaviour.

Even with recent advances in brain-scanning technology, understanding the operation of the brain and its interconnections is fraught with difficulties. Often, explanations come most clearly when dealing with parts of the brain that are known to have been damaged (Ramachandran and Blakeslee, 1998). It is perhaps, therefore, unsurprising that understanding human behaviour is a difficult process:

> When one considers the structure of the brain it is remarkable that it ever manages to come to the correct conclusion about anything. By any conventional standards neurons are an entirely unsuitable medium for computation; they die throughout the brain's life, causing random loss of stored information; they have a finite probability of firing even when they are not engaged in signal processing; the response of a neuron to any particular input is probabilistic, not fixed.
>
> (McLeod *et al.*, 1998: 32)

Thus far, the discussion has been specifically on the brain. Issues of consciousness, soul and mind are difficult ones to explain convincingly. Nathan (1987) suggests talking about mental processes rather than getting bogged down in issues of what the mind might be. A relatively simple viewpoint is to regard the mind as the emergent property of the various brain functions (e.g. Bar-Yam, 1997), although there exists considerable debate as to which functions to include so as to distinguish the human mind or consciousness from that of other animals – most, for example, insisting on the inclusion of language (Noble and Davidson, 1996). Aleksander (2000) centres his definition on imagination.

The decision-making process can be characterized as the various ways in which mental processes can be made to affect human behaviour. These processes may relate to relatively 'primitive' behaviour relating to the survival instinct (decisions on whether to run away or stand one's ground when threatened; what to eat), all the way up to the more considered ideas and plans that may develop in the frontal lobes, perhaps via discussions with others. Of course, the ever-present potential for interaction between brains complicates the matter further. As noted by Thagard (2000), it is necessary for any theory of decision-making to evaluate personal goals and constraints on them, rather than simply optimize a set of potential preferences. Decisions are not always right, nor successful! These features of mental activity and behaviour seem to have developed along with modern humans. Both Noble and Davidson (1996) and

Mithen (1996) see the onset of features characteristic of present ways of thinking around 250 000–300 000 years ago (but see Jaynes, 1980, for a significantly later date), so even environmental models that deal with the distant past may need to include significant elements relating to human activity. It is clear that our brain power (and its application to industry and technology) as well as our population have, over time, enormously increased our impact on, and (arguably) decreased our dependence on, the natural world.

Underlying our ability to model human activity is a fundamental question: can a computer be made to think? Essentially, in trying to simulate decision-making and the corresponding behaviour, we need to be able to replicate the thought process within a computational framework. This is the discipline of artificial intelligence (AI; Crevier, 1993). Searle distinguishes between 'weak' and 'strong' AI. The former assumes that principles relating to brain function can be reproduced computationally and corresponding hypotheses tested. The latter considers that 'the appropriately programmed computer really *is* a mind in the sense that computers given the right programs can literally be said to *understand* and have other cognitive states . . . the programs are not mere tools that enable us to test psychological explanations; rather, the programs are themselves the explanations' (1980: 417: emphasis in original). While he is happy to accept weak AI, Searle strongly criticizes the strong form, using the example of a non-Chinese speaker being passed a list of rules for translating and commenting (in English) upon a story given to them in Chinese. Although a native speaker might be fooled by the output (i.e. the system – essentially a replication of a computer program [the list of rules] with input [the story] and output [the translation and commentary] – passes a Turing test), Searle demonstrates that the person does not *understand* Chinese and thus the process does not represent an intelligent system. Penrose (1989) points out that the weakest point of Searle's argument is when he allows the different stages in the translation and commentary to be carried out by (large numbers of) different people, as the system then becomes more analogous to the neurons in the brain – we do not need to argue that every single neuron *understands* the outcome for intelligence to be demonstrated. The process here is one of emergence (see also Aleksander, 2000). Hofstadter (1979, 1981) argues for the possibility of strong AI in computers (but not necessarily other human attributes such as emotions). He suggests that a (large) book containing a complete description of Einstein's brain would allow us to interrogate it and give equivalent answers to those Einstein would have given. Compare this example to that of Borges' 'Library of Babel' that we discussed in the Introduction. Penrose (1989) disagrees with the strong AI view for different reasons. As a mathematician and physicist, he argues for the role of quantum phenomena in neuronal operation, allowing the flashes of insight that tend to characterize human thinking compared to the machine-based versions. This explanation also overcomes limitations such as the impossibility of designing algorithms to decide from first principles whether a result produced is correct (following Gödel's theorem – see also the excellent discussion in Hofstadter, 1979). Needless to say, Penrose's ideas have not met with universal approval (e.g. Dennett, 1991; Crick, 1994).

In summary, there are serious objections to the strong AI agenda. However, is weak AI good enough for our purposes in environmental modelling? Fortunately, the answer is yes, at least in a number of cases. For studying the development and reaction of cognitive processes, artificial neural networks (see Chapter 1) do give sufficient scope for such analysis (e.g. McLeod *et al.*, 1998; Haykin, 1999). A number of workers (e.g. Ferber, 1999) are also developing techniques of distributed AI, which allow precisely the sort of interaction in decision-making that we noted was important above. These approaches are often called agent-based or individual-based models (IBMs: see Ginot *et al.*, 2002, for a review). In the

following sections, we look at a number of problems relating to decision-making that follow more traditional modelling viewpoints, and compare them to others that account at least in part for the discussion above.

12.3 MODELLING POPULATIONS

A commonly asked question with regard to human impacts on the environment and resource management is how populations will change. Traditional approaches to modelling this question have tended to employ population-growth models similar to the simple Malthusian or carrying-capacity-limited models described in Chapter 1. These models may have the advantage of simplicity, but may also interact with a number of other modelled components (e.g. in economics-based models – see below). Studies in southern Spain by McGlade *et al.* (1994) demonstrated complex relationships between human populations, soil resources and erosion rates in relation to later Bronze Age occupations (Figure 12.2). One apparent advantage of this form of model is that they can exhibit deterministic chaos if population growth rates are increased sufficiently (McGlade, 1995). Anderies (2000) also demonstrated the development of limit cycles between human and environmental resources in his models of the Tsembaga of New Guinea and the Polynesians of Easter Island.

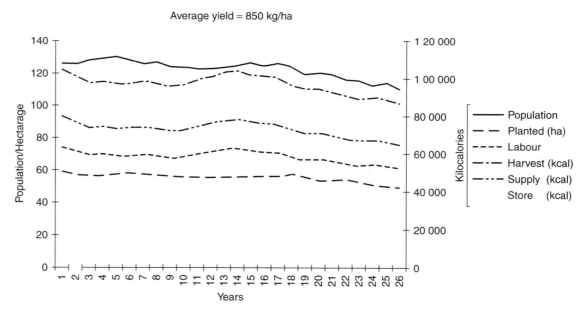

Figure 12.2 Interaction of human population, vegetation and erosion for an example simulation for the Vera Basin, Spain in the Bronze Age (after McGlade, 1995)

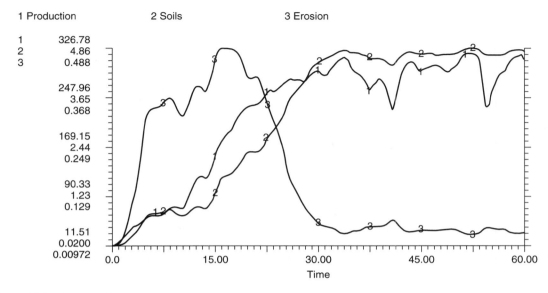

1 Production	2 Soils	3 Erosion

1	326.78
2	4.86
3	0.488

| 247.96 |
| 3.65 |
| 0.368 |

| 169.15 |
| 2.44 |
| 0.249 |

| 90.33 |
| 1.23 |
| 0.129 |

| 11.51 |
| 0.0200 |
| 0.00972 |

Figure 12.2   (*continued*)

Thus this approach provides a means of explaining apparently random population changes, relating to the rapid increases and decreases that are apparent through prehistory and history. However, many people feel unhappy with the determinism implicit in these models. A concept of free will is often thought to be central to the human condition, while these models ignore all possibility of its existence, even allowing for chaotic results.

12.4 MODELLING DECISIONS

Perhaps the most important, related, criticism of the above approaches is that they fail to account for the *reasons* why population size changes. Populations change because of the decisions taken by individuals (e.g. choice of partner, use of contraception, spacing between children, etc.) as affected by social norms (e.g. acceptability of large families, infanticide, modes of marriage, etc.) and historical and environmental contingency. (Note that a growing number of ecologists are also suggesting the use of individual-based models in their modelling of plants or animals; see Chapters 6 to 8.) Population growth is thus an emergent property or consequence of individual and social interactions and not a process in itself.

12.4.1 Agent-based modelling

Epstein and Axtell (1996) provide an agent-based approach in their Sugarscape model. Sugarscape is a simplified landscape laid out on a grid, on which a food resource (sugar) is distributed in some initial, specified pattern. The landscape is populated by a number of agents who can move around the landscape according to specified rules in order to harvest and consume the sugar. Once harvested, the sugar is able to grow back according to a simple set of rules. The agents are randomly assigned a metabolism – how fast they consume the sugar they collect, so that if their sugar resource falls to zero, they die off. They are also assigned a distance of vision allowing a form of differential planning of their movements, and are randomly distributed in space at the start of a simulation. Agents can collect more sugar than they need to survive and thus accumulate 'wealth'. The results with very simple models show important patterns. The spatial location of the agents varies according to different rates of regrowth of the harvested sugar, thus not all landscapes are equal. The landscapes converge to specific numbers of individuals for given sets of conditions. In other words, a carrying capacity (see Chapter 1) *emerges* from the agents' behaviour (Figure 12.3a). Furthermore, this carrying capacity is a function of characteristics of the agents, so that if their metabolisms and depths of vision become dynamic properties, the carrying capacity too would be variable. This result overcomes a major limitation of the carrying-capacity approach to human populations, in that the path of human history has been one of adaptation, allowing (environmentally imposed) carrying capacities to be overcome through better technology, for example. Such adaptations are impossible to include in the traditional form of the population model. Another emergent property is the unequal distribution of wealth of individuals, even if they start off with a relatively uniform initial amount (Figure 12.3b).

More complex Sugarscape rules were also developed by Epstein and Axtell to account for the processes of sexual reproduction (and genetic attributes), group affiliations, pollution, warfare, trade (of a second commodity – spice) and disease (with the possibility of overcoming it and transmitting immunity). Thus, it was possible to generate relatively complex patterns of emergent behaviour from relatively simple rules. For example, population cycles with a wavelength exceeding the average lifetime of individual agents can emerge in the presence of the spice trade, compared to the extinction under equivalent conditions without trade (Figure 12.3c). The modelling approach allows the investigation of the realism of different economic and policy models and ideas of sustainability within the landscape.

While Sugarscape looks at an artificial landscape to investigate general patterns, other attempts have been made to simulate real landscapes. Gumerman and Kohler (1995) and Axtell *et al.* (2002) have used the approach to investigate changing populations in Long

House Valley in northeastern Arizona. The area has been occupied for over 3000 years, and particular emphasis was placed on the period of agriculture under the Anasazi from *c.* 200 CE until its collapse in *c.* 1300 CE. Tree-ring records were used to derive a climate record based on drought intensity, which was then used to estimate maize yields. Agents in this case were based on family groups who make annual decisions on where to live and farm. Initial simulations track the historical pattern of household change and the spatial location of households relatively well. The simulations suggest that although population would decline in the later periods due to environmental degradation, there were probably other external factors that led to the complete abandonment of the area around 1300 CE.

Bura *et al.* (1996) and Sanders *et al.* (1997) used a model called SIMPOP to investigate patterns of urban development in a generalized area based on the southern Rhône valley in France. Their agents represented aggregation at the level of the settlement, which for obvious reasons were not allowed to move through the landscape. The landscape was divided into hexagonal cells with initial resources (agricultural, mineral or maritime) which could be occupied by settlements of different types. All sites start off as rural, but can develop into a variety of commercial, industrial or administrative centres, based

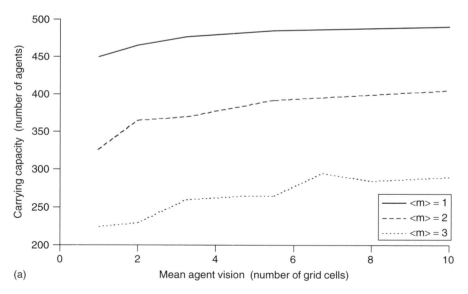

(a)

Figure 12.3 Examples of results from the Sugarscape model of Epstein and Axtell (1996). (a) Emergent carrying capacity in numbers of agents of a 50 × 50 cell landscape; (b) emergence of unequal distributions of wealth (measured in units of sugar) for similar simple rules to the results shown in (a); and (c) behaviour comparing the effect of no trade in spice, leading to extinction, with the existence of trade leading ultimately to the development of 'long wave' population oscillations

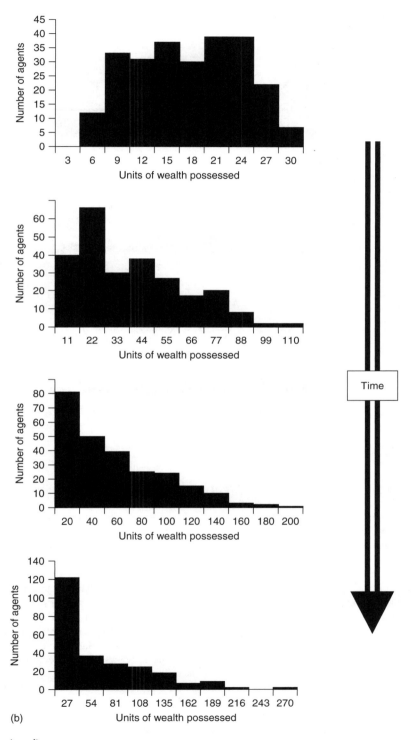

(b)

Figure 12.3   (*continued*)

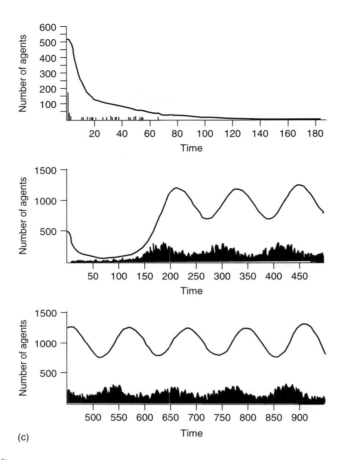

Figure 12.3   (*continued*)

on a series of economic models. Exchange is allowed between settlements and population is modelled as a series of stochastic functions based on settlement size, neighbourhood and ability to satisfy local demand. The simulations resulted in the emergence of a highly hierarchical urban landscape, with similar properties to that of the modern landscape, albeit with a poor spatial distribution of the major centres (Figure 12.4). See Chapter 14 for a discussion of more complex socio-economic interactions in agent-based models.

To summarize, we can see that environmental as well as historical contingency is important in the simulation of population change using agents. Agent-based models offer distinct explanatory power over previous approaches, in that observed phenomena emerge from interactions, rather than being imposed by the modelling framework and preconceived notions. Furthermore, they enable the testing of different hypotheses relating to social and economic phenomena, and the evaluation of

contingency in different realizations of the same set of conditions.

Of course, when interacting with the landscape, it is not just population-related properties that are important emergent features. Even more important is what is done with the land, as demonstrated in the previous two examples, for only when account is taken of changing land use are fully dynamic interactions between humans and their landscapes investigated. Many impact-related models that fail to do so are thus oversimplified to the extent of being poor representations of the system being researched. Issues relating to the modelling of land use and land-use change are covered fully by Lambin in the next chapter.

At relatively small scales, Gimblett *et al.* (2001) and Roberts *et al.* (2002) have used agent-based models to investigate the specific problem of the impact of recreational use of the Grand Canyon National Park in the US. Agents within the model plan raft trips, the use of camp sites, beaches and other recreational areas and interact with other agents, with differences reflecting commercial versus noncommercial trips. Location of the simulated trips through time matched measured data relatively well, but some features such as beach use tended to be overpredicted (although the authors suggested that this result may point to a problem in the validation data). The simulation model is actively used to support decision-making regarding conflicting uses of the national park.

This example demonstrates that it is possible to go beyond the simple rule-based approaches that typified early attempts at AI and the cellular automata approach to agent-based systems. Neural networks provide an ideal means of doing this in a way that is compatible with the discussions of the decision-making process above. Stassinopoulos and Bak (1995) have demonstrated that it is possible to design such systems

that self-organize. In other words, they produce output based on the environments in which they are localized, rather than being dependent on rules that are specifically assigned to them by the modeller. Low *et al.* (2001) have developed individual-based models to assess the implications on fisheries stocks of harvesting, and its feedback on the human populations involved. Their results showed that external variability in the resource – a factor that is rarely measured in field observations of such systems – is a critical factor in determining sustainability. Thus, we see another example of how a model can inform other aspects of our investigative methodology. Weisbuch (2000) demonstrated (using a cellular automata approach) that institutional cooperation can emerge from models of pollution-related activity. Bishop (1996) has used artificial neural networks to investigate perceived scenic beauty in western Victoria, Australia. In this example, he used a back-propagation technique

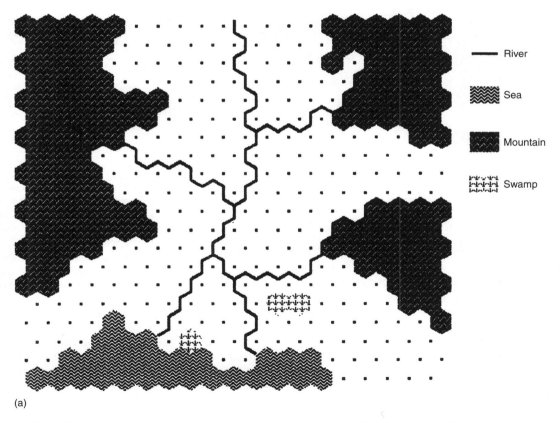

(a)

Figure 12.4 Example simulation of urban development in an area not unlike the southern Rhône valley in France. (a) initial conditions; (b) Spatial organization of the landscape after 2000 simulated years from an initially rural landscape together with corresponding emergent patterns of population and wealth
Source: Sanders *et al.* (1997)

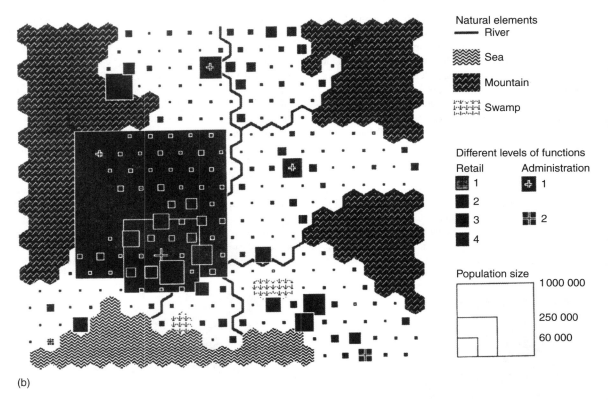

Natural elements
— River
Sea
Mountain
Swamp

Different levels of functions
Retail   Administration
1   1
2
3   2
4

Population size
1 000 000
250 000
60 000

(b)

Figure 12.4   (*continued*)

to estimate the relative importance of a number of measured variables estimated within a GIS to the mean perceived beauty of the area by a sample of 48 observers (note that this is an analytical rather than self-organizing approach, though). Bezerra *et al.* (1996) have designed a simple neural network model that takes account of factors relating to personality (distanciation from reality, knowledge, relation to time, expansiveness, ego, creativity, risk shift, anxiety and authoritarian attitude) and instantaneous state of the individual (fatigue, coping, sleep deprivation, morale and motivation as affected by the environment in which the individual is situated) in relation to the decision-making process. This process is a function of six factors in their model: perception, mental representation, data processing, problem solving, choice of solution and action. Such approaches are as yet in their infancy, and their ability to provide workable models has yet to be thoroughly evaluated (for other examples, see Gilbert and Doran, 1994; Kohler and Gumerman, 2000). Alternative approaches that may prove useful in the future include the use of virtual reality models (van Veen *et al.*, 1998).

12.4.2 Economics models

At the opposite end of the spectrum, models are required that deal with decisions at national and international levels. Models based on principles of economics are commonly used at this scale. Cantor and Yohe (1998) provide an excellent overview of the limitations of this approach, which we summarize below. Standard macroeconomic models depend on projections of factors such as income, economic activity and trade with the application of cost-benefit analysis for different environmental factors. Governmental and international organizations such as the World Bank will tend to evaluate costs and benefits relative to gross domestic product (GDP) or gross national product (GNP). Comparisons between countries are difficult because different categorizations may be used, because of the need to evaluate the relative effects of different activities and because of exchange-rate issues. Even though more reliable comparisons using purchasing-power parities are possible, these are not used by institutions such as the World Bank, leading to obvious problems in interpretation.

Cantor and Yohe divide other issues into two categories. First, measurement problems relate to difficulties in providing reasonable comparisons. It may be difficult to evaluate the true level of economic activity, particularly in less industrialized nations, where the informal economy has been estimated to be responsible for 80–98% of activity relative to the GDP, 30–57% of employment and 33–75% of credit. Other means are required for evaluating these roles. Ignoring them would significantly impact on particularly the rural sectors in less industrialized areas. There is the related issue of how to measure inequality at the international scale, so that issues of equity may be addressed within a modelling framework. Second, there are more difficult conceptual problems. How can the aggregation of individual needs and desires be addressed? This is the question of emergence that we have previously highlighted. How can social and cultural differences be accounted for? If they are ignored, any policy developed will be difficult or impossible to implement effectively (see below). How can time and the requirements of future generations be accounted for? Most economic models discount future effects so that the value of a benefit received in the future is less than one received now. The rate at which future benefits are discounted is a very sensitive model parameter – $1000 m received in 200 years would have an estimated present-day value of $137 m if discounted at 1% but only $5.27 if discounted at 10%! A 'careful' choice of rate can therefore greatly bias the outcome to benefit particular parties and is at best inappropriate but at worst unethical. How can sociopolitical aspects of growth be accounted for? The ways in which industrialization occurs may differentially affect the rates at which greenhouse gas emissions (GHG) are produced. Again, 'one size fits all' models are likely to be inappropriate. Finally, how can different ideas of sustainability be accounted for? Different viewpoints suggest highly contrasting views: optimism that technological fixes will be possible due to continued economic growth, pessimism that major thresholds will be crossed leading to catastrophic decline in environmental resources and economic growth, or intermediate viewpoints relating to adaptation (Figure 12.5). The standard approach is to produce a series of scenarios based on different assumptions of interactions, using different measurement methods, which we will discuss in more detail below.

Gielen and Yagita (2002) modelled the impacts of different policies to reduce greenhouse gases on global trade in the petrochemical industry. Using six regions – western Europe, Japan, North America, eastern Europe, the Middle East and the rest of the world – they simulated demand for CO_2 emissions at different stages in the production cycle of 50 petrochemicals. Energy and transportation costs were considered along with other constraints such as resource availability. Different scenarios were assessed using taxes on greenhouse gases applied either globally, or to specific regional subsets. Without imposition of taxes, they predicted a 67% increase in CO_2 emissions compared to decreases of 4% with taxes of ¥7500 t^{-1} in western Europe and up to 58% with global taxes of ¥15 000 t^{-1}. However, in the case of nonglobal taxes, production may only shift location.

Such an approach to abatement had led many to suggest that it would be unfeasible due to the high costs. However, Azar and Schneider (2002) used a simple macro-economic model to suggest the costs of CO_2 abatement to maintain 350 ppm in 2100 to be of the order of $18 (US) trillion ($10^{12}$) discounted to 1990 values. This figure compares to a total world economy of $20 trillion in 1990. Clearly seen by this comparison, the proportion is high. But the standard approach is to assume income growth over the same period. Put in this context, the extra cost is very marginal. This example demonstrates once again the need for extreme care in interpreting results (and ensuring that our results are not misinterpreted by others).

12.4.3 Game theory

Game theory provides one means by which such interactions can be simulated at this level. Parsons and Ward (1998) note that the standard level of analysis assumes that participants in the models can be built using individual-based approaches, but because such models would tend to be large and cumbersome, most current models assume simplification at the level of the nation, sub- or trans-national group is possible. If *a priori* assumptions can be made about the benefits to each of the groups, then game theory can provide optimal solutions for cooperation versus noncooperation for each of them. Clearly, this requirement is a problem in most approaches to future change, as it is impossible to evaluate what the benefits would be. Solutions to this problem include the use of stochastic measurements of benefits and the use of iterated 'supergame' approaches. The outcome of the latter tend to be multiple equilibrium solutions, so that some qualitative evaluation of the different outcomes is required. Other forms of simulation may thus be required to deal with the complex interactions (Parsons and Ward, 1998).

Frisvold and Caswell (2000) used game theory to investigate the need for assistance projects and technical support in the management of water supply and

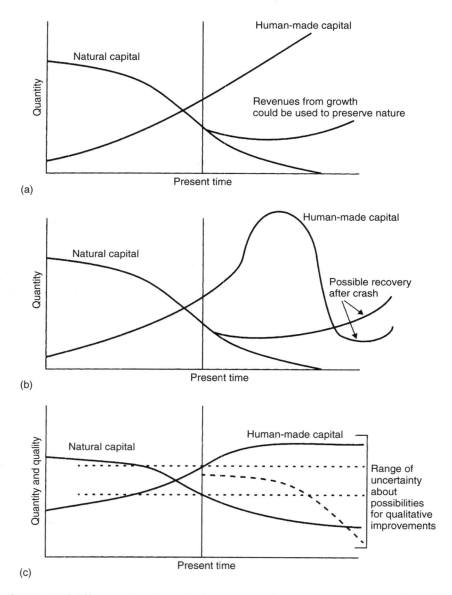

Figure 12.5 Summary of different schematic models of economic and environmental interaction. (a) 'The economic optimist'; (b) 'the environmental pessimist'; (c) 'the ecological economic vision'
Source: Cantor and Yohe (1998)

treatment issues across the border of Mexico and the United States. Negotiations within the International Boundary and Water Commission established by the 1944 Water Treaty between the two nations can be treated as a cooperation game because commitments are legally binding. Decisions such as that made in 1951 regarding the pollution of Nogales can be regarded as economically efficient within this framework as they apportioned costs to benefits – the US side benefited more in this example, and was prepared and able to pay a higher proportion of the costs. In contrast, US policy in the period of the Reagan administration stated that costs should be met equally, leading to its being highly unlikely that mutually agreeable, economically efficient solutions could be reached. After a number of project failures, equal cost sharing was abandoned as a principle in 1990. Game theoretic approaches may also prove useful in deciding the need to provide grants and technical assistance for the development of water sanitation measures. However, they point out that game theory may provide unsatisfactory solutions when one side is responsible for the problem and the other is the victim. In these cases, standard game theoretic solutions will tend towards victim-pays outcomes, which contravene the internationally accepted polluter-pays principle. They also do not allow for cases where countries do not wish to appear weak nor for the victim to offer payments to prevent the problem where it has a significantly lower income than the polluter. Frisvold and Caswell suggest that iterated games (similar to the 'supergame' discussed above) may be more appropriate techniques in such cases.

A game theoretic approach was also applied to the analysis of agreements to reduce CO_2 in the global context by Frankhauser and Kverndokk (1996). In their approach, CO_2-reduction agreements are considered as a five-player game between the USA, other OECD countries, countries of the former USSR, China and the rest of the world. Each player tries to maximize its benefits in relation to CO_2 emitted in order to increase its own wealth measured as GNP. At the same time, these emissions will lead to future effects on the wealth of the emitter and all the other players in the game, discounted at a fixed rate (see above for problems relating to these approaches), and are related to the costs in reducing emissions. In the case of noncooperation, the major costs are borne by the other OECD countries primarily and then the USA, with almost negligible impacts on the rest of the world. This outcome seemed to be due to the significantly higher costs in reducing emissions in the developing world. The results also suggested that the main beneficiaries of cooperation in reducing emissions would be the rest of the world, followed by the other OECD nations, then China and the ex-USSR countries. In the latter, there would be no financial benefit in any of the simulations assessed, implying that these countries would be reluctant to reduce emissions without other incentives. Care must still be taken, however, in interpreting these results as they are calculated as a function of GDP.

12.4.4 Scenario-based approaches

Scenarios are a common way of dealing with the effects of human decision-making on impact analyses at scales ranging from local to global. Essentially they allow us to deal with 'what if?' questions. As an example, simulations may be carried out to evaluate what happens if flood defences in particular low-lying areas are built. They may even account for predicted future changes in climate and sea-level rise. But there is a common failure to account for all the corresponding decisions that may be taken as a response to the subsequent construction of the flood defences. More homes may be built on floodplains as developers see opportunities in apparently safe areas, without accounting for the ways in which this urbanization will affect the hydrological regime, for example, perhaps producing more rapid runoff into the system and lowering the threshold flood required to cause overtopping of the defences. Developers may cynically make such decisions because they know that the costs of damage will be met through government assistance and insurance claims and therefore spread across the population and thus be bearable (although recent developments suggest that this may not be the case for much longer). Land use upstream may change, so that the production of runoff from the land surface may change. Again, this problem may arise due to urbanization, or the change in crop cycles (see Boardman, 2001, on changes relating to winter-wheat production in southern England). Such changes may even be encouraged as people believe the downstream sites can cope due to the presence of the new defences (that have been paid for out of everyone's taxes!). Even at such local scales, such 'what if?' questions are thus by no means simple.

Karjalainen *et al.* (2002) used a scenario-based approach to assess the potential impacts of forest management and climate change on the carbon budget of German forests. Their model uses scenarios of growth rate, availability, age and rate of felling of different tree species over medium timescales (up to 70 years) and at a coarse spatial resolution (10 000 ha) and climate-change scenarios based on the HadCM2 model. They simulated a decrease in the net carbon sink from $1.7\,\mathrm{Mg\,C\,ha^{-1}\,a^{-1}}$ in 1995 to $0.7\,\mathrm{Mg\,C\,ha^{-1}\,a^{-1}}$ in 2050 under current climate conditions. In contrast, for the changing climate scenario, the model simulated a later onset of the decline – some time after 2020 – and a higher sink of $0.9\,\mathrm{Mg\,C\,ha^{-1}\,a^{-1}}$ in 2050. Soil sinks remained low at around $0.2\,\mathrm{Mg\,C\,ha^{-1}\,a^{-1}}$ in both cases. Thus, although increasing tree stocks is a potential means of mitigating CO_2 emissions, its efficiency is predicted to decrease through time.

At the international scale, the so-called Blue Plan made a number of projections relating to environmental degradation in the Mediterranean (Grenon and Batisse, 1989). Population projections were made using relatively detailed economics models for 2000 and 2025 using three 'trend scenarios' assuming different continuing patterns of current (in 1985) behaviour and two 'alternative scenarios' representing more basin-wide integration and cooperation. As noted by Wainwright and Thornes (2003), comparison of the results of the 2000 estimates with FAO statistics suggests that the trend scenarios were far more realistic, so that the alternative approaches had failed to enter into the perspectives of the policy-makers in the governments involved (plus ça change!).

The Intergovernmental Panel on Climate Change (IPCC) simulations of potential future climate change are the most widely known scenario approach at the global scale. The different scenarios used in the 1992 assessment (IS92) produced a range of estimates of carbon emissions for the year 2100 ranging from $4.6\,\mathrm{Gt\,C\,a^{-1}}$ to $34.9\,\mathrm{Gt\,C\,a^{-1}}$ with such a wide range causing enormous potential for debate about the meaning of their results (Cantor and Yohe, 1998). These scenarios form the inputs of carbon-budget models which, in turn, provide the atmospheric greenhouse gas scenarios for the general circulation models themselves. Despite (or possibly because of) this potential for debate, a similar approach has been taken in the most recent assessment, the SRES of 2000, the only difference being a 'fine-tuning' of the scenarios used and an admission of the complexity involved in their determination (Houghton *et al.*, 2001). The SRES scenarios take into account changes to or improved knowledge of the carbon intensity of energy supply, the income gap between developed and developing countries, and changes to sulphur emissions (IPCC, 2000). They incorporate lower population growth than IS92, with a narrowing of income differences between regions and a greater emphasis than IS92 on technological change. The SRES scenarios were developed as four narrative storylines (A1, A2, B1 and B2) depicting divergent futures and describing the factors controlling emissions and their evolution over time in response to demographic, social, economic, technological and environmental developments. Within each storyline several models and approaches have been used to generate the emissions scenarios in order that they reflect the uncertainty in the science. The four SRES scenarios can be summarized as:

A1: Rapid economic growth, global population peaking mid-century and then declining thereafter, rapid introduction of new efficient energy technologies, convergence among regions, capacity building and reduced region to region differences in per capita income. This is further subdivided into a fossil fuel-intensive future (A1F1), an alternative energy future (A1T) and a balance across all sources (A1B).

A2: This scenario describes a very heterogeneous world of self-reliance and preservation of local identities. Fertility patterns between regions converge very slowly and this leads to a continuously increasing global population. Per capita economic growth and technological change are fragmented and slower than in other scenarios.

B1: This scenario describes a convergent world with the same population trends as A1 but with rapid changes towards a service- and information-based economy with the introduction of clean and resource-efficient technologies. The emphasis is on global solutions and increased equity.

B2: A world of continuously increasing global population (at a rate lower than A2) where the emphasis is on local solutions to economic, social and environmental sustainability. It is a world of intermediate levels of economic development and of less rapid and more diverse technological change than B1 and A1.

These rather general statements define quite specific model parameters for carbon-budget models and thus specific greenhouse-emissions scenarios and even more specific climate change outcomes. Scenarios, like model predictions, can only ever represent one of many possible futures and one will only know which, if any, of these will be *the* future when it is already here. The danger is that, while this is obvious when one is working with the scenarios themselves, it becomes less and less obvious as these scenarios are deeper and deeper hidden within models so that many read the results of GCMs with very vague or no reference to the nature of the scenarios on which their representation of greenhouse futures are based. This issue is particularly important given the difficulty in modelling human actions highlighted in this chapter (even at much smaller time and space scales than the global 100-year emissions scenarios).

Another major problem with this approach is the separation of the climate system and the human–landscape–carbon budget system. Climatic factors are a major determinant of carbon and other GHG emissions and of the human activity which underlies these. One has only to look at the global power consumption for heating and air conditioning and of the importance of temperature-dependent natural and agricultural ecosystems and ocean surfaces as sources and sinks of GHGs to see this.

Though there are now efforts to include carbon-budget models online within GCM simulations (*Cox et al.*, 2000), this is really only a recent development which, through highlighting the importance of positive feedback processes between atmospheric CO_2, global temperature and source sink dynamics, may significantly increase GCM estimates of climate change. The major problem with decoupled models is that they fail to deal with exactly the human–environment interactions that are critical in controlling the decisions that underlie the issues that are being addressed in the first place. Static projections are inappropriate in a dynamic environment and fail to deal with the adaptations, which will always take place to account for resource shortages due to pressure of the environment and the impacts of environmental change. If such adaptations did not take place (as we know they have as conscious decisions by humans for 250 000 years), then the decoupled models would still be inaccurate, because they would be representing extinct populations by those with continued growth! The lack of realism in this way encourages the use of model results by policy-makers for self-centred gains and can act to accentuate the very disparities that such models are supposedly trying to avoid (see discussion in Oreskes *et al.*, 1994). US failure to take on board the Kyoto Agreement is the most obvious and important outcome of this problem.

12.4.5 Integrated analysis

Integrated analysis of change is required to overcome such limitations. This approach allows for there to be dynamic interactions and feedbacks between the decision-making process and their impacts. Rotmans and Dowlatabadi (1998) provide a thorough overview and suggest that integrated assessment is vital because of the ability to explore such interactions and feedbacks; the production of tools that can quickly and flexibly produce results; the production of counter-intuitive results (allowing us to evaluate our preconceptions); and the provision of tools for communication. However, they also suggest eight major limitations of the technique: (a) the models produced may be overly complex, affecting their ability to communicate ideas simply; (b) they may aggregate processes to an unacceptably coarse level; (c) the crossing of disciplinary boundaries usually causes suspicion from all sides!; (d) the models generally deal with uncertainty poorly; (e) they tend not to account for stochastic behaviour; (f) they are difficult to verify and validate (in the traditional sense: see discussion in Chapter 1); (g) as with most models, our knowledge and methodology are generally limited (but

remember that model building can provide a means of assessing this issue); and (h) there are 'significant pitfalls [in] that policymakers and researchers may treat integrated assessment models as "truth machines", rather than as heuristic tools for developing a better understanding of the issues' (ibid., 1998: 302).

Because of the integration of a wide range of aspects, such models may be much more vulnerable to the propagation of errors (see Chapter 1). It is important that these errors are communicated as part of the results. As an example, Rotmans and Dowlatabadi demonstrate the large errors that can arise in integrated assessments of population, energy use, CO_2 concentration, temperature change, economic output and the economic impact of climate change (Figure 12.6). Such results clearly require a more qualitative appraisal of the potential impacts. Lund and Iremonger (2000) demonstrate that integrated assessment should be part of the methodology used to investigate environmental models from the stage of data collection. Information on different land uses, for example, are usually carried out by very different organizations, often leading to major disparities in estimates. Integrated assessment is receiving increasing attention in the development of appropriate policies at all levels (Abaza and Hamwey, 2001).

12.5 NEVER MIND THE QUANTITIES, FEEL THE BREADTH

We have presented a viewpoint in this chapter that we know will raise the hackles of many working in various fields of social science. This choice has been quite deliberate, and we have attempted to present the limitations of each approach discussed. However, a major criticism of the viewpoint remains. Human behaviour and society are not predictable, all interpretation is purely subjective and thus any results obtained using the techniques presented are highly suspect, if not totally invalid.

On one level, we agree totally that it is fundamental to bear qualitative aspects of the research question in mind. As a simple example, we consider the case of policy in relation to land-use change and tourism in Mediterranean Europe. In this environment, land degradation has generally been taken to occur when the landscape is dominated by *maquis* – open shrubland with significant areas of bare soil permitting high erosion rates. Recovery is generally thought to occur when this form of cover is replaced by more dense shrubs and trees. Yet Green and Lemon (1996) found that the local populations considered the reverse to be the case. 'Degradation' to them was what happened when the

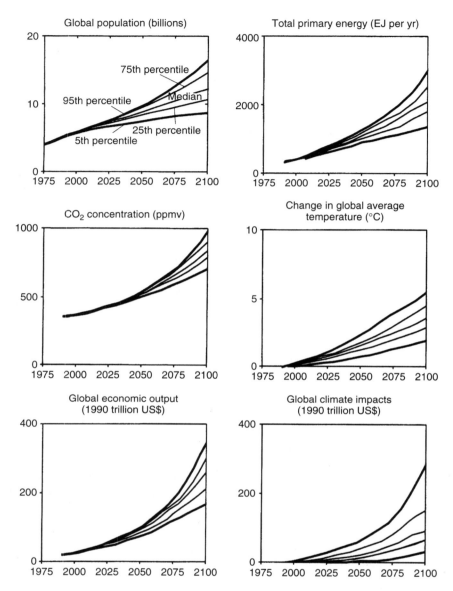

Figure 12.6   The effects of error propagation in a global estimation of climate change and human impacts using an integrated assessment model
Source: Rotmans and Dowlatabadi (1998)

landscape that had been present as far back as anyone could remember began to change. Similar attitudes were reported by Phillips (1998) with regard to the badlands of Tuscany in Italy. In contrast, O'Rourke (1999) investigated the attitudes of tourists to changing landscapes in southern France and found that those with second homes in particular attempted to make aesthetic changes to what they perceived as degraded landscapes, without contributing to the sustainability of the indigenous rural population. Clearly, a single, centralized policy aimed at reversing 'degradation' in any of these cases is doomed to failure, because it fails to deal with the qualitative aspects of what different individuals might consider the 'problem' to be!

But qualitative differences such as these do not necessarily suggest that they cannot be dealt with within a modelling framework, simply that the modelling framework must be sufficiently sophisticated to accommodate them. As noted by Casti (1994), complex systems are characterized by decentralized decision-making. Complex systems analysis therefore provides a valuable means of investigating social phenomena, particularly in these days of increasing globalization!

Many simple models of human decision-making are based on a concept of *Homo economicus* – the rational person who does everything to maximize his own personal wealth and has access to all the information required to do this. This perspective has increasingly been quite rightly criticized in recent years. Gintis (2000) points to aspects of common behaviour – drug addiction, self-interest in charitable donation, vengeance, generosity, courting of risk and present-orientated acts – that clearly run counter to this principle. Experimental data also suggest that key ideas underlying the model do not apply. For example, although individuals may discount effects of their actions on the near future quite strongly, they actually discount the effects on the more distant future less so. This hyperbolic model is fundamentally different from the exponential one commonly assumed (see the discussion above on its general limitations) and will produce significantly different consequences. Similarly, cooperative and altruistic behaviour is documented in experimental situations far more frequently than the neo-Darwinistic model – that such behaviour will be progressively removed by natural selection – would suggest. Jager *et al.* (2000) used a multi-agent approach to investigate different cognitive processes (deliberation, social comparison, repetition and imitation) in agents they defined as *Homo psychologicus* compared to *Homo economicus* agents who only engaged in deliberation. In investigating a transition from a fishing to a mining economy, they found that the former tended to produce a more complete transition than the latter, with varying levels of impact on environmental resources. To these concepts, Faber *et al.* would add a concept of *Homo politicus* as someone who 'tries to consider what is best for society' (2002: 238), which may overlap to some extent with Siebenhüner's (2000) *Homo sustinens* – a person who makes considerations with respect to future sustainability. Recent work on agent-based modelling has attempted to include emotional aspects in the cognitive process (Staller and Petta, 2001), so there is scope for further development of these concepts into our modelling approaches. Clearly, there are wide ranges of possible behaviour that need to be accounted for in our

models of the human decision-making process. As ever, the need is to provide the right sort of simplification to accommodate this variability in a meaningful way.

12.6 PERSPECTIVES

The human being is a talkative, imaginative beast, which behaves according to a whole series of mental processes involving rational and apparently irrational decisions (which may turn out to be irrational and rational, respectively). So what do we need to know to start modelling human decisions in the environment? Stern (2000) used a psychological framework to evaluate eight widely held beliefs about human–environment interactions:

1. He questioned whether *individual choice* always plays as important a role as is often suggested, given institutional and social constraints on behaviour.
2. He points out that specific forms of *consumption* by specific sections of society are more important than others in determining environmental impacts.
3. Similarly, not all *sacrifices* made to protect the environment are equal.
4. *Values and attitudes* are very often determined by situation and may not be as fixed as is often assumed; similarly, individuals' values and attitudes may not unambiguously represent the sociocultural or religious norms in which they are situated.
5. *Education* may have little beneficial effect in the short term, particularly if blocked by institutional practices.
6. *Motivation* is not always best achieved by doom-laden scenarios.
7. *Incentives* to change may not always function as intended.
8. *Emulation* of other societies in producing change may not be as clear-cut as is often assumed.

Based on these ideas, Stern developed a framework for research into environmentally relevant behaviour, in which behavioural patterns of individuals are successively constrained by higher orders of personal, social and political factors (Table 12.1). In reality, these higher orders emerge through time as a function of the discourse that occurs as a result of individual behaviour. The dynamics of this emergence is an ongoing two-way process that the theory of complex systems tells us may occur on a whole range of different timescales. Thus, we should be surprised as little by rapid change as we are by inertia.

Table 12.1 The outline of a causal model of environmentally relevant behaviour of Stern (2000). The 'level of causality' in the first column is the hierarchical order suggested by Stern so that the higher values at the top are seen as controlling the lower values below (although not necessarily in a unilinear way). Complex-systems analysis would see the factors at the top of the list emerging from the behaviours lower down (and other, unlisted ones)

Level of causality		Type of variable	Examples
	8	Social background and socialization	Socio-economic status, religion
	7	External conditions (incentives and constraints)	Prices, regulations, technology, convenience
	6	Basic values	Hedonism, power orientation, benevolence, traditionalism
	5	Environmental worldview	Belief that the environment is fragile or resilient
	4	Attitudes, beliefs and personal norms regarding environmentally relevant behaviour	Belief that recycling is good for the environment, sense of personal obligation to help prevent global warming
	3	Behaviour-specific knowledge and beliefs	Knowing which packaging is biodegradable, knowing how to petition legislators, beliefs about the personal and environmental costs and benefits of particular behaviours
	2	Behavioural commitment	Decision to travel by bus
	1	Environmentally relevant behaviour	Automobile purchase

(The left margin of the table is annotated with an upward-pointing arrow labelled "emergence".)

Traditional approaches to simplifying human behaviour in environmental models have been rightly criticized for focusing on too narrow a range of mental processes. The different mental processes emerge from complex activity within the brain as we discussed at the start of this chapter. It seems that the most promising ways of simulating these processes are from the use of complex systems models to evaluate the different roles of emergent behaviour. Stern's outline discussed above provides one means by which such studies might be structured. But as with other forms of modelling, the methodology needs to proceed hand in hand with appropriate field techniques and the collection of data that can be used to test outcomes. Uncertainty is central to the human condition and to the outcomes of models of this type. Both qualitative and quantitative models thus need to play a part in our approaches.

REFERENCES

Abaza, H. and Hamwey, R. (2001) Integrated assessment as a tool for achieving sustainable trade policies, *Environmental Impact Assessment Review* **21**, 481–510.

Aleksander, I. (2000) *How to Build a Mind*, Weidenfield and Nicolson, London.

Anderies, J. (2000) On modeling human behavior and institutions in simple ecological economic systems, *Ecological Economics* **35**, 393–412.

Axtell, R.L., Epstein, J.M., Dean, J.S., Gumerman, G.J., Swedlund, A.C., Harburger, J., Chakravarty, S., Hammond, R., Parker, J. and Parker, M. (2002) Population growth and collapse in a multiagent model of the Kayenta Anasazi in Long House Valley, *Proceedings of the National Academy of Sciences of the United States of America* **99**, 7275–7279.

Azar, C. and Schneider, S.H. (2002) Are the economic costs of stabilising the atmosphere prohibitive?, *Ecological Economics* **42**, 73–80.

Bak, P. (1997) *How Nature Works: The Science of Self-Organized Criticality*, Oxford University Press, Oxford.

Bar-Yam, Y. (1997) *Dynamics of Complex Systems*, Perseus Books, Reading, MA.

Bezerra, S., Cherruault, J., Fourcade, J. and Veron, G. (1996) A mathematical model for the human decision-making process, *Mathematical and Computer Modelling* **24** (10), 21–26.

Bishop, I.D. (1996) Comparing regression and neural net based approaches to modelling of scenic beauty, *Landscape and Urban Planning* **34**, 125–134.

Boardman, J. (2001) Storms, floods and soil erosion on the South Downs, East Sussex, autumn and winter 2000–01, *Geography* **86**, 346–355.

Bura, S., Guérin-Pace, F., Mathian, H., Pumain, D. and Sanders, L. (1996) Multiagent systems and the dynamics of a settlement system, *Geographical Analysis* **28**, 161–178.

Cantor, R. and Yohe, G. (1998) Economic analysis, in S. Rayner and E.L. Malone (eds) *Human Choice and Climate Change*, Batelle Press, Columbus, OH, Vol. 3, 1–103.

Carter, R. (1998) *Mapping the Mind*, Weidenfield and Nicolson, London.

Casti, J.L. (1994) *Complexification: Explaining a Paradoxical World Through the Science of Surprise*, Abacus, London.

Cox, P.M., Betts, R.A., Jones, C.D., Spall, S.A. and Totterdell, I.J. (2000) Acceleration of global warming due to carbon-cycle feedbacks in a coupled climate model, *Nature* **408**, 184–185.

Crevier, D. (1993) *AI: The Tumultuous History of the Search for Artificial Intelligence*, Basic Books, New York.

Crick, F. (1994) *The Astonishing Hypothesis: The Scientific Search for the Soul*, Simon and Schuster, London.

Dennett, D.C. (1991) *Consciousness Explained*, Penguin Books, Harmondsworth.

Epstein, J.M. and Axtell, R. (1996) *Growing Artificial Societies: Social Science from the Bottom Up*, Brookings Institution Press, Washington, DC. See also http://www.brookings.edu/sugarscape/

Faber, M., Petersen, T. and Schiller, J. (2002) *Homo oeconomicus* and *Homo politicus* in ecological economics, *Ecological Economics* **40**, 323–333.

Ferber, J. (1999) *Multi-Agent Systems: An Introduction to Distributed Artificial Intelligence*, Addison Wesley, Harlow.

Frankhauser, S. and Kverndokk, S. (1996) The global warming game – simulations of a CO_2-reduction agreement, *Resource and Energy Economics* **18**, 83–102.

Frisvold, G.B. and Caswell, M.F. (2000) Transboundary water management: game-theoretic lessons for projects on the US–Mexico border, *Agricultural Economics* **24**, 101–111.

Gielen, D.J. and Yagita, H. (2002) The long-term impact of GHG reduction policies on global trade: a case study for the petrochemical industry, *European Journal of Operational Research* **139**, 665–681.

Gilbert, N. and Doran, J. (eds) (1994) *Simulating Societies: The Computer Simulation of Social Phenomena*, UCL Press, London.

Gimblett, R., Daniel, T., Cherry, S. and Meitner, M.J. (2001) The simulation and visualization of complex human–environment interactions, *Landscape and Urban Planning* **54**, 63–79.

Ginot, V., Le Page, C. and Souissi, S. (2002) A multi-agent's architecture to enhance end-user individual-based modelling, *Ecological Modelling* **157**, 23–41. See also http://www.avignon.inra.fr/mobidyc

Gintis, H. (2000) Beyond *Homo economicus*: evidence from experimental economics, *Ecological Economics* **35**, 311–322. See also http://www-unix.oit.umass.edu/~gintis.

Grand, S. (2000) *Creation: Life and How to Make It*, Phoenix, London.

Green, S. and Lemon, M. (1996) Perceptual landscapes in agrarian systems: degradation processes in north-western Epirus and the Argolid valley, Greece, *Ecumene* **3**, 181–199.

Grenon, M. and Batisse, M. (1989) *Futures for the Mediterranean Basin: The Blue Plan*, Oxford University Press, Oxford.

Gumerman, G.J. and Kohler, T.A. (1995) Creating alternative cultural histories in the prehistoric Southwest: agent-based modelling in archaeology, in Proceedings of the Durango Conference on Southwest Archaeology, Durango, Co, Sept 15 1995, and at http://www.santafe.edu/sfi/publications/wpabstract/199603007

Haykin, S. (1999) *Neural Networks: A Comprehensive Foundation*, 2nd edn, Prentice Hall, Upper Saddle River, NJ.

Hofstadter, D.R. (1979) *Gödel, Escher, Bach: An Eternal Golden Braid*, Penguin Books, Harmondsworth.

Hofstadter, D.R. (1981) A conversation with Einstein's brain, in D.R. Hofstadter and D.C. Dennett (eds) *The Mind's I*, Penguin Books, Harmondsworth, 430–457.

Houghton, J.T., Ding, Y., Griggs, D.J., Noguer, M., van der Linden, P.J., Dai, X., Maskell, K. and Johnson, C.A. (eds) (2001) *Climate Change 2001: The Scientific Basis*, Cambridge University Press, Cambridge.

Hooke, R. LeB. (2000) On the history of humans as geomorphic agents, *Geology* **28**, 843–846.

IPCC (2000) *IPCC Special Report, Emissions Scenarios: Summary for Policymakers*, A Special Report of IPCC Working Group III, IPCC Working Group III, http://www.ipcc.ch/pub/sres-e.pdf.

Jager, W., Janssen, M.A., De Vries, H.J.M., De Greef, J. and Vlek, C.A.J. (2000) Behaviour in commons dilemmas: *Homo economicus* and *Homo psychologicus* in an ecological-economic model, *Ecological Economics* **35**, 357–379.

Jaynes, J. (1979) *The Origin of Consciousness in the Breakdown of the Bicameral Mind*, Houghton Mifflin, Boston.

Karjalainen, T., Pussinen, A., Liski, J., Nabuurs, G.-J., Erhard, M., Eggers, T., Sonntag, M. and Mohren, G.M.J. (2002) An approach towards an estimate of the impact of forest management and climate change on the European forest sector carbon budget: Germany as a case study, *Forest Ecology and Management* **162**, 87–103. See also http://www.efi.fi/projects/eefr/.

Kohler, T.A. and Gumerman, G.J. (eds) (2000) *Dynamics in Human and Primate Societies: Agent-Based Modelling of Social and Spatial Processes*, Oxford University Press, Oxford.

Low, B., Costanza, R., Ostrom, E., Wilson, J. and Simon, C.P. (1999) Human–ecosystem interactions: a dynamic integrated model, *Ecological Economics* **31**, 227–242.

Lund, H.G. and Iremonger, S. (2000) Omissions, commissions, and decisions: the need for integrated resource assessments, *Forest Ecology and Management* **128**, 3–10.

McGlade, J. (1995) Archaeology and the ecodynamics of human-modified landscapes, *Antiquity* **69**, 113–132.

McGlade, J., Castro, P., Courty, M.-A., Fedoroff, N. and Jones, M.K. (1994) The socioecology of the Vera basin: towards a dynamic reconstruction, in S.E. van der Leeuw (ed.) *Understanding the Natural and Anthropogenic Causes of Soil Degradation and Desertification in the Mediterranean Basin. Volume 2: Temporalities and Desertification in the Vera Basin*, Final Report on Contract EV5V-CT91-0021, EU, Brussels, 297–321.

McLeod, P., Plunkett, K. and Rolls, E.T. (1998) *Introduction to Connectionist Modelling of Cognitive Processes*, Oxford University Press, Oxford.

Mitchell, M. (1996) *An Introduction to Genetic Algorithms*, MIT Press, Cambridge, MA.

Mithen, S. (1996) *The Prehistory of the Mind: A Search for the Origins of Art, Religion and Science*, Thames and Hudson, London.

Nathan, P.W. (1987) Nervous system, in R.L. Gregory (ed.) *The Oxford Companion to the Mind*, Oxford University Press, Oxford, 514–534.

Noble, W. and Davidson, I. (1996) *Human Evolution, Language and Mind: A Psychological and Archaeological Enquiry*, Cambridge University Press, Cambridge.

O'Reilly, R.C. and Munakata, Y. (2000) *Computational Explorations in Cognitive Neuroscience: Understanding the Mind by Simulating the Brain*, MIT Press, Cambridge, MA.

Oreskes, N., Shrader-Frechette, K. and Bellitz, K. (1994) Verification, validation and confirmation of numerical models in the Earth Sciences, *Science* **263**, 641–646.

O'Rourke, E. (1999) Changing identities, changing landscapes: human-land relations in transition in the Aspre, Roussillon, *Ecumene* **6**, 29–50.

Parson and Ward, (1998) Games and simulations, in S. Rayner and E.L. Malone (eds) *Human Choice and Climate Change*, Batelle Press, Columbus, OH, Vol. 3, 105–139.

Penrose, R. (1989) *The Emperor's New Mind*, Oxford University Press, Oxford.

Phillips, C.P. (1998) The Crete Senesi, Tuscany: a vanishing landscape?, *Landscape and Urban Planning* **41**, 19–26.

Phillips, M. and Mighall, T. (2000) *Society and Exploitation Through Nature*, Prentice Hall, London.

Ramachandran, V.S. and Blakeslee, S. (1998) *Phantoms in the Brain: Human Nature and the Architecture of the Mind*, Fourth Estate, London.

Rayner, S. and Malone, E.L. (1998) *Human Choice and Climate Change*, Batelle Press, Columbus, OH.

Roberts, C.A., Stallman, D. and Bieri, J.A. (2002) Modeling complex human–environment interactions: the Grand Canyon river trip simulator, *Ecological Modelling* **153**, 181–196. Also at http://mathcs.holycross.edu/~croberts/research.

Rotmans, J. and Dowlatabadi, H. (1998) Integrated assessment modelling, in S. Rayner and E.L. Malone (eds) *Human Choice and Climate Change*, Batelle Press, Columbus, OH, Vol. 3, 291–377.

Sanders, L., Pumain, D., Mathian, H., Guérin-Pace, F. and Bura, S. (1997) SIMPOP: a multiagent system for the study of urbanism, *Environment and Planning B* **24**, 287–305.

Searle, J.R. (1980) Minds, brains and programs, *Behavioral and Brain Sciences* **3**, 417–424.

Siebenhüner, B. (2000) *Homo sustinens*: towards a new conception of human for the science of sustainability, *Ecological Economics* **32**, 15–25.

Staller, A. and Petta, P. (2001) Introducing emotions into the computational study of social norms: a first evaluation, *Journal of Artificial Societies and Social Simulation* **4** (1), http://www.soc.surrey.ac.uk/JASSS/4/1/2.html

Stassinopoulos, D. and Bak, P. (1995) Democratic reinforcement: a principle for brain function, *Physical Review E* **51**, 5033–9.

Stern, P.C. (2000) Psychology and the science of human–environment interactions, *American Psychologist* **55**, 523–530.

Thagard, P. (2000) *Coherence in Thought and Action*, MIT Press, Cambridge, MA.

Van Veen, H.A.H.C., Distler, H.K., Braun, S.J. and Bulthoff, H.H. (1998) Navigating through a virtual city: using virtual reality technology to study human action and perception, *Future Generation Computer Systems* **14**, 231–242.

Wainwright, J. and Thornes, J.B. (2003) *Environmental Issues in the Mediterranean: Processes and Perspectives from the Past and Present*, Routledge, London.

Weisbuch, G. (2000) Environment and institutions: a complex dynamical systems approach, *Ecological Economics* **34**, 381–391.

13

Modelling Land-Use Change

ERIC F. LAMBIN

13.1 THE COMPLEXITY

13.1.1 The nature of land-use change

Land-use changes are cumulatively transforming land cover at an accelerating pace, mainly in the tropics (Turner *et al.*, 1994). These changes have important implications for a range of issues such as biosphere–atmosphere interactions, species and genetic diversity associated with endangered habitats, soil conditions, water and sediment flows, vulnerability of ecosystems and social groups, and sustainable use of natural resources in the development process of human societies (Meyer and Turner, 1994). It is crucial for the modelling of the problem to distinguish between land cover and land use. The term *land cover* refers to the attributes of a part of the Earth's land surface and immediate subsurface, including biota, soil, topography, surface and groundwater, and human structures. The term *land use* refers to the purposes for which humans exploit the land cover. Common land-use types include agriculture, grazing, forestry, mineral extraction and recreation. Forest is a land cover dominated by woody species, which may be exploited for land uses as varied as recreation, timber production or wildlife conservation. Changes in land use are frequent causes of land-cover change. One generally distinguishes between land-cover *conversion* – i.e. the complete replacement of one cover type by another – and land-cover *modification* – i.e. more subtle changes that affect the character of the land cover without changing its overall classification (Turner *et al.*, 1995), such as land degradation or land-use intensification. Land-cover modifications may result in degraded ecosystems. Land-cover modifications are generally more prevalent than land-cover conversions.

The most widespread contemporary example of land-cover conversion is tropical deforestation. Forest degradation through selective logging and fires also affects large areas of forests. Forest conversion and degradation in the humid tropics account for a significant part of the carbon flux from terrestrial ecosystems and contribute to a large share of the loss of biodiversity. Another economically and demographically important process of conversion is urbanization. Land degradation is most severe and widespread in semi-arid regions. It encompasses processes such as soil erosion or soil salinization. Land degradation implies a decline in the usable natural resource base and, thus, directly affects the food supply. A synthesis of the most recent global assessments of human-induced land degradation estimated that 69.5% of the world's drylands are affected by various forms of land degradation (Dregne *et al.*, 1991). Other authors highlight the dynamic and resilient character of temperate and tropical rangelands, in response to the random interplay of human and biophysical drivers. This makes it difficult to distinguish directional change (such as loss of biological diversity or soil degradation) from readily reversible fluctuations, such that interpretations of degradation must be viewed cautiously (Lambin *et al.*, 2001). Land-use intensification may be associated with agricultural, agroforestry or grazing systems. Intensified management of land can be based on such techniques as irrigation, fertilizer use or the integration of different production activities (Netting, 1993).

13.1.2 Causes of land-use changes

There is a high variability in the land-cover types, the physical environments, the socio-economic activities and the cultural contexts which are associated with

Environmental Modelling: Finding Simplicity in Complexity. Edited by J. Wainwright and M. Mulligan
© 2004 John Wiley & Sons, Ltd ISBNs: 0-471-49617-0 (HB); 0-471-49618-9 (PB)

land-use change. The driving forces of land-use changes are categorized into the following groups: (a) factors that affect the demands that will be placed on the land, i.e. population and affluence; (b) factors that control the intensity of exploitation of the land: through technology; (c) factors that are related to access to or control over land resources: the political economy; and (d) factors that create the incentives that motivate individual decision-makers: the political structure, attitudes and values (Turner *et al.*, 1995). Identifying the causes of land-use change requires an understanding of how these different factors interact in specific environmental, historical and social contexts to produce different uses of the land (ibid.).

For example, deforestation results from slash-and-burn cultivation, both by landless migrants and by traditional shifting cultivators, government-sponsored resettlement schemes, fuelwood gathering and charcoal production, the conversion of forested areas for cattle ranching, inefficient commercial logging operations, the provision of infrastructure, and large-scale, uncontrolled forest fires of an exceptional nature. The proximate causes of deforestation are generally thought to be driven by a combination of some of the following underlying driving forces: population growth, land hunger, inequitable social conditions, property-rights regimes, misguided government policies, collective action problems, inappropriate technology, international trade relations, economic pressures afflicting debt-burdened developing countries and corruption in the forestry sector (e.g. Myers, 1989; Lambin, 1994; Kaimowitz and Angelsen, 1998). The relative importance of these causes varies widely in space and time.

The synthesis of extensive, local, case-study evidence on processes of land-use change supports the conclusion that neither population nor poverty are the primary causes of land-use change worldwide. Rather, people's responses to economic opportunities, as mediated by institutional factors, drive land-use changes. Opportunities and constraints for new land uses are created by markets and policies, increasingly influenced by global factors. Extreme biophysical events occasionally trigger further changes. Various human–environment conditions react to and reshape the impacts of drivers differently, leading to specific pathways of land-use change (Lambin *et al.*, 2001). The dependency of causes of land-use changes on historical, geographic and other factors makes it a particularly complex issue to model.

13.2 FINDING THE SIMPLICITY

The modelling of land-use change is aimed at addressing at least one of the following questions:

1. Which socio-economic and biophysical variables contribute most to an explanation of land-use changes and why?
2. Which locations are affected by land-use changes – where?
3. At what rate do land-use and land-cover change progress – when?

Four broad categories of land-use change models address some of these questions: empirical-statistical models, stochastic models, optimization models and dynamic (process-based) simulation models (Lambin *et al.*, 2000). All these model designs attempt to represent the complex land-use change pathways based on general rules, which often involve a great deal of simplification. Land-use change models offer the possibility to test the sensitivity of land-use patterns to changes in selected variables. They also allow testing of the stability of linked social and ecological systems through scenario building. While, by definition, any model falls short of incorporating all aspects of reality, it provides valuable information on the system's behaviour under a range of conditions (Veldkamp and Lambin, 2001).

13.2.1 Empirical-statistical models

Empirical-statistical models attempt to identify explicitly the causes of land-use changes using multivariate analyses of possible exogenous contributions to empirically derived rates of changes. Multiple linear regression techniques are generally used for this purpose. The finding of a statistically significant association does not establish a causal relationship. Moreover, a regression model that fits well in the region of the variable space corresponding to the original data can perform poorly outside that region. Thus, regression models cannot be used for wide-ranging extrapolations. Such models are only able to explain, in a statistical sense, patterns of land-use changes which are represented in the calibration data set.

Regression models are intrinsically not spatial. They are based on the unrealistic assumption that land-use change in an administrative unit is primarily a function of factors originating within that unit, i.e. that cross-boundary effects are minor. However, multi-level regression techniques overcome such limitations (Polsky and Easterling, 2001). In summary, regression models are mainly an exploratory tool to test for the existence of links between candidate driving forces and land-use change. Such models therefore contribute to address the why question. They are not reliable for prediction and are not spatial.

Spatial, statistical models are born from the combination of geographic information systems (GIS) and multivariate statistical models. Their emphasis is on the spatial distribution of landscape elements and on changes in land-use patterns. The goal of these models is the projection and display, in a cartographic form, of future land-use patterns which would result from the continuation of current land management practices or the lack thereof. The approach analyses the location of land-use changes in relation to maps of natural and cultural landscape variables in a GIS (Chomitz and Gray, 1996; Mertens and Lambin, 2000). A model is built to describe the relationship between the dependent variable, e.g. the binary variable forested/deforested, and the independent landscape variables. Results of spatial, statistical models demonstrate that a few landscape variables from the previous time period provide strong predictions of patterns of land-use change. Predicting the spatial pattern of land-use change is thus a much easier task than predicting future rates of change. Spatial, statistical models primarily identify predictors of the location of areas with the greatest propensity for land-use changes. They are not suited to the prediction of when land-use change will occur and only touch on proximate causes of change processes.

An example of an advanced statistical land-use change model is CLUE (Conversion of Land Use and its Effects). Veldkamp and Fresco (1996) developed a regression-based model to simulate land-use conversions as a result of interacting biophysical and human drivers. A broad spectrum of proxy variables for driving forces are incorporated. A land-use allocation module was also developed, to generate spatially explicit land-use change projections. This model has been applied at a national scale in several Latin American and Asian countries.

13.2.2 Stochastic models

Stochastic models, for land-use change, consist mainly of transition probability models such as Markov chains, stochastically describe processes that move in a sequence of steps through a set of states. For land-use change, the states of the system are defined as the amount of land covered by various types of land use. The transition probabilities can be statistically estimated from a sample of transitions occurring during some time interval. Probabilities of transitions are defined for changes from one land-use category to another. Transition probability models generally assume that land-use change is a first-order process, that is the conditional probability of land use at any time, given all previous uses at earlier times, depends at most upon the most recent

use and not upon any earlier ones (Bell and Hinojosa, 1977). Such models also rely on the assumption of the stationarity of the transition matrix, i.e. temporal homogeneity. Dynamic transition probabilities can be introduced by switching between stationary transition matrices at certain intervals or by modelling the contribution of exogenous or endogenous variables to the transitions (Baker, 1989). Transition probabilities between certain land-use categories could also be driven by ecological or socio-economic submodels, leading to process-based models (Sklar and Costanza, 1991). Spatial adjacency influences can also be taken into account in such models (Turner, 1987).

Simple, first-order transition probability models have been applied to model processes of changes in vegetation types and land use (e.g. Burnham, 1973; Robinson, 1978). However, there are few applications of Markov chain at spatial scales near or broader than the landscape scale. Thornton and Jones (1998) recently presented a conceptual model of agricultural land-use dynamics, based on Markov chains governed by a few simple decision rules. This model could be applied to assess the potential ecological impact of technological and economic change on agricultural land use, which could be of value in a range of impact assessment studies (Thornton and Jones, 1998: 519).

The main advantage of Markov chain analysis lies in its mathematical and operational simplicity. The only data requirement is for current land-use information. The stochastic nature of Markov chain masks the causative variables. Thus, the model has little explanatory power. Finally, an area is described in aggregate terms – it is a distributional (as opposed to a spatial) landscape model (Baker, 1989). In summary, transition probability models can only predict when changes in land use might take place in the short term, under a strict assumption of stationarity of the process. Such models can be used where no information on the driving forces and mechanisms of land-use changes is available.

Some other forms of stochastic models, such as spatial diffusion models, do appear to be useful in research on land-use change. Cellular automata are also increasingly used in land-use change models (White *et al.*, 1997; Wu, 1998).

13.2.3 Optimization models

In economics, many models of land-use change apply optimization techniques based either on whole-farm analyses using linear programming, at the microeconomic level, or general equilibrium models, at the macro-economic scale (Kaimowitz and Angelsen, 1998).

Many of these approaches originate in the land rent theories of von Thünen (1966) and Ricardo. Any parcel of land, given its attributes and its location, is modelled as being used in the way that yields the highest rent. Land-use change is therefore an issue of choices by land managers among alternate rents. Land conservation is an issue of investment decisions by land managers by comparing discounted benefits and costs. In econometric approaches, the supply and demand functions of the land market under investigation (assumed to be a competitive market) are estimated. Such models allow investigation of the influence of various policy measures on land allocation choices. However, models of urban and periurban land allocation appear to be much more developed than their rural counterparts (Riebsame *et al.*, 1994).

A major strength of micro-economic models compared to the other model designs reviewed here is that they usually assume that the agents whose behaviour is described within the model do have the capacity to make informed predictions and plans, so as to try to avoid disasters (Fischer *et al.*, 1996). The applicability of such models for projections is, however, limited due to unpredictable fluctuations of prices and demand factors, and to the role of noneconomic factors driving changes. For example, in the case of deforestation, it is not unusual that loggers, through their links with government officials, decide upon the economic and institutional context in which they operate. This leads to low and fixed timber prices, lack of competition and weak mechanisms to regulate access to resources. Given this inherent unpredictability of socio-economic driving forces of land-use changes, economic models usually adopt a normative approach.

The agricultural land rent theory of von Thünen (1966) explains optimal crop-production allocation following degrees of agricultural intensity. A limited number of variables used in the basic, mainly static and deterministic, model of agricultural systems explains intensity as depending on the achievable economic rent. This rent is determined by market demands in the consumer centre, transportation costs, production costs and degrees of perishability of goods produced for the central market. Price expectations and interest rates have also been shown to be important factors. Thus, these models link agricultural intensity, population density, the actual transportation network and the market economy (Jones and O'Neill, 1993, 1994).

Several partial or general equilibrium models of land use have been developed to describe the trade-offs between land clearance for agriculture and resource conservation and maintenance (Walker, 1987; Southgate, 1990; Jones and O'Neill, 1992). The work of Jones and O'Neill is particularly relevant since it models profit maximization decisions at the individual level but examines region-wide environmental outcomes. Several of these models highlight, in addition to population pressure, the institutional problems of skewed tenurial regimes and misdirected government policies (e.g. Sandler, 1993).

Optimization models suffer from other limitations, such as the somewhat arbitrary definition of objective functions and non-'optimal' behaviour of people, e.g. due to differences in values, attitudes and cultures. While, at an aggregate level, these limitations are likely to be nonsignificant, they are more important as one looks at fine-scale land-use change processes and is interested in the diversity between actors. A new generation of agent-based models simulate decisions by and competition between multiple actors and land managers (e.g. Walker, 1999; Rouchier *et al.*, 2001). Land-use change models recently developed by economists integrate spatial heterogeneity and broaden the objective function of actors from profit to utility maximization, including multiple uses of land (Irwin and Geoghegan, 2001). Behavioural models of land-use decisions by agents can be made spatially explicit thanks to cellular automata techniques. This approach accounts for the fact that landscape patterns and spatial interactions do influence land use decisions.

13.2.4 Dynamic (process-based) simulation models

Patterns of land-use changes in time and space are produced by the interaction of biophysical and socio-economic processes. Dynamic (process-based) simulation models have been developed to imitate the run of these processes and follow their evolution. Simulation models emphasize the interactions among all components forming a system. They condense and aggregate complex ecosystems into a small number of differential equations in a stylized manner. Simulation models are therefore based on an *a priori* understanding of the forces driving changes in a system. Integrated Assessment Models, e.g. IMAGE (Alcamo *et al.*, 1998), seek to integrate a wide range of sectors and process descriptions and usually operate at global scales. Integrated Assessment Models are seldom able to model the management and decision-making processes of individuals. This stems from their requirement to operate at global scales which often results in simplistic treatment of land-use change processes.

White *et al.* (1997) demonstrated the use of a land-use change model that combined a stochastic, cellular automata approach with dynamic systems models of regional economics. The approach allows spatially explicit geographic processes to be constrained by less spatially precise economic processes within the framework of a Geographic Information System (GIS). The approach has been used as a decision-support system (see Chapters 14 and 15), by allowing regional land-use planners to investigate the consequences of alternative management strategies. The combination of dynamic, process-based models with optimization techniques underpinned the development of the IMPEL model (Integrated Model to Predict European Land Use) (Rounsevell *et al.*, 1998). The requirement for IMPEL was to be able to assess modifications to the spatial distribution of agricultural land use in response to climate change.

Land-use change models that rely on a population approach and assume subsistence behaviour and limited market integration are rooted in the models of Boserup (1965) and Chayanov (1966). These models relate agricultural intensification to household needs, according to which households balance consumption and leisure and/or follow a subsistence strategy. Other simulation models integrate demands from the market. For example, in a numerical simulation model, Angelsen (1999) considers land-use intensity as one of 12 variables besides exogenously given prices and a newly introduced labour market, in that any amount of labour can be sold or hired with migrational flows in and out of the local economy. Within market-based theorizing, the dominant views used to be based on a purely market or commodity approach. This postulates that farmers accept commodity production and respond to market demands within constraints placed upon them by maximizing production to the level of maximum reward.

The strength of a simulation model depends on whether the major features affecting land-use changes are integrated, whether the functional relationships between factors affecting processes of change are appropriately represented, and on the capacity of the model to predict the most important ecological and economic impacts of land-use changes. The development of simulation models requires *a priori* knowledge of the factors and mechanisms driving land-use change. The model simply formalizes and articulates in a comprehensive manner the key relationships. Process-based ecosystem models are well suited to representing nonstationary processes since they mimic (at least in an aggregated way) the underlying processes in the system and include feedbacks and thresholds in the systems dynamics. Simulation models allow rapid exploration of probable effects of the continuation of current land-use practices or of

changes in cultural or ecological parameters. These models allow testing of scenarios on future land-use changes.

When dynamic ecosystem-simulation models are spatially explicit (i.e. include the spatial heterogeneity of landscapes), they can predict temporal changes in spatial patterns of land use. In system landscape models, the course of every individual parcel of a landscape matrix is predicted by a process-based model integrating flows between adjacent cells. The transition probabilities are generated dynamically rather than empirically. This approach allows for greater cyclic changes between land uses and longer-term prediction of landscape patterns (Sklar and Costanza, 1991). It has been successfully applied in environments driven by physical factors (Sklar *et al.*, 1985) but is difficult to translate to land-use patterns due to the need to incorporate the economic or social mechanisms driving land-use transitions (Turner, 1987). While many individual processes of decision-making cannot be modelled deterministically, it is possible to identify the main determinants of decisions on when and where to change land use. A few dynamic spatial simulations, using actor-based models, have been developed to date, e.g. for tropical deforestation in roadside areas occupied by shifting cultivators (Wilkie and Finn, 1988), in regions of official resettlement schemes associated with road development (Southworth *et al.*, 1991) or in areas dominated by shifting cultivation (Gilruth, 1995). These grid-cell models combine spatially explicit ecological information with socio-economic factors related to land-use decisions by farmers. The modelling is based on rules of behaviour determined by locational attributes, land-tenure systems and vegetation successions (see Chapter 12).

Dynamic landscape-simulation models are not pure predictive systems, even though projections of future trends in land-use change as a function of current land use practices can be performed. They are rather 'gameplaying tools' designed to understand the ecological impacts of changes in land use. The dynamic landscape simulation models developed to date are specific to narrow geographic situations and cannot easily be generalized. These models allow the testing of scenarios on changes in land-use patterns.

13.3 THE RESEARCH FRONTIER

Some of the major research challenges to improve the modelling of land-use change include better integrating the scale factor in models, representing processes of land-use modification such as agricultural intensification, and dealing with temporal heterogeneity, e.g.

as driven by endogenous factors such as technological innovation.

13.3.1 Addressing the scale issue

The scale at which land-use changes are modelled affects the type of explanation given to land-use change. At broad scales, the high level of aggregation of data obscures the variability of situations and relationships, and produces sometimes meaningless averages. Predictions based on broad-scale models are inaccurate, especially for regional and local assessments, because, at the aggregate level, key processes are masked (Turner *et al.*, 1995). On the other hand, the development of fine-scale models for every local situation would be both impractical and inadequate if there is no possibility of generalizing these models. Moreover, in scaling up several local processes and dynamics, emergent properties may appear due to synergism at a higher level of system integration, i.e. at a regional scale. Thus, regional-scale processes are more than the sum of local-scale processes.

A hierarchical approach needs therefore to be adopted in both the observation and explanation of land-use change processes. Blaikie and Brookfield (1987) suggest using a nested set of four scales to explain land degradation:

Local and site specific where individuals or small groups make the relevant decisions; the regional scale involving more generalised patterns of physiographic variation, types of land use, and property relations and settlement history; the national scale in which the particular form of class relations give the economic, political and administrative context for land-management decisions; and the international scale, which, in the most general manner, involves almost every element in the world economy, particularly through the commodization of land, labour and agricultural production.

Models of land-use change should thus be based on an analysis of the system at various spatial and temporal scales. Yet, most existing regional-scale models address neither structural nor functional complexity. Models that opt to incorporate and link a larger number of factors for a spatially heterogenous area (e.g. integrated assessment models) severely simplify the land-use system. Incorporating structural complexity becomes necessary at the coarser scales (Veldkamp and Lambin, 2001). Actually, at local scales, the direct actors of land-use change can be identified and process-based relationships can be determined. With decreasing resolution and increasing

extent, it becomes increasingly difficult to identify key processes. Thus, at these aggregated levels, the structure and assumptions of the model have to be adapted, as one cannot simply use knowledge derived from local studies and apply it at another level.

13.3.2 Modelling land-use intensification

While future food supplies will have to come mostly from agricultural intensification rather than from agricultural expansion, most modelling efforts in land-use change have been concerned with issues of land-use conversion rather than land-use modification. Note, however, that economists have a long tradition of studying agricultural intensification in relation to management practices and conditions (e.g., prices of inputs, production functions). Several studies reveal the driving factors (e.g. price differentials, tenure conditions, family cycle conditions, etc.) that cause management to change. Different modelling approaches are more or less appropriate in tackling the question of intensification at different locations. In general, however, dynamic simulation models are better suited to predict changes in land-use intensity than empirical, stochastic or static optimization models, although some stochastic and optimization methods may be useful in describing the decision-making processes that drive land management (Lambin *et al.*, 2000). The ability to predict changes in land-use intensity requires models of the process of decision-making by land managers.

13.3.3 Integrating temporal heterogeneity

Many land-use patterns have developed in the context of long-term instability, including nonlinear processes in the biophysical variables (climate, disease) and in the political economic forces (market dynamics, price fluctuations, state policy influences). Decision-making by individuals and communities has therefore to manage variable time horizons, in particular those that link the short-term management of growing seasons and cycles of consumer needs, the mid-term experience of year-to-year fluctuations (in rainfall, prices, policies), random 'chaotic' events (war, policy interventions) and the long-term horizons implied in sustainable social and ecological systems. Land-use models should therefore be built on the assumption of temporal heterogeneity rather than on the common assumption of progressive, linear trends (e.g. established models of progressive intensification of material resources). In other words, rapidly and unpredictably changing variables are as important in shaping land-use dynamics as the slowly and cumulatively changing variables. For example, technological innovations are one of the major sources of temporal heterogeneity. Given the important role of technology and management practices in driving land-use change, incorporating these factors into prognostic models of land-use change is a major challenge.

13.4 CASE STUDY

N. Stephenne and E.F. Lambin

The following case study illustrates some of the issues discussed above. It introduces a dynamic simulation model of rural land use in semi-arid Africa, to generate regional-scale projections over several decades, mostly for climate modelling applications.

13.4.1 The problem

To investigate the impact of land surface changes on regional climate, authors generally conduct experiments with General Circulation Models (GCM) at a coarse spatial resolution (e.g. Chase *et al.*, 1996). These studies require historical reconstructions of past land-cover changes and/or projections of likely future land-cover changes. While, at a local scale, part of these historical data can be generated from direct or indirect field evidence (e.g. old vegetation maps, aerial photographs, high temporal resolution pollen studies), at a regional scale, the reconstruction of past land-cover changes has to rely on backward projections using land-use change models. The SALU model (**SA**helian **L**and **U**se model) (Stephenne and Lambin, 2001a) is a dynamic simulation model which was constructed to reconstruct past land-use and land-cover changes in the African Sahel over the past 30 years, and to develop scenarios of likely future changes. Spatial resolution is an important issue in the integration of land-use change projections in Global Climatic Models (GCM). The land-use change model generates projections of land-use areas for administrative units, e.g. countries or provinces. To be integrated in a GCM, these land-use projections need to be mapped at a grid-scale level, which roughly corresponds to the size of a small Sahelian country or of an eco-climatic region of a larger country. Below, we summarize the basic principles of the SALU model. A more detailed presentation of the model is given in Stephenne and Lambin (2001a) and at http://www.geo.ucl.ac.be/Recherche/Teledetection/projects.html.

13.4.2 Model structure

In the Sudano-Sahelian region, we assume that the following land uses generate the basic resources for the population: fuelwood in natural vegetation areas, food for subsistence and market needs in cropland and fallow, livestock in pastoral land. Note that these land-use categories do not strictly coincide with land-cover types. In this model, the land-use classes 'fuelwood extraction areas' and 'pastoral lands' refer to a variety of vegetation cover types such as woodland, savannah or steppe. The exogenous variables drive yearly changes in land-use allocation. For any given year, the land-use demand is calculated under the assumption that there should be an equilibrium between the production and consumption of resources. In other words, the supply of food and energy resources derived from the areas allocated to the different land uses must satisfy the demand for these resources by the human and animal populations, given the exploitation technologies used at a given time. Fuelwood extraction areas, cropland, fallow and pastoral land compete for land through different processes of land-use change (Figure 13.1).

The competition between the different land uses takes place within the national space, which is finite. The total demand for land in a given year is the sum of demands for cropland and pastoral land. We assume that the fuelwood extraction area can be reduced on an annual basis by the expansion of cropland and pastoral land. Initially, all unused land is covered by natural vegetation. In the pastoral land, the equilibrium assumption requires that the consumption of forage is equal to the biomass production. As, in the Sahel, pastoralism is extensive, biomass production relies on the natural productivity of grasslands.

In the cultivated land, we distinguish between subsistence food crops and cash crops. The demand for subsistence food crops depends on the rural population and its basic consumption requirements. The crops which are commercialized consist mainly of food crops for the urban population, but may include some cash crops (e.g. cotton). This part of the production competes with cereal imports which are only directed toward the consumption of the urban population. In the above definitions, the cropland area includes fallow. At the most extensive level of cultivation, corresponding to a pre-intensification stage, the crop–fallow cycle is assumed to be two years of fallow for one year of cultivation (Ruthenberg, 1980).

The model simulates two processes of land-use change: agricultural expansion at the most extensive technological level, followed by agricultural intensification once some land threshold is reached. Expansion

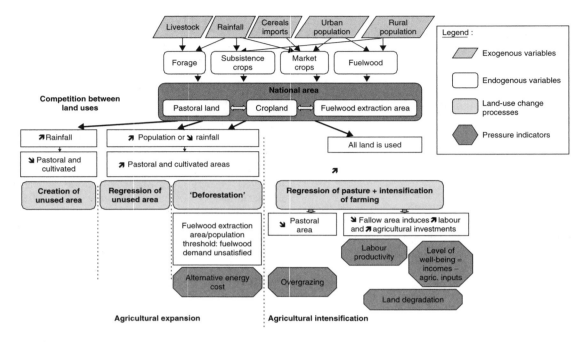

Figure 13.1   Overall structure of the SALU model

of cropland and pastoral land is driven by two sets of factors: changes in human and animal population, which increase the consumption demand for food crops and forage, and interannual variability in rainfall, which modifies land productivity and therefore increases or decreases production for a given area under a pastoral or cultivation use. Expansion of cultivation can take place into previously uncultivated areas or by migration into unsettled areas. Agricultural expansion thus leads to deforestation or to a regression of pastoral land. Pastoral land can also expand into natural vegetation areas and cropland. Expansion of cropland and pastoral land in temporarily unused land is associated with a lower environmental cost than in the case of deforestation. The unused land will only become reforested if it is left unused for a period of at least 20 years. Moreover, not all natural vegetation areas can be destroyed. We assume that the local population will always protect a certain fuelwood extraction area: a minimum area to satisfy some of the fuelwood requirements for domestic consumption, forest reserves, national parks, sacred forests, inaccessible forests or forests with a high incidence of tse-tse flies or onchocerciasis. If fuelwood needs exceed the wood production through natural regrowth of vegetation, we assume that rural households will turn to other energy sources such as kerosene. Actually, one

generally observes a sustainable use of natural vegetation resources in the Sahel, except in peri-urban areas.

Once the expansion of cropland and pastoral land has occupied all unused land, and once fuelwood extraction areas have reached their minimal area, the land is saturated and another phase of land-use change takes place. Additional demand for food crops will result mainly in agricultural intensification but also, in a lesser way, in expansion of cultivation in pastoral land. In Sahelian agriculture, intensification mostly takes place as a shortening of the fallow cycle, compensated by the use of labour and agricultural inputs such as organic or mineral fertilizers to maintain soil fertility. Agricultural intensification leads to a decrease in labour productivity and requires the use of organic or chemical inputs. A shortening of the crop–fallow cycle without input use would deplete soil fertility. As the economic value of the output per unit of area of cultivated fields is much higher than the value of the output per unit of area of pastoral land in an extensive system, thus, part of the additional demand for food crops will also lead to the expansion of cultivation in pastoral land. The increase in livestock population combined with shrinking pastoral lands (due to agricultural expansion) will result in overgrazing, which will affect the productivity of pastoral land. See Stephenne and Lambin (2001a) for more detailed results.

The model was first tested at a national scale using data from Burkina Faso, over the period 1960–97. The rates of change in cropland predicted by the model simulations for Burkina Faso were similar to rates measured for local-scale case studies by remote sensing (Stephenne and Lambin, 2001a). The application of the model to five other countries (Senegal, Mali, Niger, Chad and Nigeria) also demonstrated the robustness of the approach (Stephenne and Lambin, 2001b). For Chad and Mali, land-use projections were disaggregated per eco-climatic zone (e.g. mostly pastoral Sahelian zone versus Sudanian zone dominated by farming). The regional-scale projections generated by the model were compared to the IGBP-DISCover land-cover map of 1992/93, produced from 1-km resolution remote sensing data. A very good agreement between the two sources was found. The model was then used to test hypotheses on driving forces of land-use change via scenarios (Stephenne and Lambin, 2002). Results from these simulations were consistent with the expertise knowledge on the processes of land-use change derived from local-scale case studies in the region. This convergence suggested that the major processes of land-use change in the region were well represented in the model. The scenario simulations performed with SALU provided useful insights to better understand the processes of land-use changes in the Sudano-Sahelian countries.

REFERENCES

Alcamo, J., Leemans, R. and Kreileman, E. (1998) *Global Change Scenarios of the 21st Century*, Pergamon, Oxford.

Angelsen, A. (1999) Agricultural expansion and deforestation: modelling the impact of population, market forces and property rights, *Journal of Development Economics* **58** (1), 185–218.

Baker, W.L. (1989) A review of models of landscape change, *Landscape Ecology* **2**, 111–133.

Bell, E.J. and Hinojosa, R.C. (1977) Markov analysis of land use change: continuous time and stationary processes, *Socio-Economic Planning Science* **11**, 13–17.

Blaikie, P. and Brookfield, H.C. (1987) *Land Degradation and Society*, Methuen, London.

Boserup, E. (1965) *The Condition of Agricultural Growth*, Allen and Unwin, London.

Burnham, B.O. (1973) Markov intertemporal land use simulation model, *Southern Journal of Agricultural Economics* July, 253–258.

Chase, T.N., Pielke, R.A., Kittel, T.G.F., Nemani, R. and Running, S.W. (1996) The sensitivity of a general circulation model to global changes in leaf area index, *Journal of Geophysical Research* **101**, 7393–7408.

Chayanov, A.V. (1966) Peasant farm organization, in D. Thorner, B. Kerblay and R.E.F. Smith (eds) *A.V. Chayanov on the Theory of Peasant Economy*, Irwin, Homewood, IL, 21–57.

Chomitz, K.M. and Gray, D.A. (1996) *Roads, Land, Markets and Deforestation: A Spatial Model of Land Use in Belize*, Report from the Environment, Infrastructure, and Agricultural Division, Policy Research Department, World Bank Economic Review, Washington, DC.

Dregne, H., Kassas, M. and Rozanov, B. (1991) A new assessment of the world status of desertification, *Desertification Control Bulletin* **20**, 6–19.

Gilruth, P.T., Marsh, S.E. and Itami, R. (1995) A dynamic spatial model of shifting cultivation in the highlands of Guinea, West Africa, *Ecological Modelling* **79**, 179–197.

Fischer, G., Ermoliev, Y., Keyzer, M.A. and Rosenzweig, C. (1996) Simulating the socio-economic and biogeophysical driving forces of land-use and land-cover change: the IIASA land-use change model, *Working Paper WP-96-010*, International Institute for Applied Systems Analysis, Luxenbourg.

Irwin, E.G. and Geoghegan, J. (2001) Theory, data, methods: developing spatially explicit economic models of land use change, *Agriculture, Ecosystems and Environment* **85**, 7–24.

Jones, D.W. and O'Neill, R.V. (1992) Endogeneous environmental degradation and land conservation: agricultural land use in a large region, *Ecological Economics* **6**, 79–101.

Jones, D.W. and O'Neill, R.V. (1993) Human–environmental influences and interactions in agriculture, in T.R. Lakshmanan and P.N. Nijkamp (eds) *Structure and Change in the Space Economy*, Springer, Berlin, 297–309.

Jones, D.W. and O'Neill, R.V. (1994) Development policies, rural land use, and tropical deforestation, *Regional Science and Urban Economics* **24** (6), 753–771.

Kaimowitz, D. and Angelsen, A. (1998) *Economic Models of Tropical Deforestation: A Review*, Centre for International Forestry Research, Jakarta.

Lambin, E.F. (1994) Modelling deforestation processes: a review, *TREES Publications Series B: Research Report 1*, EUR 15744 EN, European Commission, Luxembourg.

Lambin, E.F., Rounsevell, M. and Geist, H. (2000) Are current agricultural land use models able to predict changes in land use intensity? *Agriculture, Ecosystems and Environment* **82**, 321–331.

Lambin, E.F., Turner II, B.L., Geist, H., Agbola, S., Angelsen, A., Bruce, J.W., Coomes, O., Dirzo, R., Fischer, G., Folke, C., George, P.S., Homewood, K., Imbernon, J., Leemans, R., Li, X., Moran, E.F., Mortimore, M., Ramakrishnan, P.S., Richards, J.F., Skånes, H., Steffen, W., Stone, G.D., Svedin, U., Veldkamp, T., Vogel, C. and Xu, J. (2001) The causes of land-use and land-cover change: moving beyond the myths, *Global Environmental Change* **11**, 5–13.

Mertens, B. and Lambin, E.F. (2000) Land-cover trajectories in Southern Cameroon, *Annals of the Association of American Geographers* **90** (3), 467–494.

Meyer, W.B. and Turner II, B.L. (1994) *Changes in Land Use and Land Cover: A Global Perspective*, Cambridge University Press, Cambridge.

Myers, N. (1989) *Tropical Deforestation: Rates and Causes*, Friends of the Earth, London.

Netting, R.McC. (1993) *Smallholders, Householders: Farm Families and the Ecology of Intensive, Sustainable Agriculture*, Stanford University Press, Stanford, CA.

Polsky, C. and Easterling, W.E. (2001) Adaptation to climate variability and change in the US Great Plains: a multi-scale analysis of Ricardian climate sensitivities, *Agriculture, Ecosystems and Environment* **85**, 133–144.

Riebsame, W.E., Meyer, W.B. and Turner, B.L. (1994) Modeling land-use and cover as part of global environmental change, *Climatic Change* **28** (1/2), 45–64.

Robinson, V.B. (1978) Information theory and sequences of land use: an application, *Professional Geographer* **30**, 174–179.

Rouchier, J., Bousquet, F., Requier-Desjardins, M. and Antona, M. (2001) A multi-agent model for describing transhumance in North Cameroon: comparison of different rationality to develop a routine, *Journal of Economic Dynamics Control* **25**, 527–559.

Rounsevell, M.D.A., Evans, S.P., Mayr, T.R. and Audsley, E. (1998) Integrating biophysical and socio-economic models for land use studies, *Proceedings of the ITC & ISSS Conference on Geo-information for Sustainable Land Management*, Enschede, 17–21 August (1997).

Ruthenberg, H. (1980) *Farming Systems in the Tropics*, Clarendon Press, Oxford.

Sandler, T. (1993) Tropical deforestation: markets and market failures, *Land Economics* **69**, 225–233.

Sklar, F.H. and Costanza, R. (1991) The development of dynamic spatial models for landscape ecology: a review and prognosis, in M.G. Turner and R.H. Gardner (eds) *Quantitative Methods in Landscape Ecology*, Ecological Studies Vol. 82, Springer-Verlag, New York, 239–288.

Sklar, F.H., Costanza, R. and Day, J.W. (1985) Dynamic spatial simulation modeling of coastal wetland habitat succession, *Ecological Modelling* **29**, 261–281.

Southgate, D. (1990) The causes of land degradation along 'spontaneously' expanding agricultural frontiers in the Third World, *Land Economics* **66**, 93–101.

Southworth, F., Dale, V.H. and O'Neill, R.V. (1991) Contrasting patterns of land use in Rondonia, Brazil: simulating the effects on carbon release, *International Social Science Journal* **130**, 681–698.

Stephenne, N. and Lambin, E.F. (2001a) A dynamic simulation model of land-use changes in the African Sahel (SALU), *Agriculture, Ecosystems and Environment* **85**, 145–162.

Stephenne, N. and Lambin, E.F. (2001b) Backward land-cover change projections for the Sudano-Sahelian countries of Africa with a dynamic simulation model of land-use change (SALU), in T. Matsuno and H. Kida (eds) *Present and Future of Modeling Global Environmental Change: Towards Integrated Modeling*, Terra Scientific Publishing Company, Tokyo, 255–270.

Stephenne, N. and Lambin, E.F. (2002) Scenarios of land-use change in Sudano-Sahelian countries of Africa: understanding the driving forces of environmental change, *Geojournal*.

Thornton, P.K. and Jones, P.G. (1998) A conceptual approach to dynamic agricultural land-use modelling, *Agricultural Systems* **57** (4), 505–521.

Turner II, B.L., Meyer, W.B. and Skole, D.L. (1994) Global land-use/land-cover change: towards an integrated study, *Ambio* **23**, 91–95.

Turner II, B.L., Skole, D.L., Sanderson, S., Fischer, G., Fresco, L. and Leemans, R. (1995), *Land-Use and Land-Cover Change: Science/Research Plan*, IGBP Report 35, HDP Report 7, The Royal Swedish Academy of Sciences, Stockholm.

Turner, M.G. (1987) Spatial simulation of landscape changes in Georgia: a comparison of three transition models, *Landscape Ecology* **1**, 29–36.

Veldkamp, A. and Fresco, L.O. (1996) CLUE: a conceptual model to study the conversion of land use and its effects, *Ecological Modelling* **85**, 253–270.

Veldkamp, T. and Lambin, E.F. (2001) Predicting land-use change, Special issue, *Agriculture, Ecosystems and Environment* **85**, 1–6.

van Thünen, J.H. (1966) *Der isolierte Staat in Beziehung auf Landwirtschaft und Nationalökonomie: Neudruck nach der Ausgabe letzter Hand* (1842/1850), Fischer, Stuttgart.

Walker, R.T. (1987) Land use transition and deforestation in developing countries, *Geographical Analysis* **19**, 18–30.

Walker, R.T. (1987) The structure of uncultivated wilderness: land use beyond the extensive margin, *Journal of Regional Science* **39** (2), 387–409.

White, R., Engelen, D. and Uljee, I. (1997) The use of constrained cellular automata for high-resolution modelling of urban land-use dynamics, *Environment and Planning B, Planning and Design* **24** (3), 323–343.

Wilkie, D.S. and Finn, J.T. (1988) A spatial model of land use and forest regeneration in the Ituri forest of Northeastern Zaire, *Ecological Modelling* **41**, 307–323.

Wu, F. (1998) SimLand: a prototype to simulate land conversion through the integrated GIS and cellular automata with AHP-derived transition rules, *International Journal of Geographic Information Systems* **12**, 63–82.

Part III

Models for Management

14

Models in Policy Formulation and Assessment

The WadBOS Decision-Support System

GUY ENGELEN

14.1 INTRODUCTION

In the past five to ten years, an increasing number of government organizations have started to develop rather sophisticated model-based information systems to support the policy-making process. The development of so-called Policy Support Systems is currently a booming activity. A few examples from the Netherlands only are: IMAGE (Alcamo, 1994), TARGETS (Rotmans and de Vries, 1997), Landscape Planning of the river Rhine-DSS (Schielen, 2000) and Environment Explorer (de Nijs *et al.*, 2001). This trend is propelled by the growing understanding that policy-making should be based on an integrated approach. Systems theory has clearly shown that systems and problems do not exist in isolation, rather that they are part of larger entities (see, for example, Sterman, 2000). They have dimensions that extend into other domains, other disciplines, other levels of detail and other temporal and spatial scales. Complexity and computation theory have shown that even seemingly weak linkages may have major repercussions on the behaviour of the system as a whole (Prigogine, 1981; Kauffman, 1990). Policy-makers, responsible for the management of cities, watersheds or coastal zones, are confronted with this reality on a daily basis. They are required to manage fragile systems that exhibit an extremely rich behaviour not least because of the many intelligent actors, the human inhabitants or users that steer the development in the direction of their own interest (Chapter 12). Confronted with this complexity, on the one hand, and with better-informed, agile recipients

of the policies on the other, policy-makers have to be able to rely on adequate instruments enabling them to understand better and anticipate the effects of their interventions in the system as fully as possible.

As a result, today's policy programmes strongly advocate integrated policies for land-use management, watershed management and coastal zone management among others, and today's research and development agenda strongly promotes the development of tools enabling an integrated approach. The work is propelled by the revolution in the computing hardware and software since the beginning of the 1980s, putting computation power on the desk of the individual scientist, modeller and decision-maker that could not be dreamed of 30 years ago. Information systems of growing levels of sophistication go along with the increasing capacity of the personal and micro computer. Most relevant in the field of spatial planning and policy-making has been the rapid growth of high resolution remote sensing and Geographical Information Systems (GIS) in the past two decades. Many users of the latter techniques are interested in the detail only – the precise knowledge of what is where – but there is a growing community of users and developers exploiting their potential for high resolution spatial modelling and its use as a policy-support instrument. As a result, new modelling techniques have been added to the toolbox of the spatial scientists and the policy-maker, including simulation, neural networks, genetic algorithms and cellular automata.

Environmental Modelling: Finding Simplicity in Complexity. Edited by J. Wainwright and M. Mulligan
© 2004 John Wiley & Sons, Ltd ISBNs: 0-471-49617-0 (HB); 0-471-49618-9 (PB)

However, the task ahead is still huge. Today, the scientific community cannot offer policy-makers the instruments that will solve their ill-defined problems in an absolute and indisputable manner. It probably never will. The problems encountered are too big and the knowledge available is too limited to produce unambiguous answers. But lessons are learned on how to work with models as instruments for exploration, representing a part of the complex reality with some level of certainty. These are mainly 'thinking tools' that shed light on problems that otherwise could not be manageable by the human brain alone.

In our effort to build practical instruments for planning and policy-making we have developed integrated simulation models representing the policy domain in its appropriate spatial and temporal dimensions and have embedded them in decision support systems (see http://www.riks.nl for other examples). In the remainder of this chapter we will take the reader through the design and implementation phases of the kind of system. In particular, we will dwell on the development of the WadBOS policy support system developed for and applied to the Dutch Wadden Sea. The very existence of the WadBOS system is proof of the fact that the technology and the skills exist to develop model-based policy-support tools. However, in the concluding paragraph we will contemplate briefly the many problems that still need to be solved before the kind of system presented will be a typical or a standard product used for integrated policy-making.

14.2 FUNCTIONS OF WadBOS

The Wadden Sea is part of a coastal system extending from the north of the Netherlands into northern Germany and western Denmark. In the Netherlands, the sea is a protected nature reserve because of the important ecological functions it fulfils. At the same time, the sea has important economic functions. Fishing, recreation, transportation and mining are among the main economic activities. It generates employment, income, leisure and food for many households. The management of the different activities and functions of the sea is distributed over a great number of institutions, ranging from the municipal to the European level. When decisions are to be made or policies need to be developed relative to the exploitation or protection of the area, incompatible views tend to slow down the decision-making process.

The development of WadBOS started in 1996 when CUBWAD, the association of government organizations responsible for the management of the Wadden Sea, concluded that policy-making could be enhanced

if the abundant existing knowledge about the Wadden Sea, which is generally spread among the many policy-making, management and research bodies active in the region, was gathered, ordered, combined and made available in an operational form to those responsible for policy-making. It was expected that an information system of some sort, a knowledge-based system (KBS), expert system (ES) or decision-support system (DSS), representing the Wadden Sea in a holistic manner, integrating ecological functions and human activities at the appropriate temporal and spatial scales would be a very useful instrument for this purpose. Such a system was expected to enable the exploration of the autonomous dynamics of the Wadden system as well as the analysis of effects of policy measures thereon. Thus it would boost the analytic capabilities of the policy-makers when it came to searching for solutions and alternatives to solving policy problems. It would facilitate communication when the results obtained in the different steps of the policy-making process needed to be shared and discussed with others involved: stakeholders, fellow policy-makers, or the public as the ultimate recipient of the policies. It would enable learning when it came to deepening understanding about particular topics, processes and linkages in the Wadden system. Finally, it would enable knowledge storage and retrieval when it came to amassing and integrating existing knowledge and accessing it when and where necessary. Once the available knowledge was integrated, the system would equally reveal missing or unsatisfactory elements in the knowledge base and thus give impetus to future research activities.

Thus, the scope and function of WadBOS were as a broadly defined system providing information and knowledge in support of the preparation of, and possibly the implementation of, integrated policies for the Wadden Sea. In order to do so, it has an analytic, a communication, a library and a learning function. This is an ambitious set of functions, which is not easily attained by traditional information systems such as databases or GIS systems. Rather, an information system is envisaged with the ability to manipulate and aggregate data as the result of statistical, mathematical, heuristic or algorithmic operations (Catanese, 1979). This is where models as part of decision-support systems become essential instruments.

14.3 DECISION-SUPPORT SYSTEMS

Decision-support systems (DSS) are computer-based information systems developed to assist decision-makers to address semi-structured (or ill-defined) tasks in a specific decision domain. They provide support of a formal nature by allowing decision-makers to 'access' and use 'data' and appropriate 'analytic models' (El-Najdawi and Stylianou, 1993). The term 'semi-structured' in this definition refers to the fact that DSS are typically applied to work on problems for which the scientific knowledge is missing, to solve them in an unambiguous manner, or problems for which there is insufficient consensus relative to the values, criteria or norms to define them in an unambiguous manner (van Delden, 2000). The term 'appropriate' refers to the fact that use is made of the best available solution methods to approximate heuristically the unique answer. Thus, the DSS provides the decision-maker with a suite of domain-specific 'analytic models' considered appropriate to represent the decision domain. Integrated models play a key role in any DSS in the sense that their constituting submodels are covering, at least in part, the (sub-)domains related to the decision problem, but also because integrated models explicitly include the many complex linkages between the constituting models and related domains. Thus, they provide immediate access to very rich and operational knowledge of the decision domain.

The usefulness, richness and scope of DSS are predominantly determined by the spectrum and appropriateness of the models available from its *model base* (Figure 14.1). Three more components fulfil specific tasks within the DSS (Engelen *et al.*, 1993): (1) a *user interface*, the vehicle of interaction between the user and the system; (2) a *database* containing the raw and processed data of the domain and the area being studied; and (3) a *toolbase* with the decision-theory methods, analytical techniques and software instruments required to work in an effective manner with the domain models and the data. Each of the four components has a complex internal structure.

14.4 BUILDING THE INTEGRATED MODEL

In line with the stated objectives, the approach taken in WadBOS was clearly bottom-up. It was based on a

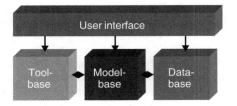

Figure 4.1 Basic functional components of the WadBOS decision-support system

reasonable understanding of the characteristic processes and problems typifying the Wadden Sea, and based on expertise, a fair amount of complementary knowledge, and modelling material available from a large number of organizations. In this way an integrated model was designed and constructed. This integrated WadBOS model is the core element of the model base of the DSS. It consists of linked submodels and represents the Wadden system as completely as possible with a view to facilitating the design and evaluation of integrated policies.

The Wadden Sea is an extensively studied coastal system and an amazing amount of data, information and knowledge is available. However, this material is very diverse in nature and spread out among the many organizations and people who produced it. With a view to carrying out a gradual systems analysis and to gather, structure and conflate the available material, three consecutive knowledge-acquisition/modelling sessions were organized (Firley and Hellens, 1991; Gonzalez and Dankel, 1993; de Kok *et al.*, 1997). Prior to these sessions, a preliminary requirements analysis had been carried out focusing strongly on the use and user of the system. The intended use and functions of WadBOS have been described earlier in this chapter. The envisaged end-users are coastal zone managers and policy-makers. Their profile is best described as high-level technicians actively involved in the design and evaluation of coastal zone management policies. They perform policy work of a formal/analytic nature in support of the administrator or politically appointed person responsible for taking the actual decision and starting the actual policy-implementation process. Thus, this policy-maker is a technician, not a politician.

14.4.1 Knowledge acquisition and systems analysis

In the first series of knowledge-acquisition sessions, an extensive group of potential end-users, all actively involved in policy-making, were interviewed. From these sessions answers were obtained to the following questions:

- What and where are the (system) boundaries of the Wadden system?
- What knowledge should be included in WadBOS? What knowledge is available, and from whom or from where can it be obtained?
- What are desirable levels of detail and accuracy of WadBOS?

The expectations were diverse, but consensus grew over a representation covering the entire Wadden Sea, not the land adjoining it, representing the natural system and all the major human activities that take place on the water. The reciprocal linkages between the natural and the human system as well as the appropriate representation of the processes in their complexly coupled multi-spatial and multi-temporal context were considered essential.

The information thus gathered, complemented by material obtained from literature research and the analysis of policy documents, was the input of an initial systems analysis: the main processes were identified and their definition in terms of measurable characteristics and state variables was carried out. Next, an initial set of system diagrams and conceptual models were drafted and for each process considered experts were identified.

In a second series of knowledge-acquisition sessions, structured interviews were organized with the selected domain experts. Most of the work with the experts was carried out in sessions involving two to five people only. Often one of the potential end-users would participate. During the interviews, visual modelling was applied: a graphical modelling tool was used to draft, discuss, upgrade and complete the system diagrams and conceptual models according to the experts' knowledge. At the end of these sessions a complete description of the Wadden system was available in the form of graphical, qualitative models. There was a very reasonable consensus relative to this representation. Ambiguity and difference of opinion had been resolved as much as possible in a number of consecutive visits and discussions in which conflicting views were clarified and resolved.

As part of the knowledge-acquisition sessions, a number of existing mathematical models had been detected and evaluated for incorporation (see below) and the translation of the qualitative models into a mathematical representation began. As far as possible, existing models were incorporated or adapted for this purpose. However, a lot of knowledge was not available in the form of readily usable mathematical models. These missing models needed to be developed.

Partway through the modelling phase, a third knowledge-acquisition round began, in which end-users, domain experts, scientists and model developers were confronted with the mathematical representation of the system. These were very intense exercises in which consensus was sought relative to the formal representations chosen, the level of integration accomplished and the simplifications introduced.

With the information thus obtained, the knowledge acquisition and modelling phases were completed, and

the technical integration of models and the technical implementation of the system began.

14.4.2 Modelling and integration of models

With a view to meeting the general objectives of the WadBOS system, a great deal of effort went into separating the detail from the essence in the domain representations. Component (sub-)models were integrated in the WadBOS model base with a view to completing the system diagram of the integrated model and to make it operational. The models, necessary for integration, are what Mulligan (1998) calls policy models as opposed to research models. The selection and incorporation of models was based on both scientific and end-user criteria.

The following scientific criteria were taken into consideration:

- *Models fitting the integration scheme.* Only models were integrated that fulfil a specific task within the WadBOS integration scheme not dealt with by any other (sub-)model. (Sub-)models compute a subset of the state-variables and exchange the necessary information among one another at the right temporal and spatial scales during the calculations.
- *Compatibility of scientific paradigms.* Only models were integrated that could be integrated from a scientific/operational point of view. The scientific assumptions and constraints underlying the models were assessed. Most of the models used in WadBOS are spatial, dynamic, nonequilibrium or quasi-equilibrium models that are solved by means of simulation. Models using both rule-based and algebraic solution methods were retained.
- *Timescales and temporal dynamics.* Only dynamic models were integrated. Models need to span a strategic time horizon (10 years) and operate at timesteps reflecting the inherent characteristics of the processes and decision-making time frame (tidal cycle, 1 month, 1 year).
- *Spatial resolution and spatial dynamics.* Only spatial models or models that can be spatialized were integrated. Models are applied to the entire Wadden Sea and operate at an appropriate spatial resolution to reflect realistically the processes represented, the spatial variability across the region, and its constituent geographical entities, subject to decision and policy-making requirements. With a view to simplifying or aggregating the model, the effect of increasing or decreasing the spatial resolution on the performance of the model is a criterion for selection.

- *Scientifically proven basis.* The process descriptions within the models should be well understood and scientifically proven. The model results should be as robust, reliable and accurate as possible.

The following list of key end-user criteria was taken as a guideline:

- *All processes represented.* The WadBOS model should adequately represent all the important processes necessary to provide the required policy outputs.
- *Spatial scale, temporal scale, time horizon.* The Wad-BOS model should provide information at a sufficient level of spatial and temporal resolution to reflect the spatial and temporal scale of variation in the most important ecological and socio-economic variables, as well as at a spatial and temporal resolution at which problems occur that need to be addressed as part of coastal zone policies. Finally, a time horizon that is relevant for policy design, implementation and assessment, is needed.
- *Routine data.* The WadBOS model should be sufficiently simple to run from routinely measured and available data. In principle, no dedicated data are collected to run WadBOS.
- *Output-centred.* The WadBOS model should be judged mostly upon the quality of its output and less upon the scientific or technical innovative character of its models.
- *Policy-centred.* The WadBOS model should provide appropriate results using indicators or variables that directly interface with the policy implementation process and focus on environmental changes, anthropic impacts and management options. More abstract, scientific or technical variables are less suitable for this purpose.
- *Interactive.* The WadBOS model should be fast, responsive and interactive and should cater for a very short attention span. A response time of less than 15 minutes per simulation run covering a period of 10 years should be aimed for. Clever models, fast algorithms and efficient software code are required to achieve this aim.

The key trade-offs in the selection process were very much between accuracy (of outputs and of process representations) and simplicity (of models and of input data). The resulting model needed to have sufficient spatial and temporal detail and model complexity to accurately represent the processes but needed to achieve this over large areas in a fast and responsive manner with a minimum of data.

14.4.3 Technical integration

Technical integration deals mostly with the hardware and software aspects of model integration: how can pieces of executable code be efficiently linked so that together they perform the operations specified in the integrated model at the right time, and so that data are exchanged in a way that is consistent with the temporal and spatial logic of the model? Is it possible to do this in a manner that enables reconfiguration of the model in a straightforward manner? And, can the material developed be re-used for DSS systems implemented elsewhere to address similar aspects? Two aspects are decisive in answering these questions: the architecture chosen to represent the model base and the integrated model and the software technology used to implement the DSS.

With a view to developing a fast, responsive system, operating stand-alone on the PCs of its end-users, an architecture was chosen for WadBOS featuring the integrated model as the core element of the DSS. Thus, the model base is equipped with an integrated model fully tailored and purposely developed to suit the precise needs of the end-users. Rather than a suite of loosely coupled submodels, it is a complex model by design. Each of its submodels is adapted to that effect and (re-)coded as a software component according to a strict template. Next, the components are coupled to one another as required in order to establish the many linkages of the integrated model. This architecture will generally result in a user-friendly system. More than in other solutions (e.g. Hahn and Engelen, 2000), it will represent most relevant processes at the same level of abstraction and detail and will give good performance because of the limited overhead in accessing and running the submodels. It is a medium- to high-cost solution because of the effort spent in the design and (re-)implementation. Unless object-oriented or component-based technology (D'Souza and Cameron Wills, 1999) is used to implement the solution, the maintenance costs can be very high, certainly if the end-user requirements change and/or if the model representation needs major repair and upgrading.

For the technical implementation of WadBOS the DSS-Generator GEONAMICA® was used. A DSS Generator is 'a package of hardware/software which provides a set of capabilities to build specific DSS[s] quickly and easily' (Sprague and Carlson, 1982). Hence, it concerns a special purpose software-development environment for the creation of new DSS applications in a more or less narrowly defined domain. GEONAMICA® is an object-oriented application framework, developed by RIKS bv. It is specially geared towards developing spatial DSS featuring models that run at multiple spatial and temporal resolutions. Typically, it will combine system dynamics models and cellular models for this purpose. In particular, use is made of spatial interaction-based models, different kinds of cellular automata models, multi-agent and other kinds of rule-based models. It is equipped with highly efficient computational techniques and algorithms for addressing spatial problems, but also with additional analytical tools visualization tools and input, import, export and output tools.

14.5 THE INTEGRATED WadBOS MODEL

The WadBOS model resulting from the exercise consists of submodels running at one of three embedded spatial scales: (1) the Wadden Sea as a whole ($+/-3000\,km^2$); (2) the 12 compartments within the sea (mostly delimited on the basis of hydrodynamic characteristics); or (3) a regular grid consisting of $\pm 11\,000$ cells of 25 ha each (see Figure 14.2). As for the temporal resolution, WadBOS integrates submodels running at (1) a yearly and (2) a monthly timestep, or (3) a timestep equal to the tidal cycle. The model represents strongly coupled social, economic, ecological, biological, physical and chemical processes (see Figure 14.3). They are grouped into three main submodels: the economy, the ecology and landscape. Enough existing GIS and statistical data are available to run its economic and ecological models. It is sufficient to generate the output required for most relevant policy questions and meets the performance criteria specified and performs a simulation run of 10 years in less than 10 minutes on a state-of-the-art PC with Windows installed.

Most economic processes run on a monthly or yearly timestep, while most of the ecological processes are

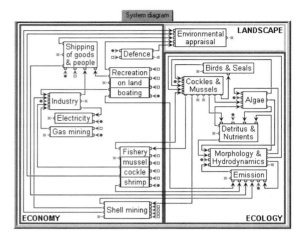

Figure 14.3 The Integrated WadBOS model represented at the highest level of abstraction. This is a screen dump of the system diagram view of the user interface (see section below). The boxes represent submodels, the arrows show linkages and flows of data between the submodels

represented at the compartment level and run on a tidal cycle timestep. Figure 14.3 shows the system diagram of the integrated model at the highest level of abstraction. This diagram is also the user interface of the model (see below). The boxes represent submodels and the arrows between the boxes show the main data flows in the model.

14.5.1 The economic submodel

In the economic submodel (see Figure 14.3), all the major economic activities present in the Wadden Sea

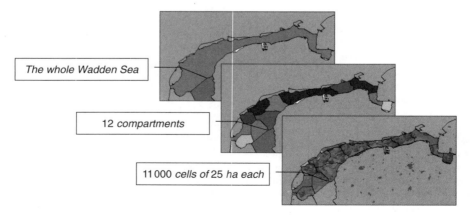

Figure 14.2 WadBOS integrates submodels running at one of three spatial scales: the whole Wadden Sea, 12 compartments, or 11 000 cells of 25 ha each

are represented at some level of detail. Shell mining, Fishery and Recreation have currently been worked out in greatest detail. Most activities carried out at sea are an input into the local Industry and generate directly or indirectly Shipping of Goods & People. Shell mining and Fishery directly extract biological resources from the Wadden Sea, while Recreation, Electricity (generation) and Defence use the open space as their prime resource. Finally, Gas mining taps underground resources and may thus have an impact on the Morphology of the seafloor. The presence and the noise generated by nearly all activities affect the landscape: its ecological state, the species present and its attractiveness to humans. Furthermore, each human activity causes some form of emission of pollutants into the waters.

Most economic activities are modelled on the basis of the Platvis economic fishery model (Salz *et al.*, 1994). Thus, they all have the same general modelling scheme, represented in Figure 14.4. This greatly increases the consistency, the learnability and the transparency of WadBOS. The calculations proceed through a series of interlinked relations on a monthly basis. In this text, the example of the Recreational boating sector (Recreatievaart) is given but the scheme of relations applies to all other sectors too, with minor adaptations in the terminology and definitions.

The Effort (Overnight stays, recreational boating) of an economic activity is determined by the available Infrastructure. Infrastructure for recreational boating is the number of mooring positions in the harbours. For other economic activities, it is, for example, the amount of equipment on board boats (expressed in horsepower)

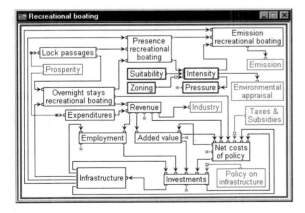

Figure 14.4 Screen dump of the model of an economic sector in WadBOS. Recreational boating is used as an example. The shaded boxes represent submodels that are not accessible from this diagram in the user interface

to drag fishing nets or suck up and wash shells, or the number of hotel beds available for visitors. In order to operate the Infrastructure, a workforce is required, thus determining the Employment in the sector. The deployment of Effort and the availability of visitors (in the recreation sector), or resources to be extracted (as calculated in the Ecology subsystem), determine the extent of the activity: Presence recreational boating (see Figure 14.5: ①) for the recreational boating sector or, for example, total catch for the mussel fishery. At a fixed rate of spending, the recreational boaters generate a particular Revenue. In WadBOS, expenditures and prices are externally determined constants. Debits, Labour costs and Technical costs are subtracted from the Revenue and thus the Added value of the sector is calculated. If the latter is positive, Investments may be made to increase the Infrastructure. Policies can intervene directly and set a maximum limit on infrastructure (Policy on infrastructure), or indirectly through Taxes & Subsidies. Another way of intervening into an economic sector is by setting a quota. These will see to it that the quantity of shells mined or fish caught is reduced and thus prevent the exhaustion of the resources.

The above relations are calculated for the entire Wadden Sea or per compartment (i.e. Quota can be set per compartment). Further to these, there are calculations relative to a more precise location of the activity at the level of the 25 ha cells. This is because the activity (Presence recreational boating (see Figure 14.5: ①) is not equally distributed over the entire sea, nor is it equally intense during the whole year (see Figure 14.5: ②). It is in particular the cell's Suitability (see Figure 14.5: ③) for an activity that determines where the Intensity (see Figure 14.5: ⑤) is high or low. However, the distribution of activities can be controlled through Zoning policies by closing areas (see Figure 14.5: ④) permanently or for particular periods during the year. The latter is an important policy lever, because a high intensity may cause Exposure (see Figure 14.6: ①) of particular species of bird or sea animals. Each activity contributes to one or more of the three types of Pressure on the natural system: presence and noise, mechanical influence and extraction.

All variables calculated at the cellular level are available as dynamic maps, updated on a monthly basis. Maps can be consulted at any time during the simulation for: Intensity for each activity, the three types of Pressure, and 11 forms of Disturbance (a combination of an activity causing the disturbance, the species affected, and the kind of disturbance, including noise, presence and mechanical).

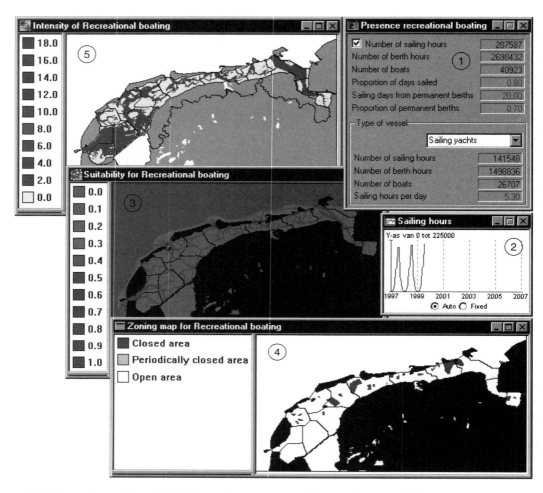

Figure 14.5 Screen dumps from WadBOS showing a map and dialogue windows relative to the recreational boating activity. The numbered items are explained in the text

For each activity the Suitability map is a composite measure calculated on the basis of physical, environmental and infrastructure elements characterizing each cell. Similarly, for each activity there is a Zoning map calculated from different factors characterizing the institutional and legal status of each cell. The Suitability will express the extent to which a cell is appropriate for carrying out the activity, while the zoning map will indicate whether or when in the course of the year the activity is allowed in the cell. Both maps can be interactively edited in the system thus enabling the user to define different kinds of spatial policy measures. The OVERLAY tool, available from the toolbase of WadBOS, enables the interactive generation of the Suitability and Zoning maps.

14.5.2 Landscape

The economic activities will affect the pristine character and landscape of the Wadden Sea. These too are localized effects that are calculated in WadBOS at the cellular level and on a monthly basis. An assessment in ecological terms (see Figure 14.6: ②) is calculated, based on the Ecotopes present in each cell and on the potential biodiversity of the ecotope. This assessment is done with and without the effect of economic Activities and Cultural elements present in the landscape. This approach enables a straightforward visualization of the human impacts. On the basis of the same information, patches of contiguous Landscape types are calculated. Types vary from 'nearly natural' to 'entirely human'.

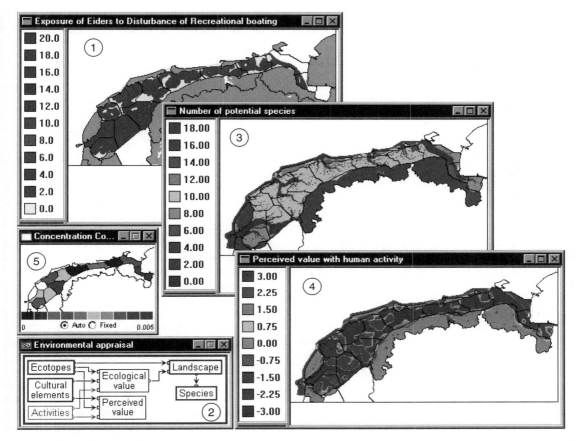

Figure 14.6 Screen dumps from WadBOS showing a few map and system diagram windows relative to the Landscape submodel. The numbered elements are explained in the text

The ecological characteristics of the type and the size of the patch will determine the kind and number of Species (see Figure 14.6: ③) potentially populating it in a sustained manner. Finally the Wadden Sea is assessed relative to the way it is perceived (see Figure 14.6: ④) by humans as 'typical and attractive'. This too is done with and without human elements present. The different characteristics mentioned – Ecological assessment, Perception, Landscape types and potential Species – are presented as dynamic maps on a monthly basis. For most of the assessment calculations there are no 'hard scientific facts' to build into the model, rather, expert rules and judgements are used. Even then, the sequence of monthly maps demonstrates in a very interesting manner how the sea loses a lot of its ecological value and its attractiveness during the summer when human activity booms. During winter, however, the sea returns to a much more natural state.

14.5.3 The ecological subsystem

Where activities are carried out some form of pollution is caused and particular Emissions (see Figure 14.6: ⑤) (copper, TBT, PAK and Oil) will end up in the water (see Figure 14.3). River-borne pollutants will enter the sea mainly from Lake Ijssel and the River Eems. A simple hydrodynamic model calculates how the water moves from one compartment to the next to eventually flow into the North Sea. Not only will pollutants move with the water, but so will other dissolved matter – Detritus & Nutrients – including phosphates, silicon and nitrogen. In a food chain model derived from the EcoWasp model (Brinkman, 1993) the Nutrients serve as an input to the Algae dynamics. Simply stated, the growth of the latter is determined by the availability of nutrients, the right climatic conditions, light, or influx from the North Sea and fresh water systems. The algae in turn are grazed by filter feeders, including the commercial species Cockles & Mussels. The latter are prey for Birds and the Fishery. Every year, policy-makers decide on the quantity of cockles and mussels needed for the survival of the species and as food for Birds. The rest can be fished commercially. Cockles are fished and cooked at sea and their shells are returned to the sea to become a resource for the Shell mining industry. The food chain dynamics are calculated on a tidal cycle timestep and mostly on the spatial level of the compartments. Some information, such as the mussel and cockle biomass and the consumption rates of birds, is calculated at the cellular level. The latter serves as a cellular input into the fishing activity.

This description is very superficial. The interested reader can find more ample and more precise descriptions of the models used in the online documentation of the WadBOS DSS itself, or in the technical reports written (Engelen, 1999; Huizing *et al.*, 1998; Uljee *et al.*, 2000). From this description, however, it should be clear how the different submodels are linked to each other in a network of mutual, reciprocal influence. It should be clear too, that the outputs are visualized by means of a large number of dynamic maps, each of which is updated at the appropriate timestep during the simulation. Outputs are also presented and stored in the form of text, MS Excel tables and time graphs. Dynamic maps generated during a simulation can be stored in files for interactive comparison and further analysis by means of the ANALYSE tool (see below). The input maps can be prepared in a commercial GIS package and imported into WadBOS. Once imported, each input map can be interactively edited by means of an appropriate editor. Thus, spatial policy measures can be localized and tried out. The interactive generation of the Zoning and Suitability maps is supported by means of the OVERLAY tool. OVERLAY and ANALYSE and many other similar instruments are part of the Toolbase of WadBOS.

14.6 THE TOOLBASE

In the DSS, it is the role of the models to provide an adequate and truthful representation of the real world system and it is the role of the tools to enable the decision-maker to work with the models. The user will hardly be aware of the fact that he is using a tool when he is editing a parameter or map, or viewing a variable graphed against time. However, without tools, the most sophisticated model is nothing but a number-cruncher, running out of control, drowning its user in a lake of numbers.

Most of the tools available in WadBOS are standard in the GEONAMICA® DSS Generator, hence were readily available for integration. In this overview the tools are ordered relative to the type of task they carry out in WadBOS. A distinction is made between Input tools, Output tools, Exploration tools and Evaluation Tools (see Table 14.1).

- Among the *Input tools* are the typical editors for changing single numbers, or series of numbers in a textual or graphical manner. As part of the latter, the table editor, which enables entering 2-D relations as a curve, and the map editors are essential and powerful instruments in WadBOS. But so are the tools to open, close, import and export files, scenarios and policy exercises.

- The *Output tools* fulfil the difficult task of presenting the massive amounts of data generated interactively in as concise and precise a manner as possible. To present dynamic spatial data interactively, fast dynamic mapping facilities are available. To store simulation results and enable their in-depth analysis at a later stage, WadBOS features recorders and players of animated maps, tools for storing dynamic map output in so-called .LOG files, and tools to write results to a linked MS Excel spreadsheet.

- *Exploration tools* enable the user to interactively search the solution space. The capacity to generate scenarios and to produce map overlays (OVERLAY tool), compare maps (ANALYSE tool), and carry out sensitivity analysis (MONTE-CARLO tool) interactively, puts great analytical power in the hands of the end-user.

- *Evaluation tools* support the user in their choice of a 'best' solution. WadBOS to that effect features Score Tables, and will be equipped with a Multi Criteria Analysis tool in its next version. A built-in Goal Seeking tool presents, based on a sensitivity analysis, the dependency of state variables on policy relevant parameters in the integrated model. When searching for potential solutions, it greatly reduces the parameter space to be analysed.

According to the function the system is to carry out in the policy-making process, the tools play a more or less pronounced role. From Table 14.1 it is clear that the analyst uses the complete set extensively. He needs both the down-to-earth instruments for entering data and viewing output as well as the sophisticated instruments for evaluating output. When the main purpose of WadBOS is communication, pertinent output instruments are essential. The learner is foremost

Table 14.1   Tools available or wanted in WadBOS and their usefulness in relation to the functions of WadBOS (on a scale from + to ++++)

Tools	Analysis	Communication	Learning	Library
Input tools				
text editor	++	+	+++	+++
value editor	+++	+	+++	+++
series editor	+++	+	+++	+++
table editor	+++	++	+++	+++
function editor	++	+	+	+++
network (points and lines) editor (*)	++++	++	++++	+++
2D map editor	++++	++	++++	+++
Open/close files	++++	++	++++	++
Import/export files	++++	++	+++	+
Output tools				
online documentation	++++	+++	++++	++++
online help	++	++	++++	++++
time graphs	++++	+	++++	+
dynamic maps (2D)	++++	++	++++	+
dynamic maps (network) (*)	++++	++	++++	+
3D representation	++	+++	+++	+
Animation tool (*)	++	++++	+++	+
Tracing tool	+++	++	+++	+
Link to Excel	++++	+	+++	+
Log to file	++++	++	+++	+
Exploration tools				
OVERLAY tool	++++	+++	+++	+
ANALYSE tool	++++	+++	+++	+
SCENARIO tool (*)	++	++++	++	+
MONTE CARLO tool (*)	+++	++	+++	+
Evaluation tools				
SCORE Table tool	++++	++++	++++	+
EVALUATE tool (MCA) (*)	+++	++++	++++	+
Goal seeking	+++	+	+	+

Note: (*) available in GEONAMICA, to be implemented in the next version of WadBOS

interested in transparent input and output tools and the documentation systems. Finally, for the library function, a good documentation system is paramount, as are facilities to quickly enter, update and retrieve the information stored in the library.

14.7 THE DATABASE

Given the fact that none of the models incorporated in WadBOS requires data from an online connection to an external database or to a monitoring system, the integration of data has been implemented by means of a stand-alone database. In fact, the original geographical information is generated, maintained and stored in the GIS system of the National Institute for Coastal

and Marine Management. A dedicated reduced version of this database is distributed with WadBOS. It is renewed when better quality data become available. The economic and ecological data are obtained from a wide variety of sources, some of which were not originally in a digital format, and have been added to the database. When used, WadBOS takes entire care of the retrieval and storage of its data: new or additional information is entered via the dialogues of the system and is stored in a hierarchically organized directory structure. In addition, all the data files are in a readable format, hence can be viewed and edited by means of standard software by the end-users if they wish to do so.

In the future, when WadBOS is installed on the machines of the many end-users of the member organizations of CUBWAD, there will be a need for a dedicated central site to organize the management and distribution of the databases. In fact, this is true for the rest of WadBOS also: its models, its tools and the application proper. Currently, options are being evaluated for a management and distribution organization that would be accessible from a page in the Interwad Internet application (www.Interwad.nl).

14.8 THE USER INTERFACE

The user interface is the vehicle of interaction between the user and the computer. It hides the complexities of the internal computer system without hampering its flexibility. It enables the user to address the different components of the DSS (tools, data, models, etc.), translates the user input into appropriate computer instructions, and reports back the results of the computations. To provide maximum user-friendliness, state-of-the-art interactive graphical techniques are applied extensively. A well-designed, intuitive and user-friendly interface will support the execution of the policy exercises to the degree considered necessary by the decision-maker and make the DSS, its models and methods as transparent as possible: at any point in time, the user should have access to the scientific information needed to understand the models, the processes represented and the numbers generated (Holtzman, 1989). Without this information, models remain black boxes and learning is impossible.

Via the user interface, the user has access to all the variables and parameters of all the submodels of Wad-BOS. Moreover, she can decide which combination of

dialogues and maps she wants to keep opened on her screen while carrying out an exercise. All opened maps and dialogues are updated instantaneously during a simulation. The parameters and variables of the submodels are organized and made accessible in a manner that clearly reflects their adherence to one or the other of the following categories: system variables and system parameters, policy parameters, scenario parameters and indicator variables. Each category is part of a dedicated window presenting a view on the integrated model. Each view shows in a graphical manner how policy-relevant features relate to the processes modelled:

1. The *system diagram view* (see Figure 14.3) contains an overview of the structure of the integrated model at the most synthetic level. It represents the linkages between the ecological, landscape and economic processes typifying the dynamics of the Wadden Sea by means of boxes connected by arrows. When a box is clicked, the details of the underlying model are shown, which is either a new system diagram with a more encompassing representation of the invoked submodel, or a dialogue window enabling the outputs generated by the submodel to be read or the required input parameters to be entered. The system diagram view is a default access to any parameter or variable of the integrated model.

2. The *impacts view* (see Figure 14.7) shows the parts of the integrated model containing the summarized information and policy indicators required to evaluate the success of scenarios and policy options tried out. To that effect, the policy-maker should begin any

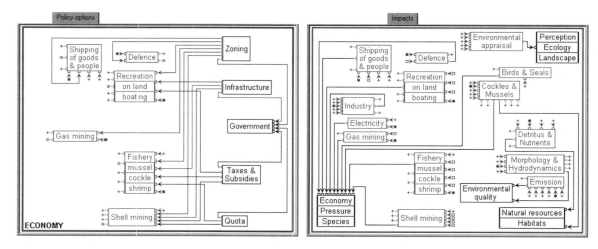

Figure 14.7 Screen dumps of the policy options view (left) and the impacts view (right) from the user interface of WadBOS

policy exercise with a clearly defined set of criteria defining the desired state at a particular point in time. In general, the set of indicators in the impacts view are taken from the management plan of the Wadden Sea. Indicators expressing the economic, social, ecological, biological, chemical and physical state of the system can be viewed. For most indicators norms are available which can be entered as a reference value in the score tables of the dialogues associated with the boxes in this view.

3. The *policy options view* (see Figure 14.7) shows the parts of the model that are most subject to policy interventions. These are elements that are under the control of the policy-maker. The user can explore different combinations of policy measures and set values for Zoning, including closure of parts of the sea; limitations on Infrastructure; Taxes & Subsidies; and Quota on fishing and mining. The costs and benefits of the policies for the Government can be assessed. In an iterative process, she can tune her policy measures in an attempt to approach the desired state with a minimum amount of effort and costs.

4. The *scenarios view* (not represented) shows the parts of the model that are most subject to external influences. These are elements not under the control of the policy-maker. In order to test the robustness of the chosen policy measures, he can impose effects on the system that in the real world are beyond his control. The user can enter a hypothesis for the following external influences: Atmospheric and Climatologic variability; Economic growth and decline in each economic sector; development of the level of Prosperity in the Netherlands; and exchanges of water and dissolved matter between the Wadden Sea, the North Sea and Lake Ijssel.

14.9 CONCLUSION

When the intention was originally formulated to develop a system 'to gather, order and link the knowledge available about the Wadden Sea to facilitate the policy-making process', it was clear that WadBOS was primarily a tool for analysis and evaluation of autonomous dynamics and the effect of policy interventions thereon. Other views coexisted relative to its function: (1) a storage tank of data, information and knowledge; (2) a tool for communication (of results); or (3) a tool for the management of information and knowledge about the Wadden system. Practice so far has shown that the role of WadBOS as an instrument for communication has been at least as important. The fact that this instrument

integrates in a formal manner the many processes that make and change the Wadden Sea gives direction to a lot of the discussions. It does not necessarily make stakeholders and policy-makers agree on issues more easily, but it helps in clarifying what they do or do not agree about. The capacity to visualize conflicting views and reduce these conflicts in an interactive session is of paramount importance.

A core element in WadBOS is the integrated model of the Wadden Sea. The use of and need for such instruments are strongly advocated in new disciplines such as Integrated Assessment (Gough *et al.*, 1998). Yet, and despite the fact that the term 'integrated model' is used to mean a wide variety of things, there are very few operational definitions and recipes or procedures for model integration available from the scientific literature. Thus, the development of integrated models seems more an art than a science at present. It is a deep scientific problem but also a pragmatic multi-criteria multi-objective problem as it requires dealing with end-use aspects: what is appropriate to be integrated in view of the intended use; scientific aspects: what can and cannot be integrated on scientific grounds; and technical aspects: how the integrated model will be assembled and run.

The WadBOS example has shown that integrated models used to support the policy-making process come with a set of requirements of their own, distinguishing them clearly from research models (see also Mulligan, 1998). Policy-makers are best served by models in which the time horizon, the spatial and the temporal resolution are policy problem-oriented and not as process-oriented as research models. They need adequate rather than accurate representations of the processes modelled and sketchy but integral rather than in depth and sectorial models. While research models are as complicated as necessary and scientifically innovative, the policy-maker is better served with an instrument that is as simple as possible and scientifically proven, and which produces usable results as fast as possible, but certainly within 10 minutes. If these differences are ignored in the integration process, the result will often be a large-scale sluggish model not tailored to the needs and expectations of its end-user. Clearly, a fast, interactive model will do much better for policy exploration.

A model will only serve the policy end-user if it is presented in a format that enables it to be worked with. In decision-support systems, models are supplemented with sets of tools to structure and carry out the analysis in a manner that makes intuitive sense to the policy-maker. One of the crucial elements in the DSS is the user interface. Attaining the level of user-friendliness and flexibility that policy-makers seem to desire remains a very big challenge for computer scientists, model developers and domain specialists alike. Indeed, it is very difficult to package in a single application a system with the level of interactivity, flexibility and the fast response times wanted, the user-friendliness, simplicity and the transparency desired, and, the level of accuracy and certainty expected. A lot more innovative research, design and implementation work will need to be carried out to get to this point, if ever we will. Certainly, the demand for the kind of instruments is real. Policy questions have reached a level of complexity that can no longer be dealt with by politicians alone. High-level technicians are playing an ever-increasing role, and the revolution in hardware and software technologies has equipped them with very powerful multimedia calculators.

The technology presented in this chapter is not an alternative for calibration and good modelling practices. On the contrary, as long as a model is not calibrated, it should stay in the hands of the model developer and should be used for policy-making purposes under very strict conditions. Calibration remains a very difficult and time-consuming task. It is more so for a complex integrated model than for a simple and small one. However, linking individual models into an integrated model does not necessarily increase the level of uncertainty and error in the final product. Rather, the strong connections and loops between the individual models and the fact that the values of variables are passed from submodel to submodel at every simulation timestep, bring to the surface mistakes in the model formulations and calculations much more easily than when the same models are applied in isolation (de Kok *et al.*, 2001). Nevertheless, it should be emphasized that even extensively calibrated WadBOS-like models will only generate potential developments rather than predictions of future states or changes. This is not only due to the problem of calibration, but because of the inherent uncertainty in the processes represented. Consequently, the model and the encompassing DSS should be used for explorative rather than predictive purposes at all times.

From the WadBOS project it can be concluded that a policy-support tool can be developed within very reasonable constraints relative to budget, human resources and development time. This development is much easier when good base material and expertise are available and when a stimulating collaboration between visionary end-users and competent DSS developers is propelling the development. However, the development phase described is only the first one in the life of a decision-support system. It needs to be followed by an in-depth evaluation of the technical contents of the system: its constituent models, its coverage of the decision domain(s), its way of dealing with the spatial and temporal dynamics of the processes and the policies. Also an evaluation on behalf of the end-users is pertinent. How do they believe their decision domain has been represented? Does the system speak their language? Does it work in a way they find useful and pleasant? With the answers to all these questions a more or less major redesign of the DSS might be possible or necessary.

Introducing and institutionalizing the DSS in the end-user organization is the ultimate factor determining its success (Alter, 1980; Klein and Methie, 1995). Even the best DSS will go unnoticed or will fail if the wrong implementation and introduction strategy are chosen. Generally speaking, the development and acceptance of the DSS will be much easier if the development is initiated by its end-users right from the beginning. The more they feel the need for a change and the more they are involved in defining the precise role of the instrument in the organization, the more likely it is that the DSS will be accepted and used (Marakas, 1998). The more the product itself fulfils tasks that are perceived as real, in ways that are transparent to the end-user, and in ways that solve the problem in an obviously better way than before its introduction, the better the product's chances of survival. It is therefore crucial to determine the right moment in the development phase of the DSS to have it change state from a prototype into an operational version. At this stage, the gradual introduction, good technical documentation, hands-on training and prolonged technical assistance during usage become of paramount importance. A hasty, unprepared introduction should be avoided in all circumstances.

ACKNOWLEDGEMENTS

WadBOS was developed by a consortium of Dutch R&D institutes consisting of the Research Institute for Knowledge Systems bv (Maastricht), Infram bv (Zeewolde), Delft Hydraulics (Delft), Institute for Environmental Studies (Amsterdam), Resource Analysis (Delft) and DHV (Amersfoort). The work was supported by the Land Water Information Technology Programme of the Dutch Government as well as the Ministry of Transport, Public Works and Water Management. A more elaborate, earlier version of this text was produced for the National Institute for Coastal and Marine Management/RIKZ, The Hague, The Netherlands as part of Contract 42002555 (Engelen, 2000).

REFERENCES

Alcamo, J. (1994) IMAGE 2.0: *Integrated Modeling of Global Climate Change*, Kluwer, Dordrecht.

Alter, S.L. (1980) *Decision Support Systems: Current Practices and Continuing Challenges*, Addison-Wesley, Reading, MA.

Brinkman, A. (1993) *Biological Processes in the EcoWasp Ecosystem Model*, IBN-DLO report 93/6, Texel.

Catanese, A.J. (1979) Information for planning, in *The Practice of Local Government Planning*, International City Management Association, Washington, DC.

de Kok, J.L., Arifin, T., Noor, A., Wind, H.G. and Augustinus, P.G.E.F. (1997) Systems analysis as a methodology for sustainable coastal-zone management in tropical countries, *Torani, Marine Science and Technology Bulletin* **8**, 31–41.

de Kok, J.L., Engelen, G., White, R. and Wind, H.G. (2001) Modeling land-use change in a decision-support system for coastal-zone management, *Environmental Modeling and Assessment* **6**, 123–132.

de Nijs, T., Engelen, G., White, R., van Delden, H. and Uljee, I. (2001) *De LeefOmgevingsVerkenner. Technische documentatie*, RijksInstituut voor Volksgezondheid en Milieuhygiene, Bilthoven, Report 408505007/2001.

D'Souza, D.F. and Cameron Wills, D. (1999) *Objects, Components, and Frameworks with UML: The Catalysis Approach*, Addison-Wesley, Reading, MA.

El-Najdawi, M.K. and Stylianou, A.C. (1993) Expert support systems: integrating AI technologies, *Communications of the ACM* **36** (2), 55–65.

Engelen, G. (ed.) (1999) *BOS Integraal beheer van Estuariene en Waddensystemen. Deelproject: Case Waddenzee*, Stichting LWI, Gouda.

Engelen, G. (2000) *The Development of the WadBOS Decision Support System: A Bridge between Knowledge and Policy in the Wadden Sea*, National Institute for Coastal and Marine Management, The Hague.

Engelen, G., White, R. and Uljee, I. (1993) Exploratory modelling of socio-economic impacts of climatic change, in G. Maul (ed.) *Climate Change in the Intra-Americas Sea*, Edward Arnold, London, 306–324.

Fayad, M.E., Schmidt, D.C. and Johnson, R.E. (1999) *Building Application Frameworks: Object-Oriented Foundations of Framework Design*, Wiley Computer Publishing, Chichester.

Firley, M. and Hellens, D. (1991) *Knowledge Elicitation: A Practical Handbook*, Prentice Hall, New York.

Gonzalez, A.J. and Dankel, D.D. (1993) *The Engineering of Knowledge-based Systems: Theory and Practice*, Prentice Hall, Englewood Cliffs, NJ.

Gough, C., Castells, N. and Funtowics, S. (1998) Integrated assessment; an emerging methodology for complex issues, *Environmental Modelling and Assessment* **3**, 19–29.

Hahn, B. and Engelen, G. (2000) Concepts of DSS systems, *Veranstaltungen* **4**, Bundesanstalt für Gewässerkunde, Koblenz, 9–44.

Huizing, J.J., Engelen, G., van de Ven, K., Pothof, I. and Uljee, I. (1998) *WadBOS: Een prototype van een kennissysteem voor beleidsanalyse van de Waddenzee*, Eindrapport, Directie Noord Nederland, Rijkswaterstaat, Leeuwarden.

Holtzman, S. (1989) *Intelligent Decision Systems*, Addison-Wesley, Reading, MA.

Kauffman, S.A. (1990) Requirements for evolvability in complex systems: orderly dynamics and frozen components, *Physica D* **42**, 135–152.

Klein, M.R. and Methie, L.B. (1995) *Knowledge-Based Decision Support Systems*, John Wiley & Sons, New York.

Marakas, G.M. (1998) *Decision Support Systems in the 21st Century*, Prentice Hall, Upper Saddle River, NJ.

Mulligan, M. (1998) *Modelling Desertification: EU Concerted Action on Mediterranean Desertification*, thematic report, King's College London, London.

Prigogine, I. (1981) *From Being to Becoming*, Freeman, San Francisco.

Rotmans, J. and de Vries, B. (1997) *Perspectives on Global Futures: The TARGETS Approach*, Harvard University Press, Cambridge, MA.

Salz, P., Dol, W. and Smit, W. (1994) *Schol Case: Economisch Model*, LEI-DLO, Den Haag.

Schielen, R.M.J. (2000) Recent history and future developments of decision support systems for the Dutch rivers, *Veranstaltungen* **4**, Bundesanstalt für Gewässerkunde, Koblenz, 77–87.

Sprague, R.H. Jr. and Carlson, E.D. (1982) *Building Effective Decision Support Systems*, Prentice Hall, Englewood Cliffs, NJ.

Sterman, J.D. (2000) *Business Dynamics: Systems Thinking and Modeling for a Complex World*, Irwin/McGraw-Hill, Boston, MA.

Uljee, I., Hahn, B., van der Meulen, M. and Engelen, G. (2000) *WadBOS Gebruikershandleiding*, National Institute for Coastal and Marine Management/RIKZ, Haren.

van Delden, H. (2000) *A Generic Approach for the Design of Decision Support Systems for River Basin Management*, University of Twente, Enschede.

15

Decision-Support Systems for Managing Water Resources

SOPHIA BURKE

15.1 INTRODUCTION

Chapter 14 addresses how decision-support systems can be used for policy making in the coastal zone and this chapter identifies how models are used within a decision-support tool for optimal management of water resources. Optimal water-resource management involves the identification of the best possible protection and efficient utilization of available water resources, whether for hydropower, water supply, agricultural use (such as irrigation and drainage) or for environmental management of freshwater or marine resource management. This chapter will outline why decision-support systems (DSS) are needed, how they are designed, and will give some brief examples of their use in water resource management. Further reviews of the use of DSS in water resources can be found in Loucks and Gladwell (1999), Simonovic (2000) and Watkins and McKinney (1995).

15.2 WHY ARE DSS NEEDED?

Water-resource managers are charged with the long-term optimal management, regulation and protection of water resources. However, it is recognized that water-resource managers must take into account the multitude of water resources development preferences that are put forward by stakeholders such as agriculturalists, water suppliers and environmental groups. As a result, conflicts over the utilization of a particular resource may occur. In recent years, there has been increasing interest in the potential of using integrated catchment management (ICM) as a means of sustainable management of water resources on

a catchment scale. ICM is a holistic, natural, resources-management system comprising interrelated elements of water, soils and other environmental and socio-economic factors, in a river basin, managed on an environmental and economic basis. The rationale of managing water resources on a basin scale instead of on a project-by-project basis at a subcatchment scale is based on the recognition that the water and land resources of a basin are integrated together and hence must be treated as such. Given the complexity of the hydrological system, the development and use of mathematical modelling are often necessary. However, to facilitate the identification of the best development strategy of catchments we need to identify and evaluate alternative water management plans and policies and to predict and understand their short- and long-term physical, economic and social impacts. An adequate understanding of the complex hydrological function of the resource and the limits of resource exploitation is necessary, given the environmental (soil, land use, climate), human (socio-economics, policy) and infrastructural constraints. In addition, we need to understand the impacts of additional pressures on water in the future due to population growth and land use development, climate variability and regulatory requirements. For example, the supply of water often does not keep up with demand, and in the face of increasing population and an uncertain climate, computer models can be used to simulate future scenarios for water-supply systems. The complexity of land-surface controls such as soils, climate and human activity on water-resource quality and quantity and the needs of the users for adequate and predictable supplies of water means that an improved approach to modelling

Environmental Modelling: Finding Simplicity in Complexity. Edited by J. Wainwright and M. Mulligan
© 2004 John Wiley & Sons, Ltd ISBNs: 0-471-49617-0 (HB); 0-471-49618-9 (PB)

is needed. Additionally, the use and interpretation of the hydrological model output can be difficult for the nonspecialist, particularly for nondeterministic models or where complex calibration procedures are needed. To incorporate the complexity into the decision-making process and to facilitate the practical use of existing complex models and technologies for decision-making, the development and use of decision-support systems is often necessary.

DSSs can be defined as 'interactive computer programs that utilize analytical methods, such as decision analysis, optimization algorithms, program scheduling routines, and so on, for developing models to help decision makers formulate alternatives, analyse their impacts, and interpret and select appropriate options for implementation' (Adelman, 1992). They are interactive frameworks built around established technologies that address the questions or issues pertaining to problems and contribute to option selection. They allow the user to use existing modelling technology from a range of disciplines and investigate the impacts of future scenarios. Traditional modelling assessments of water quality and quantity are joined by assessments of ecology, biodiversity, geomorphology and environmental economics, and the DSS should be able to integrate the output of these previously independent models of processes to aid decision-making.

Moreover DSS are based on formalized knowledge, and their application in the decision-making process provides the additional benefit that the final decisions are by definition as rational as possible, and the process of arriving at that decision is documented and reproducible (Vacik and Lexer, 2001).

15.3 DESIGN OF A DECISION-SUPPORT SYSTEM

The needs of the end-users must be taken into account during the early stages of design of the DSS (Wilson and Droste, 2000). Good communication between developers and end-users throughout development is imperative in order that it truly reflects the needs of the users (Loucks and da Costa, 1991; Argent and Grayson, 2001). In addition, advice is needed on the criteria of the post-model evaluation and optimization tools.

The availability and quality of appropriate data also need to be established at an early stage. The development of a DSS has the advantages of bringing together existing available environmental and economic data so that they can usefully be analysed together. One of the main problems in inter-disciplinary, regional decision-making where there is an environment of entrenched

bureaucracy is the lack of synthesis of even basic spatial information. The DSS is therefore also useful for the display and analysis of existing resource data as a user-friendly Geographic Information System (GIS), database or website. The DSS provides a synoptic view of the main elements of water resource management in a particular catchment, combining rainfall, land use and water quality with infrastructure, economic data and other sectors.

The DSS comprises a network of models, such as water-balance models and economic forecasting models. Often simpler models are favoured where there are a large number of models in the network so that design, computing and data demands are kept to a minimum. For example, Georgopoulou *et al.* (2001) used a DSS in the Mediterranean to investigate the physical and economic feasibility of extracting brackish groundwater for desalination together with artificial recharging the groundwater using treated wastewater to prevent further saline intrusion. Suitable models were therefore needed to describe each main process that affects the system or forms part of the proposed scheme. The elements described were therefore the hydrologic budget, natural recharge (infiltration) of the aquifer, a groundwater model that included seawater intrusion, water demands on the aquifer, a desalinization model, and recharge of the treated reclaimed water. The models ranged from complex, well-tested existing models (for the groundwater) to simple mathematical expressions (for cost functions) and historical data functions (for water demand).

The models in a DSS are linked to the database of existing information for the parameterization and validation of the models in the target region. The DSS may demand spatial and temporal data of different scales, such as hydrological model data at a hillslope scale and data for upscaling to the catchment scale. GIS are useful databases as these hold geographical data as well as tabular data and they are often used for the output of results for the user too. Data for possible future scenarios are also held in the database such as climate data from GCM (General Circulation Model) output, or forecasts of future water demands.

The DSS should have a user-friendly interface if it is to be readily adopted by the decision-maker. The complexity of this interface must be balanced with the ease of application. An example of a fairly sophisticated DSS is where the user can interact with both the model parameters and optimization models (see later discussion on optimization), such as that described and illustrated in Ostfeld *et al.* (2001). This DSS is called 'HANDSS' and was designed to assist in the operation of the canal network fed from the River Jordan in Israel to the Lake Hula wetlands. HANDSS consists of the MODFLOW groundwater model, visualization layer, a simulation layer and a management layer. The visualization layer uses GIS to provide spatial maps of solutions, the simulation layer is linked to the selection of model equations and the management layer allows the user to control the optimization procedure.

The user interface must be flexible enough to allow interaction with as many parameters as possible so that a series of scenarios can be fully tested by the user. The parameters that are affected by the end-users' decisions could be altered within limits, for example, land-use decisions or economic markets. A number of future scenarios for factors that are not directly influenced by the decision-maker, such as external markets or climate change, could also be selected. For example, Staudenrausch and Flugel (2001) have developed an application for Southern Africa, a region of complex political economy where unfavourable climates mean that renewable water resources are scarce. Scenarios of land-use and climate change were coupled with scenarios of population growth and policy forecasts and applied to a hydrological and socio-economic model linked by a rule-based expert system.

Users are encouraged to perform sensitivity analyses of the impacts of change, which greatly enhances the understanding the system behaviour. Through an iterative process, the user can explore the multiple impacts of various management options. For example, the user may investigate the sensitivity of the river regime to anthropic influences within the catchment such as irrigation, and changes in the sensitivity in the face of different climate change scenarios. The DSS is often menu-driven and uses graphics to display data input and output, which facilitates their use. The output is displayed in a meaningful way, for example, using maps and animations, and these can be quickly assimilated by the decision-maker and communicated to a variety of stakeholder groups (Welp, 2001).

The final step is to find the optimal solution for water resource management. Automated optimization tools are used in many DSS, and the optimization model is one of the most important components of the DSS. It comprises rule-based or linguistic algorithms that interrogate the DSS models. For example, the rules can be set to optimize the allocation of water for a given objective or set of objectives. The model is run many times within given constraints (economic, hydrological, cultural). The optimal solutions will fulfil the maximum number of objectives and the model run result will allow for sensitivity testing of the solutions and for

proposing and illustrating the trade-offs and the less than optimal solutions. For example, Ghoneim *et al.* (1996) use a simple linear programming optimization simulation model to find the best possible allocation of water in part of the Nile Valley. The optimality is based upon the planner's preferences for using different resources, the physical constraints, and the desired operation rules that constrain the model. Finding the optimal solution for multiple nonlinear objectives is explored by Eskandari *et al.* (1995), who describe a DSS of a pine-forested catchment in north-central Arizona. The DSS was used to test four forestry strategies to maximize objectives of water production, wildlife habitat, aesthetics, livestock and wood production and to minimize costs. Distance-minimizing methods are used, which are computed by first maximizing each objective individually and then calculating the feasible solution with minimal distance from the ideal point. However, the objectives will vary according to interest groups (water users, livestock producers, foresters, environmentalists and land-use planners), and so after consultation with these groups the objectives were defined by a distribution function, the form of which varies according to its relative importance to each group. Five different methods of optimization by distance minimization are used. They found that for each optimization method, different alternative management techniques appeared to be the best.

DSS is most useful where there are a large number of objectives for assimilation. Optimal solutions therefore demand the exact preference of each objective. Twery and Hornbeck (2001) analysed forests in the northeastern United States with respect to water (quantity and quality), wildlife, timber production, visual quality and general ecological goals. Model results showed that some of the desired objectives were linked, for example, water and ecological goals, and others were not compatible with each other (see also Chapter 17). Vacik and Lexer (2001) use an algorithm for comparing goals. They apply a DSS to forest management in Vienna, also using a large number of management objectives, including water production, timber production, conservation of biodiversity, recreation objectives and rock stability. The decision model decomposes the objectives into several hierarchically structured decision criteria based on partial objectives. Qualitative expert knowledge is used to compute preference values, and the preferences are aggregated additively at each hierarchical level.

Improved optimization techniques have come to the fore in water management models. One example is Dynamic Programming, which is not dependent on the shape of the objective function as it is discretized into a finite set of stages, with an optimal solution at each stage, although this is computationally expensive. Second, there is a growing interest in evolutionary or genetic algorithms (for example, Booty *et al.*, 2001), based on Darwinian evolutionary theory. The solution space is evaluated for fitness and the fittest of the populations are selected and combined to produce a new generation of solution points with increased likelihood of increased fitness. The process is iterated until the 'gene' with the best fitness value meets the pre-defined convergence criteria.

15.4 CONCLUSION

Decision-support systems are a useful platform for allowing nonexperts to exploit state-of-the-art hydrological modelling technology. They allow the visualization of the holistic water-management framework and can aid policy decision-making where complexities of the water resources management demand a number of factors to be taken into account. In the face of climatic, environmental and socio-economic change, water resources management will become more complex and DSS allows the impacts of decisions on the system under a number of scenarios to be analysed. The DSS must be tested and robust enough to be passed safely into the hands of nonexperts but also be flexible enough to future-proof the decision-making environment.

15.5 DSS ON THE WEB

There are water resources DSS freely available on the Internet for academic/research purposes. One such package was released by IIASA (International Institute for Applied Systems Analysis) and IWEP (Institute for Water and Environmental Problems) in 1996 and is called DESERT (DEcision Support system for Evaluation of River basin sTrategies). The current website address is http://www.iiasa.ac.at/Research/WAT/docs/desert.html. This was designed for decision support for water-quality management on a river-basin scale. More information can be found in the manual, which is downloadable from this website.

REFERENCES

Adelman, L. (1992) *Evaluating Decision Support and Expert Systems*, John Wiley & Sons, New York.
Argent, R.M. and Grayson, R.B. (2001) Design of information systems for environmental managers: an example using interface prototyping, *Environmental Modelling and Software with Environment Data News* **16** (5), 433–438.

Booty, W.G., Lam, D.C.L., Wong, I.W.S. and Siconolfi, P. (2001) Design and implementation of an environmental decision support system, *Environmental Modelling and Software with Environment Data News* **16** (5), 453–458.
Eskandari, A., Ffolliott, P. and Szidarovsky, F. (1995) Decision support system in watershed management, in T.J. Ward (ed.) *Watershed Management: Planning for the 21st Century*, American Society of Civil Engineers, New York.
Georgopoulou, E., Kotronarou, A., Koussis, A., Restrepo, P.J., Gomez-Gotor, A. and Rodriguez Jimenez, J.J. (2001) A methodology to investigate brackish groundwater desalination coupled with aquifer recharge by treated wastewater as an alternative strategy for water supply in Mediterranean areas, *Desalination* **136** (1), 307–315.
Ghoneim, G.A., El-Terzy, A.I., Bahjat, A.M. and Fontane, D.G. (1996) Optimal operational guidelines for conjunctive use of multi-water resources system, in M.A. Abu-Zeid and A.K. Biswas (eds) *River Basin Planning and Management*, Water Resources Management Series 4, Oxford University Press, Calcutta.
Loucks, D.P. and da Costa, J.R. (1991) Computer aided decision support in water resources planning and management, in D.P Loucks and J.R. da Costa (eds) *Decision Support Systems, Water Resources Planning*, NATO ASI Series G, Vol. 26, Springer-Verlag, Berlin.
Loucks, D.P. and Gladwell, J.S. (1999) *Sustainability Criteria for Water Resource Systems*, International Hydrology Series, Cambridge, Cambridge University Press.
Ostfeld, A., Muzaffar, E. and Lansey, K. (2001) HANDSS: The Hula Aggregated Numerical Decision Support System, *Journal of Geographic Information and Decision Analysis* **5** (1), 16–31 (also available for download at http://www.geodec.org/Ostfeld.pdf).
Simonovic, S.P. (2000) Tools for water management: one view of the future, *Water International* **25** (1), 76–88.
Staudenrausch, H. and Flugel, W.A. (2001) Development of an integrated water resources management system in southern African catchments, *Physics and Chemistry of the Earth, Part B: Hydrology, Oceans and Atmosphere* **26** (7), 561–564.
Twery, M.J. and Hornbeck, J.W. (2001) Incorporating water goals into forest management decisions at a local level, *Forest Ecology and Management* **143** (1), 87–93.
Vacik, H. and Lexer, M.J. (2001) Application of a spatial decision support system in managing the protection forests of Vienna for sustained yield of water resources, *Forest Ecology and Management* **143** (1), 65–76.
Watkins, D.W. and McKinney, D.C. (1995) Recent developments associated with decision support systems in water resources, *Reviews of Geophysics Supplement 33, American Geophysical Union* (also available at http://earth.agu.org/revgeophys/watkin00/watkin00.html).
Welp, M. (2001) The use of decision support tools in participatory river basin management, *Physics and Chemistry of the Earth, Part B: Hydrology, Oceans and Atmosphere* **26**, 535–539.
Wilson, D.J. and Droste, R.L. (2000) Design considerations for watershed management decision support systems, *Water Quality Research Journal of Canada* **35**, 2–26.

16

Soil Erosion and Conservation

MARK A. NEARING

16.1 THE PROBLEM

Accelerated soil erosion induced by human activities is the principal cause of soil degradation across the world. The main culprit behind the problem is agriculture, and at stake is the long-term viability of the agricultural production capacity of the planet. Barring major, unknown scientific advances in the future, and if soil erosion and population growth remain unchecked from their current rates, humanity will eventually lose the ability to feed itself. Another significant problem associated with soil erosion is off-site sediment pollution. Costs associated with the delivery of sediment to streams and other water bodies worldwide are huge (e.g. Pimental, 1995). This chapter will focus on models of soil erosion as they are used for purposes of soil conservation. In particular, we focus here exclusively on soil erosion by water (see also Chapter 9). Models of other agricultural erosion processes, such as wind erosion and tillage erosion, are certainly important, but they will not be addressed here.

Models can be used in conservation work for three primary purposes: (a) to help a land owner or manager choose suitable conservation practices from among alternatives; (b) to make large-scale erosion surveys in order to understand the scope of the problem over a region and to track changes in erosion over time; and (c) to regulate activities on the land for the purposes of conservation compliance.

In selecting or designing an erosion model, a decision must be made as to whether the model is to be used for on-site or off-site concerns, or both. On-site concerns are generally associated with degradation or thinning of the soil profile in the field, which may become a problem of crop-productivity loss. Conservationists refer to this process as soil loss, referring to the net loss of soil over only the portion of the field that experiences net loss over the long term. Areas of soil loss end where net deposition begins. Off-site concerns, on the other hand, are associated with the sediment that leaves the field, which we term here sediment yield. In this case, we are not necessarily concerned with the soil loss, or for that matter the amount of sediment deposited prior to leaving the field, although estimation of both of these may be used to estimate sediment yields. Ideally, a model will compute soil loss, deposition and sediment yield, and thus have the capability to address both on-site and off-site issues.

Data variability and model uncertainty are two related and important issues associated with the application of erosion models. Data from soil-erosion plots contain a large amount of unexplained variability, which is an important consideration for using erosion data to evaluate soil-erosion models, as well as for interpreting erosion data. This variability is due both to natural causes and measurement errors. When comparing measured rates of erosion to predicted values, a portion of the difference between the two will be due to model error, but a portion will also be due to unexplained variance of the measured sample value from the representative, mean value for a particular treatment.

Knowledge of variability in soil-erosion data, however, is somewhat limited, though recent studies have enlightened us to some degree. Only one experimental erosion study to date has been conducted with a sufficient number of replicated erosion plots to allow an in-depth analysis of variability. Wendt *et al.* (1986) measured soil-erosion rates on 40 cultivated, fallow, experimental plots located in Kingdom City, MO, in 1981. All the 40 plots were cultivated and in other ways treated identically. The coefficients of variation for the

Environmental Modelling: Finding Simplicity in Complexity. Edited by J. Wainwright and M. Mulligan
© 2004 John Wiley & Sons, Ltd ISBNs: 0-471-49617-0 (HB); 0-471-49618-9 (PB)

25 storms ranged from 18% to 91%, with 15 of the storms falling in the range of less than 30%. The more erosive storms tended to show the lesser degree of variability. Of the 15 storms with mean erosion rates of greater than $0.1\,kg\,m^{-2}$ ($1.0\,Mg\,ha^{-1}$), 13 showed coefficients of variation of less than 30%. The results of the study indicated that 'only minor amounts of observed variability could be attributed to any of several measured plot properties, and plot differences expressed by the 25 events did not persist in prior or subsequent runoff and soil loss observations at the site'.

Ruttimann *et al.* (1995) reported a statistical analysis of data from four sites, each with five to six reported treatments. Each treatment had three replications. Reported coefficients of variation of soil loss ranged from 3.4% to 173.2%, with an average of 71%. The authors concluded by suggesting 'as many replications as possible' for erosion experiments.

Nearing *et al.* (1999) studied erosion variability using data from replicated soil-loss plots from the Universal Soil Loss Equation (USLE) database. Data from replicated plot pairs for 2061 storms, 797 annual erosion measurements and 53 multi-year erosion totals were used. They found that the relative differences between replicated plot pair measurements tended to decrease as the magnitude of the measured soil loss increased. Using an assumption that soil-loss magnitude was the principal factor in explaining variance in the soil-loss measurements, the authors were able to calculate the coefficient of variation of within-treatment, plot replicate values of measured soil loss. Variances between replicates decreased as a power function ($r^2 = 0.78$) of measured soil loss, and were independent of whether the measurements were event-, annual- or multi-year values. Values of the coefficient of variability ranged from nearly 150% for a soil loss of $0.1\,kg\,m^{-2}$ to as low as 18% or less for soil-loss values greater than $10\,kg\,m^{-2}$.

One important question for scientists is: 'How do we know when an erosion model is working adequately?' Given that the data are highly variable, when we ask the question about how well a model works, the answer is not so simple. One cannot just compare the model output to an erosion rate. One must simultaneously ask the question: 'How variable is nature?'

Risse *et al.* (1993) applied the USLE to 1700 plot years of data from 208 natural runoff plots. Annual values of measured soil loss averaged $3.51\,kg\,m^{-2}$ with an average magnitude of prediction error of $2.13\,kg\,m^{-2}$, or approximately 60% of the mean. Zhang *et al.* (1996) applied the Water Erosion Prediction Project (WEPP) computer simulation model to 290 annual values and

obtained an average of $2.18\,kg^{-2}$ for the measured soil loss, with an average magnitude of prediction error of $1.34\,kg^{-2}$, or approximately 61% of the mean. In both cases the relative errors tended to be greater for the lower soil-loss values. Given these results and others from similar types of studies (Liu *et al.*, 1997; Rapp, 1994; Govers, 1991), the question may be asked: are the predictions 'good enough' relative to measured data? What is an acceptable and expected level of model prediction error?

One manner in which we can address this problem is to think of the replicated plot as the best possible 'real-world, physical model' of soil erosion. As such, one might further consider that the physical model represented by the replicate plot represents essentially a 'best-case' scenario in terms of erosion prediction, which we can use as a baseline with which the performance of erosion prediction models might be compared. Using, as discussed above, data from natural runoff plots from the USLE plot database, Nearing (2000) suggested a basis for an erosion-model evaluation method using the idea of the replicate plot as a physical model of the replicated plot. He suggested that if the difference between the model prediction and a measured plot data value lies within the population of differences between pairs of measured values, then the prediction is considered 'acceptable'. A model 'effectiveness' coefficient was defined for studies undertaken on large numbers of prediction vs. measured data comparisons. The method provides a quantitative criterion for taking into account natural variability and uncertainty in measured erosion-plot data when those data are used to evaluate erosion models.

Nearing (2000) outlines the specific procedures for how erosion model evaluation can be done in the presence of data uncertainty. The method is straightforward, but requires some detail in the computations. Using similar arguments with the erosion plot replicates data, but using a slightly less complex analysis, we can achieve a rule-of-thumb measure of model validity simply by looking at the coefficient of determination for the regression line between measured and predicted soil loss values. Using measured soil loss data pairs from 3007 storms (overlapping with some of the same data used in the pre-mentioned studies), Nearing (1998) obtained a coefficient of determination between measured and predicted soil loss of 0.77. One certainly would not expect (on uncalibrated data) to obtain results between model predictions and measured data substantively better than this, and for all practical purposes expectations of fit must be less. In the study by Risse *et al.* (1993) using the USLE and 1700+ plot years of data, the overall coefficients of determination were 0.58 for annual values and 0.75 for annual average soil-loss data. In the study by Zhang *et al.* (1996), the WEPP model was applied using data from 4124 storm events, the coefficients of determination were 0.36 for the individual storms, 0.60 for annual values, and 0.85 for annual average soil loss values. The observation that the fit improves from storm to annual to average annual predictions reflects the trend that data variability decreases with increasing soil-loss magnitudes, as discussed above.

Given that we know, based on the data from erosion plots, that soil erosion is highly variable, and then using the information on variability to set limits on the ability of models to predict soil-erosion rates, the question then becomes one of utility. Is the model accurate enough to solve our problems? We will address this question later in this chapter. But first we need to look at the models themselves, and look at an example of how an erosion model might be used to solve a problem.

16.2 THE APPROACHES

Erosion models used in application for conservation planning fall into two basic categories: empirical and process-based. Undoubtedly the prime example of an empirically based model is the Universal Soil Loss Equation (USLE), which was developed in the United States during the 1950s and 1960s (Wischmeier and Smith, 1965, 1978). This equation has been adapted, modified, expanded and used for conservation purposes throughout the world (e.g. Schwertmann *et al.*, 1990; Larionov, 1993).

The USLE was originally based on statistical analyses of more than 10 000 plot-years of data collected from natural runoff plots located at 49 erosion research stations in the United States, with data from additional runoff plots and experimental rainfall simulator studies incorporated into the final version published in 1978 (Wischmeier and Smith, 1978). The large database upon which the model is based is certainly the principal reason for its success as the most used erosion model in the world, but its simplicity of form is also important:

$$A = R\,K\,L\,S\,C\,P \qquad (16.1)$$

where A ($t\,ha^{-1}\,a^{-1}$) is average annual soil loss over the area of hillslope that experiences net loss, R ($MJ\,mm\,h^{-1}\,ha^{-1}\,a^{-1}$) is rainfall erosivity, K ($t\,hr\,MJ^{-1}\,mm^{-1}$) is soil erodibility, L (unitless ratio) is the slope length factor, S (unitless ratio) is the slope steepness factor, C (unitless ratio) is the cropping factor, and P (unitless ratio) is the conservation practices factor. Terminology is important here. Note first that the

USLE predicts soil loss (see discussion above) and not sediment yield. Second, the word *erosivity* is used to denote the driving force in the erosion process (i.e., rainfall in this case) while the term *erodibility* is used to note the soil resistance term. These two terms are not interchangeable. Third, the model predicts average annual soil loss: it was not intended to predict soil loss for storms or for individual years. Conservationists often describe the predictions as long term, while from the geomorphic perspective the predictions would be referred to as medium term (Govers, 1996).

The units of the USLE appear a bit daunting as written (Equation 16.1), but become somewhat clearer with explanation. The units were originally written, and are still used in the United States, as imperial, but conversion to metric is generally straightforward (Foster *et al.*, 1981). The key to understanding the dimensional units lies with the definition of rainfall erosivity and the concept of the unit plot. Wischmeier (1959) found for the plot data that the erosive power of the rain was statistically best related to the total storm energy multiplied by the maximum 30-minute storm intensity. Thus, we have the energy term (MJ) multiplied by the intensity term ($mm\,h^{-1}$) in the units of R, both of which are calculated as totals per hectare and per year. The unit plot was defined as a standard of 9% slope, 22.13 m length,[1] and left fallow (cultivated for weed control). The K value was defined as A/R for the unit plot. In other words, erodibility was the soil loss per unit value of erosivity on the standard plot. The remaining terms, L, S, C and P, are ratios of soil loss for the experimental plot to that of the unit plot. For example, the C value for a particular cropped plot is the ratio of soil loss on the cropped plot to the value for the fallow plot, other factors held constant.

The USLE reduced a very complex system to a quite simple one for the purposes of erosion prediction. There are many complex interactions within the erosional system which are not, and cannot be, represented within the USLE. We will illustrate a few of these interactions below. On the other hand, for the purposes as stated above for which an erosion model is used, the USLE has been, and still can be, very successful. This issue is also discussed below in more detail.

The USLE was upgraded to the Revised Universal Soil Loss Equation (RUSLE) during the 1990s (Renard *et al.*, 1997). RUSLE is a hybrid model. Its basic structure is the multiplicative form of the USLE, but it also has many process-based auxiliary components. It is computer based, and has routines for calculating time-variable soil erodibility, plant growth, residue management, residue decomposition and soil surface roughness as a function of physical and biological processes. RUSLE also has updated values for erosivity (R), new relationships for L and S factors which include ratios of rill and inter-rill erosion, and additional P factors for rangelands and subsurface drainage, among other improvements. RUSLE has the advantage of being based on the same extensive database as is the USLE, with some of the advantages of process-based computations for time-varying environmental effects on the erosional system. It still has the limitations, however, in model structure that allows only for limited interactions and inter-relationships between the basic multiplicative factors of the USLE (Equation 16.1).

Various process-based erosion models have been developed in the past ten years including EUROSEM in Europe (Morgan *et al.*, 1998), the GUEST model in Australia (Misra and Rose, 1996), and the WEPP model in the United States (Flanagan and Nearing, 1995). We will focus here on the example of the WEPP model, largely because it is the technology most familiar to the author.

The WEPP profile computer model includes seven major components, including climate, infiltration, water balance, plant growth and residue decomposition, surface runoff, erosion and channel routing for watersheds. The climate component of the profile computer model (Nicks, 1985) generates daily precipitation, daily maximum and minimum temperature, and daily solar radiation based on a statistical representation of weather data at a particular location. The climate model has been tested for erosion and well parameterized for the United States (Baffaut *et al.*, 1996). The infiltration component of the hillslope model is based on the Green and Ampt equation, as modified by Mein and Larson (1973), with the ponding time calculation for an unsteady rainfall (Chu, 1978). The water balance and percolation component of the profile model is based on the water balance component of SWRRB (Simulator for Water Resources in Rural Basins) (Williams and Nicks, 1985; Arnold *et al.*, 1990), with some modifications to improve estimation of percolation and soil evaporation parameters. The plant-growth component of the model simulates plant growth and residue decomposition for cropland and rangeland conditions. The residue and root decomposition model simulates decomposition of surface residue (both standing and flat), buried residue, and roots for the annual crops specified in the WEPP User Requirements (Flanagan and Livingston, 1995) plus perennial crops of alfalfa and grasses. Surface runoff is calculated using a kinematic wave equation. Flow is partitioned into broad sheet flow for inter-rill erosion calculations and concentrated

flow for rill erosion calculations. The erosion component of the model uses a steady-state sediment continuity equation that calculates net values of detachment or deposition rates along the hillslope profile (Nearing *et al.*, 1989). The erosion process is divided into rill and inter-rill components where the inter-rill areas act as sediment feeds to the rills, or small channel flows. The model is applicable to hillslopes and small watersheds.

Because the model is based on all of the processes described above, and more, it is possible with WEPP to have an enormous array of possible system interactions represented in the simulations. Just to name a very few examples, slope-length and steepness effects are functions of soil consolidation, surface sealing, ground residue cover, canopy cover, soil water content, crop type and many other factors. Ground residue cover is a function of biomass production rates, tillage implement types, residue type, soil moisture, temperature and solar radiation, previous rainfall and many other factors. Rill-erosion rates are a function of soil-surface roughness, ground cover, consolidation of the soil, soil physical and chemical properties, organic matter, roots, inter-rill erosion rates, slope, and runoff rates, among other factors. The lists continue *ad infinitum*. These are interactions which are simply not possible to represent by an empirical model. WEPP is a very complex model in this sense.

The disadvantage of the process-based model is also the complexity of the model. Data requirements are huge, and with every new data element comes the opportunity to introduce uncertainty, as a first-order error analysis would clearly indicate. Model structure interactions are also enormous in number, and with every structural interaction comes the opportunity for error, as well. In a sense, the goal in using the process-based model is to capture the advantages of the complexity of model interactions, while gaining the accuracy and dependability associated with the simpler empirically based model. This goal can be achieved, and was achieved with the WEPP model, using a combination of detailed sensitivity analyses and calibration of the model to the large database of natural runoff-plot information used to develop the USLE and RUSLE. Without the tie between model and database, and without knowledge of the sensitive input variables so as to know where to focus efforts, turning a complex model such as WEPP into a useful conservation tool would not be possible. Thus in a sense, even though WEPP routines are process-based descriptors of various components of the erosional system, ultimately the model must be empirically based on the same type of data as was used to develop the USLE and RUSLE, along with additional experimental data collected specifically for WEPP.

16.3 THE CONTRIBUTIONS OF MODELLING

The accuracy of the three models introduced above has been tested using measured soil-loss data from plots. We mentioned above the study by Risse *et al.* (1993) using the USLE and 1700+ plot years of data, and the study by Zhang *et al.* (1996) of the WEPP model using data from 4124 storm events. The data of Risse *et al.* (1993) were also applied to the RUSLE model with very similar levels of accuracy as obtained with the USLE (Rapp, 1994). These three models all produced essentially equivalent levels of accuracy for prediction of soil loss, and the level was somewhat less than the level of fit obtained with the 'best-case' replicate plot-model discussed above. The results suggest that we have approached with these models the maximum level of possible soil-loss accuracy for ungauged, uncalibrated sites.

This does not imply, however, that the three models are equivalent in usage. RUSLE has certain advantages over the USLE because its database and internal relationships have been expanded beyond that of the USLE for particular applications such as rangelands in the western United States and no-till cropped lands in the eastern United States. The data comparisons reported in the three studies above included no rangeland data and very little no-till data, so these advantages were not apparent from those studies. The USLE may have advantages, however, in other applications. In areas where data are few, or computations need to be kept simple, the USLE has distinct advantages over both RUSLE and WEPP.

Another category of differences between the models is the type of information provided, rather than the accuracy of the information. USLE provides essentially only average annual soil loss over the area of the field experiencing net loss. RUSLE also provides only average annual values of erosion, however, it provides estimates of off-slope sediment delivery in addition to estimates of on-slope soil loss. RUSLE can also provide estimates of certain auxiliary system variables, such as residue amounts and crop yields. The WEPP model provides a massive array of system information to the user, if such information is desired. The model predicts both on-site soil loss and off-site sediment delivery, including ephemeral gully erosion, which neither USLE nor RUSLE attempts to predict. Sediment delivery information includes not just the amount of sediment yield, but the particle-size distribution information

for that sediment, which can be important in terms of chemical transport by sediment. WEPP also provides a detailed description of the spatial and temporal distributions of soil loss, deposition and sediment yields, both along the hillslopes and across the watershed. Auxiliary system information from WEPP is enormous, and is available on a daily basis. Information includes soil-water content with depth, surface residue amounts and coverage in both rill and inter-rill areas separately, buried residue and root masses, canopy cover and leaf area index, evapotranspiration rates, soil-surface roughness, soil bulk density, changes in hydraulic conductivities of the soil surface layer, changes in soil erodibility with consolidation and surface sealing, crop biomass and yields, subsurface interflow of water, tile drainage, and surface runoff amounts and peak rates, among others.

USLE, RUSLE and WEPP (or other process-based models) constitute a complementary suite of models to be chosen to meet the specific user need. To illustrate this idea, we will take a look at recent applications of the USLE and WEPP to address the question of the potential impact of climate change on erosion rates in the United States. As we will see, we are able to use the USLE to provide certain information which WEPP simply cannot provide because of the restrictions of model complexity, and we are able to use the WEPP model in a way where only the complex model interactions will provide us with the information we want regarding system response.

In the first study we used the RUSLE R-factor to estimate the potential changes during the next century for rainfall erosivity across the whole of the United States, southern Canada and northern Mexico. In this case, we do not want to become embroiled in the subtle differences between effects of various soils, slopes, cropping systems and other system variables. Instead, we are looking for the primary effects over regions. With the USLE and RUSLE we can do this, because RUSLE uses an R-factor that was derived from a wide array of plot conditions, and it is not interdependent with the other system variables. Also, there have been developed, as we will see, statistical relationships between general precipitation data and erosivity. If we attempted to conduct such a large-scale study with the WEPP model, we would quickly find ourselves with complicated sets of analyses, which we would then need to compose back to the general trends that RUSLE and the USLE provide directly. There would also be a data problem in this case, because WEPP requires certain details of precipitation which are not available from the Global Circulation Models used to predict future climate change.

In the second study we review here, the objective was to determine the specific effects of changes in rainfall erosivity that might occur as a function of changes in the number of rain days in the year versus erosivity changes that are expected to occur when precipitation amounts per day and associated rainfall intensities change. In this study, the USLE and RUSLE would have been largely ineffective, because these changes are related to process changes within the system which USLE and RUSLE do not take into account. We shall find that in this case the detailed process interactions within WEPP enable us to see some quite interesting and important system interactions which significantly affect the results.

16.3.1 Potential changes in rainfall erosivity in the USA during the twenty-first century

Soil-erosion rates may be expected to change in response to changes in climate for a variety of reasons, including, for example, changes in plant-biomass production, plant-residue-decomposition rates, soil microbial activity, evapotranspiration rates, soil-surface sealing and crusting, as well as shifts in land use necessary to accommodate a new climatic regime (Williams *et al.*, 1996). However, the direct, and arguably the most consequential, effect of changing climate on erosion by water can be expected to be the effect of changes in the erosive power, or erosivity, of rainfall. Studies using WEPP (Flanagan and Nearing, 1995) have indicated that erosion response is much more sensitive to the amount and intensity of rainfall than to other environmental variables (Nearing *et al.*, 1990). Warmer atmospheric temperatures associated with potential greenhouse warming of the Earth are expected to lead to a more vigorous hydrological cycle, with the correspondent effect of generally more extreme rainfall events (IPCC, 1995). Such a process may already be taking place in the United States. Historical weather records analysed by Karl *et al.* (1996) indicate that since 1910 there has been a steady increase in the area of the United States affected by extreme precipitation events (>50.8 mm in a 24-hour period). According to statistical analyses of the data, there is less than one chance in a thousand that this observed trend could occur in a quasi-stationary climate. Karl *et al.* (1996) also observed in the weather records an increase in the proportion of the country experiencing a greater than normal number of wet days.

Atmosphere–Ocean Global Climate Models (see Chapter 2) also indicate potential future changes in rainfall patterns, with changes in both the number of wet days and the percentage of precipitation coming in intense convective storms as opposed to longer duration, less intense storms (McFarlane *et al.*, 1992; Johns *et al.*, 1997).

Rainfall erosivity is known to be strongly correlated to the product of the total energy of a rainstorm multiplied by the maximum 30-minute rainfall intensity during a storm (Wischmeier, 1959). The relationship first derived by Wischmeier has proved to be robust for use in the United States, and is still used today in the Revised Universal Soil Loss Equation (Renard *et al.*, 1997).

A direct computation of the rainfall erosivity factor, R, for the RUSLE model requires long-term data for rainfall amounts and intensities. Current global circulation models do not provide the requisite details for a direct computation of R-factors (McFarlane *et al.*, 1992; Johns *et al.*, 1997). However, the models do provide scenarios of monthly and annual changes in total precipitation around the world. Renard and Freimund (1994) recently developed statistical relationships between the R-factor and both total annual precipitation at the location and a modified Fournier coefficient (Fournier, 1960; Arnoldus, 1977), F, calculated from monthly rainfall distributions.

The case study which we want to examine here was conducted by Nearing (in press), who recently used the erosivity relationships developed by Renard and Freimund (1994) to estimate the potentials for changes in rainfall erosivity in the United States during the twenty-first century under global climate change scenarios generated from two coupled Atmosphere–Ocean Global Climate Models. The two coupled Atmosphere–Ocean Global Climate Models from which results were used were developed by the UK Hadley Centre and the Canadian Centre for Climate Modelling and Analysis.

The most current UK Hadley Centre model, HadCM3 (Wood *et al.*, 1999; Gordon *et al.*, 2000; Pope *et al.*, 2000), is the third generation of Atmosphere–Ocean Global Climate Models produced by the Hadley Centre. It simulates a 1% increase in greenhouse gases for the time period studied, as well as the effects of sulphate aerosols. The model also considers the effects of the minor trace gases CH_4, N_2O, CFC-11, CFC-12 and HCFC-22 (Edwards and Slingo, 1996), a parameterization of simple background aerosol climatology (Cusack *et al.*, 1998) and several other improvements over the previous Hadley Centre model, HadCM2. Results from the model are reported on a 2.5° latitude by 3.75° longitude grid.

The Canadian Global Coupled Model, CGCM1 (Boer *et al.*, 2000), is composed of an atmospheric component based on the model GCMII (McFarlane *et al.*, 1992) coupled with an ocean component based on the model GFDL MOM1.1 (Boer *et al.*, 2000). For the current study we used results from the simulation GHG + A1, which incorporated an increase of atmospheric concentration of greenhouse gases (GHG) corresponding to an increase of 1% per year for the time period studied, as well as the direct forcing effect of sulphate aerosols (Reader and Boer, 1998). The data from this model were presented on a Gaussian 3.75° by 3.75° grid.

Changes in rainfall erosivity for the two models were computed for two time intervals, 40 and 80 years. In the first case the values of erosivity for the 20-year period from 2040 to 2059 were compared with the period 2000–2019, and in the second case, the values of erosivity for the 20-year period from 2080 to 2099 were compared with the period 2000–2019. Erosivity changes were computed in two ways: (a) as a function of change in average annual precipitation for the 20-year periods using equations 11 and 12 from Renard and Freimund (1994); and (b) as a function of the Fournier coefficient for the 20-year periods using equations 13 and 14 from Renard and Freimund (1994).

The erosivity results calculated from the Hadley Centre model analyses indicated a general increase in rainfall erosivity over large parts of the eastern United States, including most of New England and the mid-Atlantic states as far south as Georgia, as well as a general increase across the northern states of the USA and southern Canada (see maps in Nearing, 2000). The Hadley Centre results also indicated a tendency for erosivity increases over parts of Arizona and New Mexico. Decreases in erosivity were indicated in other parts of the south-western USA, including parts of California, Nevada, Utah and western Arizona. Decreases were also shown over eastern Texas and a large portion of the southern central plains from Texas to Nebraska.

The erosivity results calculated from the Canadian Centre for Climate Modelling and Analysis model also showed an increase in erosivity across the northern states of the USA, including New England, and southern Canada (see maps in Nearing, 2001). The Canadian Centre model results also indicated a reduction in erosivity across much of the southern plains, again from Texas to Nebraska, but extending somewhat west of the corresponding area shown in the Hadley Centre results. The Canadian Centre model did not show consistent results for the south-eastern United States. Results of computations using the annual precipitation (see maps in Nearing, 2000) indicate changes in parts of the south-east USA tending toward lower erosivity, corresponding to a tendency toward a decrease in the annual precipitation in that region. Results of the erosivity computations using the Fournier coefficient indicate the possibility of little change or increases over part of the region for the 80-year comparison (see maps in Nearing, 2001). Calculated increases in erosivity using the Fournier coefficient suggest a change in the distribution of rainfall patterns through the year.

Erosivity results calculated from the Canadian Centre for Climate Modelling and Analysis and the Hadley Centre models show major differences in the south-western United States, including California, Arizona, Nevada and Utah. Whereas the Hadley Centre model results suggest a definite trend towards lower erosivity in this area, the Canadian Centre for Climate Modelling and Analysis model results suggest a definite, strong trend toward greater erosivity through the twenty-first century.

The amount of inconsistency in the calculations from the two methods of calculating erosivity trends was, for the most part, similar between the two models (Table 16.1). Overall, between 16 and 20% of the calculations resulted in negative values of the R-factor calculated from total annual rainfall, RP, when the R-factor calculated from the Modified Fournier coefficient, RF, was positive, or vice versa. For the cases where both RP and RF were large, i.e., greater than 10%, those

Table 16.1 Percentages of map grid cells in which changes over time in erosivity values, RP, calculated using precipitation were inconsistent in sign with changes in the values of erosivity, RF, calculated using the Fournier coefficient

Model scenario	Inconsistencies in erosivity between RP and RF			
	For all data		Where also both |RP| and |RF| < 10%	
	40-yr time interval (%)	80-yr time interval (%)	40-yr time interval (%)	80-yr time interval (%)
HadCM3	17.2	16.2	1.0	1.5
CGCM1 HG + A1	17.4	19.4	0.7	7.6

percentages were much smaller, although 7.6% of the pairs were inconsistent in this case for the Canadian model results for the 80-year time interval (2000–2019 to 2080–2099). It is not out of the question to expect inconsistencies between results of RP and RF, since RP is based on total annual precipitation and RF is based on the monthly distributions of precipitation. Both relationships are statistically based, and we have no reason to favour one over the other.

One might expect a consistent trend for the change of erosivity as a function of time, and in general this was true (Table 16.2). In this case, the Canadian model exhibited more inconsistency as a function of time when using the monthly precipitation values to calculate erosivity, though it was consistent temporally in terms of the erosivity calculated using the annual precipitation.

The RF values tended to show a somewhat greater magnitude, in terms of the average of the absolute value of percentage erosivity change, than did the RP values (Table 16.3). The difference between the two models in this regard was striking. The Canadian model indicated a much greater level of erosivity

changes overall as compared to the Hadley Centre model (Table 16.3). Both models suggested erosivity changes which generally increased in magnitude from the 40-year to the 80-year comparison.

16.3.2 Effects of precipitation intensity changes versus number of days of rainfall

Now we take a look at another study of the effects of precipitation changes on soil-erosion rates, but this time we use the WEPP model. As we mentioned above, historical weather records analysed by Karl *et al.* (1996) indicate that since 1910 there has been a steady increase in the area of the United States affected by extreme precipitation events as well as an increase in the proportion of the country experiencing a greater than normal number of wet days. The results given by Nearing (2001) discussed above provide a broad view of expected changes in erosivity based on the statistical models, but an important question not addressed is the expected differences in erosivity that come about relative to rainfall intensity versus a simple increase in the average number of rain days in a year. Erosion is not

Table 16.2 Percentages of map grid cells in which changes over time in erosivity values calculated over the 40-year time interval were inconsistent in sign with changes in the values of erosivity calculated over the 80-year time interval

| Model scenario | Inconsistencies in erosivity between 40- and 80-year time intervals | | | |
| | For all data | | Where also both the 40 yr \|R\| and the 80 yr \|R\| < 10% | |
	RP (%)	RF (%)	RP (%)	RF (%)
HadCM3	16.2	15.2	1.5	1.0
CGCM1 HG + A1	7.6	23.6	0	5.6

Table 16.3 Average magnitudes (absolute values) of erosivity change calculated

| Model scenario | Average magnitude of change | | | |
| | 40-yr time interval | | 80-yr time interval | |
	RP (%)	RF (%)	RP (%)	RF (%)
HadCM3	11.8	16.5	15.9	20.9
CGCM1 HG + A1	23.4	29.1	53.4	58.3

linearly proportional to rainfall intensity (Wischmeier and Smith, 1978; Nearing *et al.*, 1990).

Pruski and Nearing (2002) recently performed computer simulations to obtain estimates of potential runoff and soil loss changes as a function of precipitation changes. In particular, they studied the different responses of the erosional system to changes in precipitation as they occurred with changes in rainfall intensities, including the amount of rainfall which occurs on a given day of rain, versus responses to changes in simply the average number of days of rain. Assessments were made using WEPP for several combinations of geographic locations, soils, crops and slopes. Geographic locations included West Lafayette, IN, Temple, TX, and Corvallis, OR. Soils were sandy clay loam, silt loam and clay loam. Crops included grazing pasture, corn and soybean rotation, winter wheat and fallow. Slopes were 3, 7 and 15%. Three scenarios of precipitation changes were considered: (a) all precipitation change occurring as number of days of rainfall; (b) all precipitation change occurring as amount of rainfall in a given day; and (c) half of the precipitation change occurring from each source. Under these scenarios, and using the climate generator for WEPP, changes in the number of days of rainfall do not influence rainfall intensity, while changes in the amount of rainfall on a given day increase the duration, peak intensities and average intensities of rain. Levels of changes considered in each case were approximately zero, $\pm 10\%$ and $\pm 20\%$ of total precipitation, with the same relative proportion of precipitation for the year maintained as a function of month.

Erosion rates changed much more with changes in the amount of rainfall per precipitation event, which also implies changes in the rainfall durations and intensities for the events. When total precipitation in this case was increased by 10%, soil loss increased by an average of 26%. Realistically, we can expect that any changes in precipitation will come as a combination of both changes in the number of wet days as well as in changes in the amount and intensities of rainfall. As we discussed earlier, historical changes in rainfall over the past century have occurred in both of these terms (Karl *et al.*, 1996). For the combined case of both changes in wet days and changes in rainfall per day, Pruski and Nearing (2002) found that erosion response was intermediate to the two extremes. For a 10% increase in total precipitation, simulated erosion increased by an average of 16%.

The average results for the combined case of changes in both number of days of precipitation and changes in amount of rain per day from the study by Pruski and Nearing (2002) are similar to those for the empirical relationship proposed by Renard and Freimund (1994) between erosivity and total annual precipitation for the RUSLE model as discussed above. Using Renard and Freimund's first equation for erosivity results in a 17% change as a function of a 10% change in total annual precipitation. However, it is important to note that regardless of this fact, obtaining the large-scale erosivity change information similar to the information we obtained from the study discussed in the previous section (Nearing, 2001) would have been extremely difficult using WEPP.

Now let's look at some of the details of the results from the WEPP erosivity study. Greater amounts and rates of runoff, other factors being equal, will generally tend to cause an increase in erosion. Increased runoff causes increased energy of surface flow, which increases the detachment capability and the sediment-transport capacity of the flow. Inter-rill erosion also increases with increased rain.

The simulation results of Pruski and Nearing (2002) showed a general increase in soil loss with increase in precipitation, and vice versa (Table 16.4), however, the changes were generally not as great as for runoff (Table 16.5). One major reason for the difference between the sensitivity results for runoff and those for soil loss is related to biomass production. Both runoff and soil loss are sensitive to biomass, but soil loss is more so. Soil loss is affected by plant canopy, which reduces the impact energy of rainfall; by crop residues which protect the soil from raindrop impact and reduce rill detachment rates and sediment transport capacities; and from subsurface roots and decaying residue, which mechanically hold the soil in place and provide a medium for micro-organisms to live. Thus, the increase of biomass production with increased rainfall tends to counteract to some degree the increased erosivity of the rain. This argument is supported by the results of the simulations for fallow conditions in comparison to the other treatments. The sensitivity values for the three precipitation scenarios for fallow conditions average 1.63 for soil loss and 1.55 for runoff. Thus fallow was the only crop treatment for which the sensitivities for runoff were less than for soil loss.

The difference between a sensitivity of 0.95 for soil loss and 1.06 for runoff for the fallow scenario of change only in the number of days of rainfall (Tables 16.4 and 16.5) can be explained in terms of surface sealing and consolidation processes. Surface sealing and consolidation occur as a function of rainfall amount in nature and in the WEPP model (Flanagan and Nearing, 1995), so that any increase in rainfall will increase soil resistance to erosion via consolidation. This

Table 16.4   Sensitivities of changes in soil loss to changes in average annual precipitation. Sensitivity values are calculated as the ratio of the percentage change in soil loss to the percentage change in precipitation. Values represent averages for all simulation runs associated with the soil, crop, slope or location listed in the first column. Values greater than zero indicate that soil loss increases with increased annual precipitation. A value of greater than one indicates a greater percentage change in soil loss than the percentage change in precipitation

Scenarios	Normalized sensitivity of soil loss to changes in average annual precipitation		
	Change in number of wet days	Change in amount of rain per day	Combined changes in both
Silt loam soil	0.90	2.45	1.72
Sandy loam soil	0.89	2.60	1.82
Clay soil	0.79	2.10	1.46
Grazing pasture	1.02	2.66	1.96
Fallow	0.95	2.22	1.71
Corn and soybean	0.70	2.46	1.48
Wheat winter	0.77	2.18	1.50
S-shape (0%–3%–1%) 40 m	0.92	2.47	1.71
S-shape (0%–7%–1%) 40 m	0.84	2.40	1.67
S-shape (0%–15%–1%) 40 m	0.82	2.27	1.61
West Lafayette, IN	0.74	2.35	1.56
Temple, TX	0.88	2.10	1.50
Corvallis, OR	0.92	2.69	1.93
Overall average	0.85	2.38	1.66

process also acts as a feedback effect, similar to, but in less degree, the effect of biomass on partially offsetting the impact of increased rainfall on erosion. This explains the lesser sensitivity of 0.95 for soil loss as compared to 1.06 for runoff.

The soil-loss sensitivity value for fallow conditions for the scenario of change in amount of rainfall per day was greater (2.22) than that for runoff (1.99), whereas for the other crops the trend was reversed (Tables 16.4 and 16.5). Although the effect of surface sealing and consolidation as discussed above is present in this case also, that effect is apparently superseded by yet another process when rainfall amounts and intensities per day are increased. These processes were related to rill- and inter-rill soil-detachment processes. Inter-rill erosion rates are represented in the WEPP model as proportional to the rainfall intensity and the runoff rate (Flanagan and Nearing, 1995), which are relationships based on experimental data (Zhang *et al.*, 1996). Since both of these variables increase with increased rainfall intensity, the effect of increased

rainfall intensity on inter-rill erosion is greater than unity. Also, rill erosion occurs as a threshold process. Rill detachment occurs proportional to the excess shear stress of water flow above the threshold critical shear stress of the soil, rather than to the shear stress of the flow itself. The overall effect is that the sensitivity of the rill erosion rate to runoff rate will be somewhat more than unity, other factors being constant. The effect is not present in the precipitation scenario of changes in the number of rainfall days because in that case, the average runoff rate is essentially not changing, but rather only the frequency of runoff events changes.

These are only a portion of the interactions discussed by Pruski and Nearing (2002) that were evident in the results of this study, but they provide a flavour of the types of information which the process-based model provides that the empirical model cannot address. Hopefully the above discussions of these two model applications will provide the reader with a sense of how each type of model might be used to advantage depending upon the desired application.

Table 16.5 Sensitivities of changes in runoff to changes in average annual precipitation. Sensitivity values are calculated as the ratio of the percentage change in runoff to the percentage change in precipitation. Values represent averages for all simulation runs associated with the soil, crop, slope or location listed in the first column. Values greater than zero indicate that runoff increases with increased annual precipitation. A value of greater than one indicates a greater percentage change in runoff than the percentage change in precipitation

Scenarios	Normalized sensitivity of runoff to changes in average annual precipitation		
	Change in number of wet days	Change in amount of rain per day	Combined changes in both
Silt loam soil	1.32	2.57	2.00
Sandy loam soil	1.31	2.80	2.17
Clay soil	1.15	2.17	1.75
Grazing pasture	1.54	3.09	2.41
Fallow	1.06	1.99	1.60
Corn and soybean	1.32	2.51	1.97
Wheat winter	1.21	2.43	1.91
S-shape (0%–3%–1%) 40 m	1.32	2.59	2.03
S-shape (0%–7%–1%) 40 m	1.29	2.49	1.98
S-shape (0%–15%–1%) 40 m	1.23	2.42	1.91
West Lafayette, IN	1.16	2.61	1.94
Temple, TX	1.19	2.25	1.73
Corvallis, OR	1.50	2.64	2.23
Overall average	1.28	2.50	1.97

16.4 LESSONS AND IMPLICATIONS

At the start of this chapter we listed three primary uses for soil-erosion models: (a) to help a land owner or manager choose suitable conservation; (b) to make large-scale erosion surveys in order to understand the scope of the problem over a region and to track changes in erosion over time; and (c) to regulate activities on the land for purposes of conservation compliance. Let us look at each of these goals in turn.

Choosing how to manage land, from the practical perspective, is often a matter of choosing between an array of potential options. Often, therefore, what we need to know is not necessarily the exact erosion rate for a particular option to a high level of accuracy, but rather we want to know how the various options stack up against one another. We may certainly be interested in having a general quantitative idea of the erosion rate, but for purposes of land management, this is not critical. Choosing which model to use then becomes a matter of (a) what type of information we would like

to know, and (b) what information (data) we have for the particular site of application. We know from our discussions above that the USLE provides only estimates of average annual soil loss on the portion of the field that experiences a net loss of soil. If we have an interest in off-site impacts, then we probably want to choose either RUSLE, which will provide us with a rough idea of the sediment leaving the profile, or WEPP, if we want more comprehensive sediment-yield information or if we are modelling a small watershed. If we have an interest in obtaining other, auxiliary information about our choice of management strategy, such as soil moisture or crop yields, we might also decide to use WEPP. On the other hand, if data are limited for the situation to be modelled, then the USLE might be the best option in any case, and one would be forced to move to other options for assessing information not supplied by the USLE. At the current time most applications of WEPP are possible in the United States because of the availability of soil, climate and crop information, but in other areas this might not be the case.

Making large-scale erosion surveys in order to understand the scope of the erosion problem over a region and to track changes in erosion over time can be done with any of the models discussed above. Often what is done is to use a statistical sampling scheme to take random points over the area of interest, and to apply the erosion model to each point (USDA, 1996). In this case, too, we are not so concerned about the individual prediction for each point of application, but rather the ability of the model to predict overall averages of soil loss in a quantitatively accurate manner. While we know that none of these models will necessarily predict erosion for a particular site to the quantitative level of accuracy we would like to see for survey assessment purposes (Nearing, 2000), each of the three models does predict the averages for treatments quite effectively (Risse, *et al.*, 1993; Rapp, 1994; Zhang, *et al.*, 1996). As with the case discussed above, the issues related to choosing the correct model are related to the information desired and the available data.

Conservation compliance, governmental policy-making, and regulation of land-users' actions follow the same guidelines as for the other two applications: information desired and data availability are again the keys to choice of model. In this case, however, the argument is often given, most often by the farmer who is being regulated, that if we know that there are uncertainties in the erosion predictions for individual applications, how can we be sure that his field is being evaluated accurately? The answer is, of course, that we cannot be sure. If the model predicts that the farmer's field is eroding at a rate in excess of what our society's policy indicates to be acceptable, the model could well be wrong for this particular field. This problem is really no different from that faced by insurance companies as they set rates for insurance coverage. My home may be more secure from the possibility of fire than my neighbour's home because I am more careful than my neighbour. But unless my home falls in a different category (e.g. better smoke-alarm protection), I will not have much luck in going to my insurance company and asking for a lower payment rate. Likewise, if I am the farmer, I cannot expect to give a coherent argument for lower soil loss than the model predicts unless I conduct some practice, such as reduced tillage or buffers, which arguably reduces erosion.

Complexity and uncertainty are key issues relative to the development, understanding and use of erosion models for conservation purposes. They are inevitable considerations because of the many complex interactions inherent in the erosional system as well as the enormous inherent variability in measured erosion data.

These issues do not, however, prevent us from using models effectively for conservation planning. In fact, the scientific evidence indicates that choice of models, which implies choice of model complexity, is more a matter of the type of information desired and the quality and amount of data available for the specific application. If our goal is to know to a high level of accuracy the erosion rate on a particular area of ungauged land, we cannot rely upon the models. Natural variability is too great, and uncertainty in predictions is too high (see Nearing, *et al.*, 1999, 2000). For appropriate and common uses, such as those discussed above, models can be effective conservation tools.

ACKNOWLEDGEMENTS

The precipitation data from the HadCM3 model for the period 2000–2099 were supplied by the Climate Impacts LINK Project (DETR Contract EPG 1/1/68) on behalf of the Hadley Centre and the UK Meteorological Office. The precipitation data from the CGCM1 model, GHG + A1 scenario, for the period 2000–2099 was supplied by the Canadian Centre for Climate Modelling and Analysis.

NOTE

[1] Most of the early erosion plots were 1.83 m (6 feet) wide. A length of 22.13 m (72.6 feet) and a width of 1.83 m (6 feet) resulted in a total area of 1/100 of an acre. Prior to the days of calculators and computers this was obviously a convenient value for computational purposes.

REFERENCES

Arnold, J.G., Williams, J.R., Nicks, A.D. and Sammons, N.B. (1990) *SWRRB: A Basin Scale Simulation Model for Soil and Water Resource Management*, Texas A&M University Press, College Station, TX.

Arnoldus, H.M.L. (1977) Methodology used to determine the maximum potential average annual soil loss due to sheet and rill erosion in Morocco, *FAO Soils Bulletin* **34**, 39–51.

Baffaut, C., Nearing, M.A. and Nicks, A.D. (1996) Impact of climate parameters on soil erosion using CLIGEN and WEPP, *Transactions of the American Society of Agricultural Engineers* **39**, 447–457.

Boer, G.J., Flato, G., Reader, M.C. and Ramsden, D. (2000) A transient climate change simulation with greenhouse gas and aerosol forcing: experimental design and comparison with the instrumental record for the 20th century, *Climate Dynamics* **16**, 405–425.

Chu, S.T. (1978) Infiltration during an unsteady rain, *Water Resources Research* **14** (3), 461–466.

Cusack, S., Slingo, A., Edwards, J.M. and Wild, M. (1998) The radiative impact of a simple aerosol climatology on the Hadley Centre GCM, *Quarterly Journal of the Royal Meteorological Society* **124** (551), 2517–2526.

Edwards, J.M. and Slingo, A. (1996) Studies with a flexible new radiation code. 1: choosing a configuration for a large scale model, *Quarterly Journal of the Royal Meteorological Society* **122**, 689–719.

Flanagan, D.C. and Livingston, S.J. (1995) *USDA-Water Erosion Prediction Project: WEPP User Summary*, NSERL Report No. 11, USDA-ARS National Soil Erosion Research Laboratory, West Lafayette, IN, 47097–1196.

Flanagan, D.C. and Nearing, M.A. (1995) *USDA–Water Erosion Prediction Project: Hillslope Profile and Watershed Model Documentation*, NSERL Report No. 10, USDA-ARS National Soil Erosion Research Laboratory, West Lafayette, IN, 47097–1196.

Foster, G.R., McCool, D.K., Renard, K.G. and Moldenhauer, W.C. (1981) Conversion of the universal soil loss equation to SI metric units, *Journal of Soil and Water Conservation* **36**, 355–359.

Fournier, F. (1960) *Climat et Erosion*, Universitaires de France, Paris.

Gordon, C., Cooper, C., Senior, C.A., Banks, H., Gregory, J.M., Johns, T.C., Mitchell, J.F.B. and Wood, R.A. (2000) The simulation of SST, sea ice extents and ocean heat transports in a version of the Hadley Centre coupled model without flux adjustments, *Climate Dynamics* **16** (2–3), 147–168.

Govers, G. (1991) Rill erosion on arable land in central Belgium: rates, controls, and predictability, *Catena* **18**, 133–155.

Govers, G. (1996) Soil erosion process research: a state of the art, Academie voor Wetenschappen, *Letteren en Schone Kunsten van Belgie, Klasse der Wettenschappen* **58** (1).

IPCC (1995) *Second Assessment Synthesis of Scientific-Technical Information Relevant to Interpreting Article 2 of the UN Framework Convention on Climate Change*, Intergovernmental Panel on Climate Change, Geneva, Switzerland.

Johns, T.C., Carnell, R.E., Crossley, J.F., Gregory, J.M., Mitchell, J.F.B., Senior, C.A., Tett, S.F.B. and Wood, R.A. (1997) The second Hadley Centre coupled ocean–atmosphere GCM: model description, spin-up and validation, *Climate Dynamics* **13**, 103–134.

Karl, T.R., Knight, R.W., Easterling, D.R. and Quayle, R.G. (1996) Indices of climate change for the United States, *Bulletin of the American Meteorological Society* **77** (2), 279–292.

Larionov, G.A. (1993) *Erosion and Wind Blown Soil*, Moscow State University Press, Moscow.

Liu, B.Y., Nearing, M.A., Baffaut, C. and Ascough II, J.C. (1997) The WEPP watershed model: III. Comparisons to measured data from small watersheds, *Transactions of the American Society of Agricultural Engineers* **40** (4), 945–951.

McFarlane, N.A., Boer, G.J., Blanchet, J.-P. and Lazare, M. (1992) The Canadian Climate Centre second-generation general circulation model and its equilibrium climate, *Journal of Climate* **5**, 1013–1044.

Mein, R.G. and Larson, C.L. (1973) Modeling infiltration during a steady rain, *Water Resources Research* **9** (2), 384–394.

Misra, R.K. and Rose, C.W. (1996) Application and sensitivity analysis of process-based erosion model GUEST, *European Journal of Soil Science* **47**, 593–604.

Morgan, R.P.C., Quinton, J.N., Smith, R.E., Govers, G., Poesen, J.W.A., Auerswald, K., Chisci, G., Torri, D. and Styczen, M.E. (1998) The European Soil Erosion Model (EUROSEM): a dynamic approach for predicting sediment transport from fields and small catchments, *Earth Surface Processes and Landforms* **23**, 527–544.

Nearing, M.A. (1998) Why soil erosion models over-predict small soil losses and under-predict large soil losses, *Catena* **32**, 15–22.

Nearing, M.A. (2000) Evaluating soil erosion models using measured plot data: accounting for variability in the data, *Earth Surface Processes and Landforms* **25**, 1035–1043.

Nearing, M.A. (2001) Potential changes in rainfall erosivity in the U.S. with climate change during the 21st century, *Journal of Soil and Water Conservation* **56**, 229–232.

Nearing, M.A., Ascough, L.D. and Laflen, J.M. (1990) Sensitivity analysis of the WEPP hillslope profile erosion model, *Transactions of the American Society of Agricultural Engineers* **33**, 839–849.

Nearing, M.A., Foster, G.R., Lane, L.J. and Finkner, S.C. (1989) A process-based soil erosion model for USDA–Water Erosion Prediction Project technology, *Transactions of the American Society of Agricultural Engineers* **32**, 1587–1593.

Nearing, M.A., Govers, G. and Norton, L.D. (1999) Variability in soil erosion data from replicated plots, *Soil Science Society of America Journal* **63** (6), 1829–1835.

Nicks, A.D. (1985) Generation of climate data, in D.G. DeCoursey (ed.) *Proceedings of the Natural Resources Modeling Symposium*, Pingree Park, CO, 16–21 October 1983, USDA-ARS ARS-30.

Ott, L. (1977) *An Introduction to Statistical Methods and Data Analysis*, Duxbury Press, North Scituate, MA.

Pimentel, D.E. *et al.* (1995) Environmental and economic costs of soil erosion and conservation benefits, *Science* **267**, 1117–1123.

Pope, V.D., Gallani, M.L., Rowntree, P.R. and Stratton, R.A. (2000) The impact of new physical parametrizations in the Hadley Centre climate model – HadAM3, *Climate Dynamics* **16** (2–3), 123–146.

Pruski, F.F. and Nearing, M.A. (2002) Runoff and soil-loss responses to changes in precipitation: a computer simulation study, *Journal of Soil and Water Conservation* **57**, 7–16.

Rapp, J.F. (1994) Error assessment of the Revised Universal Soil Loss Equation using natural runoff plot data, MS thesis, School of Renewable Natural Resources, University of Arizona, Tucson, AZ.

Reader, M.C. and Boer, G.J. (1998) The modification of greenhouse gas warming by the direct effect of sulphate aerosols, *Climate Dynamics* **14**, 593–607.

Renard, K.G., Foster, G.R., Weesies, G.A., McCool, D.K. and Yoder, D.C. (1997) *Predicting Soil Erosion by Water:*

A Guide to Conservation Planning with the Revised Universal Soil Loss Equation (RUSLE), Agricultural Handbook No. 703, US Government Printing Office, Washington, DC.

Renard, K.G. and Freimund, J.R. (1994) Using monthly precipitation data to estimate the R-factor in the revised USLE, *Journal of Hydrology* **157**, 287–306.

Risse, L.M., Nearing, M.A., Nicks, A.D. and Laflen, J.M. (1993) Assessment of error in the Universal Soil Loss Equation, *Soil Science Society of America Journal* **57**, 825–833.

Ruttimann, M., Schaub, D., Prasuhn, V. and Ruegg, W. (1995) Measurement of runoff and soil erosion on regularly cultivated fields in Switzerland – some critical considerations, *Catena* **25**, 127–139.

Schwertmann, U., Vogl, W. and Kainz, M. (1990) *Bodenerosion durch Wasser*, Eugen Ulmer GmbH and Co., Stuttgart.

USDA (1996) *Summary Report 1992: National Resource Inventory*, US Government Printing Office, Washington, DC.

Walpole, R.E. and Myers, R.H. (1993) *Probability and Statistics for Engineers and Scientists*, 5th edn, Prentice Hall, Englewood Cliffs, NJ.

Wendt, R.C., Alberts, E.E. and Hjelmfelt, A.T. Jr. (1986) Variability of runoff and soil loss from fallow experimental plots, *Soil Science Society of America Journal* **50**, 730–736.

Williams, J., Nearing, M.A., Nicks, A., Skidmore, E., Valentine, C., King, K. and Savabi, R. (1996) Using soil erosion models for global change studies, *Journal of Soil and Water Conservation* **51** (5), 381–385.

Williams, J.R. and Nicks, A.D. (1985) SWRRB, a simulator for water resources in rural basins: an overview, in D.G. DeCoursey (ed.) *Proceedings of the Natural Resources Modeling Symposium*, Pingree Park, CO, 16–21 October 1983, USDA-ARS, ARS-30, 17–22.

Wischmeier, W.H. (1959) A rainfall erosion index for a universal soil loss equation, *Soil Science Society of America Proceedings* **23**, 322–326.

Wischmeier, W.H. and Smith, D.D. (1965) *Predicting Rainfall Erosion Losses in the Eastern US: A Guide to Conservation Planning*, Agricultural Handbook No. 282, US Government Printing Office, Washington, DC.

Wischmeier, W.H. and Smith, D.D. (1978) *Predicting Rainfall Erosion Losses: A Guide to Conservation Planning*, Agricultural Handbook No. 537, US Government Printing Office, Washington, DC.

Wood, R.A., Keen, A.B., Mitchell, J.F.B. and Gregory, J.M. (1999) Changing spatial structure of the thermohaline circulation in response to atmospheric CO_2 forcing in a climate model, *Nature* **399**, 572–575.

Zhang, X.C., Nearing, M.A., Risse, L.M. and McGregor, K.C. (1996) Evaluation of runoff and soil loss predictions using natural runoff plot data, *Transactions of the American Society of Agricultural Engineers* **39** (3), 855–863.

17

Modelling in Forest Management

MARK J. TWERY

17.1 THE ISSUE

Forest management has traditionally been considered management of trees for timber. It really includes vegetation management and land management and people management as multiple objectives. As such, forest management is intimately linked with other topics in this volume, most especially those chapters on ecological modelling and human dimensions. The key to responsible forest management is to understand both how forest ecosystems work and how to use this understanding to satisfy society's expectations and values. The key to forest modelling is to portray accurately the dynamics of forests. Successful forest-management modelling finds a means to improve management through accurate representation of all parts of the system. In this chapter I will review both modelling approaches and the types of applications that various modelling techniques have been used to address.

The complexity of forest management does not stop with the intricate details of the biological system found in a forest. Because all forest management is by people to meet goals or objectives desired by people, forestry is at its core a social activity. Thus, it demands that we understand the relationships among land owners, professional forest managers, forest-dependent communities and other stakeholders if we are to model the results of decisions regarding forest management. Individuals and communities have broad interests in the physical, biological and social goods and services that forest ecosystems provide. To meet the challenges of today we need to know as much about the people as about the physical and biological conditions, and to deepen our understanding of the social goods and services that we expect our forest ecosystems to supply.

Models used to assist in forest management consist of several types. First and most prevalent are growth and yield models, which predict the development of stands of trees through time. Initially, these models focused on single species, single age stands, such as one would find in plantations. Little wonder, because in plantations were the highest investments of forest managers who primarily sought timber value from the forest. As modelling sophistication increased, so did the models. Multiple species and multiple age and size classes in an individual stand have been included in growth models with varying success. Further developments along a different track have seen modelling techniques used to help schedule activities, such as harvesting or thinning in forests. Linear programming and other techniques that allow a modeller to specify constraints on resources have allowed models to find solutions to allocation problems, again primarily for those interested in extracting timber value from forests.

As various public and private groups have come to recognize the importance of forests for values beyond timber and focused research into those topics, further modelling efforts have begun to characterize these resources. Wildlife habitat, recreation opportunities and watershed protection are just a few of the benefits that now must be modelled and presented to decision-makers for their evaluation of alternative forest management strategies. These efforts face many challenges due to the inherent complexities of the systems they attempt to model and the lack of good quantitative date for many of the factors involved. In addition, the most complex and fuzziest of the factors is that of the role of humans in the forest. Models capable of predicting human needs and desires, and their implications for the state of forested ecosystems, are a great need that is still largely unmet.

Environmental Modelling: Finding Simplicity in Complexity. Edited by J. Wainwright and M. Mulligan
© 2004 John Wiley & Sons, Ltd ISBNs: 0-471-49617-0 (HB); 0-471-49618-9 (PB)

Managers need to predict effects of implementing different alternatives. Although a large body of scientific knowledge exists on relations between forest structure and pattern and ecosystem attributes, this information is frequently difficult to interpret and apply. To have and to use models that both incorporate the best knowledge available about the biological system and present the results in an interpretable and useful fashion is a significant challenge. Some such integrated models have been developed, and more are in various stages of design. Further, the incorporation of stakeholders' knowledge and goals, and the integration of social disciplines, theories and measurement techniques are difficult for many traditionally trained resource managers. One of today's greatest challenges is the development and testing of new theories and tools that describe the multiple ramifications of management decisions and that provide a practical, understandable decision process. Developing, evaluating and adapting new decision processes and their supporting software tools is a critically important endeavour in forest management and elsewhere.

17.2 THE APPROACHES

17.2.1 The empirical approach

Traditionally, simulation models of tree growth have been developed to enable projection of future timber production. Economic values drove the effort, and the models were built to do a good job of estimating growth of trees that have already reached moderate size. Limitations include the fact that these models were built from measurements on stands that have intentionally included only 'undisturbed, fully stocked' stands, rendering them of questionable use when trying to predict responses to disturbances such as thinnings or windstorms.

In a review of the earliest forest yield studies, Assmann (1970) describes several eighteenth- and nineteenth-century efforts to predict expected yields in German forests of beech and other important species. In North America, forest growth and yield prediction models began as simple estimates of standing volume (Pinchot, 1898) and progressed to estimates using increment cores (Fenska and Lauderburn, 1924). These studies and others led to stand table projection systems still in use as a basic form of empirical forest-growth modelling. Other reviews of the development of this type of modelling can be found in Chapman and Meyer (1949), Spurr (1952) and Clutter et al. (1983). See also http://sres.anu.edu.au/associated/mensuration/xref.htm.

17.2.2 The mechanistic approach

Models based on presumed or observed mechanisms of growth, theoretical controlling factors, and interactions among elements of a system are what I refer to here as mechanistic models. JABOWA (Botkin et al., 1972) is among the earliest of these models used to simulate forest growth, which it predicts for individual trees based on available light, nutrients, temperature and other parameters. Successors to the JABOWA model include Fortnite (Aber and Melillo, 1982) and Zelig (Urban and Shugart, 1992), among others, all of which use basic ecological principles to predict the response of an individual tree to its surroundings and grow a forest by accumulating the sum of the individuals. Sortie (Pacala et al., 1993, 1996) is a spatially explicit model developed using similar ecological variables to the earlier gap models. The stochastic element added to gap models complicates their use in management applications because of the need to run many simulations and average their outcomes. On the other hand, because of their reliance on basic principles, these models do a better job of predicting multiple generations of forest development than empirical models based on growth within the lifetime of an existing stand of trees, and thus are more useful for modelling long-term succession under existing or changing conditions.

A different approach to modelling tree growth mechanistically is to model individual trees primarily based on inherent physiological characteristics. For example, the pipe-model theory presumes that a given unit of foliage requires a given unit of sapwood area to supply water (Shinozaki et al., 1964; Waring et al., 1982). One such model, TREGRO (Weinstein and Yanai, 1994), predicts growth and carbon allocation patterns based on various levels of ozone, nutrient stress and water availability. Hybrid (Friend et al., 1997) is a growth model using a combination of techniques, including competition between plants that is modelled with a gap model approach, while physiological knowledge is used to predict plant growth. Pipestem[*1] (Valentine et al., 1997) is a pipe-model-based stand growth model that projects even-aged, single-species stands using production and loss rates of leaves, stems and roots.

17.2.3 The knowledge-based approach

Knowledge-based or rule-based systems are a special case of modelling in which the components being modelled and the interactions between them are not necessarily represented mathematically. Approaches such as these use a symbolic representation of information to model systems by effectively simulating the logical processes of human experts (Reynolds *et al.*, 1999). Knowledge-based systems have the advantages that they do not necessarily require the specific, detailed data that many simulation models do, and they can be adapted to situations in which some information may be lacking entirely. As such, they can be very useful in providing assistance to decision-makers who must analyse situations and choose actions without complete knowledge. Schmoldt and Rauscher (1996) point out that knowledge-based systems also prove useful as agents to codify institutional memory, manage the collection and delivery of scientific knowledge, and train managers through their ability to provide explanations of their reasoning processes. All these characteristics make knowledge-based models extremely useful in forest management.

17.3 THE CONTRIBUTION OF MODELLING

17.3.1 Models of the forest system

17.3.1.1 Growth and yield models

Predicting growth and yield has long been at the heart of simulating the future of forests. Growth and yield models were classified by Clutter *et al.* (1983) as for natural forests (either even-aged or uneven-aged) or plantations (either thinned or unthinned). Early modelling efforts, restricted by lack of computational power, typically resulted in the publication of yield tables that presented basal area or volume of a stand at regular intervals of development. See Hann and Riitters (1982) for a summary of such models in the western United States or Schumacher and Coile (1960) for an example from the southeastern United States. Assmann (1970) presents a comprehensive description of the history of European forest-yield modelling. Often, such yield tables are still adequate for managers interested primarily in timber-volume production and who have only extensive data on size class and basal area in their stands. Without more detailed data to run computer-based simulation models, yield tables still prove useful.

Computer-based simulation models now dominate the field. Simulators may be divided between stand-level models that project stand-level summary variables such as basal area and number of stems, or individual tree models that project individual trees from a tree list or by species and size class. Individual tree models can be further classified as distance-independent or distance-dependent. Clutter (1983) provides a detailed review of the various techniques. Ritchie (1999) has compiled an exhaustive description of the numerous growth and yield

simulators available for the western United States and an extensive bibliography of publications using the various simulators. Ritchie (1999)* describes some newer techniques in individual tree modelling, including disaggregative techniques in which stands grown as a whole are disaggregated among trees in a list, which is maintained to allow more detailed stand parameters needed in predicting other variables. His analysis includes an evaluation of the suitability of the models for management planning. Pretzsch (2001) reviews forest growth modelling from a European perspective, including recently developed simulators of growth and visualization such as SILVA.

Examples of growth models in use in the eastern United States include FIBER, SILVAH and TWIGS. FIBER* (Solomon *et al.*, 1995) is a two-stage matrix model using dynamic transition probabilities for different ecological classifications to obtain the growth of trees between diameter classes. These transition probabilities are a function of diameter, initial and residual stand basal area, proportion of hardwoods, and elevation. SILVAH* (Marquis and Ernst, 1992) is an expert system for making silvicultural decisions in hardwood stands of the Allegheny Plateau and Allegheny Mountain region that recommends appropriate treatments based on user objectives and overstorey, understorey and site data provided by the user. SILVAH also contains a forest-stand growth simulator, provides the ability to test alternative cuts, enables development of a forest-wide inventory database, and facilitates other forest management planning functions. TWIGS* (Miner *et al.*, 1988) is a computer program used to simulate growth and yield for forests in the North Central region of the United States. It grows individual tree lists, has a regeneration component, and also includes management and economic analyses. Two variants are available: Central States (Indiana, Illinois, and Missouri) and Lake States (Michigan, Minnesota, and Wisconsin).

17.3.1.2 Regeneration models

Models of forest regeneration that provide reasonable estimates of tree species composition and density after a disturbance have been difficult to develop. Gap-dynamics models in the JABOWA family tend to use an approach of generating many small individuals in a predetermined proportion based on their prevalence in the seed bank or in the overstorey before disturbance and letting them die in early steps of the simulation. Empirical stand models typically have no regeneration function or a crude one that applies ingrowth to the smaller size classes based on proportions of a previous stand (e.g. Solomon *et al.*, 1995).

Recent developments using knowledge-based models to predict the composition of understorey after a minor disturbance or a newly regenerated stand after a major disturbance show some promise. Rauscher *et al.* (1997a) have developed a rule-based regeneration-prediction program for the southern Appalachians. Yaussy *et al.* (1996) describe their efforts to catalogue ecological characteristics of various species of the central hardwood forest of the United States and the individual-tree regeneration model developed from those characteristics. Ribbens (1996) developed a spatially explicit, data-intensive regeneration model called RECRUITS, which calculates the production and spatial dispersion of recruited seedlings in reference to the adults and uses maximum likelihood analysis to calibrate functions of recruitment. Because this program requires map data of adults and transect sampling of seedlings, it is unlikely to be useful in management applications.

17.3.1.3 Mortality models

Mortality of trees is an important process in forest development. Empirical simulation models usually calculate mortality through generating probabilities based on species and relative sizes, densities and ages of trees measured in the data sets used to generate the model parameters. Mechanistic models typically set a minimum level of growth parameters for survival, and a tree dies if it does not reach the minimum level. Mortality has been important to forest-management models as an indication of timber loss, so typically trees that die are removed from the model in further projections. A comprehensive review of the state of mortality models can be found in Hawkes (2000).

In recent years, dead trees, both standing and fallen, have become more widely recognized as important parts of the forest in their own right, and models are being developed to simulate creation, duration and decomposition of standing snags, fallen logs and coarse woody debris in general. Forest-fire models have recognized that dead wood within the forest is an important factor as fuel for potential fires, but forest managers are seeing the need to estimate dead wood in its various forms to feed wildlife habitat, visual quality and water quality models as well. Models to predict the longevity of dead wood include systems as diverse as subalpine Colorado (Brown *et al.*, 1998), and coastal Oregon (Spies *et al.*, 1988). The subject is well addressed by Parminter at http://www.for.gov.bc.ca/hre/deadwood/. Other references can be found at http://www.tws-west.org

/deadwoodabstracts.html, a summary of a symposium sponsored by the Wildlife Society in 1999.

17.3.1.4 Habitat models

Providing wildlife habitat has long been one of the objectives of forest management. Often the availability of habitat that has been assumed in the forest is managed to maximize timber. Recent controversies such as those over the spotted owl and salmon habitat in the Pacific Northwest have shown that sometimes forest practices need to be altered to meet multiple objectives, and sometimes objectives other than timber are of overriding importance. Habitat-suitability models have been a common technique for formulating descriptions of the conditions needed to provide habitat for individual species. These models typically are generated from expert knowledge and expressed in terms of ranges and thresholds of suitability for several important habitat characteristics. Models that use such techniques lend themselves to adaptation to the use of fuzzy logic in a knowledge-based computer system.

Recent developments using general habitat information in a Geographic Information System (GIS) coupled with other techniques have produced a number of promising approaches to integrating timber and wildlife habitat modelling in a spatially explicit context. Hof and Joyce (1992, 1993) describe the use of mixed linear and integer programming techniques to optimize wildlife habitat and timber in the context of the Rocky Mountain region of the western United States. Ortigosa *et al.* (2000) present a software tool called VVF, which accomplishes an integration of habitat suitability models into a GIS to evaluate territories as habitat for particular species.

17.3.2 Models of human responses and interactions

17.3.2.1 Harvest-scheduling models

Large-scale analyses are necessary for policy and for including ecosystem processes that include a greater area than a stand. Spatially explicit techniques are important and valuable because we know that patterns and arrangements affect the interactions of components.

Forest managers need to plan activities across a landscape in part to maintain a reasonable allocation of their resources, but also to include considerations of maintenance of wildlife habitat and to minimize negative effects on the aesthetic senses of people who see the management activities. Gustafson (1999) has developed such a model, HARVEST*, to enable analysis of such activities across a landscape, including an educational version, HARVEST Lite. The model has now been combined with LANDIS* (Mladenoff *et al.*, 1996) to integrate analyses of timber harvesting, forest succession and landscape patterns (Gustafson and Crow, 1999; Gustafson *et al.*, 2000). Hof and Bevers (1998) take a mathematical optimization approach to a similar problem, to maximize or minimize a management objective using spatial optimization given constraints of limited area, finite resources and spatial relationships in an ecosystem.

17.3.2.2 Recreation-opportunity models

Providing recreation opportunities is an important part of forest management, especially on public lands. Indeed, the total value generated from recreation on National Forests in the United States competes with that from timber sales, and may well surpass it soon (USDA Forest Service, 1995). Forest managers have long used the concept of a 'recreation opportunity spectrum' (Driver and Brown, 1978) to describe the range of recreation activities that might be feasible in a particular area, with the intention of characterizing the experience and evaluating the compatibility of recreation with other activities and goals in a particular forest or other property.

RBSim* (Gimblett *et al.*, 1995, 1996) is a computer program that simulates the behaviour of human recreationists in high-use natural environments using GIS to represent the environment and autonomous human agents to simulate human behaviour within geographic space. In RBSim, combinations of hikers, mountain bikers and jeep tours are assigned individual characteristics and set loose to roam mountain roads and trails. The behaviours and interactions of the various agents are compiled and analysed to provide managers with evaluations of the likely success of an assortment of management options.

17.3.2.3 Visualization

Many people tend to respond to visual images, leading to the adage, 'a picture is worth a thousand words'. Much information generated by forest models is in the form of data tables, which are intelligible to the well initiated, but meaningless to many, including public stakeholders and many forest managers. Photographs of a forest may be nearly as good at conveying an image of the conditions as actually visiting a site, but models are used to project conditions that do not yet exist. The best available means of providing an image of potential future conditions is a computer representation of the data. One such system, the Stand Visualization System (SVS*) (McGaughey, 1997), generates graphic images depicting stand conditions represented by a list of individual stand components, e.g., trees, shrubs and down material. It is in wide use as a secondary tool, connected to growth models such as FVS* (Stage, 1973, 1997), LMS (McCarter *et al.*, 1998) and NED (Twery *et al.*, 2000). UTOOLS* and UVIEW are geographic analysis and visualization software for watershed-level planning (Ager and McGaughey, 1997). The system uses a Paradox database to store spatial information and displays landscape conditions of a forested watershed in a flexible framework. Another similar visualization tool is SmartForest* (Orland, 1995), which is also an interactive program to display forest data for the purposes of visualizing the effects of various alternative treatments before actually implementing them. Different versions have been developed on a variety of platforms, many of them requiring more data or computer power than is practical for management activities, but SmartForest II (Orland *et al.*, 1999) is designed to run on a PC and display either stand-level or landscape data. Recently, McGaughey has developed an advanced landscape-scale visualization program addressing the same issues, entitled EnVision*.

17.3.3 Integrating techniques

17.3.3.1 Decision-support systems

Adaptive management has recently been viewed as a very promising and intuitively useful conceptual strategic framework for defining ecosystem management (Rauscher, 1999). Adaptive management is a continuing cycle of four activities: planning, implementation, monitoring and evaluation (Walters and Holling, 1990; Bormann *et al.*, 1993). Planning is the process of deciding what to do. Implementation is deciding how to do it and then doing it. Monitoring and evaluation incorporate analysing whether the state of the managed system was moved closer to the desired goal state or not. After each cycle, the results of evaluation are provided to the planning activity to produce adaptive learning. Unfortunately, this general theory of decision analysis is not specific enough to be operational. Further, different decision-making environments typically require different, operationally specific decision processes. Decision-support systems are combinations of tools designed to facilitate operation of the decision process (Oliver and Twery, 1999).

Mowrer *et al.* (1997) surveyed 24 of the leading ecosystem-management decision-support systems

(EM-DSS) developed in the government, academic and private sectors in the United States. Their report identified five general trends: (1) while at least one EM-DSS fulfilled each criterion in the questionnaire used, no single system successfully addressed all important considerations; (2) ecological and management interactions across multiple scales were not comprehensively addressed by any of the systems evaluated; (3) the ability of the current generation EM-DSS to address social and economic issues lags far behind biophysical issues; (4) the ability to simultaneously consider social, economic and biophysical issues is entirely missing from current systems; (5) group consensus-building support was missing from all but one system – a system which was highly dependent upon trained facilitation personnel (Mowrer *et al.*, 1997). In addition, systems that did offer explicit support for choosing among alternatives provided decision-makers with only one choice of methodology.

There are few full-service DSSs for ecosystem management (Table 17.1). At each operational scale, competing full-service EM-DSSs implement very different decision processes because the decision-making environment they are meant to serve is very different. For example, at the management-unit level, EM-DSSs can be separated into those that use a goal-driven approach and those that use a data-driven approach to the decision support problem. NED (Rauscher *et al.*, 1997a; Twery *et al.*, 2000) is an example of a goal-driven EM-DSS where goals are selected by the user(s). In fact, NED is the only goal-driven, full-service EM-DSS operating at the management-unit level. These goals define the desired future conditions, which define the future state of the forest. Management actions should be chosen that move the current state of the forest closer to the desired future conditions. In contrast, INFORMS (Williams *et al.*, 1995) is a data-driven system that begins with a list of actions and searches the

Table 17.1   A representative sample of existing ecosystem management decision-support software for forest conditions of the United States arranged by operational scale and function

Full service EM-DSS		Functional service modules	
Operational scale	Models	Function	Models
Regional Assessments	EMDS LUCAS*	Group negotiations	AR/GIS IBIS*
		Vegetation dynamics	FVS
Forest-level planning	RELM SPECTRUM WOODSTOCK ARCFOREST SARA TERRA VISION EZ-IMPACT* DECISION PLUS* DEFINITE*		LANDIS CRBSUM SIMPPLLE
		Disturbance simulations	FIREBGC GYPSES UPEST
		Spatial visualization	UTOOLS/UVIEW SVS* SMARTFOREST*
Management-unit-level planning	NED INFORMS MAGIS KLEMS TEAMS LMS*	Interoperable system architecture	LOKI CORBA* IMPLAN
		Economic impact analysis	SNAP
		Activity scheduling	

Note: * References for models not described in Mowrer *et al.* (1997): EZ-IMPACT (Behan, 1994); DECISION PLUS (Sygenex, 1994); IBIS (Hashim, 1990); DEFINITE (Janssen and van Herwijnen, 1994); SMARTFOREST (Orland, 1995); CORBA (Otte *et al.*, 1996); SVS (McGaughey, 1997); LMS (Oliver and McCarter, 1996); LUCAS (Berry *et al.*, 1996).

existing conditions to find possible locations to implement those management actions.

Group decision-making tools are a special category of decision support, designed to facilitate negotiation and further progress toward a decision in a situation in which there are multiple stakeholders with varied perspectives and opinions of both the preferred outcomes and the means to proceed. Schmoldt and Peterson (2000) describe a methodology using the analytic hierarchy process (Saaty, 1980) to facilitate group decision-making in the context of a fire-disturbance workshop, in which the objective was to plan and prioritize research activities. Faber *et al.* (1997) developed an 'Active Response GIS' that uses networked computers to display proposed options and as intermediaries to facilitate idea generation and negotiation of alternative solutions for management of US national forests.

17.4 LESSONS AND IMPLICATIONS

17.4.1 Models can be useful

Models of various kinds have been very useful to forest management for a long time. The most basic models provide at least an estimate of how much timber is available and what it may be worth on the market, so that managers can determine the economic feasibility of timber cutting. More sophisticated modelling techniques provide better estimates of timber, include other forest characteristics, and project likely developments into the future. Reliability of empirical models tends to be restricted to the current generation of trees, for which they are very good.

Other forest-growth models use ecological and physiological principles to make projections of growth. Theoretical, mechanistically based models tend to be better for general pictures of forest characteristics in a more distant future projection, but may be less reliable for near-term forecasts. They tend to require more data than managers are capable of collecting for extensive tracts, and thus are often restricted to use in scientific research contexts, rather than management decisions directly. Still, such research-oriented models are still very useful in the long term, as they help increase understanding of the system and direct further investigations.

With greater and greater computing power in recent years, modelling techniques have expanded to include spatially explicit models of landscape-level change. These models now help provide the context in which a stand-level forest management decision is made, giving a manager a better understanding of the implications one action has on other areas. Positive effects are being seen in wildlife management, fire management,

watershed management, land-use changes and recreation opportunities.

Other improvements in computing power and collaboration between forestry and landscape architecture have resulted in greatly enhanced capabilities to display potential conditions under alternative management scenarios before they are implemented. This capability enhances the quality of planning and management decisions by allowing more of the stakeholders and decision-makers to understand the implications of choosing one option over another. As computing power increases and digital renderings improve, care must be taken to ensure that viewers of the renderings do not equate the pictures they see with absolute certainty that such conditions will occur. We are still subject to considerable uncertainty in the forest system itself, and there is considerable danger that people will believe whatever they see on a computer screen simply because the computer produced it.

17.4.2 Goals matter

Forestry practice in general and silviculture in particular are based on the premise that any activity in the forest is intended to meet the goals of the land owner. Indeed, identification of the landowner's objectives is the first step taught to silviculturists in forestry schools (Smith, 1986). However, there has always been societal pressure for management practices, even on private lands, to recognize that actions on any particular private tract influence and are influenced by conditions on surrounding lands, including nearby communities and society at large. This implies that decision-makers need to be cognizant of the social components and context of their actions. Forest-management models intended to help land owners or managers determine appropriate actions must focus on meeting the goals defined by the user if they are to be used. Models that predetermine goals or constrain options too severely are unlikely to be useful to managers.

17.4.3 People need to understand trade-offs

There are substantial and well-developed theory and methodological tools of the social sciences to increase our understanding of the human element of forest ecosystem management (Burch and Grove, 1999; Cortner and Moote, 1999; Parker *et al.*, 1999). Models of human behaviour, social organizations and institutional functions need to be applied to forest planning, policy and management. Existing laws, tax incentives and best management practices provide some context for delivering social goods, benefits and services from forest management (Cortner and Moote, 1999). In addition, recent forest initiatives such as sustainable forestry certification through the forest industry's Sustainable Forestry Initiative (SFI) and the independent Forest Stewardship Council's (FSC) 'Green Certification' programmes include explicit, albeit modest, social considerations (Vogt *et al.*, 1999). Unfortunately, these sideboards to forest management fail to deal with the complexity of forest ecosystem management. Indeed, new modelling approaches are needed to identify effectively, collect and relate the social context and components of forest ecosystem management in order to enhance and guide management decisions (Burch and Grove, 1999; Villa and Costanza, 2000). One of today's greatest challenges is the development and testing of new theories and tools that describe the multiple ramifications of management decisions and that provide a practical, understandable decision process. Developing, evaluating and adapting new decision processes and their supporting software tools are critically important endeavours.

NOTE

[1] Models with an asterisk next to their name have URLs given in the list at the end of the chapter.

REFERENCES

Ager, A.A. and McGaughey, R.J. (1997) *UTOOLS: Microcomputer Software for Spatial Analysis and Landscape Visualization.* Gen. Tech. Rep. PNW-GTR-397. Portland, OR: Pacific Northwest Research Station, Forest Service, US Department of Agriculture, Washington, DC.

Assmann, E. (1970) *The Principles of Forest Yield Study,* (translated by S.H. Gardner), Pergamon Press, Oxford.

Behan, R.W. (1994) Multiresource management and planning with EZ-IMPACT, *Journal of Forestry* **92** (2), 32–36.

Berry, M.W., Flamm, R.O., Hazen, B.C. and MacIntyre, R.M. (1996) The land-use change and analysis (LUCAS) for evaluating landscape management decisions, *IEEE Computational Science and Engineering* **3**, 24–35.

Bormann, B.T., Brooks, M.H., Ford, E.D., Kiester, A.R., Oliver, C.D. and Weigand, J.F. (1993) *A Framework for Sustainable Ecosystem Management,* USDA Forest Service Gen. Tech. Rep. PNW-331, Portland, OR.

Botkin, D.B., Janak, J.F. and Wallis, J.R. (1972) Some ecological consequences of a computer model of forest growth, *Journal of Ecology* **60**, 849–872.

Brown, P.M., Shepperd, W.D., Mata, S.A. and McClain, D.L. (1998) Longevity of windthrown logs in a subalpine forest of central Colorado, *Canadian Journal of Research* **28**, 1–5.

Burch, W.R. and Morgan Grove, J. (1999) Ecosystem management – some social and operational guidelines for practitioners, in W.T. Sexton, A.J. Malk, R.C. Szaro and N.C.

Johnson (eds) *Ecological Stewardship: A Common Reference for Ecosystem Management,* Oxford: Elsevier Science, Vol. 3, 279–295.

Chapman, H.H. and Meyer, W.H. (1949) *Forest Mensuration,* McGraw-Hill, New York.

Clutter, J.L., Fortson, J.C., Pienaar, L.V., Brister, G.H. and Bailey, R.L. (1983) *Timber Management: A Quantitative Approach,* John Wiley & Sons, Chichester.

Cortner, H. and Moote, M.A. (1999) *The Politics of Ecosystem Management,* Island Press, New York.

Driver, B.L. and Brown, P.J. (1978) The opportunity spectrum concept and behavioral information in outdoor recreation resource supply inventories: a rationale, in *Integrated Inventories of Renewable Natural Resources,* USDA Forest Service General Technical Report RM-55, Fort Collins, CO, 24–31.

Faber, B.G., Wallace, W., Croteau, K., Thomas, V. and Small, L. (1997) Active response GIS: an architecture for interactive resource modeling, in *Proceedings of GIS '97 Annual Symposium,* GIS World, Inc., Vancouver, 296–301.

Fenska, R.R. and Lauderburn, D.E. (1924) Cruise and yield study for management, *Journal of Forestry* **22** (1), 75–80.

Friend, A.D., Stevens, A.K., Knox, R.G. and Cannell, M.G.R. (1997) A process-based, biogeochemical, terrestrial biosphere model of ecosystem dynamics (Hybrid v3.0), *Ecological Modelling* **95**, 249–287.

Gimblett, H.R., Durnota, B. and Itami, R.M. (1996) *Spatially-Explicit Autonomous Agents for Modelling Recreation Use in Wilderness: Complexity International,* Vol. 3, http://www.csu.edu.au/ci/vol03/ci3.html

Gimblett, H.R., Richards, M., Itami, R.M. and Rawlings, R. (1995) Using interactive, GIS-based models of recreation opportunities for multi-level forest recreation planning, *Geomatics 1995,* Ottawa, Ontario, Canada, June 1995.

Gustafson, E.J. (1999) Harvest: a timber harvest allocation model for simulating management alternatives, in J.M. Klopatek and R.H. Gardner (eds) *Landscape Ecological Analysis Issues and Applications,* Springer-Verlag, New York, 109–124.

Gustafson, E.J. and Crow, T.R. (1999) HARVEST: linking timber harvesting strategies to landscape patterns, in D.J. Mladenoff and W.L. Baker (eds) *Spatial Modeling of Forest Landscapes: Approaches and Applications,* Cambridge University Press, Cambridge, 309–332.

Gustafson, E.J., Shifley, S.R., Mladenoff, D.J., He, H.S. and Nimerfro, K.K. (2000) Spatial simulation of forest succession and timber harvesting using LANDIS, *Canadian Journal of Forest Research* **30**, 32–43.

Hann, D.W. and Riitters, K. (1982) *A Key to the Literature of Forest Growth and Yield in the Pacific Northwest: 1910–1981: Research Bulletin 39,* Oregon State University, Forest Research Laboratory, Corvallis, OR.

Hashim, S.H. (1990) *Exploring Hypertext Programming,* Windcrest Books, Blue Ridge Summit, PA.

Hawkes, C. (2000) Woody plant mortality algorithms: description, problems, and progress, *Ecological Modelling* **126**, 225–248.

Hof, J. and Bevers, M. (1998) *Spatial Optimization for Managed Ecosystems*, Columbia University Press, New York.

Hof, J. and Joyce, L. (1992) Spatial optimization for wildlife and timber in managed forest ecosystems, *Forest Science* 38 (3), 489–508.

Hof, J. and Joyce, L. (1993) A mixed integer linear programming approach for spatially optimizing wildlife and timber in managed forest ecosystems, *Forest Science* 39 (4), 816–834.

Janssen, R. and van Herwijnen, M. (1994) *DEFINITE: Decisions on a Finite Set of Alternatives*, Institute for Environmental Studies, Free University Amsterdam, The Netherlands, Kluwer Academic Publishers, software on 2 disks.

Marquis, D.A. and Ernst, R.L. (1992) *User's Guide to SILVAH: Stand Analysis, Prescription, and Management Simulator Program for Hardwood Stands of the Alleghenies*, Gen. Tech. Rep. NE-162, US Department of Agriculture, Forest Service, Northeastern Forest Experiment Station, Radnor, PA.

McCarter, J.B., Wilson, J.S., Baker, P.J., Moffett, J.L. and Oliver, C.D. (1998) Landscape management through integration of existing tools and emerging technologies, *Journal of Forestry*, June, 17–23 (also at http://lms.cfr.washington.edu/lms.html).

McGaughey, R.J. (1997) Visualizing forest stand dynamics using the stand visualization system, in *Proceedings of the 1997 ACSM/ASPRS Annual Convention and Exposition*, April, 7–10, American Society for Photogrammetry and Remote Sensing, Seattle, WA, 4, 248–257.

Miner *et al.* (1988) *A Guide to the TWIGS program for the North Central United States*, Gen. Tech. Rep. NC-125, USDA Forest Service, North Central Forest Experiment Station, St Paul, MN.

Mladenoff, D.J., Host, G.E., Boeder, J. and Crow, T.R. (1996) Landis: a spatial model of forest landscape disturbance, succession, and management, in M.F. Goodchild, L.T. Steyaert, B.O. Parks, C.A. Johnston, D.R. Maidment and M.P. Crane (eds) *GIS and Environmental Modeling: Progress and Research Issues*, Fort Collins, CO, 175–179.

Mowrer, H.T., Barber, K., Campbell, J., Crookston, N., Dahms, C., Day, J., Laacke, J., Merzenich, J., Mighton, S., Rauscher, M., Reynolds, K., Thompson, J., Trenchi, P. and Twery, M. (1997) *Decision Support Systems for Ecosystem Management: An Evaluation of Existing Systems*, USDA Forest Service, Interregional Ecosystem Management Coordination Group, Decision Support System Task Team, Rocky Mountain Forest and Range Experiment Station, RM-GTR-296, Fort Collins, CO.

Oliver, C.D. and McCarter, J.B. (1996) Developments in decision support for landscape management, in M. Heit, H.D. Parker and A. Shortreid (eds) *GIS Applications in Natural Resource Management 2*, Proceedings of the 9th American Symposium on Geographic Information Systems, Vancouver, BC, 501–509.

Oliver, C.D. and Twery, M.J. (1999) Decision support systems/models and analyses, in W.T. Sexton, A.J. Malk, R.C. Szaro and N.C. Johnson (eds) *Ecological Stewardship: A Common Reference for Ecosystem Management*, Oxford: Elsevier Science, Vol. III, 661–685.

Orland, B. (1995) SMARTFOREST: a 3-D interactive forest visualization and analysis system, in J.M. Power, M. Strome and T.C. Daniel (eds) *Proceedings of the Decision Support-2001 Conference*, 12–16 September 1994, Toronto, Ontario, 181–190.

Orland, B., Liu, X., Kim, Y. and Zheng, W. (1999) *SmartForest-II: Dynamic forest visualization*, software release to US Forest Service, Imaging Systems Laboratory, University of Illinois, Urbana (http://www.imlab.psu.edu/smartforest/)

Ortigosa, G.R., de Leo, G.A. and Gatto, M. (2000) VVF: integrating modelling and GIS in a software tool for habitat suitability assessment, *Environmental Modelling & Software* 15 (1), 1–12.

Otte, R., Patrick, P. and Roy, M. (1996) *Understanding CORBA: The Common Object Request Broker Architecture*, Prentice-Hall, Englewood Cliffs, NJ.

Pacala, S.W., Canham, C.D. and Silander, J.A.J. (1993) Forest models defined by field measurements: I. the design of a northeastern forest simulator, *Canadian Journal of Forest Research* 23, 1980–1988.

Pacala, S.W., Canham, C.D., Silander, J.A.J., Kobe, R.K. and Ribbens, E. (1996) Forest models defined by field measurements: Estimation, error analysis and dynamics, *Ecological Monographs* 66, 1–43.

Parker, J.K. *et al.* (1999) Some contributions of social theory to ecosystem management, in N.C. Johnson, A.J. Malk, W.T. Sexton and R. Szaro (eds) *Ecological Stewardship: A Common Reference for Ecosystem Management*, Elsevier Science, Oxford, 245–277.

Pinchot, G. (1898) *The Adirondack Spruce*. Critic Co, New York.

Pretzsch, H. (2001) *Modelling Forest Growth*, Blackwell Verlag, Berlin.

Rauscher, H.M. (1999) Ecosystem management decision support for federal forests of the United States: a review, *Forest Ecology and Management* 114, 173–197.

Rauscher, H.M., Kollasch, R.P., Thomasma, S.A., Nute, D.E., Chen, N., Twery, M.J., Bennett, D.J. and Cleveland, H. (1997b) NED-1: a goal-driven ecosystem management decision support system: technical description, in *Proceedings of GIS World '97: Integrating Spatial Information Technologies for Tomorrow*, 17–20 February, Vancouver, BC, 324–332.

Rauscher, H.M., Loftis, D.L., McGee, C.E. and Worth, C.V. (1997a) Oak regeneration: a knowledge synthesis, *The COMPILER* 15, 52 + 3 disks.

Reynolds, K., Bjork, J., Riemann Hershey, R., Schmoldt, D., Payne, J., King, S., DeCola, L., Twery, M. and Cunningham, P. (1999) Decision support for ecosystem management, in W.T. Sexton, A.J. Malk, R.C. Szaro and N.C. Johnson (eds) *Ecological Stewardship: A Common Reference for Ecosystem Management*. Oxford: Elsevier Science, Vol. III, 687–721.

Ribbens, E. (1996) Spatial modelling of forest regeneration: how can recruitment be calibrated?, in J.P. Skovsgaard and V.K. Johannsen (eds) *IUFRO Conference on Forest Regeneration and Modelling*, Danish Forest and Landscape Research Institute, Hoersholm, 112–120.

Ritchie, M.W. (1999) *A Compendium of Forest Growth and Yield Simulators for the Pacific Coast States*, Gen. Tech. Rep. PSW-GTR-174, Pacific Southwest Research Station, Forest Service, S Department of Agriculture, Albany, CA.

Saaty, T.L. (1980) *The Analytic Hierarchy Process*, McGraw-Hill, New York.

Schmoldt, D.T. and Peterson, D.L. (2000) Analytical group decision making in natural resources: methodology and application, *Forest Science* **46** (1), 62–75.

Schmoldt, D.T. and Rauscher, H.M. (1996) *Building Knowledge-Based Systems For Resource Management*, Chapman & Hall, New York.

Schumacher, F.X. and Coile, T.S. (1960) *Growth and Yield of Natural Stands of the Southern Pines*, T.S. Coile, Durham, NC.

Shinozaki, K., Yoda, K., Hozumi, K. and Kiro, T. (1964) A quantitative analysis of plant form – the pipe model theory, I. Basic analysis, *Japanese Journal of Ecology* **14**, 97–105.

Smith, D.M. (1986) *The Practice of Silviculture*, 8th edn, John Wiley & Sons, New York.

Solomon, D.S., Herman, D.A. and Leak, W.B. (1995) *FIBER 3.0: An Ecological Growth Model for Northeastern Forest Types*, USDA Forest Service Northeastern Forest Experiment Station, GTR-NE-204, Radnor, PA.

Solomon, D.S., Hosmer, R.A. and Hayslett, Jr. H.T. (1987) A two-stage matrix model for predicting growth of forest stands in the Northeast, *Canadian Journal of Forest Research* **16** (3), 521–528.

Spies, T.A., Franklin, J.F. and Thomas, T.B. (1988) Coarse woody debris in Douglas-fir forests of western Oregon and Washington, *Ecology* **69**, 1689–1702.

Spurr, S.H. (1952) *Forest Inventory*, Roland Press, New York.

Stage, A.R. (1973) *Prognosis Model for Stand Development*. Research Paper INT-137, US Department of Agriculture, Forest Service, Intermountain Forest and Range Experiment Station, Ogden, UT.

Stage, A.R. (1997) Using FVS to provide structural class attributes to a forest succession model (CRBSUM), in R. Teck, M. Moeur and J. Adams (eds) *Proceedings of Forest Vegetation Simulator Conference, February 3–7 1997*, Fort Collins, CO, Gen. Tech. Rep. INT-GTR-373, Ogden, UT: US Department of Agriculture, Forest Service, Intermountain Research Station, Ogden, UT, 139–147.

Sygenex (1994) *Criterium Decision Plus: The Complete Decision Formulation, Analysis, and Presentation for Windows: User's Guide*, Sygenex Inc., 15446 Bel-Red Road, Redmond, WA 98052.

Twery, M.J., Rauscher, H.M., Bennett, D., Thomasma, S., Stout, S., Palmer, J., Hoffman, R., DeCalesta, D.S., Gustafson, E., Cleveland, H., Grove, J.M., Nute, D., Kim, G. and Kollasch, R.P. (2000) NED-1: integrated analyses for forest stewardship decisions, *Computers and Electronics in Agriculture* **27**, 167–193.

Urban, D.L. and Shugart, H.H. (1992) Individual-based models of forest succession, in D.C. Glenn-Lewin, R.K. Peet and T.T. Veblen (eds) *Plant Succession*, Chapman & Hall, New York, 249–292.

USDA Forest Service. (1995) *The Forest Service Program for Forest and Rangeland Resources: A Long-Term Strategic Plan*, Washington: US Department of Agriculture, Forest Service.

Valentine, H.T., Gregoire, T.G., Burkhart, H.E. and Hollinger, D.Y. (1997) A stand-level model of carbon allocation and growth, calibrated for loblolly pine, *Canadian Journal of Forest Research* **27**, 817–830.

Villa, F. and Costanza, R. (2000) Design of multi-paradigm integrating modelling tools for ecological research, *Environmental Modelling & Software* **15** (2), 169–177.

Vogt, K.A., Larson, B.C., Gordon, J.C., Vogt, D.J. and Franzeres, A. (1999) *Forest Certification: Roots, Issues, Challenges, and Benefits*, CRC Press, New York.

Walters, C.J. and Holling, C.S. (1990) Large-scale management experiments and learning by doing, *Ecology* **71**, 2060–2068.

Waring, R.H., Schroeder, P.E. and Oren, R. (1982) Application of the pipe model theory to predict canopy leaf area, *Canadian Journal of Forest Research* **12**, 556–560.

Weinstein, D.A. and Yanai, R.D. (1994) Integrating the effects of simultaneous multiple stresses on plants using the simulation model TREGRO, *Journal of Environmental Quality* **23**, 418–428.

Williams, S.B., Roschke, D.J. and Holtfrerich, D.R. (1995) Designing configurable decision-support software: lessons learned. *AI Applications* **9** (3), 103–114.

Yaussy, D.A., Sutherland, E.K. and Hale, B.J. (1996) Rule-based individual-tree regeneration model for forest simulators, in J.P. Skovsgaard and V.K. Johannsen (eds) *Modelling Regeneration Success and Early Growth of Forest Stands: Proceedings from the IUFRO Conference*, Copenhagen, 10–13 June 1996, Danish Landscape Research Institute, Horsholm, 176–182.

Models available on the Web

EnVision	–	http://forsys.cfr.washington.edu/envision.html
FIBER	–	http://www.fs.fed.us/ne/durham/4104/products/fiber.shtml
FVS	–	http://www.fs.fed.us/fmsc/fvs/index.php
HARVEST	–	http://www.ncrs.fs.fed.us/products/Software.htm
LANDIS	–	http://landscape.forest.wisc.edu/Projects/LANDIS_overview/landis_overview.html
NED	–	http://www.fs.fed.us/ne/burlington/ned
Pipestem	–	http://www.fs.fed.us/ne/durham/4104/products/pipestem.html
RBSim	–	http://www.srnr.arizona.edu/~gimblett/rbsim.html
Ritchie	–	http://www.snr.missouri.edu/silviculture/tools/index.html for numerous useful links to models of various kinds that are available for downloading
SILVAH	–	http://www.fs.fed.us/ne/warren/silvah.html
SmartForest	–	http://www.imlab.psu.edu/smartforest/
SVS	–	http://forsys.cfr.washington.edu/svs.html
TWIGS	–	http://www.ncrs.fs.fed.us/products/Software.htm (this site also contains other related models)
UTOOLS	–	http://faculty.washington.edu/mcgoy/utools.html

18

Stability and Instability in the Management of Mediterranean Desertification

JOHN B. THORNES

18.1 INTRODUCTION

In this chapter, the main aim is to introduce the problems of desertification and to develop some ideas based on the complexity of response, as a basis for a fresh theoretical approach to understanding and managing desertification. The emphasis will be on the stability of soil and vegetation, the relationships between them and the behaviour of this relationship in the light of varying climatic conditions. Surprisingly, in the early 1990s there were some who believed that the issue of desertification in the Mediterranean required not much more than the solution of the differential equations of atmospheric motion over flat regions, provided that the boundary conditions and the model parameters could be properly identified and satisfied.

The approach here is heuristic reasoning, revitalized by Polya (1964). Heuristic reasoning is not regarded as final and strict, but as provisional and plausible only, the purpose of which is to discover the solution of the present problem. Before obtaining certainty, we must be satisfied with a more or less plausible guess. We may need the provisional before we obtain the final (ibid.: 113). This approach is not intended to provide an excuse for lack of rigour, but rather to admit the possibility of a creative approach to the problem, where 'technical fixes' (solutions) have largely failed. The creative–technical dichotomy of approaches is an alternative expression of the culture/nature dichotomy so carefully exposed and discussed by van der Leeuw (1998), following Evernden (1992). Hudson (1992) also identifies the contrast between the technical and land-husbandry approaches to land degradation. The problem

is not simply to identify and apply technical fixes to solve soil erosion. Rather, it is to develop a deeper consciousness of the problem and cultivate the notion of social responsibility to the land, so well encapsulated in the term 'land husbandry'.

Nevertheless, at the end of the day, efforts have to be focused in the right direction on a common basis, especially in a multidisciplinary context, to avoid repetition, replication and poor return for the effort applied. To this end, 'desertification' is understood here to be: 'land degradation in arid, semi-arid and dry subhumid areas, resulting from various factors, including climatic variations and human activities'. 'Land' means the terrestrial bio-productive system that comprises soil, vegetation, other biota and the ecological and hydrological processes that operate within the system. 'Land degradation' means reduction and loss of the biological or economic productivity caused by land-use change, from a physical process. These include processes arising from human activities and habitation patterns such as soil erosion, deterioration of the physical, chemical and biological or economic properties of the soil and long-term loss of vegetation.

In this chapter, the priority is to engage with the physical processes that involve the interaction between soil erosion and vegetation cover. The chapter starts by examining this interaction in a complex systems context, developing the argument that soil erosion results from the instability between soil production and removal rates, which in turn is a function of the stability of the vegetation cover. This interaction in turn leads to the concept of competition between soil and plant cover that

Environmental Modelling: Finding Simplicity in Complexity. Edited by J. Wainwright and M. Mulligan

leads to erosional outcomes and to a deeper discussion of the concept of complexity and how it can be identified.

Since the great drought crisis of the Sahel in the 1960s and 1970s, desertification has been strongly identified with climate change. The third part of the chapter attempts to demonstrate how change in climatic gradients can lead to critical instabilities at different places along the gradient, and how to conceptualize the recovery trajectories in both space and time. This emphasis on climate change belies the fact that plant productivity is strongly affected both by human and animal impacts on the plant cover and by the intrinsic changes within the plant cover itself through time. These issues form the final section of the chapter, where we emphasize that the emergence of instability in the plant cover itself results not only from climatic and anthropic variation but also from genetic drift among the hillside plant communities.

18.2 BASIC PROPOSITIONS

The fundamental core of what follows is the proposition that plant cover reduces the rate of erosion. Although hypothesized for over 100 years, this was first empirically demonstrated in a simple but important experiment by Elwell and Stocking, published in 1976 and re-affirmed several times since (e.g. Shaxson *et al.*, 1989). They showed that there is a steep decay in the rate of soil erosion, compared to that on bare soil, as the plant cover increases. The erosion rate is scaled so that the erosion rate for a bare soil (i.e. no vegetation) = 100% and vegetation is scaled between 0 and 100% (Figure 18.1a). Notice that the curve is not linear but negative exponential. With 30% cover, the erosion rate has reduced to

30% of the bare soil value. Francis and Thornes (1990) further showed that this relationship depends on the intensity of rainfall. With higher mean intensity of rainfall, the curve is steeper than with lower mean rainfall intensity (Figure 18.1b).

Another variant of the exponential delay (Figure 18.1c) is that the erosion rate increases slowly above zero as the cover increases. This is thought to represent the effect of shrubs or bunch grasses breaking up sheet flow into more concentrated rill flows as the flow becomes reticular between the scattered and concentrated vegetation cover. Rills form and lead to higher erosion rates.

The second basic proposition is that erosion is caused by either Hortonian or saturated overland flow. The former occurs when the rainfall intensity is greater than the soil-infiltration rate. The latter occurs when the soil is already so full of water that no more can enter. Soil-water storage is thus crucial to the occurrence of erosion. Soil storage is the product of the soil moisture per unit depth and the depth over which the soil moisture is stored. Deep soils with low moisture content have greater storage. Conversely, shallow soils with high moisture content are close to saturation. According to Musgrave's equation, erosion rate (E, MT^{-1}) is a function of the power of unit discharge and slope in the form (Musgrave, 1947):

$$E = Kq^m s^n \qquad (18.1)$$

where K is an erodibility parameter, q is the discharge per unit width (LT^{-1}), s is the slope angle (dimensionless), and m and n are exponents shown empirically and theoretically for sheet wash to be 2 and

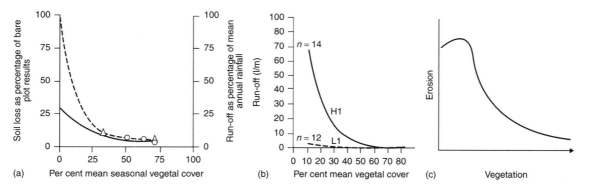

Figure 18.1 Exponential relationships between erosion rate and vegetation cover (a) the exponential decrease of erosion as vegetation cover increases (after Elwell and Stocking, 1976); (b) the effect of high and low intensity rainfall on exponential decrease of erosion (after Francis and Thornes, 1990); and (c) the effect of intermittent bushes on the exponential decrease of erosion at low densities

1.66 respectively (Carson and Kirkby, 1972). This relationship means that thin soils with high soil-moisture content have much erosion, while deep soils with low soil-moisture content have low erosion rates and that intense rainfall produces more runoff and erosion than less intense rainfall. K in the Musgrave equation is the soil erodibility, as measured by the shear strength and cohesion of the soil. Most activities that change the soil erodibility, water-holding capacity or infiltrability of the soil also thereby change the likelihood of erosion.

Since 1976, the role of the stone content of soils has also been identified as being involved in runoff production, though the evidence is not unequivocal (Poesen and van Waesemael, 1995). When particles are lying on the surface, infiltration tends to increase as the fraction of stone fragments increases. When particles are embedded within the soil, the infiltration and storage capacities decrease. Surface stones reduce runoff and protect soil against erosion. Embedded stones reduce storage, increase runoff and may lead after erosion to an armoured layer. In the models described below, stone content is considered an important component of erosion control.

The third basic proposition needed is that, after vegetation has been cleared, for example, by grubbing up, by deliberate removal by machinery, or perhaps by fire, the recovery rate is assumed to be logistic. That is, the rate of re-growth takes place slowly at first,

then very quickly, then it converges on the carrying capacity of the environment. This process is illustrated in Figure 18.2, where the carrying capacity is at level Kv. The rate of growth converges asymptomatically on to Kv. If too much vegetation is produced, then the system is said to overshoot and it will fall again until it converges on Kv. This pattern of growth has been demonstrated empirically. The steep part of the growth, between t_1 and t_2, results from positive feedback. In this time period, more vegetation encourages growth, so that the rate of growth depends on the existing vegetation biomass. As vegetation growth occurs, it provides organic matter to the soil, which provides more nutrients and better soil water-holding capacity. There is, we suppose, a limit to the maximum possible growth, determined by the available moisture. This limit is, generally, of the order of $1-5 \, \text{kg m}^{-2}$ of biomass in semi-arid areas, so that in Almeria, Spain, the standing biomass is approximately $3 \, \text{kg m}^{-2}$, with a rainfall of 300 mm (Francis and Thornes, 1990).

Obando (1997) studied a range of abandoned fields of different ages in Murcia Province, Spain. By measuring the biomass, she was able to plot the amount of biomass after different periods of regeneration. The results confirmed a logistic pattern of growth. Similarly Godron *et al.* (1981) estimated the recovery of Mediterranean vegetation after fire and again this recovery was shown to follow the logistic form.

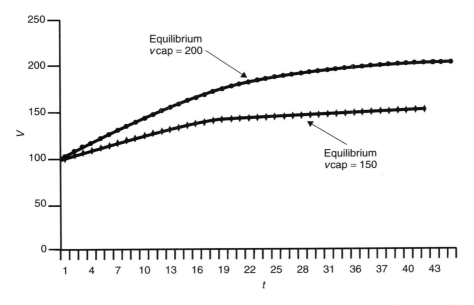

Figure 18.2 The form of the logistic curves of vegetation growth with different maximum possible vegetation biomass values

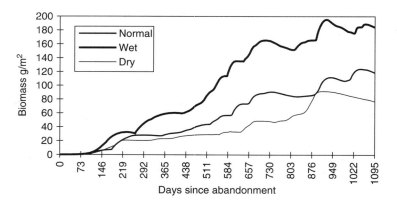

Figure 18.3 The modelled recovery of vegetation cover from bare ground, starting at different times and therefore having different rainfall histories in the recovery period
Source: Obando (1997)

Obando (1997) also modelled plant production on abandoned land, using a plant growth model developed by Thornes (1990) on the basis of Tilman's (1982) resource-based model using essentially a daily growth period and making growth proportional to water efficiency and then partitioning the growth to leaf, stem and root according to partition coefficients. Respiration needs are also budgeted in the model. An important result emerged – that the rate of recovery was heavily determined by the amount and pattern of rainfall (as modelled of course), so that generation that started at different times and therefore having different rainfall histories during the regeneration period has different rates of recovery. This process is shown in Figure 18.3, where each curve of regeneration starts at a different year and years with large extreme rainfalls usually show faster regeneration.

The same logistic growth is assumed to hold for soil erosion. That is, the more erosion there is, the more there will be. Erosion starts slowly, accelerates and then slows down as equilibrium is reached. The positive feedback occurs because shallower soils produce more runoff, more erosion and therefore shallower soils. The rate of erosion slows down eventually as more and more soil is removed. This slowing is mainly because lower horizons are denser and have a higher stone content (Thornes, 1990). Ultimately, when there is no soil left, there can be no more erosion, so the maximum 'carrying capacity' that can be reached (Ks) is the total soil depth. Unfortunately, in soil erosion formulae, such as the Universal Soil Loss Equation (USLE: see Chapter 16), the erosion is assumed to proceed to potential depth, even where this might be deeper than the actual soil depth. Sometimes it is assumed by the

unwitting model user that bedrock can be consumed at that rate! Claims for very high soil-erosion rates based on the USLE should be treated with caution. Although the logistic equation has not been validated for soil erosion, the 'slowing down' effect is often recorded from suspended sediment data and is ascribed to soil 'depletion'. This depletion is usually taken to indicate that all the 'available' soil has been used up.

The logistic assumption is very important because it underpins many ecological models developed by Lotka and Voltera (Lotka, 1925; Voltera, 1926) and described by Maynard Smith (1974) and May (1973). These models form the basis of the following sections of this chapter. A more detailed discussion of the technique and the details of the modelling approach which follows is contained in Thornes (1985).

18.3 COMPLEX INTERACTIONS

There are constraints to the logistic growth described above and these give rise to some very interesting and complex behaviour. The interactions can be set up as competition between erosion and plant growth. As in the above section, more plant cover means less erosion, so the tendency to erosion is inhibited by vegetation cover. Thinner soils mean less available soil moisture and so less plant growth. Following Maynard Smith's analysis (1974), Thornes (1985) set up the differential growth equations for plant logistic growth and erosion logistic growth, constrained by erosion and vegetation respectively. The mathematical argument is not repeated here but, by solving the partial differential equations, the behaviour of the erosion–vegetation interactions can be

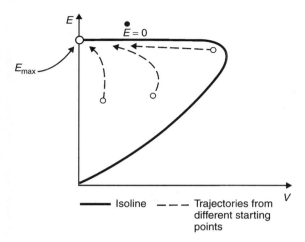

—— Isoline — — — Trajectories from different starting points

Figure 18.4 System behaviour in terms of erosion in the absence of vegetation. The solid line is the isocline along which there is no change in the erosion ($dE/dt = 0$)

explored. By this procedure, we can start the system at any value of E and V and follow the behaviour as solved in time by the equations. Figure 18.4 shows how the erosion behaves in the absence of vegetation. The line $\dfrac{dE}{dt} = \dot{E} = 0$ is called an isocline. It shows all the values along which there is no change in the erosion rate (i.e. it is in equilibrium). For a point below the isocline, the erosion rate increases (moves up) until the isocline is reached. For points above the isocline, their time trajectories move down to the line of equilibrium. Figure 18.5 shows the equivalent diagram for the vegetation system behaviour in the absence of erosion. In this case, the isocline is the point along which $\dfrac{dV}{dt} = \dot{V} = 0$ and if the system is in any state in the plane (given by E, V values), the trajectories within the isocline carry the system towards higher V until they eventually reach the isocline. Notice that, outside the isocline, the system moves to the left until

it again reaches the isocline $\dot{V} = 0$. For the left-hand rising limb, the system always flows away from the equilibrium. Thus, at point a, the vegetation will move to full cover at point a^1 if it is just to the right of the isocline. If it is just to the left of the isocline, it will move towards zero vegetation cover at point a^{11}. So, all the points on the left-hand, rising limb of the isocline are unstable equilibria because a slight perturbation will carry the system away from the equilibrium to either zero or complete cover (nearly 100%). Because there is convergence on the right-hand set of equilibria (e.g. point a^{111}), these are stable equilibria. If the system starts on the equilibrium, it is impossible for it to move away and, in fact, if it is artificially displaced (e.g. by grazing) the dynamics of the system force it to return to the isocline.

A few general comments are in order here. First, the equilibria we have shown here are both unstable and stable. In unstable equilibria, slight perturbations carry the system away from the equilibrium, sometimes far away and even a tiny perturbation can cause a big change. This situation is like a ball balanced on the nose of a sea lion. A tiny wobble (perhaps caused by a breeze) can easily upset the balance. In the stable equilibrium, considerable effort is needed to drive the system away from the isocline and if it is driven away, then the complex behaviour forces it back to the equilibrium, like a marble displaced from its resting place at the bottom of a tea cup and then rolling back to the lowest point. Stable equilibria which draw in the system after perturbations are called attractors. Figure 18.6 shows another metaphor for this behaviour. Ball A is in unstable equilibrium and a slight perturbation will push it into local attractor basins at B or C. Further perturbations can force it into D or E. F is called the global equilibrium because the ball will remain stable for all perturbations and, following a set of sequential perturbations, the system (ball) will end up at F. There is another point here. The deeper the attractor basin, the bigger the perturbation needed to displace the system from the equilibrium.

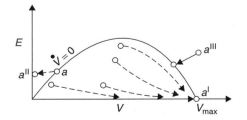

Figure 18.5 Isocline for vegetation in the absence of erosion ($dV/dt = 0$)

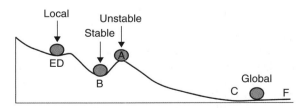

Figure 18.6 Metaphor for different types of equilibria: local, stable, unstable and global

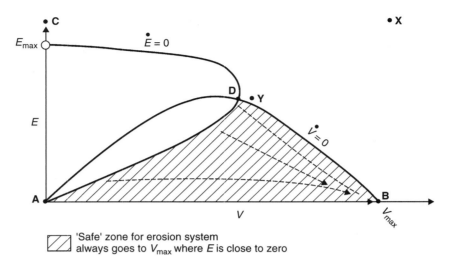

Figure 18.7 Combined effects of vegetation and erosion. The intersection of the isoclines is where both variables are unchanging. The shaded area is the 'safe' zone for erosion

In Figure 18.7, the combined effect of both the erosion and vegetation components are shown. The points where $\dot{E}$ and $\dot{V}$ intersect are the joint equilibria – the point where neither erosion nor vegetation is changing. At A, both vegetation and erosion are zero. At B, vegetation is at a maximum and erosion is 0. At C, erosion is at a maximum and vegetation is 0. The behaviour at D is a mixture of unstable and stable. If the system is perturbed in the areas with shading, it will move in towards D, so now the point D is acting as a repeller. In the triangle ADB, the system behaviour will carry it to the attractor at B, which is very attractive for soil conservationists, because it has zero erosion and maximum vegetation cover. In the shaded area ADC, the system is forced to the attractor at C, which is very undesirable because it represents zero vegetation cover and maximum erosion. In fact, we can go further. The line A–D (if we can locate it) separates all trajectories to attractor C from those to attractor B. In conservation terms, this is a most desirable line. Ideally, we want to keep the system to the right of this line, or move it there by management actions. The triangle ABD is the area we wish our system to be in. Any combined values of $\dot{E}$ and $\dot{V}$ in this area take the system to B. Conversely, in ACD the system will move to the attractor (stable) equilibrium at C with maximum erosion and minimum vegetation cover and it will then require a great effort to move it away. At D, it could go either way and A, C and B are possible destinations. Another feature emerges. If the system is at point X according to its values of E and V, then it has a long way to go before the system

dynamics take over. In this sense, point X is fairly stable. By contrast, if it is in condition (state) Y, then quite a small perturbation will carry it to C, D or B. So, in some cases, very small adjustments to the system can quickly take the result to stable conditions (states) or to unstable conditions. In complex systems, a little change goes a long way, principally because of the positive feedbacks that were described above. The difficulty is to know (a) where are the isoclines (thresholds) over which the behaviour changes very rapidly with only small perturbations?; (b) where are the attractors and are they stable or unstable? and (c) what are the equilibria and are they stable or unstable or saddles? Thornes (1990) examined these questions with more elaborate systems, including those with stony soils.

There are two catches to this optimistic picture. First, descent to the attractor is not smooth; the approach is very fuzzy in the sense that there may be a variability that 'captures' the system for the attractor before it gets there. If the 'external' forces perturbing the system are also noisy (vary a lot in their strength), then the system may be forced onto a trajectory from an unexpected position. Thornes and Brandt (1993) explored the behaviour of the erosion–vegetation dynamical system for stochastic (unpredictable in time) variations in rainfall for a system in the dry Mediterranean in southeast Spain. Although this fuzziness produced some quite unexpected behaviour, the vegetation showed a strong oscillation of high and low vegetation covers and corresponding low and high erosion rates. This kind of behaviour in dynamical systems usually indicates that there is no strongly preferred attractor and that the system is moving between attractors. An alternative explanation is that the vegetation–moisture–erosion system is a 'hypercycle' (Eigen and Schuster, 1979) that can also produce nonlinear oscillations. This alternative explanation has been explored very recently in the context of grazing (Thornes, in preparation). The results of Thornes and Brandt also showed that occasionally there would be overshoots. With a lot of rainfall, the vegetation grew well in excess of carrying capacity and in subsequent years had to 'die' back to the carrying capacity through negative feedback. It was also found that the stability of the system, as formulated, was significantly affected by the stone content.

Note that two controlling influences have been identified in this section, internal or intrinsic and external or extrinsic. The intrinsic influences are thresholds and behaviour within the system itself, the trajectories, thresholds and equilibria (repellers and attractors) that determine the courses of behaviour, their final destinations (unless they go off the graph, far away from equilibrium) and their stability. Extrinsic forces are those that perturb the system away from the internal controls towards new equilibria. If we had a good understanding of both sets of mechanisms, and there was not too much fuzziness or randomness, then we might be in a position to use this knowledge to determine the future trajectories or the reactions to deliberate perturbations, such as 'over-grazing' or torrential rainfall. Obtaining absolute values for this behaviour, such as the trajectories in space and time, requires not only an understanding of the system couched in this framework, but also a good knowledge of the parameters of the system and the initial conditions (what are the values of E and V at the start?). These are demanding requirements, though if some of the many plot experiments were reconfigured, the data could be easily obtained.

Because there are several (sometimes many) thresholds in the system separating attractor basins, an important question is, how close to the threshold can we get before the system is captured by an attractor basin? In Figure 18.7, how close can we get to the A–D line before we go to either complete cover and no erosion (attractor B) or no cover and very high erosion (attractor C)? Obviously this is a crucial question for management. If we accidentally tip the system across the boundary, the resulting outcome could be the exact opposite of what is required. The judgement as to what is a good outcome and a bad outcome is independent of the dynamics. But in the case of the management implications, it might be a question of how many $kg\,m^{-2}$ of biomass should be harvested or animals should be allowed to graze.

This decision will determine the horizontal movement in Figure 18.7 and hence depends on two things: (a) how far the system is away from the boundary; and (b) how 'noisy' the process is. This last point is very important.

18.3.1 Spatial variability

Because the value of E depends on the soil and rainfall characteristics over space, the runoff prediction is extremely variable. On two adjacent soil patches the erosion rate could be quite different, putting us either side of the threshold. So bare and strongly vegetated patches could co-exist and indeed they do! Through time, the intensity varies, changing the overland flow production. Although generally one uses a mean intensity to fix the position in the space $E-V$, if this variation is very large, then it is likely to straddle the threshold. In fact, the distribution of intensity is exponential (Kosmas *et al.*, 1999). So if the mean is located on the threshold between two attractors, we can define a 'belt' either side of the threshold, in which there is a risk of the system moving to another attractor. The larger the deviation, the larger the risk. The 'noisier' the system, the larger the risk. This 'noise' varies spatially, as does the runoff production, vegetation cover and erosion. Consequently, spatial outcomes could be quite varied and it is interesting to ask where the thresholds exist in geographical space. In a patchy landscape with different soil types and different runoff productions, the system is differentially located with respect to the thresholds. Just crossing a soil boundary or the edge of a vegetation patch could trip us towards another attractor. Human-induced patchiness is, in this respect, critically important in trying to foresee the location of desertification in space (Thornes, 1990). What has not yet been studied or modelled is any 'contagion' effect of desertification, though given the catena concept and patch dynamical growth approaches (Tilman and Karieva, 1997), this is clearly an important problem for the near future. Often in Mediterranean environments, one can observe a whole hillside fairly uniformly covered with a single shrub species (such as *Anthyllis cytisoides*) and the following question arises. If a small locus of erosion breaks out somewhere in the middle, will it propagate outwards in an unstable fashion, moving the whole hillside to the erosion attractor? This evolution has been suggested by Kosmas for the case of the bush *Saceropoterium*, on the island of Lesbos, Greece (Kosmas, personal comm.). Alternatively, given that different species have hydrologically different characteristics, if a different species invades such a slope, could it shift the system across a threshold to a different attractor? We shall refer back to this question in the fourth section of this chapter, arguing that species mixtures may lead to conditions that are 'unstably contagious' by virtue of their genetic evolutionary potential and population dynamics.

18.3.2 Temporal variability

Of course, the very word dynamical implies that the analytical models are evolutionary (changing through time) according to their internal mechanism, but the external controls are also changing through time. Theoretically, it should be possible to use the equations of dynamical behaviour to determine how long the system will take to adjust to a perturbation or external change (the relaxation time: Brunsden and Thornes, 1979). This delay has important impacts in dynamical systems and external changes can be scaled. High frequency changes (sudden storms) may have significantly different impacts from lower frequency changes (such as internal variations in storm intensities).

In desertification studies, these continue to be a preoccupation, with climate change as a major cause in spite of the overwhelming evidence that socio-economic controls are probably more important. In this respect, Allen and his colleagues developed the human dimensions aspect of dynamical systems models in desertification studies (Allen, 1988). There also, thresholds, attractors, intrinsic and extrinsic forces all have their roles.

Climate variation can also be regarded as perturbations operating over the long timescale (tens of years). The main effects, in terms of the models outlined above, are on the overall carrying capacities (V_{cap}), the instantaneous growth rate coefficient in the logistic equations (k_1), the soil moisture constraining plant growth and the direct production of overland flow through varying rainfall intensities and therefore the rate of erosion. The variability in rainfall is also important, as indicated in the previous section. Temperature and CO_2 concentrations also affect plant growth and are therefore an integral component of modelling the impacts of climate change on desertification, as used by Diamond and Woodward (1998).

There are several ways of attempting to model the impacts of climate change on desertification. The most common is direct digital simulation of the impacts of change. For the Mediterranean, the MEDALUS team at Leeds University under Mike Kirkby has produced a variety of simulation models for different space and time scales. The MEDALUS hillslope model (Thornes *et al.*, 1996) investigates field-scale erosion in semi-arid environments and operates at an hourly event scale. It is essentially based on the principles of TOP-MODEL (Beven and Kirkby, 1979). The MEDRUSH model (Kirkby *et al.*, 2002) operates at the large catchment scale and uses the technique of a statistical ensemble of a 'characteristic' flow strip whose hydrology, vegetation growth and erosion are modelled by physical process laws and applied to the representative strips which thus change its morphology. The theory developed permits this changed morphology to cause the strips to migrate within a basin and the sediment is routed down the channel. In this model, the initial vegetation cover is provided in the catchment GIS from remotely sensed imagery and the models can be used to predict the impacts of different physical (rainfall) and human (land-use) perturbations on runoff and sediment yields from the catchment.

Empirical models have also been used to examine the effects of climate change on land degradation and runoff, both over space (Kirkby and Cox, 1995) and over time (Arnell, 1996). These examples have the usual problems of such models. Time series have been parameterized by particular data sets and therefore are not readily transportable. They often have (and should have) error bands attached to them that may be quite wide. This rather ancient technology is highly speculative and often misleadingly simple.

Even further down the line are analogue studies in which the authors seek contemporary or historical analogues for supposed climate changes among existing or past environments. Thus one might expect parts of the south shore of the Mediterranean to be suitable analogues for the northern shore. The problem is obvious. Obtaining suitable analogues is very difficult especially when the systems are only weakly understood. This is the kind of approach that enables us to hope that, 70 years from now, southern Britain might be like southern Spain and therefore the farmers of Sussex can change their potato fields for vineyards.

A special version of this argument that has become popular in recent years is the gradient analogue. Where there is a strong climatic gradient, detailed studies of geomorphological and ecological processes with their respective responses can be used to demonstrate how the changing climatic parameters along the gradient control the erosion rates. When these are coupled with plot studies, the linkages so made can be confirmed. An outstanding study of this type was carried out by Lavee and Pariente (1998) between Jerusalem and Jericho, a gradient of strongly decreasing rainfall from the uplands of central Israel to the Dead Sea. The methodology is to seek along the gradients for analogues of expected future climate conditions. This might reveal what the future

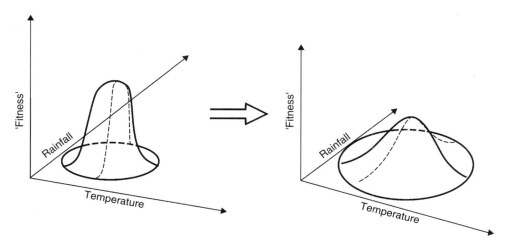

Figure 18.8 Pianka's representation of the hypothesized impact of climate change on plant fitness

processes and responses will be, but does not reveal the trajectories of change. As was shown earlier, these trajectories are crucial in determining the final state, especially if they carry the system across thresholds between quite different attractor basins. The path to equilibrium need be neither steady nor smooth. This problem often occurs in mathematical modelling of change when, by setting the rate of change equal to zero and solving for equilibrium, the equilibrium state is obtained but not the trajectories towards it. Kirkby (1971) used this technique in his characteristic slope-form paper, in which the characteristic forms are derived from the values of the parameters in the Musgrave erosion equation (Musgrave, 1947). The equilibrium forms obtained are in fact attractors in the $m-n$ space though neither the trajectories nor the relaxation times to equilibrium were obtained in the analysis. This example appears to be one of the earliest analytical applications of dynamical systems analysis in the field of geomorphology, though it is rarely recognized as such.

Another approach (Pianka, 1978) is to provide the fitness distribution for plants in a climate phase space and then judge the climatic change impacts by shifting the curves in the phase space (Figure 18.8) and adjusting the fitness response.

18.4 CLIMATE GRADIENT AND CLIMATE CHANGE

Using the tools described earlier, we consider in this section how the gradient-change problem can be approached heuristically using the dynamical systems approach. A novel treatment of the gradient problem

was developed by Pease *et al.* (1989). They envisaged the production of the vegetation at ground level by the intersection of an atmospheric gradient with a planar surface over which the change is passing (Figure 18.9). The response at equilibrium is controlled by Lotka–Voltera competitive growth (cf. Thornes, 1988) whereby not only is there competition between the plants, but also the stability of the outcomes can be specified from a study of the modelled interactions, as described below (Lotka, 1925; Voltera, 1926). As the climate changes, the atmospheric zones shift across the land, generating new vegetation in response to this change. The species, in fact, track the moving environment by adjustment of the patches through a critical patch size dynamic, in which the population within the patch is counter-balanced by dispersal into an unsuitable habitat (Skellam, 1951). The dispersal changes the mean phenotype and, because of the environmental gradient, natural selection will favour different phenotypic values in different parts of the environment. The model also incorporates additive genetic variance.

Changing the climate also changes the carrying capacity (V_{cap}) in the logistic equations for plant response (see below). But soil moisture is the control over the above-ground biomass. Hence the cover and soil-moisture content are not a simple function of mean annual rainfall, but also of the magnitude and frequency of rainfall and the evapotranspiration conditions forced mainly by temperature. So rainfall (R) and temperature (T) variations offer a good approximation to plant cover (Eagleson, 1979), given all other co-linearities. Different coefficients of water use efficiencies may also determine the main functional vegetation type (bare, grass, shrubs or

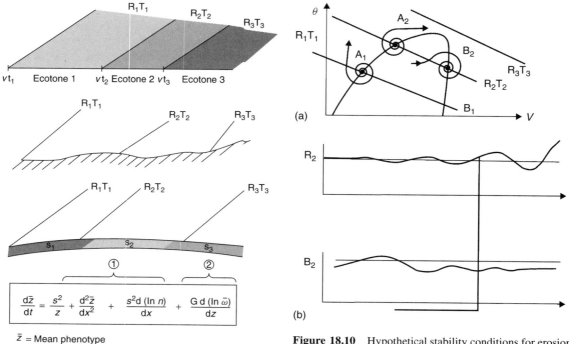

$$\frac{d\bar{z}}{dt} = \frac{s^2}{z} + \frac{d^2\bar{z}}{dx^2} + \frac{s^2 d (\ln n)}{dx} + \frac{G d (\ln \bar{\omega})}{dz}$$

$\bar{z}$ = Mean phenotype
G = Additive genetic variance in character
① How migration changes phenotype
② Adjustment of adaptive capacity

Figure 18.9 The formulation by Pease *et al.* (1989) of a climate change moving across a landscape. Upper: on a flat plane; middle: on a topographically varying surface; lower: on a surface with different soil types. Box and text equations used for modelling behaviour

Figure 18.10 Hypothetical stability conditions for erosion under changing rainfall (R) and temperature (T) along a climatic gradient. See text for detailed explanation. Upper part represents stability conditions on a plot of soil moisture (θ) and vegetation biomass (V). Lower part shows time trajectories of system against time

trees) (Tilman, 1982). Using the logistic equation

$$\frac{dV}{dt} = k_1 V \left(1 - \frac{V}{V_{cap}}\right) \qquad (18.2)$$

with $V_{cap} = 200$ and $k_1 = 6.1$, the equilibrium biomass is higher and reached later than if $V_{cap} = 150$ and $k_1 = 6.1$ as indicated in Figure 18.2. This expression can be re-written to allow for positive growth, as θ (soil moisture) increases linearly with coefficient k_3 as follows:

$$\frac{dV}{dt} = k_2 V \left(1 - \frac{V}{V_{cap}}\right) + k_3 \theta. \qquad (18.3)$$

This equation can then be solved for $dV/dt = 0$ with the resulting isocline in the plane as shown in the earlier section (and by the 'hump' in the upper part of

Figure 18.10) and the trajectories calculated. If now θ is reduced by interception, then:

$$\frac{d\theta}{dt} = -k_4(R, T) - k_5\theta + k_6 R. \qquad (18.4)$$

θ is assumed to increase as a linear function of rainfall and decreases in relation to water content θ (by drainage proportional to conductivity, k_5) and vegetation V (by water consumption and interception). The consumption rate is also assumed to be a function of T, the mean annual ambient air temperature (for evapotranspiration). Setting $d\theta/dt = 0$, the isoclines for a given value of T can be obtained. As a consequence, the isoclines run from high θ at low V to low θ at high V (Figure 18.10). The separate lines T_1–T_3 represent the equilibrium between θ and V and they can be represented on the same diagram as the $\dot{V}$ isocline. The family of R–T isoclines now intersect the $\dot{V}$ isocline at points A and B, where there are joint equilibria corresponding to vegetation cover, rainfall, soil moisture and air temperature. As before, the system dynamics can be obtained for any point on the $V-\theta$ plane. For the plane, points above the isocline will have $\dot{V}$ decreasing until they reach the isocline and because the system is moving towards the isocline from both higher and lower values of V, the $\dot{V}$ isocline is stable on the right-hand limb. Once the system comes to lie on the isocline, there can be no further change. At the intersection of the two isoclines (points A and B), there is a joint equilibrium of V and θ whose behaviour is resultant of the behaviour of both (Figure 18.10). Intersections on the rising limb are unstable equilibria. Perturbations from equilibrium A will be driven away from A. By contrast, perturbations from equilibrium B will be driven back towards B, which is an attractor. This point is shown by the time graphs in the lower part of Figure 18.10. Moreover, an oscillatory damped approach to equilibrium characterizes perturbations spiralling into B, whereas there will be an undamped explosive behaviour spiralling out from A. In practice, fluctuations about A and B are likely to occur at random, so that there is a bivariate probabilistic distribution of values about A and B of bifurcation points.

18.5 IMPLICATIONS

The implications are that shifts in either $\dot{\theta}$ or $\dot{V}$ isoclines not only identify changes in equilibrium that may result from systematic or random perturbations in the controlling variables, but also the very character of future responses to perturbations, and accommodate the 'boom or bust' adjustments reported in May (1973) or in Thornes (1985, 1990).

To the extent that the behaviour of vegetation and soil moisture is represented by the equations described, then their responses to climate change or perturbations in the controlling variables can be modelled tentatively by the evolutionary modelling strategy described in this section. Different values and combinations of the values of coefficients k_1-k_6 will determine the stability conditions following climate change. The very existence of the climate gradient implies that the stability conditions of the response at different places and different times will vary in kind rather than in magnitude, that the responses cannot be assumed to be simple and linear, but will be complex and nonlinear, and that the gradient-change methodology is a tool of dubious value, even when the local processes and their responses are carefully defined by field studies. Here is the rub. Which strategy, field experiment and observation or heuristic modelling, can we most rely on? The limitations of models are identified by Roughgarden (1979: 298) 'Most models in ecology are intended for other ecologists, cooked up by scientists for other scientists.'

18.6 PLANTS

So far in this chapter, we have concentrated on plant cover as the key property in relation to desertification. Nevertheless, different plant functional types may be a more appropriate level of analysis because something at least is known about the relative efficacy of trees, shrubs and grass with respect to erosion and this is mainly through their morphology. Eventually the problem will devolve to species. Here, prompted by the work of Pease *et al.* (1989), some complex aspects of the reduction of vegetation through quite evolutionary population dynamics are discussed, and thoroughly explored in a complex dynamical systems context by Roughgarden (1979). It is raised here as a promising area for future research in the vegetation-erosion paradigm. Essentially, plants survive in these environments through their capacity to resist pestilence and drought. This resistance of the parent plant is basically conferred by the genetic constitution of the plant, the genotype. If the genotype is inadequate, then the plant is unable to survive. The character of the genotype determines the fitness of plants and thus is an internal complexity that controls the fitness of the species. The genes have variants that make up the fibrous blueprint. These variants combine and recombine in a complex fashion as a result of reproduction.

Roughgarden (1979) demonstrated how the evolution of the gene pool can determine the population of the species. Essentially, the combinatorial possibilities can produce complex behaviour that exhibits the usual characteristics of complexity, convergence and divergence to and from stable and unstable equilibria and the corresponding fluctuations in the population numbers. He also showed that three 'external' factors control the actual gene pool. These are natural selection, mutation and genetic drift. In the first, the fittest plants (and their contribution to the gene pool) survive. This is a fast process, ranging from a single to several thousand generations according to the make-up of the gene pool and the external stresses that the phenotype (reproductive parent plant) is exposed to. The second is the effect of random chemical changes in the gene pool that change its basic composition and therefore the possible recombinations in reproduction. The third is the sum of all the chance variations that occur in the reproductive processes. These three processes can all lead to unexpected outcome as the 'internal' and 'external' forces operate through time. From the desertification point of view, they can lead to the disappearance of unfit populations in a mixture of populations. As new plants are introduced, they may destabilize existing communities and to this extent they could induce the type of 'contagious' desertification that is referred to above. Furthermore, climatic fluctuations (Thornes, 2003) and grazing densities (Thornes, in preparation) are implicated in desertification contagion.

18.7 CONCLUSION

In desertification, vegetation and land degradation are inextricably mixed and strongly interactive. This interaction can be modelled using the predator–prey strategy that produces clear separations (bifurcations) between stable attractors and unstable repeller equilibria. By identifying the bifurcations in phase space, with properly parameterized equations, we can anticipate how the behaviour of the system will lead to one of the equilibria. As with all complex systems, small variations in initial conditions can lead to strongly different outcomes and, because natural systems are very noisy, there are many possibilities for diversion along the routes to stability. The beauty of the approach is that a basic knowledge of the underlying dynamics provides a good insight into the likely possible trajectories. Another important aspect is the separation of the 'intrinsic' or 'internal' controls of evolution and the external controls that push or force the trajectories. A further dichotomy is between the deterministic modelling represented by the differential equations of change and the manipulations that explore the nature of the equilibria, on the one hand, and the probabilistic modelling exposed by the combinatorial possibilities within the genetic modelling and the genetic drift as a stochastic process, on the other. These are the intrinsic probabilistic elements. The 'external' forcing, climate, human activities and natural spatial variability also lead to probabilistic behaviour and modelling near the bifurcation sets (thresholds) and at the attractors themselves.

The implications for desertification are beginning to emerge. The first and most important is that the intrinsic complexity has to be appreciated before mitigation actions are undertaken. Tiny changes can have major outcomes. They can set the system along trajectories that have far-reaching and (usually) unpredictable outcomes. In this chapter, some of the major bifurcations have been identified but, for desertification, many more wait to be revealed. The second major implication for management and history is that observed outcomes are not necessarily a good guide to process mechanisms. The inference of process from form is still in its infancy. The best hope is that following the trajectories of specific changes from known perturbations will lead us to a better appreciation of the vegetation-degradation system.

It is too naïve always to seek climate-change explanations for land degradation. This outlook is a residue from the denudation-chronology preoccupation of early to mid-twentieth-century geomorphology, compounded with the contemporary concerns with global climate change. Semi-arid environments are often located at or near to the isoclines and are therefore especially susceptible to small perturbations of both physical and socio-economic origins. Recognition and acceptance of the complexity (in the technical sense intended in this chapter) are the first step towards better management of desertification. Further research is needed into complexity for more fruitful management.

REFERENCES

Allen, P.M. (1988) Dynamic models of evolving systems, *Systems Dynamics Review* **4**, 109–130.

Arnell, N. (1996) *Global Warming, River Flows and Water Resources*, John Wiley & Sons, Chichester.

Beven, K.J. and Kirkby, M.J. (1979) A physically based, variable contributing area model of basic hydrology, *Hydrological Sciences Bulletin* **24**, 43–69.

Brunsden, D. and Thornes, J.B. (1979) Landscape sensitivity and change, *Transactions of the Institute of British Geographers* **4** (4), 463–484.

Carson, M.A. and Kirkby, M.J. (1972) *Hillslope Form and Process*. Cambridge University Press, Cambridge.

Diamond, S. and Woodward, F.E.I. (1998) Vegetation modelling, in P. Mairota, J.B. Thornes and N. Geeson (eds) *An Atlas of Mediterranean Europe: The Desertification Context*, John Wiley & Sons, Chichester, 68–69.

Eagleson, P.S. (1979) Climate, soil and vegetation, 4: the expected value of annual evapotranspiration, *Water Resources Research* **14** (5), 731–764.

Eigen, M. and Schuster, P. (1979) *The Hypercycle*, Berlin, Springer-Verlag.

Elwell, H.A. and Stocking, M.A. (1976) Vegetative cover to estimate soil erosion hazard in Rhodesia, *Geoderma* **15**, 61–70.

Evernden, N. (1992) *The Social Creation of Nature*, Johns Hopkins University Press, Baltimore, MD.

Francis, C.F. and Thornes, J.B. (1990) Runoff hydrographs from three Mediterranean cover types, in J.B. Thornes (ed.) *Vegetation and Erosion*, John Wiley & Sons, Chichester, 363–384.

Godron, M., Guillerm, J.L., Poissonet, J., Poissonet, P., Thiault, M. and Trabaud, L. (1981) Dynamics and management of vegetation, in F. di Castri, D. Goodall and R.L. Specht (eds) *Ecosystems of the World 11: Mediterranean-type Shrublands*, Elsevier Scientific Publishing Co, Amsterdam, 317–445.

Hudson, N. (1992) *Land Husbandry*, B.T. Batsford Limited, London.

Kirkby, M.J. (1971) Hillslope process-response models based on the continuity equation, *Institute of British Geographers Special Publication* **3**, 15–30.

Kirkby, M.J., Abrahart, R.J., Bathurst, J.C., Kilsby, C.G., McMahon, M.L, Osborne, C.P., Thornes, J.B. and Woodward, F.E.I. (2002) MEDRUSH: a basin-scale physically based model for forecasting runoff and sediment yield, in N.A. Geeson, C.J. Brandt, and J.B. Thornes (eds) *Mediterranean Desertification: A Mosaic of Processes and Responses*, John Wiley & Sons, Chichester, 203–229.

Kirkby, M.J. and Cox, N.J. (1995) A climatic index for soil erosion potential (CSEP) including seasonal and vegetation factors, *Catena* **25**, 333–352.

Kosmas, C., Kirkby, M.J. and Geeson, N. (1999) *The MEDALUS Project Mediterranean Desertification and Land Use: Manual on Key Indicators of Desertification and Mapping Environmentally Sensitive Areas to Desertification*, Official Publications of the European Communities, Luxembourg, EUR 18882.

Lavee, H. and Pariente, S. (1998) Resilience of soil structures to environmental changes, in J. Brandt (ed.) *MEDALUS III Project 1: Core Project, Final Report covering the period 1 January 1996–31 December 1998*, EU, Brussels, 433–485.

Lotka, A.J. (1925) *Elements of Physical Biology*, Williams and Wilkins, Baltimore, MD.

May, R.L. (1973) *Stability and Complexity in Model Ecosystems*, University Press, Princetown, NJ.

Maynard Smith, J. (1974) *Models in Ecology*, Cambridge University Press, Cambridge.

Musgrave, G.W. (1947) Quantitative evaluation of factors in water erosion: a first approximation, *Journal of Soil and Water Conservation* **2**, 133–138.

Obando, J.A. (1997) Modelling the impact of land abandonment on runoff and soil erosion in a semi-arid catchment, PhD thesis, King's College London.

Pease, C.M., Land, R. and Bull, J.J. (1989) A model of population growth, dispersal and evolution in a changing environment, *Ecology* **70** (6), 1657–1664.

Pianka, E.R. (1978) *Evolutionary Ecology*, 2nd edn, Harper and Row, New York.

Poesen, J. and van Waesemael, B. (1995) Effects of rock fragments on the structure of tilled topsoil during rain, in E. Derbyshire, T. Dikstra and I.J. Smalley (eds) *Genesis and Properties of Collapsible Soils*, NATO Advanced Studies Institute Series, Series C: Mathematical and Physical Sciences, 486, Kluwer Academic Publishers, Dordrecht, 333–343.

Polya, G. (1957) *How to Solve It: A New Aspect of Mathematical Method*, 2nd edn, Princeton University Press, Princeton, NJ.

Roughgarden, J. (1979) *The Basics of Population Genetics*, Macmillan, New York.

Shaxson, T.F.N., Hudson, W., Sanders, D.W., Roose, E. and Moldenhauer, W.C. (1989) *Land Husbandry: A Framework for Soil and Water Conservation*, Soil and Water Conservation Society, Ankeny, IO.

Skellam, J.G. (1951) Random dispersal in theoretical populations, *Biometrica* **38**, 196–218.

Thornes, J.B. (1990) The interaction of erosional and vegetational dynamics in land degradation: spatial outcomes, in J.B. Thornes (ed.) *Vegetation and Erosion*, John Wiley & Sons, Chichester, 41–53.

Thornes, J.B. (1985) The ecology of erosion, *Geography*, **70** (3), 222–236.

Thornes, J.B. (1988) Erosional equilibria under grazing, in J. Bintliff, D. Davidson and E. Grant (eds) *Conceptual Issues in Environmental Archaeology*, Edinburgh University Press, Edinburgh, 193–211.

Thornes, J.B. (2003) (in press) Exploring the grass-bush transitions in South Africa through modelling the response of biomass to environmental change, *Geographical Journal*.

Thornes, J.B. (in preparation) Vegetation, erosion and grazing: a hypercycle approach. Paper to be presented at the COM-LAND Conference on Land Degradation and Mitigation, Iceland, August 2003.

Thornes, J.B. and Brandt, C.J. (1993) Erosion-vegetation competition in an environment undergoing climatic change with stochastic rainfall variations, in A.C. Millington and K.T. Pye (eds) *Environmental Change in the Drylands: Biogeographical and Geomorphological Responses*, John Wiley & Sons, Chichester, 305–320.

Thornes, J.B, Shao, J.X., Diaz, E., Roldan, A., Hawkes, C. and McMahon, M. (1996) Testing the MEDALUS Hillslope Model, *Catena* **26**, 137–160.

Tilman, D. (1982) *Resources Competition and Community Structure*, Princeton University Press, Princeton, NJ.

Tilman, D. and Karieva, P. (1997) *Spatial Ecology: The Role of Space in Population Dynamics and Interspecific Interactions*, Princeton University Press, Princeton, NJ.

Van der Leeuw, S. (ed.) (1998) *Understanding the Natural and Anthropogenic Causes of Land Degradation and Desertification in the Mediterranean Basin*, Office for Official Publications of the European Communities, Luxembourg, EUR18181 EN.

Voltera, B. (1926) Fluctuations in the abundance of a species considered mathematically, *Nature* **118**, 558–560.

Part IV

Current and Future Developments

19

Scaling Issues in Environmental Modelling

XIAOYANG ZHANG, NICK A. DRAKE AND JOHN WAINWRIGHT

19.1 INTRODUCTION

Modelling is useful for understanding and predicting environmental changes at various times and areas. It can incorporate descriptions of the key processes that modulate system performance or behaviour with varying degrees of sophistication (Moore *et al.*, 1993). Over the past few hundred years, we have invested considerable effort to understand the processes of the environmental system, to build up a long record of environmental parameters, and to create a number of reasonable models at local (plot) scales. For example, land-surface and hydrological processes are usually monitored and experimented on an isolated uniform plot or under laboratory control with a spatial scale between one metre and one hundred metres and a time period of hours or days. Analysis of an ecological system is built up by observing the regulation of stomatal conductance, photosynthesis, and the flux of water and carbon at the scale of a leaf or a canopy (plant) with a corresponding timescale from seconds to hours. Our knowledge and corresponding physically realistic models of environmental processes have been extensively developed and validated, based on field measurements at local scales. The underlying processes occurring at these scales reveal critical causes of environmental changes.

Many pressing environmental problems affecting humans are, however, always to be found at regional and global scales. The assessments of regional and global environmental changes force us to address large-scale and long-term issues. Thus, we must focus on how our Earth system is changing and how the future changes of the Earth system can be estimated, for example, in terms of carbon cycling and climate change at regional and global scales. Hence we primarily use our knowledge at

local scales to address hydrology, climatic change and the carbon dioxide cycle at regional and global scales using grid data ranging from $0.5° × 0.5°$ to $5° × 5°$. For example, a dataset with a resolution of $0.5° × 0.5°$ was used in the Vegetation/Ecosystem Modelling and Analysis Project (VEMAP members, 1995). Although environmental processes at large scales are to a great extent the result of processes at local scales, the models representing these processes can vary considerably from one scale to another (Heuvelink, 1998).

Global databases are gradually being built up with good quality information for environmental modelling. High spatial and temporal resolution data are not always available at a regional or global scale. In contrast, the coarse spatial and fine temporal resolution datasets are widely available. These databases include $0.5°$ and $1°$ soil types, vegetation types and land use in the Global Ecosystems Database (GED), daily 1 kilometre global Advanced Very High Resolution Radiometer (AVHRR), 1 kilometre MODerate Resolution Imaging Spectroradiometer (MODIS) products including land cover, temperature, albedo, leaf-area index (LAI) and Fraction of Photosynthetically Active Radiation (FPAR), and daily climatic data in the Global Climate Data Center. Nevertheless, it is still not clear how these data change with scales and to what extent they fit both environmental processes and models from local to global scales.

Because many environmental processes and patterns are scale-dependent, we currently face two pairs of contradictions. One is that although the local environmental processes can be simulated accurately, many environmental managers require the environmental assessments at a regional or global scale. Another is that most physically based models have been developed and validated

Environmental Modelling: Finding Simplicity in Complexity. Edited by J. Wainwright and M. Mulligan
© 2004 John Wiley & Sons, Ltd ISBNs: 0-471-49617-0 (HB); 0-471-49618-9 (PB)

at uniform field-plot scales or under laboratory conditions while the widely available data for model inputs are very coarse and heterogeneous. The scaling issues of environmental parameters and modelling restrict the applicability of the principles and theories learned under plot conditions to the assessment of environmental risk at a regional or global scale. They arise because little understanding has been achieved about the linkage between well-developed models at fine scales and environmental processes operating at large scales.

This chapter discusses the scaling issues facing environmental modelling and the currently used methods reducing the scaling effects on both models and their parameters. A case study is then presented on the approaches to scale up a soil-erosion model established at a plot scale to regional or global scales by scaling land-surface parameters. A primary focus is to develop a scale-invariant soil-erosion model by scaling the slope and vegetation-cover factors that control erosion to fit the modelling scale.

19.2 SCALE AND SCALING

19.2.1 Meanings of scale in environmental modelling

'Scale is a confusing concept often misunderstood, and meaning different things depending on the context and disciplinary perspective' (Goodchild and Quattrochi, 1997). 'The various definitions for scale are often used interchangeably and it is not always clear which one is used. The term "scale" refers to a rough indication of the order of magnitude rather than to an accurate figure' (Blöschl and Sivapalan, 1995). Although there are over 30 meanings of the word (scale) (Curran and Atkinson, 1999), five meanings are commonly used in environmental analysis (Lam and Quattrochi, 1992; Blöschl and Sivapalan, 1995). A *cartographic map scale* refers to the proportion of a distance on a map to the corresponding distance on the ground. A large-scale map covers a smaller area generally with more detailed information, while a small-scale map covers a larger area often with brief information about the area. In contrast, a *geographic scale* is associated with the size or spatial extent of the study. A large geographic scale deals with a larger area, as opposed to a small geographic scale, which covers a smaller area.

An *operational scale* (process scale, characteristic scale) is defined as the scale at which a process operates in the environment. This scale is associated with the spatial extent and temporal duration (lifetime and cycle) depending on the nature of the process. The property of

an object is identified at this scale. An environmental phenomenon is best observed at its operational scale.

Measurement scale (observational scale) is the spatial resolution that is used to determine an object. At this scale a construct is imposed on an object in an attempt to detect the relevant variation (Collins, 1998). Measurement scale can be defined as the spatial or temporal extent of a dataset, the space (resolution) between samples, and the integration time of a sample. The spatial scale is related to the size of the smallest part of a spatial dataset, such as the plot size in the field investigation, pixel size of remotely sensed data and digital elevation data. With the change of measurement scale, the environmental parameters may represent different information in concerned disciplines. For example, the observed characteristics in biosphere processes may represent leaf, plant and ecosystem respectively when the spatial data are aggregated from a small scale to a large scale.

Modelling (working) scale, building up an environmental model, is partly related to processes and partly to the application models. Typical hydrological modelling scales in space include the local (plot) scale (1 m), the hillslope (research) scale (100 m), the catchment scale (10 km), and the regional scale (1000 km). The corresponding temporal scales are the event scale (1 day), the seasonal scale (1 year), and the long-term scale (100 years: Figure 19.1; Dooge, 1986). In the assessment of biospheric ecosystem function, data for fine-scale physiology models are measured at leaf level over seconds. The prediction of canopy photosynthesis for forest stands is operated with daily climatic information and a spatial scale from 1 to 10 ha. On the contrary, coarse-scale biogeochemical models are designed to estimate biosphere processes over regional scale extrapolating to a 10 to 100 km^2 spatial resolution with monthly climate data (McNulty *et al.*, 1996).

Application of environmental modelling always involves four different scales. These are geographic scale of a research area, temporal scale related to the time period of research, measurement scale of parameters (input data resolution) and model scale referring to both temporal and spatial scales when a model was established.

19.2.2 Scaling

Scaling focuses on what happens to the characteristics of an object when its scale (size/dimension) is changed proportionately. An object in the real world can be considered as the composition of line, area and volume. To explain the scaling issue, we consider a unit side of

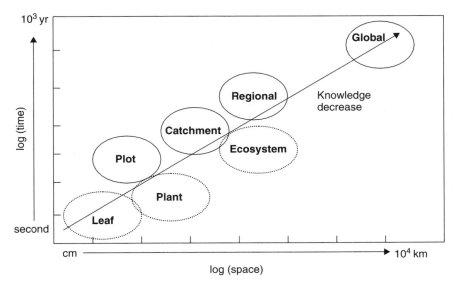

Figure 19.1 Typical scales of environmental models

1 centimetre. When using this side to build up a square and a cube, the square surface area is 1 square centimetre and the volume of the cube is 1 cubic centimetre. We can see all three objects with the same value as one. When we double the side, the surface area becomes 4 square centimetres, and the volume of the cube 8 cubic centimetres (http://science.kennesaw.edu/~avesperl). It clearly shows that when the side scale changes by a factor of two, the surface area increases by a factor of four and the volume by a factor of eight. There is clearly a nonlinear process of scaling properties between the side length, the square surface and the cube volume. Considering such a problem in hydrology, we should find how topographic attributes change if we double the spatial resolution of a topographic map, or how the drainage area changes if we double the length of a stream (Dodds and Rothman, 2000).

The parameter values in heterogeneous environments are usually dependent on the measurement scales. This dependence implies that the value of a parameter in a large measurement scale cannot be linearly averaged from a small measurement scale. In order to reduce or increase measurement scale, scaling studies in environmental modelling are related to such a fundamental question as how a model changes with the variation of parameter measurement scales. Hence it becomes crucial to determine the linkage of both environmental parameters and models across scales only depending on the knowledge at one scale by interpolation/extrapolation. Upscaling refers to the environmental issues at a higher scale based on the knowledge obtained from a lower scale, whereas downscaling determines the issues at a lower scale using knowledge at a higher scale (Figure 19.2).

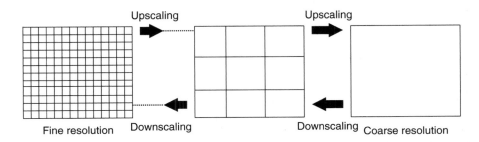

Figure 19.2 Linkages of parameters across scales

19.3 CAUSES OF SCALING PROBLEMS

There are a number of conceptually distinct reasons resulting in scaling problems (Heuvelink, 1998; Harvey, 2000; Peterson, 2000). First of all, the most common and fundamental cause of scaling problems is the existence of both spatial heterogeneity and relevant process nonlinearities. Environmental models more or less require both climatic/weather input data and land-surface parameters, which are derived from measurements including precipitation, temperature, topography, soil physical and chemical properties, land cover/land use and hydrological properties. These data are considerably variable both spatially and temporally, creating difficulties in aggregating large-scale behaviour from local processes.

Second, different processes are primarily dominated at different scales. This means the correlations derived at one scale might not be applicable at another. Because of the existence of different dominant processes at various scales, the effects of an increasing number of processes need to be incorporated as a scaling method attempts to span over a wider range of scale. Third, feedback is associated with the interaction between small-scale parameters of a system and large-scale variables since processes at different scales affect each another. Processes operating at small and fast scales are constrained by processes operating at slow and large scales, while large and slow processes are constructed and organized by the interactions of many small fast processes (Ahl and Allen, 1996). These cross-scale connections suggest that scaling should only be applied over a limited range of scales and in specific situations. Fourth, emergent properties arise from the mutual interaction of small-scale components among themselves rather than some outside force. The processes can abruptly reorganize themselves through time, rendering previously developed scaling relationships invalid as the structure and processes that they incorporate cease to exist. Fifth, when there is a temporal lag in the response of a system to perturbation, a scaling problem may arise because of the lack of information about process linkages in a dynamic environment.

19.4 SCALING ISSUES OF INPUT PARAMETERS AND POSSIBLE SOLUTIONS

19.4.1 Change of parameters with scale

Data required for environmental modelling are measured either at a point such as precipitation, temperature and soil properties, or over a given area such as remotely sensed data. The reliability of the data value in an area is affected by both its neighbours and internal heterogeneity. Pixels in remote sensing imagery, which are the smallest element of an electronic image, have been widely used to measure environmental properties. Nevertheless, sensors are commonly centre-biased such that the reflectance towards the centre of the field of view has most influence on the reflectance (Fisher, 1997). It is poorly understood whether the reflectance from one location on the ground only influences the corresponding pixel, or whether it may have an effect on values for surrounding pixels. On the other hand, information in one pixel, especially a large pixel, is usually the mixture of different ground objects rather than presenting a true geographical object (Settle and Drake, 1993; Fisher, 1997). The environmental parameter extracted from a large grid only indicates a representative value rather than a real physical meaning. For example, if there are more than three types of soil or land cover/land use, the corresponding categorical value in a large grid is only a domain value which may represent a proportion less than 50% of reality. It is a snare and a delusion to take pixels with various resolutions as a homogenous reality on the ground (Fisher, 1997).

Changing the scale of measurement has a significant impact on the variability of object quantities. A land-surface parameter at various scales represents the different amount of details referring to both spatial patterns and observed values. The most well-known example is the measurement of the length of Britain's coastline, in which the values change greatly from one measurement to another when different spatial resolution maps are used. If the whole of Britain were reduced to one pixel, the length measurement of the coastline would be no more than the sum of four pixel sides. It would be ignored if the pixel size were as large as the whole of Europe. In contrast, it would be huge if a measurement scale were less than a millimetre. Obviously, all these values are not able to represent reality. In many cases of the natural world, a single value of a parameter is not meaningful when the measurement scale is too small or too large.

It has been recognized that measured data are an explicit representation, abstraction or model of some properties varying in N-dimensional Euclidean space instead of reality itself (Gatrell, 1991; Burrough and McDonnell, 1998; Atkinson and Tate, 1999). Since all data are a function of the underlying reality and sampling framework, the data we obtained can generally be described as the following equation (Atkinson, 1999):

$$Z(x) = f(Y(x), d) \qquad (19.1)$$

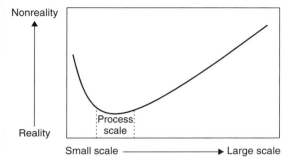

Figure 19.3 Changes of parameter values with measurement scales

where $Z(x)$ is the observed variable, $Y(x)$ is the underlying property and d presents the measurement scale (sampling framework).

The reality of a variable is associated with its corresponding processes. If a variable is measured at its process scale, its value can approximate the reality. Otherwise, the larger (or smaller) the measurement scales than the process scale, the more meaningless the variable values (Figure 19.3). For example, when using remotely sensed data to classify land cover, the resultant land-cover map will be most accurate when using approximately field-sized pixels. If pixels either much smaller or much larger than a typical field are used, classification accuracy will consequently decrease (Woodcock *et al.*, 1988a, 1988b; Curran and Atkinson, 1999).

19.4.2 Methods of scaling parameters

Techniques are often required to relate information at one spatial scale to another. In some cases, this process involves the removal of data points to produce a coarser resolution, often so that data from different sources have a compatible resolution. Commonly used techniques include the averaging method, the thinning method and the dominant method, which are simple and easy to operate but do not deal with the intrinsic scaling problem. In other cases, we need to find ways of reproducing the finer details, by a process of interpolation. The methods of fractal and geostatistics, however, are more intricate but robust than those mentioned above. They can also be used for interpolation of data to finer resolutions.

For a grid-based dataset, the *averaging method* makes use of the average value over an N by N pixel window to form a dataset of coarse resolution, which smoothes the variance of the dataset and increases spatial autocorrelation. It may reveal the basic relationships

involved, but simple averaging is unrealistic and may have an effect on the outcome because there might be nonlinear correlations among different grids. Currently, this is a commonly used technique for degrading fine-scale remotely sensed data and digital elevation models (DEMs) (e.g. Hay *et al.*, 1997; De Cola, 1997; Helmlinger *et al.*, 1993). The *thinning method* is used to construct a dataset by subsampling data at fixed intervals, taking every Nth pixel to create a series of coarser data. The frequency distribution of the sampled data may be very similar to the original. This method can maximally remain the variance of the dataset but the information between every Nth pixel is lost.

The *dominant method* is used to create a coarse resolution dataset on the basis of the dominant values in an N by N window. The variance will be reduced in the coarser data since the low frequency values are excluded during aggregation. The dominant procedure retains the values of the original data but may alter the spatial pattern (including spatial autocorrelation) at coarse resolutions (Bian, 1997). Categorical data are usually aggregated using this method. It should be noted that this procedure creates bias since types with low proportions will diminish or disappear while others will increase with scales depending on the spatial patterns and probability distributions of the cover types.

The *linear convolution* method is a useful technique for degrading remotely sensed data. The degraded radiance value $Y(t)$ at point t is the convolution of the scene radiance $Z(t)$ with the system point spread function $p(u)$ (Rosenfeld and Kak, 1982; Collins, 1998):

$$Y(t) = \int_u p(u)Z(t-u)\,du \qquad (19.2)$$

Generally, all parameters are vectors representing position in a two-dimension space. The system point-spread function can be approximated by a probability distribution, such as a Gaussian function.

The *fractal method* provides the possibility of interpolating an object with self-similarity at various scales (both upscaling and downscaling). In general, a scale-dependent parameter can be defined as:

$$F(d) = f(r)g(d) \qquad (19.3)$$

where d is a scaling factor, $g(d)$ is a scaling function, $f(r)$ represents the reality of an object at scale r and $F(d)$ is a parameter value at measurement scale d.

If an object has fractal (unifractal) properties, the scaling of the object F can be described as a power law (Mandelbrot, 1982):

$$F(d) = Ad^\alpha \qquad (19.4)$$

with

$$\alpha = D - L + 1 \qquad (19.5)$$

where D is the fractal dimension, L is a Euclidean dimension and A is the amplitude or prefactor which is related to the lacunarity (patchiness) of an object.

This universal scaling law has been widely applied in biological and environmental research. For example, the metabolic rate for a series of organisms ranging from the smallest microbes to the largest mammals is a power function of mass (West *et al.*, 2000). The number of species found in a censused patch of habitat on the area of that patch can be described by a power function of area (Harte, 2000). Hydrological parameters and morphological attributes are clear special scaling features with this power law (Braun *et al.*, 1997; Zhang *et al.*, 1999). Note that the spatial distribution of the fractal dimension has to be determined if we need to scale a parameter spatially (Zhang *et al.*, 1999). The limitation of the fractal method is that multifractal properties of objects may exist over a broad range of scales (e.g. Evens and McClean, 1995), and the unifractal relationship may break down when the measurement scales are very small or very large.

The *kriging* method denotes a set of techniques to predict (extrapolate) data quantities and their spatial distribution. Commonly used methods are simple kriging, universal kriging and co-kriging. Simple kriging is a heterogeneous best linear unbiased predictor in terms of both the mean function and the covariance of sample data, whereas universal kriging only requires determining a variogram for the extrapolation of point data. Cokriging provides a consistent framework to incorporate auxiliary information, which is easy to obtain, into the prediction. This technique can predict spatial data more precisely (Papritz and Stein, 1999). Other robust measurements of spatial interpolation include the fitting of splines (e.g. De Boor, 1978) and the modelling process itself (see Chapter 20).

19.5 METHODOLOGY FOR SCALING PHYSICALLY BASED MODELS

19.5.1 Incompatibilities between scales

A major source of error in environmental models comes from the incompatibilities between model scale, input-parameter (database) scale and the intended scale of model outputs. The most common issue is that the parameters are measured at one scale (or several different scales) and are input into a model constructed at another scale. When we build an environmental model, at least two groups of measurements at the same or a similar scale are required. One is used to establish a model while another is employed to test the model. When changing measurement scales, the models based on these measurements may vary considerably. For example, Campbell *et al.* (1999) found that effective dispersion values estimated using data from each type of probe were systematically different when using different devices to obtain solute transport parameters for modelling. This difference may be a scale issue resulting from the probe sampling volumes. Even if a model operates linearly across the range of model-input values, the aggregation of the model-parameter values with increased scale still biases model predictions because of the heterogeneity. Both the model constants and the relationship significance between dependent and independent variables are usually controlled by the scale of processing of the model. For example, when investigating a linear model between elevation and biomass index using different spatial resolution of parameters, Bian (1997) found the constants in the model and the correlation coefficient between independent variables and the dependent variable to be very changeable. The same result was also identified by Walsh *et al.* (1997) when analysing a linear regression model between Normalized Difference Vegetation Index (NDVI) and elevation on a mountain. Therefore, if a model is established at a process scale and input parameters measured at the same scale, the model output should be accurate and acceptable. The relationship between process and measurement scale is that processes larger than the coverage appear as trends in the data, whereas processes smaller than the resolution appear as noise (Figure 19.4; Blöschl and Sivapalan, 1995).

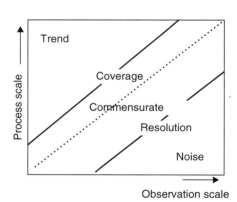

Figure 19.4   Process scale versus observation scale
Source: Blöschl and Sivapalan (1995). Reproduced by permission of John Wiley & Sons

19.5.2 Methods for upscaling environmental models

If we could take the Earth as a unit to build up a global model, it would produce a very accurate output of our interests when each input and output is only one single measured value at the global scale. However, global-scale model operation should rely on scale linkages of environmental objects to certain conceptual frameworks due to the difficulties in making direct measurements at a large scale. Although it is very hard to scale a complex assemblage of parameters and models to corresponding process scales, it is feasible to identify a few scaleable parameters within the complex environment in an attempt to reduce the scale uncertainties of modelling. Any methods reducing the scaling effects would be a great improvement to modelling results since it seems impossible to completely eliminate the impact of scale on environmental modelling (Luo, 1999). There are several methods available to translate ecological and environmental models across spatial scales (King, 1991; Rastetter *et al.*, 1992; Harvey, 2000; Bugmann *et al.*, 2000).

19.5.2.1 Calibration of a model

This approach is to adopt a model, which is applicable at small scales, for application at large scales using 'calibration' values. The calibration technique is to set up an empirically calibrated relationship between fine- and coarse-scale data. This approach requires that both fine- and coarse-scale data are available to perform the calibration. The calibrated models may only be treated as valid within the ranges of input and output data used for the calibration, and the incorporation of new model variables requires recalibration of the relationship (Rastetter *et al.*, 1992; Friedl *et al.*, 1997). In practical terms, it is difficult to carry out measurements for calibration at various scales, especially very large scales. However, because of the nonlinearity of the processes involved, together with the heterogeneity of the natural system, there is no reason to suppose that the use of calibration values should be any more successful in reproducing the areal average result (Beven, 1995).

19.5.2.2 Multi-scale models

Since the processes change with various scales, a set of different models are required for various particular scales, such as specifically used at field plot, watershed, regional and global scales (David *et al.*, 1996; Kirkby *et al.*, 1996; Poesen *et al.*, 1996). Separate scaling methods could be developed for each individual process,

and the rules for transitions between them could be used to synthesize these separate scaling methods into an aggregate scaling method which incorporates the behaviour of the ensemble of scales (Peterson, 2000). The plot-scale environmental models may represent physical and other processes in detail while large-scale models can provide a potential value, such as potential cumulative soil erosion at the global scale (Kirkby and Cox, 1995; Kirkby *et al.*, 1996). In doing so, the scale effects are included in each individual model and the environment could be accurately simulated. Unlike plot-scale models, however, regional- and global-scale models are rarely available.

19.5.2.3 Lumped models

The lumping method is carried out by integrating the increased heterogeneity that accompanies the change in model extent by aggregating across heterogeneity in environmental parameters. Since an area is taken as a whole to calculate the model result, a lumped model is like a zero-dimensional representation of spatial features (Maidment, 1993). A major concern in using a lumped model is the possibility of obtaining a single value of a spatially changeable parameter that allows a model to predict the mean response of the modelled area (Moore and Gallant, 1991). To address this need in upscaling studies, the most common way to treat a lumped system is that the related properties are spatially averaged and the mean values are calculated for the modelling parameters. This approach is based on the assumption that a model is linear in its structure, and no attempt is made to describe the spatial heterogeneity. Alternatively, a more effective way of processing a lumped model is to choose aggregated representative parameters accounting for the heterogeneity. A noticeable example is that the representative parameters across scales are calculated by using self-similar fractals, such as topographic properties for hydrological models (Braun *et al.*, 1997; Zhang *et al.*, 1999). Since one single set of model calculations is applied across the whole area, the spatial distribution of model results is not taken into consideration.

19.5.2.4 Distributed (partitioning) method

A research area is divided into many grid cells (or subareas, patches and strata) with relatively homogeneous environmental conditions. Each grid cell is represented by mean values of environmental processes in the computation of regional and global estimates, such as in a grid-based approach. The heterogeneity (variance) within a grid cell is minimized while it is maximized between cells. The environmental behaviour in a large area is determined as the areally weighted average of the behaviour of all those cells after a local model is employed in each grid cell. The resultant value in a large area can be estimated in Equation 19.6:

$$F = \sum_{i=1}^{i=k} p_i f(\overline{x}_i) \qquad (19.6)$$

where p_i is the proportion of the total area that stratum i occupies, and $\overline{x}_i$ is the mean of x in the stratum i.

A number of practical approaches can be used to determine homogeneous areas. First, a straightforward method is to run the model in each grid cell across a region. It is assumed that each grid cell is homogeneous and fits the modelling scale no matter what the size of the cell is. For example, land vulnerability to water erosion was assessed directly by running a plot model on a 0.5° dataset by Batjes (1996). There are several more reasonable ways to apply distributed models.

Second, when studying rainfall and runoff, Wood *et al.* (1988, 1990) created the representative elementary area (REA) concept and gave it a definition of the 'smallest discernible point which is representative of the continuum' (see also Chapter 3). The REA concept is employed for finding a certain preferred time and spatial scale over which the process representations can remain simple and at which distributed catchment behaviours can be represented without the apparently undefinable complexity of local heterogeneity. This concept provides a motivation for measurements of spatial variability as it highlights the interaction between scale and variability. It indicates that the variances and covariances of key variables are invariant in land units above a certain threshold size. It is difficult in practice, however, to determine the size of the REA because it is strongly controlled by environmental characteristics (Blöschl *et al.*, 1995).

Third, in regional hydrological modelling, the concept of a hydrological response unit (HRU) has been developed. The HRU is a distributed, heterogeneously structured entity having a common climate, land use and underlying pedo-topo-geological associations controlling their hydrological dynamics. The crucial assumption for each HRU is that the variation of the hydrological process dynamics within the HRU must be small compared with the dynamics in a different HRU (Flügel, 1995).

Fourth, geostatistics, such as measures of spatial autocorrelation between data values, provides a methodology of determining a relative homogeneous size for environmental modelling. Homogeneous areas are seen as having high spatial autocorrelation such that a coarse spatial resolution would be appropriate, and heterogeneous areas are seen as having low spatial autocorrelation so that a fine spatial resolution would be needed (Curran and Atkinson, 1999). Analysing the variogram change with spatial resolution can help to identify the predominant scale of spatial variation, which might be an optimal scale for environmental modelling. Several statistical models can provide ways to identify the optimum scale of a variable. The average local variance estimating from a moving window is a function of measurement scale (Woodcock and Strahler, 1987). The scale at which the peak occurs may be used to select the predominant scale of variables. Similar to the local variance method, the semi-variance at a lag of one pixel plotted against spatial resolution is an effective means of determining the optimal scale. For example, when using this technique with remote sensing data, Curran and Atkinson (1999) found that a spatial scale of 60 m was suitable for the analysis of urban areas, whereas 120 m was suitable for agricultural areas. The dispersion variance of a variable defined on a spatial scale within a specified region may also be applied to link information pertinent to the choice of optimal scale (Van der Meer, 1994). All the above approaches are acceptable for dealing with scaling problems that arise due to spatial heterogeneity combined with process nonlinearity, but are not valid when there are interactions between adjacent grids (or areas).

19.5.2.5 Routing approach

The model structure remains the same across scales while the variables (fluxes) for the models are spatially and temporally modified. Such interactions always occur in hydrological processes, where the model output at large scales is far from being the simple sum of each grid cell. When water flow and sediment move from the divide to the outflow of a catchment and from a small catchment to a large catchment, the values of related hydrological variables in a grid are affected not only by the environmental characteristics within this grid but also by properties in the surrounding grid cells. The runoff in a pixel is determined by the precipitation and infiltration within this grid, and water inputting from and discharging to neighbouring grids. The amount of sediment deposited or detached in a grid cell is strongly dependent on the movement of both water and sediment in surrounding neighbouring grids. Hence the routing approach is usually employed after calculating the drainage direction on the basis of the steepest

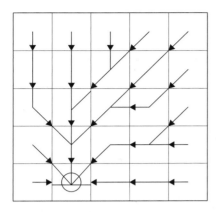

Figure 19.5 The routing of hydrological models according to the steepest topographic slope

descent of topography (Figure 19.5). This technique has been effectively employed in runoff and sediment transport models at large scales (e.g. Richards, 1993; Pilotti and Bacchi, 1997).

19.5.2.6 Frequency-distribution-based method (expected value or partial transformations)

This approach is defined as the calculation of the expected value of the model output based on the joint frequency distribution of the variables describing environmental heterogeneity (King, 1991; Rastetter *et al.*, 1992). Probability density functions that describe the heterogeneity at subgrid scales in the input data have to be determined to estimate the expected model output value over the region of interest. The equation used may be expressed as:

$$F = E[f(X)] = \int_{-\infty}^{+\infty} f(x)p(x)dx \qquad (19.7)$$

where $f(X)$ is a local model, $E[f(X)]$ is the expectation value of $f(X)$, X is a random variable (or vector of random variables), and $p(x)$ is the probability density function (or the joint frequency distribution) describing the subgrid scale variance in X. In an environmental model, parameters always include attributes derived from factors such as climate, landscape and soil. Clearly, the joint frequency distribution of such parameters is rarely known if there are a large number of parameters for a model. Nevertheless, using such an approach to account for one or two of the most important nonlinear model input parameters can substantially improve model results (Avissar, 1992; Friedl, 1997).

To solve this equation simply, Bresler and Dagan (1988) and Band (1991) use a Taylor series expansion about the vector mean of model parameters, where the distribution information required is reduced to the variance-covariance matrix of the key model variables. Taylor series expansions can transform models from fine to coarse scales. The equation becomes:

$$F = \sum \frac{f^{(i)}(\mu_x)}{i!} E[(X - \mu_x)^i] \qquad (19.8)$$

where μ_x is the mean of X, $f^{(i)}(\mu_x)$ is the ith derivative of $f(X)$ evaluated at μ_x with respect to X, and $E(X - \mu_x)^i$ is the expected value of the ith moment of X about μ_x. To implement this type of solution, the model must either have higher-order derivatives all equal to zero, or the expansions must be estimated from sample data. In either case, this approach is typically limited to an approximation of the exact solution provided by partial transformations (Friedl, 1997). Using this method, it is only possible to derive areal average values of environmental variables rather than their spatial distribution. Such a result is a very useful means of approximating values over large areas, for example in estimating the average global carbon storage, whereas it would be meaningless to talk in these terms in relation to land-surface properties.

19.5.2.7 Analytical integration

The method of analytical integration is used to extrapolate a model employing explicit, analytical integration of the local model in time and space, which of course provides not only an accurate, but even an 'exact estimate':

$$F = \int_x \int_y \int_t f(x, y, t) \, dt \, dy \, dx \qquad (19.9)$$

where $f(x, y, t)$ is the local model in time t and space (x, y). This method is typically not feasible because closed-form analytical integrals cannot be found for many environmental models (Bugmann *et al.*, 2000).

19.5.2.8 Parameterizing interaction

Fluxes in the natural world not only interfere but also have feedback with each other (Harvey, 2000). It is necessary either to parameterize the interaction effects between grid cells directly, or to create a whole new model which incorporates these interactions and their effects. The parameterization approach of representing inter-grid interactions can be used when the small-scale effects do not significantly affect the process at large scales.

19.5.3 Approaches for downscaling climate models

Downscaling is concerned with taking the output of global change models and deducing the changes that would occur at finer scales than those resolved by such a model as a general circulation model (GCM). This approach has subsequently emerged as a means of interpolating regional-scale atmospheric predictor variables (such as a mean sea level pressure or vorticity) to station-scale meteorological series or agricultural production (Karl *et al.*, 1990; Wigley *et al.*, 1990; Hay *et al.*, 1992; Wilby and Wigley, 1997: see also Chapter 2). The fundamental assumption is that relationships can be established between atmospheric processes occurring at disparate temporal and/or spatial scales. Wilby and Wigley (1997) reviewed four categories of downscaling methods in GCM. First, regression methods involve establishing linear or nonlinear relationships between fine-resolution parameters and coarse-resolution predictor variables. Second, weather-pattern approaches of downscaling typically involve statistically relating observed station or area-average meteorological data to a given weather classification scheme, which may be either objectively or subjectively derived. Third, stochastic weather generators are designed to simulate daily time-series of climate data for each successive day governed by the characteristics of previous days based on first or multiple-order Markov renewal processes. These models are commonly used for climate-impact studies. Fourth, limited-area climate models are high-resolution limited-area climate models embedded with GCM and use GCM to define the boundary conditions. These models have the ability to simulate small-scale atmospheric features and may provide atmospheric data for impact assessments that reflect the natural heterogeneity of the climate at regional scales.

19.6 SCALING LAND-SURFACE PARAMETERS FOR A SOIL-EROSION MODEL: A CASE STUDY

Soil erosion is recognized as a major problem arising from agricultural intensification, land degradation and possibly global climatic change (see Chapters 9 and 16). However, the extent of the problem is hard to quantify as field measurements of erosion are rare, time-consuming, and are only acquired over restricted temporal and spatial scales. Global scale soil-erosion

modelling can provide a quantitative and consistent approach to estimating erosion. Such a methodology can be used to define regions where erosion is high and management is needed (Drake *et al.*, 1999).

Current erosion models are developed from the analysis of the results of plot-scale erosion experiments (1 m^2 to 30 m^2). One of the physically based models developed by Thornes (1985, see also Chapter 18) is defined as:

$$E = k \ OF^2 s^{1.67} e^{-0.07v} \qquad (19.10)$$

where E is erosion (mm day^{-1}), k is a soil erodibility coefficient, OF denotes overland flow (mm day^{-1}), s is the slope (m m^{-1}) and v the vegetation cover (%).

In order to overcome the scaling problem, methods have been developed to upscale this plot-scale model by downscaling topography and vegetation cover from the global scale (10 arc minutes) to the plot scale (30 m). Global soil erosion has then been calculated by implementing this upscaled erosion model.

19.6.1 Sensitivity of both topographic slope and vegetation cover to erosion

Both the nonlinearity between model output and parameters and the spatial heterogeneity of parameters can be identified using sensitivity analysis. In order to investigate the sensitivity of the soil-erosion model to the scale of the slope measurement, a global 30″ DEM was degraded to lower resolutions using pixel thinning (see above). The slope values were calculated from these DEMs and then erosion was calculated using average monthly values of both overland flow and vegetation cover. Thus, the only changing parameter is the spatial resolution of the slope. Average slope is reduced from 3.93% at 30″ to 0.39% at 30′ in the area of Eurasia and northern Africa. Estimated erosion is reduced exponentially by two orders of magnitude (from 0.03 to 0.0007 mm month^{-1}) as slope is reduced by the scaling effect over this range (Zhang *et al.*, 1997).

When changing spatial distributions and scales of vegetation cover, the amount of soil loss predicated by using the erosion model varies considerably (Drake *et al.*, 1999). This variability occurs because the negative exponential relationship between cover and erosion combined with the heterogeneity of vegetation cover means that erosion is very high in bare areas but very low once cover is greater than 30%. If we consider a high-resolution image of individual plants with 100% cover surrounded by bare soil, each pixel is homogenous with either vegetation or bare soil. Erosion is then very high in bare areas and nonexistent under the plants.

When the spatial resolution of the image is reduced to a point exceeding the size of the field plant, heterogeneity within a pixel occurs and predicted erosion is reduced because some of the vegetation cover of the field is assigned to the bare areas. Thus a scaling technique must be employed to account for the original structure of the vegetation cover.

19.6.2 A fractal method for scaling topographic slope

When analysing the regular decrease of slope values with the increase of spatial resolution of DEMs, Zhang *et al.* (1999) directly link the scaling of slope measurements with the fractal dimension of topography. When focusing on the difference in elevation between two points and the distance between them, the variogram equation used to calculate the fractal dimension of topography (Klinkenberg and Goodchild, 1992) can be converted to the following formula:

$$\left| \frac{Z_p - Z_q}{d} \right| = \gamma d^{1-D} \qquad (19.11)$$

where Z_p and Z_q are the elevations at points p and q, d is the distance between p and q, γ is a coefficient and D is fractal dimension. Because the equation $\left| \dfrac{Z_p - Z_q}{d} \right|$ represents the surface slope it can be assumed that the slope value s is associated with its corresponding scale (grid size) d by the equation:

$$s = r d^{1-D} \qquad (19.12)$$

This result implies that if topography is unifractal in a specified range of measurement scale, slope will then be a function of the measurement scale. If a research area were taken as a whole to determine the parameters γ and D, there would be only one value of scaled slope. It is necessary in practice to keep the spatial heterogeneity of slope values. After analysing the spatial variation of γ and D in different subareas, it was discovered that both γ and D are mainly controlled by the standard deviation of the elevations (Zhang *et al.*, 1999). Hence correlation functions are established between γ and D and local standard deviation. When the standard deviation in each pixel is determined by moving a 3×3 pixel window, the spatial distributions of both γ and D are then calculated. Therefore, the slope values for each pixel with finer measurement scales can be successfully estimated on the basis of coarse resolution DEMs.

19.6.3 A frequency-distribution function for scaling vegetation cover

The entire information required for environmental modelling is saved in distribution functions. Instead of assuming a Gaussian distribution, different distribution modes are discovered after analysing multi-scale data of vegetation cover (Zhang *et al.*, 2002). The Polya distribution, a mixture distribution of a beta and a binomial distribution, is very effective in simulating the distribution of vegetation cover at numerous measurement scales (Zhang *et al.*, 2002).

$$f(x) = {}^nC_x \int_0^1 \theta^x (1-\theta)^{n-x} \left[\frac{\Gamma(\alpha + \beta)}{\Gamma(\alpha)\Gamma(\beta)} \right]$$
$$\times \, \theta^{\alpha-1}(1-\theta)^{\beta-1} d\theta \quad 0 \le x \le n \le N \quad (19.13)$$

where n is the number of sampling events, α and β are the parameters defined by variance and expected value, x represents a random variable between 0 and n, and θ is a variable that ranges between 0 and 1.

Both the expected value and the variance in this function have to be calculated for the determination of parameters α and β. When changing the spatial resolution of vegetation cover, it is seen that the expectation is stable across scales while the variance is reduced at increasingly smaller scales. Therefore a method of predicting this reduction in variance is needed in order to employ the Poyla function to predict the frequency distribution at the fine scales from coarse resolution data. When a multi-scale dataset degrading the high resolution vegetation cover (0.55 m) derived from aerial photography is analysed, it can be seen that a logarithmic function between variance and scale in all the subimages can effectively describe the decline of subimage variance with increasing spatial resolution (Zhang *et al.*, 2002):

$$\sigma^2 = a + b \ln(d) \qquad (19.14)$$

where σ^2 represents variance, d represents measurement scale, and a and b are coefficients.

After the subimage variance is calculated, the distribution of vegetation cover in a subimage (or a pixel) can be estimated across (up or down) scales. Using this technique, the frequency distributions of vegetation cover at a measurement scale of 30 m are predicted on the basis of vegetation cover derived from AVHRR-NDVI at a resolution of 1 km (Figure 19.6).

19.6.4 Upscaled soil-erosion models

It is clear that for accurately calculating soil erosion the global scale, the Thornes erosion model nee

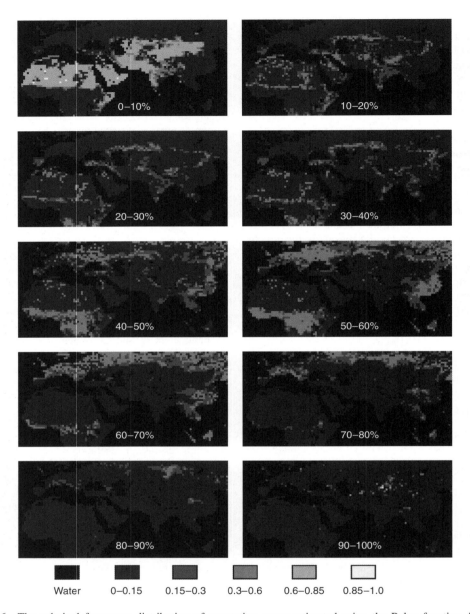

Figure 19.6 The subpixel frequency distribution of vegetation cover estimated using the Polya function. This figure shows the probability (values in the legend) of each level of percentage vegetation cover if the cover within each large pixel is measured using a scale (subpixel size) of 30 m

be scaled up. Therefore, the slope s is scaled down to the plot scale using Equation 19.12. The vegetation cover is represented as the subpixel frequency distribution by using Equation 19.13 with 101 levels from 0% to 100%. The monthly overland flow is scaled to a rain day by using the exponential frequency distribution of rainfall distribution described by Carson and Kirkby (1972). Thus the scaled erosion model becomes:

$$E = 2klOF_0^2(\gamma d^{1-D})^{1.67} \sum_{v=0}^{100} f(v, d)e^{-0.07v} \quad (19.15)$$

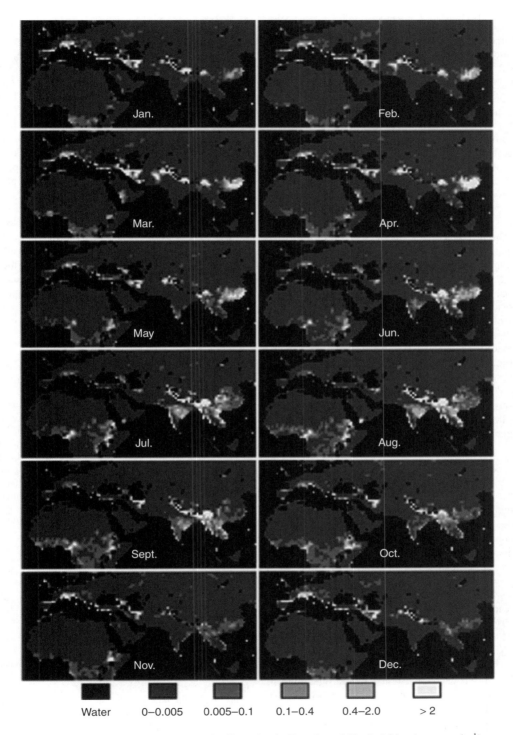

Figure 19.7   Spatial and temporal distributions of soil erosion in Eurasia and North Africa (mm month^{-1})

where E is monthly erosion (mm), k is a soil erodibility coefficient, OF is average daily overland flow (mm), l is the monthly number of rainfall events (days), D is the local topographic fraction dimension, γ is a local topographic parameter, v is the percentage vegetation cover, $f(v, d)$ is the Polya mass function, d represents the original modelling scale and s is the slope (m m^{-1}).

This upscaled model is employed to calculate monthly global erosion in Eurasia and North Africa based on a dataset of 1 km-resolution (Figure 19.7). The required dataset includes monthly overland flow calculated using a modified Carson and Kirkby (1972) model (Zhang *et al.*, 2002), monthly vegetation cover estimated in terms of AVHRR-NDVI (Zhang *et al.*, 1997), and a DEM selected from http://edcwww.cr.usgs.gov/landdaac/. The resultant spatial pattern of soil erosion suggests that high rates of soil erosion can be accounted for in terms of the steep unstable terrain, highly erodible soil, high monthly precipitation, and vegetation removal by human activity or seasonal factors. Conversely, low soil-erosion rates reflect the lower relief, the greater density of the vegetation canopy, and the areas of low precipitation. Erosion is serious in southern East Asia, India, along the Himalayas and Alps, and in West Africa. In seasonal terms, the erosion rate is high from spring to autumn in south-east China, summer and autumn in India, and from April to October in West Africa. Relatively high erosion rates occur in Western Europe during the winter months while very low soil loss occurs in summer.

19.7 CONCLUSION

The importance of scaling issues in environmental modeling has been widely recognized. A number of techniques have been developed to reduce the scaling effects on modelling outputs. A very effective approach is to identify scaleable parameters that characterize the relevant intrinsic environmental processes. When scaling several important environmental parameters to fit the scales of environmental models, the resultant values can be greatly improved. Since an environmental model is generally complex and related to several types of parameters, it may be necessary to combine several scaling techniques of scaling both parameters and models for the reduction of scale uncertainties. This procedure has been demonstrated in the modelling of soil erosion at the global scale. However, we need to pay more attention to the following issues.

1. Few scaling methods have effectively dealt with intrinsic environmental behaviour at regional and global scales since our solid knowledge remains on the local processes.

2. There are mainly two purposes of upscaling a model to regional or global scales. One is to derive the spatial distribution of our interests while the second is to calculate a single value in a large area. The output modelled on the basis of a distributed dataset may just represent the spatial distribution of the result in the resolution of the dataset. In many cases, the sum of the spatially distributed output cannot be taken as the value in a whole region (globe) since the spatial and temporal interaction and feedback of the processes may occur at large scales.

3. An applicable technique is required for a reasonable validation of scaled models at regional and global scales. It is insufficient to only verify the reliability of global scale results by comparing the difference among several typical spatial areas such as the Amazon tropical forest area and the Sahara desert, since any kind of model can produce several obviously different patterns at a global scale.

REFERENCES

Ahl, V. and Allen, T.F.H. (1996) *Hierarchy Theory: A Vision, Vocabulary, and Epistemology*, Columbia University Press, New York.

Atkinson, P.M. (1999) Spatial statistics, in A. Stein, F. van der Meer and B. Gorte (eds) *Spatial Statistics for Remote Sensing*, Kluwer Academic Publishers, Dordrecht, 57–81.

Atkinson, P.M. and Tate, N.J. (1999) Extracting information from remotely sensed and GIS data, in P.M. Atkinson and N.J. Tate (eds) *Advances in Remote Sensing and GIS Analysis*, John Wiley & Sons, Chichester, 263–268.

Avissar, R. (1992) Conceptual aspects of a statistical-dynamical approach to represent landscape subgrid scale heterogeneities in atmospheric models, *Journal of Geophysical Research-Atmospheres* **97**, 2729–2742.

Band, L.E. (1991) Distributed parameterization of complex terrain, in E.F. Wood (ed.) *Land Surface–Atmosphere Interactions for Climate Modelling*, Kluwer Academic, Dordrecht, 249–270.

Batjes, N.H. (1996) Global assessment of land vulnerability to water erosion on a 1/2 degrees by 1/2 degrees grid, *Land Degradation & Development* **7** (4), 353–365.

Beven, K. (1995) Linking parameters across scales: subgrid parameterizations and scale dependent hydrological models, in J.D. Kalma and M. Sivapalan (eds) *Scale Issues in Hydrological Modelling*, John Wiley & Sons, Chichester, 263–282.

Bian, L. (1997) Multiscale nature of spatial data in scaling up environmental models, in D.A. Quattrochi and M.F. Goodchild (eds) *Scale in Remote Sensing and GIS*, Raton Lewis Publishers, Boca Raton, FL, 13–26.

Blöschl, G., Grayson, R.B. and Sivapalan, M. (1995) On the representative elementary area (REA) concept and its utility for distributed rainfall-runoff modelling, in J.D. Kalma and M. Sivapalan (eds) *Scale Issues In Hydrological Modelling*, John Wiley & Sons, Chichester, 71–88.

Blöschl, G. and Sivapalan, M. (1995) Scale issues in hydrological modeling: a review, in J.D. Kalma and M. Sivapalan (eds) *Scale Issues in Hydrological Modelling*, John Wiley & Sons, Chichester, 9–48.

Braun, P., Molnar, T. and Kleeberg, H.-B. (1997) The problem of scaling in grid-related hydrological process modelling, *Hydrological Processes* **11**, 1219–1230.

Bresler, E. and Dagan, G. (1988) Variability of yield of an irrigated crop and its causes, 1, Statement of the problem and methodology, *Water Resources Research* **24**, 381–388.

Bugmann, H., Linder, M., Lasch, P, Flechsig, M., Ebert, B. and Cramer, W. (2000) Scaling issues in forest succession modelling, *Climatic Change* **44**, 265–289.

Burrough, P.A. and McDonnell, R.A. (1998) *Principles of Geographical Information Systems. Spatial Information Systems and Geostatics*, Oxford University Press, New York.

Campbell, C.G., Ghodrati, M. and Garrido, F. (1999) Comparison of time domain reflectometry, fiber optic mini-probes, and solution samplers for real time measurement of solute transport in soil, *Soil Science* **164** (3), 156–170.

Carson, M.A. and Kirkby, M.J. (1972) *Hillslope Form and Process*, Cambridge University Press, Cambridge.

Collins, J.B. (1998) Geostatistical methods for analysis of multiple scales of variation in spatial data, PhD thesis, Boston University.

Curran, P.J. and Atkinson, P.M. (1999) Issues of scale and optimal pixel size, in A. Stain, F. van der Meer and B. Gorte (eds) *Spatial Statistics for Remote Sensing*, Kluwer Academic Publishers, Dordrecht, 115–1333.

David, T.F., John, N.Q. and Dickinson, W.T. (1996) The GCTE validation of soil erosion models for global change studies, *Journal of Soil and Water Conservation* **15** (5), 397–403.

De Boor, C. (1978) *A Practical Guide to Splines*, Springer Verlag, Berlin.

De Cola, L. (1997) Multiresolution covariation among Landsat and AVHRR vegetation indices, in D.A. Quattrochi and M.F. Goodchild (eds) *Scale in Remote Sensing and GIS*, Lewis Publishers, Boca Raton, FL, 72–91.

Dodds, P.S. and Rothman, D.H. (2000) Scaling, university, and geomorphology, *Annual Review of Earth and Planetary Sciences* **28**, 571–610.

Dooge, J.C.I. (1986) Looking for hydrologic laws, *Water Resources Research* **22**, 465–585.

Drake, N.A., Zhang, X., Berkhout, E., Bonifacio, R., Grimes, D., Wainwright, J. and Mulligan, M. (1999) Modelling soil erosion at global and regional scales using remote sensing and GIS techniques, in P. Atkinson and N.J. Nate (eds) *Spatial Analysis for Remote Sensing and GIS*, John Wiley & Sons, Chichester, 241–261.

Evans, I.S. and McClean, C.J. (1995) The land surface is not unifractal: variograms, cirque scale and allometry, *Zeitschrift für Geomorphologie, N.F., Supplementband* **101**, 127–147.

Fisher, P. (1997) The pixel: a snare and a delusion, *International Journal of Remote Sensing* **18** (3), 679–685.

Flügel, W.-A. (1995) Delineating hydrological response units by geographical information system analyses for regional hydrological modelling using PRMS/MMS in the drainage basin of the river Bröl, Germany, in J.D. Kalma and M. Sivapalan (eds) *Scale Issues in Hydrological Modelling*, John Wiley & Sons, Chichester, 181–194.

Friedl, M.A. (1997) Examining the effects of sensor resolution and sub-pixel heterogeneity on spectral vegetation indices: implications for biophysical modelling, in D.A. Quattrochi and M.F. Goodchild (eds) *Scale in Remote Sensing and GIS*, Lewis Publishers, Boca Raton, FL, 113–139.

Gatrell, A.C. (1991) Concepts of space and geographical data, in D.J. Maguire, M.F. Goodchild and D. Rhind (eds) *Geographical Information Systems: Principles and Applications*, Longman, London, 119–134.

Goodchild, M.F. and Quattrochi, D.A. (1997) Scale, multiscaling, remote sensing, and GIS, in D.A. Quattrochi and M.F. Goodchild (eds) *Scale in Remote Sensing and GIS*, Lewis Publishers, Boca Raton, FL, 1–11.

Harte, J. (2000) Scaling and self-similarity in species distributions: implication for extinction, species richness, abundance, and range, in J.H. Brown and G.B. West (eds) *Scaling in Biology*, Oxford University Press, Oxford, 325–342.

Harvey, L.D.D. (2000) Upscaling in global change research, *Climatic Change* **44**, 225–263.

Hay, G.J., Niemann, K.O. and Gooddenough, D.G. (1997) Spatial thresholds, images-objects, and upscaling: a multiscale evaluation, *Remote Sensing of Environment* **62**, 1–19.

Hay, L.E., McCabe, G.J., Wolock, D.M. and Ayers, M.A. (1992) Use weather types of disaggregate general circulation model predictions, *Journal of Geophysical Research* **97**, 2781–2790.

Helmlinger, K.R., Kumar, P. and Foufoula-Georgiou, E. (1993) On the use of digital elevation model data for Hortonian and fractal analyses of channel networks, *Water Resources Research* **29** (8), 2599–2613.

Heuvelink, G.B.M. (1998) Uncertainty analysis in environmental modelling under a change of spatial scale, *Nutrient Cycling in Agroecosystems* **50** (1–3), 255–264.

Karl, T.R., Wang, W.C., Schlesinger, M.E., Kniht, R.W. and Portman, D. (1990) A method of relating general circulation model simulated climate to the observed local climate. Part I. Seasonal statistics, *Journal of Climate* **3**, 1053–1079.

King, A.W. (1991) Translating models of across scales in the landscape, in M.G. Turnere and R.H. Gardner (eds) *Quantitative Methods in Landscape Ecology: Ecological Studies*, Springer, New York, 479–517.

Kirkby, M.J. and Cox, N.J. (1995) A climatic index for soil erosion potential (CSEP) including seasonal and vegetation factors, *Catena* **25**, 333–352.

Kirkby, M.J., Imeson, A.C., Bergkamp, G. and Cammeraat, L.H. (1996) Scaling up process and models, *Journal of Soil and Water Conservation* **15** (5), 391–396.

Klinkenberg, B. and Goodchild, M.F. (1992) The fractal properties of topography: a comparison of methods, *Earth Surface Processes and Landforms* **17** (3), 217–234.

Lam, N. and Quattrochi, D.A. (1992) On the issue of scale, resolution, and fractal analysis in the mapping sciences, *Professional Geographer* **44**, 88–98.

Luo, Y. (1999) Scaling against environmental and biological variability: general principles and A case study, in Y. Luo and H.A. Mooney (eds) *Carbon Dioxide and Environmental Stress*, Academic Press, San Diego, 309–331.

Maidment, D.R. (1993) GIS and hydrological modeling, in M.F. Goodchild, B.O. Parks and L.T. Steyaert (eds) *Environmental Modeling with GIS*, Oxford University Press, New York, 147–167.

Mandelbrot, B.B. (1982) *The Fractal Geometry of Nature*, Freeman, New York.

McNulty, S.G., Vose, J.M. and Swank, W.T. (1996) Scaling predicted pine forest hydrology and productivity across the southern United States, in D.A. Quattrochi and M.F. Goodchild (eds) *Scale in Remote Sensing and GIS*, Lewis Publishers, Boca Raton, FL, 187–209.

Moore, I.D. and Gallant, J.C. (1991) Overview of hydrologic and water quality modeling, in I.D. Moor (ed.) *Modeling the Fate of Chemicals in the Environment*, Center for Resource and Environmental Studies, the Australian National University, Canberra, 1–8.

Moore, I.D., Turner, A.K., Wilson, J.P., Jenson, S.K. and Band, L.E. (1993) GIS and land surface-subsurface process modeling, in M.F. Goodchild, B.O. Parks and L.T. Steyaert (eds) *Environmental Modelling with GIS*, Oxford University Press, New York, 196–229.

Papritz, A. and Stein, A. (1999) Spatial prediction by linear kriging, in A. Stain, F. van der Meer and B. Gorte (eds) *Spatial Statistics for Remote Sensing*, Kluwer Academic Publishers, Dordrecht, 83–113.

Peterson, G.D. (2000) Scaling ecological dynamics: self-organization, hierarchical structure, and ecological resilience, *Climatic Change* **44**, 291–309.

Pilotti, M. and Bacchi, B. (1997) Distributed evaluation of the contribution of soil erosion to the sediment yield from a watershed, *Earth Surface Processes and Landforms* **22**, 1239–1251.

Poesen, J.W., Boordman, J., Wilcox, B. and Valentin, C. (1996) Water erosion monitoring and experimentation for global change studies, *Journal of Soil and Water Conservation* **15** (5), 386–390.

Rastetter, E.B., King, A.W., Cosby, B.J., Hornberger, G.M., O'Neill, R.V. and Hobbie, J.E. (1992) Aggregating fine-scale ecological knowledge to model coarser-scale attributes of ecosystems *Ecological Application* **2**, 55–70.

Richards, K. (1993) Sediment delivery and the drainage network, in K. Beven and M.J. Kirkby (eds) *Channel Network Hydrology*, John Wiley & Sons, Chichester, 221–254.

Rosenfeld, A. and Kak, A.C. (1982) *Digital Picture Processing*, Academic Press, New York.

Settle, J.J. and Drake, N.A. (1993) Linear mixing and the estimation of ground cover proportions, *International Journal of Remote Sensing* **14** (6), 1159–1177.

Thornes, J.B. (1985) The ecology of erosion, *Geography* **70**, 222–234.

Van der Meer, F. (1994) Extraction of mineral absorption features from high-spectral resolution data using non-parametric geostatistical techniques, *International Journal of Remote Sensing* **15**, 2193–2214.

VEMAP Members (1995) Vegetation/ecosystem modelling and analysis project: comparing biogeography and biogeochemistry models in a continental-scale study of terrestrial ecosystem response to climate change and CO_2 doubling, *Global Biogeochemical Cycles* **4**, 407–437.

Walsh, S.J., Moody, A., Allen, T.R. and Brown, D.G. (1997) Scale dependence of NDVI and its relationship to mountainous terrain, in D.A. Quattrochi and M.F. Goodchild (eds) *Scale in Remote Sensing and GIS*, Lewis Publishers, Boca Raton, FL, 27–55.

West, G.B., Brown, J.H. and Enquist, B.J. (2000) The origin of universal scaling laws in biology, in J.H. Brown and G.B. West (eds) *Scaling in Biology*, Oxford University Press, Oxford, 87–112.

Wigley, T.M.L., Jones, P.D., Briffa, K.R. and Sith, G. (1990) Obtaining sub-grid scale information from coarse resolution general circulation model output, *Journal of Geophysical Research* **95**, 1943–1953.

Wilby, R.L. and Wigley, T.M.L. (1997) Downscaling general circulation model output: a review of methods and limitations, *Progress in Physical Geography* **21** (4), 530–548.

Wood, E.F., Sivapalan, M. and Beven, K.J. (1990) Similarity and scale in catchment storm response, *Review of Geophysics* **28**, 1–18.

Wood, E.F., Sivapalan, M., Beven, K.J. and Band, L. (1988) Effects of spatial variability and scale with implications to hydrologic modelling, *Journal of Hydrology* **102**, 29–47.

Woodcock, C.E. and Strahler, A.H. (1987) The factor of scale in remote sensing, *Remote Sensing of Environment* **21**, 311–332.

Woodcock, C.E., Strahler, A.H. and Jupp, D.L.B. (1988a) The use of variograms in remote sensing I: Scene models and simulated images, *Remote Sensing of Environment* **25**, 323–348.

Woodcock, C.E., Strahler, A.H. and Jupp, D.L.B. (1988b) The use of variograms in remote sensing I: Real digital images, *Remote Sensing of Environment* **25**, 349–379.

Zhang, X., Drake, N.A., Wainwright, J. and Mulligan, M. (1997) Global scale overland flow and soil erosion modelling using remote sensing and GIS techniques: model implementation and scaling, in G. Griffiths and D. Pearson (eds) *RSS97 Observations and Interactions: 23rd Annual Conference and Exhibition of the Remote Sensing Society*, The Remote Sensing Society, Nottingham, 379–384.

Zhang, X.Y., Drake, N.A., Wainwright, J. and Mulligan, M. (1999) Comparison of slope estimates from low resolution DEMs: scaling issues and a fractal method for their solution, *Earth Surface Processes and Landforms* **24** (9), 763–779.

Zhang, X., Drake, N.A. and Wainwright, J. (2002) Scaling land-surface parameters for global scale soil-erosion estimation, *Water Resources Research* **38** (10), 1180. Doi: 10.1029/2001 WR000356.

Environmental Applications of Computational Fluid Dynamics

NIGEL G. WRIGHT AND CHRISTOPHER J. BAKER

20.1 INTRODUCTION

Computational Fluid Dynamics (CFD) has been in use in various fields of science and engineering for over 40 years. Aeronautics and aerospace were the main fields of application in the early days, but its application has much to offer in other fields if due consideration is given to their particular requirements. CFD offers:

- Full-scale simulation as opposed to the model scale of many physical simulations. In environmental applications this can be of vital importance as the domain of interest may be of considerable size.
- Interpolation between measured data. Measured data are often both temporally and spatially sparse. As long as sufficient data are available, they can be used to calibrate a CFD model that will provide information at many time and space points.
- Excellent visualization as results are easily amenable to graphical representation.
- Repeatability: a CFD simulation can be run again and again with the same parameters or with many variations in parameters. As computer power develops, this approach also opens the route to optimization and parameter-uncertainty studies for more complex problems.
- CFD simulations can be carried out in situations where a real-life simulation is impossible such as the release of toxic substances into the natural environment.

As well as drawing attention to the advantages and possibilities of CFD, it is also necessary to explain the drawbacks and issues that must be addressed in any CFD study. These are:

- The results output by a CFD package are not necessarily a valid solution for a particular fluid flow problem. Knowledge and experience of fluid mechanics are needed to evaluate the results critically. Furthermore, some understanding of how a CFD code works is needed to evaluate whether it has been applied appropriately in a given situation. This combination of skills is not widely available and is not quickly learnt.
- Although hardware costs are much lower now that most software can be run on PCs, the cost is still reasonably high given that a high-specification PC is required. If a commercial CFD package is used, the cost of a licence will be much higher than the cost of a PC. This, combined with the cost of employing someone with the skills previously described, can make CFD an expensive undertaking. However, this must be weighed against the cost of more traditional methods.
- There are several guidelines available on the validation of CFD simulations (ASME, 1993; AIAA, 1998; Casey and Wintergerste, 2000) and these should be carefully studied (especially in view of the comments on skills above). In some cases there may be no data available to validate or calibrate the CFD model and in this situation care must be taken not to interpret too much into the results (Lane and Richards, 2001).

20.2 CFD FUNDAMENTALS

20.2.1 Overview

Computational Fluid Dynamics is concerned with taking the physical laws of fluid flow and producing solutions

Environmental Modelling: Finding Simplicity in Complexity. Edited by J. Wainwright and M. Mulligan
© 2004 John Wiley & Sons, Ltd ISBNs: 0-471-49617-0 (HB); 0-471-49618-9 (PB)

in particular situations with the aid of computers. Soon after the advent of computers, mathematicians began to investigate their use in solving partial differential equations and by the 1970s the use of computers to solve nonlinear partial differential equations governing fluid flow was under active investigation by researchers. During the 1980s, CFD became a viable commercial tool and a number of companies were successfully marketing CFD software.

While in general the term CFD is taken to imply a full three-dimensional calculation of a fully turbulent flow field, in environmental calculations it is often possible to use calculations that are either one- or two-dimensional, or include significant simplifications of the flow field in other ways. Thus, while in this section CFD is described in its conventional three-dimensional form, many of the applications discussed below utilize models that are much simpler, which gives significant advantages in terms of practicality and ease of use.

There are many books devoted to CFD and it is not possible to cover the field in a short chapter. In view of this, the following discussion highlights the major issues in environmental applications. For more detail, readers are referred to other texts: Versteeg and Malalasekera (1995) for a general treatment and Abbott and Basco (1989) for applications to the shallow water equations.

20.2.2 Equations of motion

CFD is based on solving the physical laws of Conservation of Mass and Newton's 2nd Law as applied to a fluid. These are encapsulated in the Navier–Stokes equations for incompressible flow:

$$\frac{\partial u_i}{\partial x_i} = 0 \qquad (20.1)$$

$$\frac{\partial(\rho u_i)}{\partial t} + \frac{\partial(\rho u_i u_j)}{\partial x_j} = -\frac{\partial p}{\partial x_i}$$

$$+ \frac{\partial}{\partial x_j}\left(\mu \frac{\partial u_i}{\partial x_j}\right) + S_i \qquad (20.2)$$

where x_i ($i = 1, 2, 3$) are the three coordinate directions [L], u_i ($i = 1, 2, 3$) are the velocities in these directions [L T^{-1}], p is pressure [M L^{-1} T^{-2}], ρ is density [M L^{-3}] and μ is viscosity [M L^{-1} T^{-2}]. The first equation encapsulates conservation of mass. The second is a momentum equation where the first term is the time variation and the second is the convection term: on the right-hand side the terms are, respectively, a pressure gradient, a diffusion term and the source term (i.e. Coriolis force, wind shear, gravity). In most cases these

equations cannot be solved to give an explicit function. CFD opens up the possibility of providing approximate solutions at a finite number of discrete points.

20.2.3 Grid structure

Initial investigations into the solutions of PDEs used the finite difference method and this approach was carried through into early CFD. The finite difference method approximates a derivative at a point in terms of values at that point and adjacent points through the use of a Taylor series approximation (see Chapter 1). It is simple to calculate and implement for basic cases. However, it was found to have limitations in its application to fluid-flow problems. Engineering CFD researchers developed and favoured an alternative, more versatile method called the finite volume method. This divides the whole area of interest into a large number of small control volumes. These control volumes are distinct and cover the whole domain. The physical laws are integrated over each control volume to give an equation for each law in terms of values on the face of each control volume. These face values or fluxes are then calculated from adjacent values by interpolation. This method ensures that mass is conserved in the discrete form of the equations just as it is in the physical situation. The finite volume technique is the most widely used in 3D CFD software.

The initial CFD work with finite differences and finite volumes used what is known as a structured mesh. Essentially, this is a mesh that can be represented in a Cartesian grid system as shown in Figure 20.1. As CFD developed, there was a desire to apply it to geometries that did not conform to this rectilinear form. One solution is to deform the grid as shown in Figure 20.2. This approach was successful, but cases arose where the grid was so deformed that it was difficult to obtain solutions to the problem. In view of this problem, attention was drawn to the finite element method that had originally been developed for structural analysis (Zienkiewicz and Taylor, 2000) which allowed for an irregular splitting up of the domain (see Figure 20.3). This proved successful in dealing with complex geometries. It is therefore popular for software solving the Shallow Water Equations (TELEMAC: Hervouet, 1994, and RMA2: King *et al.*, 1975), but less so for 3D CFD packages. Because the method minimizes error globally rather than locally, it does not guarantee conservation of physical quantities without extremely fine grids or special treatments. In view of this problem, the method has limits for applications to sediment transport or water quality.

A significant development in tackling complex geometries was a combination of the finite element and

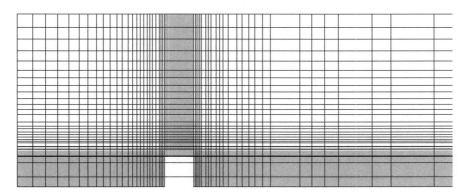

Figure 20.1 Cartesian grid

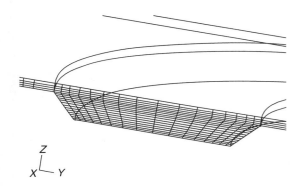

Figure 20.2 Curvilinear grid for a river cross-section (after Morvan, 2001)

finite volume approach. Researchers (Schneider and Raw, 1987) took the unstructured grids of finite elements, but applied the finite volume methodology. This approach now forms the basis for the main commercial 3D CFD packages (CFX, FLUENT, STAR-CD). It has the advantages of ease of application to complex geometries (including natural geometries) and implicit conservation of physical quantities.

Constructing a suitable mesh is often the most demanding part of a CFD simulation both in terms of operator time and expertise. Adequate details of the physical situation must be included in the mesh, but too great a level of detail will lead to a mesh that contains more cells than can be accommodated on the computer.

Figure 20.3 Example of an unstructured grid

An appropriate spatial scale must be selected. Further, a mesh should be refined in certain key areas such as boundary layers where velocity gradients are highest.

20.2.4 Discretization and solution methods

As described above, there are several methodologies for converting the continuous equations to a set of algebraic equations for the discrete values. Even within each methodology there are different ways of making the approximations. These different approximations for producing discrete values are known as discretization techniques. Within the finite volume framework the equations are integrated over a control volume and the problem is reduced to finding a face value from values in surrounding volumes. A popular method for this is the upwind method (Versteeg and Malalasekera, 1995) which uses the intuitive idea that the face values are more closely related to upstream than downstream values and therefore approximates them by the value immediately upstream. This scheme is the default in many packages, but it must be used with caution. An analysis of the error in the approximation reveals that it has the form of a false, numerically generated diffusion. This numerical diffusion gives the scheme excellent stability, but also poor accuracy. Other schemes exist that are more accurate (Versteeg and Malalasekera, 1995) and each of these is a balance between robustness and accuracy.

Once a particular discretization method has been adopted, a set of nonlinear algebraic equations is obtained. This nonlinearity and the lack of an explicit equation for pressure are the main difficulties that have to be addressed in devising a robust solution algorithm for these equations. The SIMPLE algorithm of Patankar and Spalding (1972) used a derived, approximate pressure equation that was solved separately from the velocity equations. It rapidly became the most popular technique and was the mainstay of commercial packages until recently. During the 1980s research was conducted on alternatives that solved for velocity and pressure simultaneously and removed the necessity of deriving a pressure equation (Schneider and Raw, 1987; Wright, 1988). Unsegregated solvers have now been adopted for some commercial codes and are delivering significant increases in computational efficiency and robustness.

20.2.5 The turbulence-closure problem and turbulence models

One of the fundamental phenomena of fluid dynamics is turbulence. As the Reynolds number of a flow increases, random motions are generated that are not suppressed by viscous forces. The resulting turbulence consists of a hierarchy of eddies of differing sizes. They form an energy cascade which extracts energy from the mean flow into large eddies and in turn smaller eddies extract energy from these which is ultimately dissipated via viscous forces.

In environmental flows, turbulence is particularly important. The rough boundaries usually found in surface water and atmospheric flows generate significant turbulence. Further, an understanding of turbulence is vital in studying dispersion. Turbulence is perhaps the most important remaining challenge for CFD. In theory, it is possible to predict all the eddy structures from the large ones down to the smallest. This approach is known as Direct Numerical Simulation (DNS). However, for practical flows it requires computing power that is not available at present and may not be available for many years. A first level of approximation can be made through the use of Large Eddy Simulations (LES). These use a length scale to differentiate between larger and smaller eddies. The larger eddies are predicted directly through the use of an appropriately fine grid that allows them to be resolved. The smaller eddies are not directly predicted, but are accounted for through what is known as a subgrid scale model (Smagorinsky, 1963). This methodology can be justified physically through the argument that large eddies account for most of the effect on the mean flow and are highly anisotropic whereas the smaller eddies are less important and mostly isotropic. Care is needed in applying these methods as an inappropriate filter or grid size and low accuracy spatio-temporal discretization can produce spurious results. If this is not done, LES is not much more than an inaccurate laminar flow simulation. Although less computationally demanding than DNS, LES still requires fine grids and consequently significant computing resources that still mean it is not a viable, practical solution.

In view of the demands of DNS and LES, most turbulence modelling still relies on the concept of Reynolds averaging where the turbulent fluctuations are averaged out and included as additional modelled terms in the Navier–Stokes equations. The most popular option is the $k - \varepsilon$ model which is usually the default option in CFD software. k represents the kinetic energy in the turbulent fluctuations and ε represents the rate of dissipation of k. Equations for k and ε can be derived from the Navier–Stokes equations and solved in a manner similar to the momentum equations. To represent the effect of turbulence on the momentum equation, the eddy-viscosity hypothesis of Boussinesq (1877) is used. This hypothesis assumes an analogy between the turbulent stresses and the viscous stresses and calculates an additional 'turbulent' viscosity based on values of k and ε. Interested readers are referred to CFD texts (Versteeg and Malalasekera, 1995) for further details and details of variants of this standard model such as RNG $k - \varepsilon$ and nonlinear $k - \varepsilon$ models. Such texts also give details of Reynolds Stress models that do not use the Boussinesq hypothesis and can calculate anisotropic turbulence. In some situations, these models represent a significant advantage over the isotropic $k - \varepsilon$ model.

20.2.6 Boundary conditions

Whether computational or purely analytic, any solution to the Navier–Stokes equations requires that values are specified at the boundaries of the domain of interest. These can take a variety of forms and have various levels of computational complexity. The most common and easiest to implement specify a fixed value of a variable at the boundary which is known mathematically as a Dirichlet condition (Smith, 1978). In fluid flow this is commonly a fixed velocity at an inlet although pressure may sometimes be specified. At a flow outlet it is more difficult to specify conditions. It can be said that, for a steady state problem, the outflow must equal the inflow or that the flow profile must be uniform in the along stream direction. The latter would require that the outlet is sufficiently downstream of the area of interest and, if this is so, a condition on the derivative of the along stream velocity may be imposed. This condition must be approached carefully, in order to prevent poor convergence or unphysical solutions. Another condition, common in CFD, is a symmetry condition which may be used to allow for solution of only half of a symmetric domain by imposing zero derivatives on all variables except velocity into the symmetry plane which has a value of zero. Before making use of a symmetry plane it must be certain that the flow solution will be symmetric – a symmetric domain and boundary conditions do not guarantee a stable, symmetric solution. Other boundary conditions occur, such as periodic and shear-free, and readers are referred to CFD texts (e.g. Versteeg and Malalasekera, 1995) for details of these.

In environmental flows, care must be taken to ensure that the turbulence quantities are correctly specified at inlets and that the inlet profiles are consistent with the definition of roughness along the boundaries of the domain.

In most flow situations the velocity gradients in the boundary layer are so high that an adequate grid size to resolve the flow would result in an unfeasibly large mesh. In these cases a coarser mesh is used and the Navier–Stokes equations are replaced by a model of boundary layer flow known as 'the Law of the Wall'. Care must be taken in setting a grid size that is appropriate for the Law of the Wall and guidance can be found in most CFD software manuals and textbooks (e.g. Versteeg and Malalasekera, 1995). The Law of the Wall can be amended to take account of surface roughness which is obviously important in many environmental flows. Again, guidance can be found elsewhere on appropriate values for this.

20.2.7 Post-processing

Visualization is one of the great strengths of CFD, but can also be one of its downfalls in inexperienced hands. The ability to predict results at positions throughout the domain offers a great advantage over experimental methods that often have a restricted number of measuring points. The many forms of graphics techniques such as shaded contours, velocity vectors, iso-surfaces, particle trajectories and 3D virtual reality allow CFD users to gain tremendous and unprecedented insights into the calculated solutions. However, these powerful visualization tools can give false confidence in the results. Incorrect results are still incorrect, however well they are visualized.

20.2.8 Validation and verification

The rapid development of CFD has not brought it to the level of a 'black-box' analysis tool. Experience and understanding are still needed on the part of the operator both in setting up a problem and analysing its results. A number of efforts have been made to assist and the recently produced ERCOFTAC Guidelines (Casey and Wintergerste, 2000) are an excellent example of these. Guidance is also available from reports such as that produced by the ASME (1993) and AIAA (1998). Lane and Richards (2001) have also discussed the more scientific questions of what is meant by validation and verification and their arguments should be borne in mind by any CFD user (see also the discussion in Chapter 1).

20.3 APPLICATIONS OF CFD IN ENVIRONMENTAL MODELLING

20.3.1 Hydraulic applications

As mentioned above, the term CFD is often taken to cover solutions to the Navier–Stokes equations in three dimensions. However, in many environmental applications such as rivers, estuaries and coastal areas this approach is not appropriate. Flows in these situations usually have one or two dimensions considerably larger than the other. Fully 3D solutions are neither feasible nor necessary. For example, in studying the entire length of a river (maybe up to 1000 km) it is not necessary to have details of the velocity profile across the channel. In these situations, approximations are made that allow the 3D equations to be reduced to a simpler, more tractable set. These are referred to as the shallow water equations (SWE) and they assume that there are no significant vertical accelerations and consequently that a hydrostatic pressure exists in the vertical direction. This assumption is reasonable in many cases, but the limitations must be borne in mind in examining the results of simulations. The computational solution of these equations was covered comprehensively in Abbott (1979).

The most widely used version of the shallow water equations is the 1D version known as the St Venant equations (Cunge *et al.*, 1980) which has been implemented in a number of commercial codes including ISIS from Halcrow/HR Wallingford, MIKE11 from the Danish Hydraulic Institute and HEC-RAS from the US Army Corps of Engineers. These have limitations in their application in that:

- they assume small variations in the free surface inclination;
- they assume that the friction laws from steady state apply to unsteady flow;
- they have no allowances for sinuosity.

In some situations where the cross-channel distance or topography variation is significant, a 1D model may not be adequate. In these situations 2D models are used. These still assume small variations in the free surface, but predict velocities in both horizontal directions in addition to the depth of water. This procedure is referred to as depth averaging which relates to the assumed vertical profile. These codes are particularly useful in floodplains, coastal areas and estuaries. There are a number of academic or research codes (Falconer, 1980; Olsen and Stokseth, 1995; Sleigh *et al.*, 1998) and an increasing number of commercially available codes: TELEMAC2D (Hervouet 1994), MIKE21(DHI 1998), SMS (http://emrl.byu.edu/sms.htm). In some cases these codes have been extended to include a number of vertical layers (Falconer and Lin, 1997). This inclusion allows for the prediction of secondary circulations, that is circulations with an axis along the channel. These layered models can also be adjusted to

account for nonhydrostatic pressure variations (Stansby and Zhou, 1998) thereby overcoming the limitations of the assessment of vertical hydrostatic pressure in the SWE.

The commercial and academic codes in use can be classified into those based on the finite element methodology and those based on the alternative finite volume methodology. A finite element algorithm minimizes errors globally and does not necessarily ensure local conservation of mass. Mathematically, the lack of conservation merely reflects the fact that the solution is an approximation. While in flood-extent studies precise conservation may not be crucial, when solving for sediment transport or water quality parameters, it is important to have a conservative scheme.

The finite volume method does guarantee mass conservation and codes have been developed for sediment transport and water-quality studies both for Cartesian grids (Falconer, 1980) and boundary fitted grids (Falconer and Lin, 1997). In the past decade there has been significant development of unstructured finite volume codes (Anastasiou and Chan, 1997; Sleigh *et al.*, 1998; Olsen, 2000).

The issue of wetting and drying is a perennially difficult one for 2D models. As water levels drop, areas of the domain may become dry and the calculation procedure must remove these from the computation in a way that does not compromise mass conservation or computational stability. Most available codes accommodate this phenomenon with varying degrees of success, but they all compromise between accuracy and stability. This issue must be carefully examined in results from any 2D simulation where wetting and drying are significant.

Ultimately an application that has significant 3D features can be solved by a general CFD code that solves the full Navier–Stokes equations and either uses a prior estimate of the free surface position or uses a technique such as Volume of Fluid (VoF) (Hirt and Nichols, 1981) to predict the position of the air/water interface. Such 3D codes are significantly different from codes that solve for the SWEs within a series of vertical layers. These are often referred to as '3D codes', but should, perhaps, be more appropriately referred to 'quasi-3D' codes.

20.3.1.1 Coastal and estuarine applications

The application of numerical methods in this field is extensive and readers are referred elsewhere (e.g. Dean and Dalrymple, 1991) for further information.

Coastal zones and estuaries have significant variations in all lateral directions and computer-modelling uses the 2D shallow water equations or more complex models. Early examples are given by Abbott (1979). The DIVAST code was developed by Falconer (1980) and has been applied successfully in both coastal and estuarial situations by a number of different researchers. Other models have been developed from this basis and a review of earlier work is given by Cheng and Smith (1989). More recently models have been extended to whole regions of the continental shelf (Sauvaget *et al.*, 2000).

In addition to the solution of the depth-averaged equations, computational solutions in this field require account to be taken of the effects of wind shear and Coriolis forces. The former is accommodated as a momentum source on the free surface and the latter as a source term.

The motion of waves is a significant factor in coastal studies. The modelling of waves is not possible with the shallow water equations because of the neglect of vertical acceleration and the consequent assumption of no significant free surface curvature. Models have been developed based on the Boussinesq equations for waves in shallow water (Berzins and Walkley, 2003). Li and Fleming (2001) have developed fully 3D models for coastal zones that predict wave motions directly.

In estuaries there can be significant effects due to thermal and saline stratification. Adequate resolution of these effects necessitates the use of software that solves the shallow water equations within a number of vertical layers. This approach allows for different temperatures and salinities in the vertical and makes some allowance for vertical velocities. Falconer and Lin (1997) have developed this technique and applied it to studies of morphological changes and the transport of heavy metals in the Humber Estuary.

20.3.1.2 *Rivers*

1D models are in widespread use in commercial consultancy and are the main tool for assessing flood risk, water quality and construction impact in rivers and channels. In addition to modelling the discharge and depth at river cross-sections, these codes have units that model hydraulic structures such as sluice gates and weirs. Through use of these units a model of a complex network of rivers and artificial channels can be built up. In this way current computer power is beginning to allow for models of whole catchments and real-time flood prediction. Cunge *et al.* (1980) contains an excellent background to the theory and application of river models. Predominantly, the 1D equations are solved by the Preissmann or other 'box' method. These

methods allow for the simple inclusion of units such as sluice gates and weirs. However, they do not take account of the fact that the equations are hyperbolic (see Smith, 1978, for a definition). Other techniques have been proposed (Garcia-Navarro *et al.*, 1992; Alcrudo and Garcia-Navarro, 1993; Garcia-Navarro *et al.*, 1995; Crossley, 1999), but these have yet to be put into common use.

2D modelling of rivers is now being used commercially, although the recent work on flood extent prediction undertaken by the Environment Agency for England and Wales used 1D models indicating that this is not yet accepted as standard practice. Various research groups have developed and validated these techniques (Bates *et al.*, 1996; Sleigh *et al.*, 1998; Bates *et al.*, 1999). A significant amount of work has been carried out on the use of remotely-sensed data in 2D modelling (Horritt *et al.*, 2001) which offers the advantages of increased accuracy and faster model development.

In many cases 2D solutions are only required in parts of a river model and thus in the majority of the river system a 1D model is adequate. Dhondia and Stelling (2002) have implemented a combined 1D/2D approach that allows for a complete 1D model to be augmented, rather than replaced, in key areas.

The use of fully 3D CFD codes within rivers is still a predominantly research activity. However, this work can give new understanding of the fundamentals of fluid flow in channels. The first attempts at using CFD considered simplified channels (Rastogi and Rodi, 1978; Leschziner and Rodi, 1979; Rodi, 1980; Naot and Rodi, 1982; Demuren and Rodi, 1986; Gibson and Rodi, 1989), but did demonstrate the potential application. Bradbrook *et al.* (2000) have also published LES results for a laboratory-type channel. Although these results require substantial computing resources and may be some way off practical application, they offer a means of understanding the fundamental processes of turbulence in river channels.

One of the first 3D models to attempt to model natural river geometry in full was reported by Olsen and Stokseth (1995) who modelled a short reach (20 × 80 m) of the Skona River in Norway. The use of 3D modelling for the design of river rehabilitation schemes has been developed (Swindale and Wright, 1997; Swindale, 1999; Wright *et al.*, 2000) with the free surface being determined from 1D and 2D models. In this situation the extra information available from 3D is of great benefit in analysing the varied habitat required for different species. Lane *et al.* (1999) carried out 3D simulations and compared the results with a 2D simulation for flow at a river confluence for which high quality field data was available. They found that 3D simulations offered better predictions for secondary circulations. The 3D model also gave better estimates of shear stress and provided better information for calculating mixing processes. For more detail on the use of higher dimensional models in river modelling, readers are referred to a review paper commissioned by HR Wallingford for the Environment Agency of England and Wales (Wright, 2001).

20.3.1.3 Small-scale river works

3D CFD software can also be used for analysing flows at small-scale river works such as sluice gates, weirs, outfalls and fish passes (Rodríguez *et al.*, 2001). This work is neither straightforward nor rapid, but with further verification may become an acceptable alternative to scale models. In these cases CFD has the advantage over physical models of being able to work at full scale.

20.3.2 Atmospheric applications

In this section we consider the application of CFD techniques to two categories of atmospheric flows. As above, the appropriate length scale is used to differentiate them. First, we discuss work on predicting the wind flow over large topography (with a scale of the order of 0.1 to 10 km). Second, we consider the use of CFD in small-scale engineering and environmental problems with scales of the order of 10 to 100 m. We do not consider the application of CFD techniques (and in particular the LES technique) to the prediction of large-scale atmospheric circulations. For details of such applications, the reader is referred to Mason (1994).

20.3.2.1 Wind over large-scale topography

The calculation of wind speeds and other atmospheric variables, such as temperature and moisture, over large-scale topography such as hills, is of some practical importance for assessing local wind climate for planning purposes, the construction of wind farms, the dispersion of atmospheric pollutants, etc., and also for providing a means by which larger-scale numerical weather predictions can be used to predict smaller-scale effects. An excellent historical review of the subject has been presented by Wood (2000) and the following few paragraphs draw heavily on this work.

Work in this field began in the 1940s and 1950s with inviscid flow calculations over a range of large-scale

topography. The work of Scorer (1949) on trapped lee waves stands out as being of considerable importance. While these calculations were of fundamental importance, they could not of course take into account the real surface boundary condition, nor the turbulent nature of the boundary layer. To assess the flow regions in which these methods were applicable, the concepts of different layers of the flow were derived – in particular, the idea that there was a relatively thin region close to the surface of the topography where the effects of the topography and of turbulence were important, and a region above that that could be regarded as inviscid to a first approximation. This concept was central to the work of Jackson and Hunt (1975) who formalized it in their linear analysis of turbulent flow over low hills, defining an inner layer close to the hill and an outer layer above that. This theory led to quantifiable predictions of wind speed-up and the change of turbulent structure over a variety of topographical shapes. When tested against actual full-scale experimental data obtained from controlled measurements over simple topography (such as the Black Mountain and Askervein Hill), this theory was shown to perform more than adequately. The theory continued to be developed over the next decade or so (Hunt *et al.*, 1988; Belcher, 1990; Belcher and Hunt, 1998) and forms the basis of a number of calculation methods that are currently in use for the prediction of wind fields and dispersion over arbitrary topography. An example of this is provided in the work of Inglis *et al.* (1995) for the prediction of wind conditions over Kintyre (which was carried out to assess the impact of wind on forests). Typical comparisons of the predictions with theory are shown in Figure 20.4.

Although the linear type of model is a CFD model in that it requires a numerical solution to the equations of fluid flow, it is not a CFD model in the terms described in previous sections. So what of the adequacy of standard CFD models to predict flow over large topography? At this point it is worth directly quoting the words of Wood (2000):

Using standard closure schemes, simulated flow over hills exhibits a strong dependence on the specific closure employed and yet for various reasons (known deficiencies, dependence on unknown tuneable constants, misrepresentation of the near wall behaviour), none of the currently available closures are satisfactory in the complex geometry to which they are applied.

Wood then goes on to advocate the continued development of LES models in this field, and this seems to be

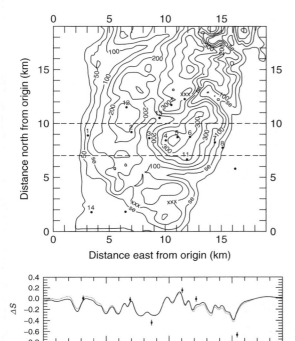

Figure 20.4 Comparison of observed (filled circles) and modelled fractional speed-ups (continuous and dotted lines – with and without modelling of atmospheric stability) along a line connecting sites 1 to 8
Source: Inglis *et al.* (1995)

generally accepted as the way forward for the use of CFD techniques for such problems. It has the particular advantage of being able to predict flow fluctuations, extreme values, etc., which are often the parameters required in practical situations. An example of the current use of LES is the RANS model developed at Colorado State University, which has been used extensively in predicting topographical flows (see, for example, Cai *et al.*, 1995; Cai and Steyn, 1996).

20.3.2.2 Small-scale wind modelling

The use of CFD techniques at relatively small engineering scales is becoming increasingly common both within the research community and within an industrial context. The proceedings of the three recent conferences on computational wind engineering (in Tokyo in 1992, Fort Collins, Colorado in 1996 and Birmingham in 2000) indicate the considerable effort that is being put into developing and using models to predict wind effects on pedestrians, pollution dispersion, and wind

loads on buildings and other structures. The motivation for this is often the reduced cost compared with physical model testing. This point will be discussed further below.

The main type of model that is used is standard RANS models, usually in the form of one of the standard software packages, such as FLUENT, CFX, STAR-CD, etc. Thus, in general, only time-mean averages of the flow field or of contaminant transport can be predicted, although there is an increasing use of unsteady RANS modelling. The major problem with the use of these methods is the general inability to predict the extent of separated flow fields. This problem is associated with the over-prediction of turbulent kinetic energy in stagnation regions by such models (see Murakami, 1997).

Examples of the type of results that can be obtained are given in Figure 20.5, which shows calculations of pedestrian-level wind velocities in Auckland (Richards *et al.*, 2000), pollutant transport from a storage tank (Fothergill *et al.*, 2000) and forces on a wall (plotted from data in Robertson *et al.*, 1997). The latter investigation only gives mean pressures and forces of course. In reality, in design, extreme values of the parameters are needed. These cannot be predicted by RANS models, which thus have a limited application in the prediction of structural forces due to turbulent winds.

Other types of modelling technique are also being used. The discrete vortex method, in which vorticity is fed into the flow field from surfaces, has been shown to be of particular use in predicting the flow around bridge decks, and can be incorporated into dynamic models of bridge movement in high winds (Walther, 1998). Typical results are given in Figure 20.6. Also the LES technique is beginning to be applied for small-scale engineering problems. Again, this technique is likely to become of increasing use in the future because of its ability to predict unsteady flows and forces, which are often design requirements (Murakami, 1997).

Having described the techniques that are in use, it is necessary to urge caution in their use. First, as has been stated above, the standard RANS techniques currently in use cannot predict unsteady fluctuating flows and forces. Whilst this might be adequate for some applications where only time-averaged conditions are required (average pedestrian-level wind speeds, average plume trajectories), it is not adequate for most structural calculations. Second, the paper by Cowan *et al.* (1997) is of some interest. The trials reported in that paper were of calculations of pollutant concentrations for a variety of well-documented cases, most of which have experimental verification. They were carried out independently by a number of different organizations, using the same

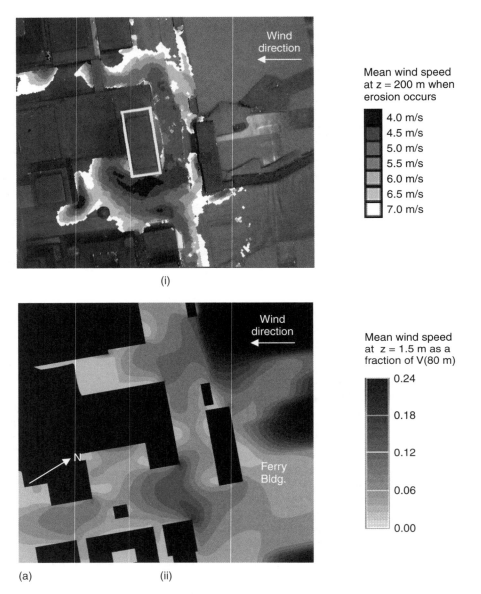

Figure 20.5   Comparison of CFD calculations and experimental data. (a) Pedestrian-level wind velocities (from Richards *et al.*, 2000) – (i) wind tunnel scour test, (ii) CFD calculations of low level wind speeds; (b) flow around oil storage tank (from Fothergill *et al.*, 2000) – (i) wind tunnel result for normalized wind speeds, (ii) $k-\varepsilon$ CFD calculations for normalized wind speeds; (c) variation of force coefficient with wind angle for a free-standing 2 m high wall (from Robertson *et al.*, 1997). Values are shown for three panels – 0-*h* is panel of one wall height at free end of wall, *h*-2*h* is second panel from end, and 2*h*-3*h* is third panel from end; wind direction is measured relative to wall normal; solid lines are experimental results, dotted lines are CFD calculations

Source: (b) Reproduced with permission of Shell Research Ltd

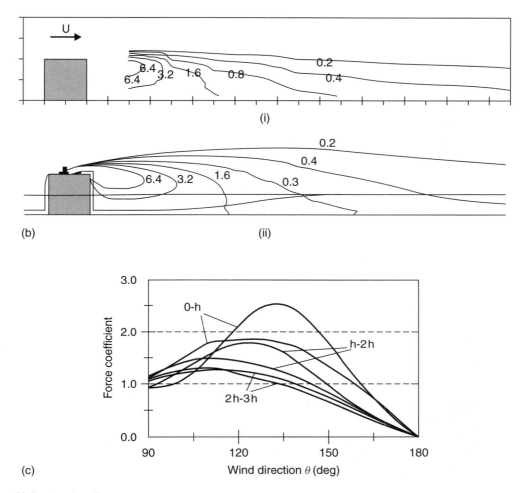

Figure 20.5 (*continued*)

computer code. The differences between the calculations were thus in the realms of grid generation, numerical schemes, etc. The results produced by different investigators were found to vary very significantly (often by an order of magnitude or more). Typical comparisons are given in Figure 20.7. These results serve as a warning against placing too great a reliance on the accuracy of any one calculation that does not have some sort of experimental validation. The unverified use of CFD codes to reduce the costs of physical model tests may, if great care is not used, simply produce results that are unreliable. These points are further discussed and emphasized by Castro and Graham (1999). It must be concluded that CFD and physical modelling should be seen as complementary technologies that should be used in conjunction with one another to varying degrees for any particular situation.

20.4 CONCLUSION

While CFD has many applications in environmental flows as outlined here, there are still challenges to be faced to increase its accuracy and ability to deal with more complex situations. It is also vital that users understand the difficulties and pitfalls in CFD. This requires experience in CFD, but also an understanding of the particular applications area.

In future, new techniques will become feasible for practical modelling which will open up new areas of

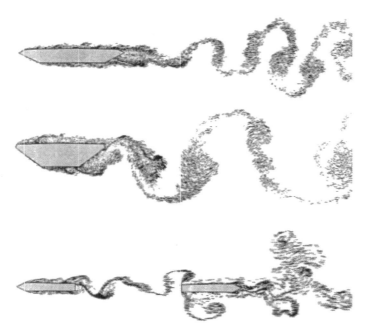

Figure 20.6   Calculations of the flow around bridge decks
Source: Walther (1998)

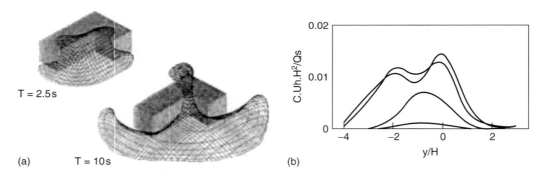

Figure 20.7   Pollutant dispersion from an L-shaped building
Source: Cowan *et al*. (1997) Wind direction bisects the L and the release is near the inner corner of the building:
(a) predicted concentration distributions after times of 2.5 s and 10 s; (b) dimensionless concentration profiles in the
cross-wind direction predicted by different investigators at half a building height upstream of the source, one building
height above the ground after 40 s. The same CFD code and turbulence model is used in each case

application. These may well include:

- large eddy simulations for a more realistic representation of turbulence;
- improved roughness implementation;
- unstructured meshes with automatic adaptivity;
- increased computing power that will allow for:
  - much larger meshes (a billion cells?);
  - multiple runs opening up the possibility of automatic calibration, numerical optimization and parametric uncertainty analysis.

Although there are many valid reservations about the use of CFD, it is a technique that is established as an engineering tool and whose use will become more and more commonplace.

REFERENCES

Abbott, M. and Basco, D. (1989) *Computational Fluid Dynamics: An Introduction for Engineers*, Longman Scientific, Singapore.

Abbott, M.B. (1979) *Computational Hydraulics: Elements of the Theory of Free Surface Flows*, Pitman, London.

AIAA (1998) *Guide for the Verification and Validation of Computational Fluid Dynamics Simulations*, AIAA.

Alcrudo, F. and Garcia-Navarro, P. (1993). A high-resolution Godunov-type scheme in finite volumes for the 2D shallow-water equations, *International Journal for Numerical Methods in Fluids* **16** (6), 489–505.

Anastasiou, K. and Chan, C. (1997) Solution of the 2D shallow water equations using the finite volume method on unstructured triangular meshes, *International Journal for Numerical Methods in Fluids* **24**, 1225–1245.

ASME (1993) Statement on the control of numerical accuracy, *Journal of Fluids Engineering* **115** (3), 339–340.

Bates, P., Anderson, M., Price, D., Hardy, R. and Smith, C. (1996) *Analysis and Development of Hydraulic Models for Floodplain Flows in Floodplain Processes*, John Wiley & Sons, Chichester.

Bates, P., Wilson, C., Hervouet, J.-M. and Stewart, M. (1999) Modélisation à deux dimensions par éléments finis d'un écoulement en plaine, *La Houille Blanche* **3/4**, 61–67.

Belcher, S. (1990) *Turbulent Boundary Layer Flow over Undulating Surfaces*, Cambridge University Press, Cambridge.

Belcher, S. and Hunt, J. (1998) Turbulent flow over hills and waves, *Annual Review of Fluid Mechanics* **30**, 507–538.

Berzins, M. and Walkley, M. (2003) A finite element method for the two-dimensional extended Boussinesq equations, *International Journal for Numerical Methods in Fluids* **39** (10), 865–885.

Boussinesq, J. (1877) Essai sur la théorie des eaux courantes, *Mémoires à l'Académie des Sciences*, XXIII, Paris.

Bradbrook, K., Lane, S., Richards, K., Biron, P. and Roy, A. (2000) Large eddy simulation of periodic flow characteristics at river channel confluences, *Journal of Hydraulic Research* **38** (3), 207–215.

Cai, X.-M. and Steyn, D. (1996) The Von Karman constant determined by large eddy simulation, *Boundary Layer Meteorology* **78**, 143–164.

Cai, X.-M., Steyn, D. and Gartshore, I. (1995) Resolved scale turbulence in the atmospheric surface layer from a large eddy simulation, *Boundary Layer Meteorology* **75**, 301–314.

Casey, M. and Wintergerste, T. (2000) *Best Practice Guidelines*, ERCOFTAC Special Interest Group on Quality and Trust in CFD, Swiss Federal Institute of Technology, Lausanne.

Castro, I. and Graham, M. (1999) Numerical Wind Engineering: The Way Ahead?, *Proceedings of the Institute of Civil Engineers, Structures & Buildings* **134**, 275–277.

Cheng, R. and Smith, P. (1989) A Survey of Three-Dimensional Numerical Estuarine Models, in M. Spaulding (ed.) *Estuarine and Coastal Modeling: American Society of Civil Engineers Specially Conference*, 1–15, ASCE, Reston, VA.

Cowan, I.R., Castro, I.P. and Robins, A.G. (1997) Numerical considerations for simulations of flow and dispersion around buildings, *Journal of Wind Engineering and Industrial Aerodynamics* **67–68**, 535–545.

Crossley, A. (1999) *Accurate and Efficient Numerical Solutions for the Saint Venant Equations of Open Channel Flow*, School of Civil Engineering, University of Nottingham, Nottingham.

Cunge, J., Holly Jr, F. and Verwey, A. (1980) *Practical Aspects of Computational River Hydraulics*, Pitman, London.

Dean, R. and Dalrymple, R. (1991) *Water Wave Mechanics for Engineers and Scientists*, World Scientific, Singapore.

Demuren, A. and Rodi, W. (1986) Calculation of flow and pollutant dispersion in meandering channels, *Journal of Fluid Mechanics* **172** (1), 65–92.

DHI (1998) *User Guide and Scientific Documentation*, Danish Hydraulic Institute.

Dhondia, J. and Stelling, G. (2002) *Application of One Dimensional – Two Dimensional Integrated Hydraulic Model for Flood Simulation and Damage Assessment*, Hydroinformatics 2002, IWA Publishing, Cardiff.

Falconer, R. (1980) Numerical modelling of tidal circulation in harbours, *Proceedings of the Institution of Civil Engineers, Waterway, Port and Coastal Division* **106**, 32–33.

Falconer, R. and Lin, B. (1997) Three-dimensional modelling of water quality in the Humber Estuary, *Water Research, IAWQ* **31** (5), 1092–1102.

Fothergill, C., Roberts, P. and Packwood, A. (2000) *Flow and Dispersion around Storage Tanks: Comparison between Wind Tunnel Studies and CFD Simulation*, CWE2000, Birmingham.

Garcia-Navarro, P., Alcrudo, F. and Saviron, J.M. (1992) 1-D open-channel flow simulation using Tvd-McCormack scheme, *Journal of Hydraulic Engineering-Asce* **118** (10), 1359–1372.

Garcia-Navarro, P., Hubbard, M.E. and Preistley, A. (1995) Genuinely multidimensional upwinding for the 2D shallow water equations, *Journal of Computational Physics* **121**, 79–93.

Gibson, M. and Rodi, W. (1989) Simulation of free surface effects on turbulence with a Reynolds stress model, *Journal of Hydraulic Research* **27** (2), 233–244.

Hervouet, J.-M. (1994) *Computation of 2D Free Surface Flows: The State of the Art in the TELEMAC System*, Hydroinformatics 94, Balkema.

Hirt, C. and Nichols, B. (1981) Volume of fluid methods for the dynamics of free boundaries, *Journal of Computational Physics* **39**, 201–225.

Horritt, M., Mason, D. and Luckman, A. (2001) Flood boundary delineation from synthetic aperture radar imagery using a statistical active contour model, *International Journal of Remote Sensing* **22**, 2489–2507.

Hunt, J., Leibovich, S. and Richards, K. (1988) Turbulent shear flow over low hills, *Quarterly Journal of the Royal Meteorological Society* **114**, 1435–1470.

Inglis, D., Choularton, T., Stromberg, I., Gardiner, B. and Hill, M. (1995) Testing of a linear airflow model for flow over complex terrain and subject to stable structured stratification, in J. Grace (ed.) *Wind and Trees*, Cambridge University Press, Cambridge, 71–87.

Jackson, P. and Hunt, J. (1975) Turbulent wind flow over a low hill, *Quarterly Journal of the Royal Meteorological Society* **101**, 929–955.

King, I., Norton, W. and Iceman, K. (1975) A finite element solution for two-dimensional stratified flow problems, in O.C. Zienkiewicz (ed.) *Finite Elements in Fluids: International Conference on the Finite Element Method in Flow Analysis*, John Wiley & Sons, New York, **1**, 133–156.

Lane, S., Bradbrook, K., Richards, K., Biron, P. and Roy, A. (1999) The application of computational fluid dynamics to natural river channels: three-dimensional versus two-dimensional approaches, *Geomorphology* **29**, 1–20.

Lane, S.N. and Richards, K.S. (2001) The 'validation' of hydrodynamic models: some critical perspectives, in M.G. Anderson and P.D. Bates, *Model Validation: Perspectives in Hydrological Science*, John Wiley & Sons, Chichester, 413–438.

Leschziner, M. and Rodi, W. (1979) Calculation of strongly curved open channel flow, *Journal of Hydraulic Engineering* **105** (10), 1297–1314.

Li, B. and Fleming, C. (2001) Three-dimensional model of Navier–Stokes equations for water waves, *Journal of Waterway, Port, Coastal and Ocean Engineering* **127** (1), 16–25.

Mason, P. (1994) Large eddy simulation: a critical review of the technique, *Quarterly Journal of the Royal Meteorological Society* **120**, 1–26.

Murakami, S. (1997) Current status and future trends in computational wind engineering, *Journal of Wind Engineering and Industrial Aerodynamics* **67–68**, 3–34.

Naot, D. and Rodi, W. (1982) Calculation of secondary currents in channel flow, *Journal Hydraulic Division ASCE* **108** (HY8), 948–968.

Olsen, N. (2000) *Unstructured Hexahedral 3D Grids for CFD Modelling in Fluvial Geomorphology*, 4th International Conf. Hydroinformatics 2000, Iowa, USA.

Olsen, N.R.B. and Stokseth, S. (1995) 3-dimensional numerical modelling of water flow in a river with large bed roughness, *Journal of Hydraulic Research* **33** (4), 571–581.

Patankar, S. and Spalding, D. (1972) A calculation procedure for heat, mass and momentum transfer in three-dimensional parabolic flows, *International Journal of Heat Mass Transfer* **15**, 1787–1806.

Rastogi, A. and Rodi, W. (1978) Predictions of heat and mass transfer in open channels, *Journal of Hydraulic Division* **104**, 397–420.

Richards, P., Mallinson, G. and McMillan, D. (2000) *Pedestrian Level Wind Speeds in Downtown Auckland*, CWE 2000, Birmingham.

Robertson, A., Hoxey, R., Short, J. and Ferguson, W. (1997) Full-scale measurements and computational predictions of wind loads on free standing walls, *Journal of Wind Engineering and Industrial Aerodynamics* **67–68**, 639–646.

Rodi, W. (1980) *Turbulence Models and Their Application in Hydraulics: State of the Art Paper*, IAWR, Delft.

Rodríguez, J., Belby, B., Bombardelli, F., García, C., Rhoads, B. and García, M. (2001) *Numerical and Physical Modeling of Pool-Riffle Sequences for Low-Gradient Urban Streams*, 3rd International Symposium on Environmental Hydraulics, Tempe, Arizona, USA.

Sauvaget, P., David, E. and Guedes Soares, C. (2000) Modelling tidal currents on the coast of Portugal, *Coastal Engineering* **40**, 393–409.

Schneider, G. and Raw, M. (1987) Control-volume finite element method for heat transfer and fluid flow using co-located variables – 1 computational procedure, *Numerical Heat Transactions* **11**, 363–390.

Scorer, R. (1949) Theory of waves in the lees of mountains, *Quarterly Journal of the Royal Meteorological Society* **76**, 41–56.

Sleigh, P., Gaskell, P., Berzins, M. and Wright, N. (1998) An unstructured finite-volume algorithm for predicting flow in rivers and estuaries, *Computers & Fluids* **27** (4), 479–508.

Smagorinsky, J. (1963) General circulation experiments with primitive equations Part 1: basic experiments, *Monthly Weather Review* **91**, 99–164.

Smith, G. (1978) *Numerical Solution of Partial Differential Equations: Finite Difference Methods*, Oxford University Press, Oxford.

Stansby, P. and Zhou, J. (1998) Shallow-water flow solver with non-hydrostatic pressure: 2D vertical plane problems, *International Journal of Numerical Methods in Fluids* **28**, 541–563.

Swindale, N. (1999) *Numerical Modelling of River Rehabilitation Schemes*, Department of Civil Engineering, University of Nottingham, Nottingham.

Swindale, N. and Wright, N. (1997) *The Analysis of River Rehabilitation Schemes Using One and Two Dimensional Hydraulic Modelling*, IAHR Congress, 27th Congress of the IAHR, San Francisco, August.

Versteeg, H. and Malalasekera, W. (1995) *An Introduction to Computational Fluid Dynamics – the Finite Volume Method*, Longman, London.

Walther, J. (1998) Discrete vortex methods in bridge aerodynamics and prospects for parallel computing techniques, in *Bridge Aerodynamics*, Esdahl, Rotterdam, Balkema.

Wood, N. (2000) Wind flow over complex terrain: a historical perspective and the prospect for large eddy modelling, *Boundary Layer Meteorology* **96**, 11–32.

Wright, N. (1988) *Multigrid Solution of Elliptic Fluid Flow Problems*, Department of Mechanical Engineering, University of Leeds, Leeds.

Wright, N. (2001) *Conveyance Implications for 2D and 3D Modelling*, Environment Agency, London.

Wright, N., Swindale, N., Whitlow, C. and Downs, P. (2000) *The Use of Hydraulic Models in the Rehabilitation of the River Idle, UK*, Hydroinformatics 2000, Iowa, USA.

Zienkiewicz, O. and Taylor, R. (2000) *The Finite Element Method*, McGraw-Hill, London.

21

Self-Organization and Cellular Automata Models

DAVID FAVIS-MORTLOCK

21.1 INTRODUCTION

Models are one of the main tools available to help us understand the environment, and to attempt to predict what will happen in it. Indeed, 'The control [which] science gives us over reality, we normally obtain by the application of models' (Ackoff *et al.*, 1962: 108). Many aspects of the environment, however, are too complex for mere humans to grasp fully: if this is the case, we are forced to simplify, in effect to fall back on making analogies. The aim of an analogy is to highlight similarities between the things compared, and in the same way environmental models aim to link the more easily understood system (the model) with the less comprehensible system (some aspect of the environment). Analogies, though, are never perfect, and in the same way models are always less than perfect in making this link. To return to the original point, it could hardly be otherwise, since a model which is as complicated as the thing it represents would be no easier to understand. 'The best material model of a cat is another, or preferably the same, cat' (Rosenblueth and Wiener, 1945; see also the discussion in the Introduction and Chapter 1).

The resulting tension, between the robustness of our models and their complexity, is the focus of this chapter. It deals with recent theoretical developments in self-organization of complex systems, and how these may be used in a practical way to construct spatial models. The resulting approach (cellular automata modelling) is one which appears to have great promise for lessening this tension in future spatial models of all kinds.

While many of the examples presented here are drawn from physical geography, since that is the area of science in which work, the ideas and approaches discussed are of wide applicability (cf. Wolfram, 2002).

21.1.1 The ever-decreasing simplicity of models?

Models used by geographers have steadily become more and more complex, from the 'quantitative revolution' in geography in the 1960s (e.g. Harvey, 1969) when models first began to be used widely, to the present. This is a trend that looks set to continue. Figure 21.1 uses the number of lines of programming code as a surrogate for model complexity[1] for a selection of computer implementations of US-written field-scale models of soil erosion by water. Model complexity in this particular domain has clearly increased in a somewhat nonlinear way.

21.1.1.1 The quest for 'better' models

The reason for this increased complexity is, of course, the desire for 'better' models. Although environmental models differ widely, two general criteria have come to dominate in assessment of any model's 'success' (or otherwise).[2] First, the model is usually ranked more highly if it is physically based, i.e. if it is founded on the laws of conservation of energy, momentum and mass; and if it has its parameters and variables defined by means of equations that are at least partly based on the physics of the problem, such as Darcy's law and the Richards equation (Kirkby *et al.*, 1992; see

---

Environmental Modelling: Finding Simplicity in Complexity.   Edited by J. Wainwright and M. Mulligan
© 2004 John Wiley & Sons, Ltd   ISBNs: 0-471-49617-0 (HB); 0-471-49618-9 (PB)

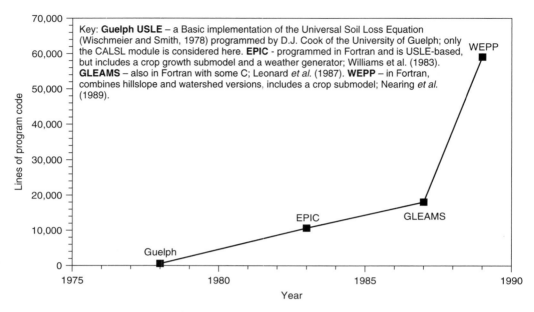

Figure 21.1   An estimate of the complexity of some North American field-scale erosion models using the number of lines of programming source code. The date given is of a 'representative' publication and not the release analysed here
Source: Redrawn from data in Favis-Mortlock *et al.* (2001)

also Chapter 2). The presumed universal applicability of the laws of physics is the reason for preferring physically based models. Thus, the more a model is rooted in these laws (and, conversely, the less it depends upon empirically derived relationships), the more widely applicable – i.e. less location-specific – it is assumed to be. By contrast, the second criterion is pragmatic: how well the model does when model results are compared against measured values. For hydrological models, a time series of simulated discharges might be compared with an observed time series, or for erosion models simulated soil loss compared with measured values (e.g. Favis-Mortlock *et al.*, 1996; Favis-Mortlock, 1998a). The inevitable differences between computed and observed time series are usually attributed to the failure of the model to describe some aspect of the real world adequately. Thus in an evaluation of catchment-scale soil-erosion models (Jetten *et al.*, 1999), the models were largely able to predict correctly sediment delivery at the catchment outlet but were much less successful at identifying the erosional 'hot spots' within the catchment which supplied this sediment. The failure was in part attributed to deficiencies in modelling within-catchment flow paths. The standard remedy for such shortcomings is the addition of detail, preferably physics-based, to the model or its successors. Thus the empirical Universal Soil Loss

Equation (USLE: Wischmeier and Smith, 1978) made no distinction between rainsplash-dominated interrill soil loss, and flow-dominated rill erosion; whereas these processes are separately modelled, in a more physically based way, in two subsequent models: the Water Erosion Prediction Project (WEPP: Nearing *et al.*, 1989; see also Chapter 16) model and the European Soil Erosion Model (EUROSEM: Morgan *et al.*, 1998; see also Chapter 9). This strategy is often – though not always – successful; however it inevitably leads to an explosion in model complexity (cf. Figure 21.1) and data requirements (Favis-Mortlock *et al.*, 2001).

Despite great strides, our still incomplete knowledge of the physics of several environmental processes (e.g. for soil erosion, the details of soil-surface crusting) gives rise to an associated modelling problem. These poorly understood processes can only be described in current models in a more-or-less empirical way; this means that some model parameters essentially fulfil the function of curve-fitting parameters, adjusted to provide a match between the observed and computed time series rather than measured independently. This inclusion of empirical elements in otherwise physics-based models is to the dismay of authors such as Klemeš, who wrote: 'For a good mathematical model it is not enough to work well. It must work well for the right reasons. It must reflect, even if only in a simplified form, the essential features of the physical prototype' (Klemeš, 1986: 178S). Model parameterization under such conditions then becomes more and more a curve-fitting exercise (Kirchner *et al.*, 1996). As an example, a recent evaluation of field-scale erosion models (Favis-Mortlock, 1998a) found calibration to be essential for almost all models involved, despite the supposed physical basis of the models.

Additionally, results from a more complex model may not necessarily improve upon those of a simpler model if interactions between processes are inadequately represented within the model (Mark Nearing, personal communication, 1992; Beven, 1996). Since the addition of more model parameters increases the number of degrees of freedom for the model, adding extra free parameters to a model means that changes in the value of one input parameter may be compensated by changes in the value of another. Therefore unrealistic values for individual input parameters may still produce realistic results (in the sense of a close match between the observed and computed time series). The model is 'unidentified' with its parent theory[3] in the sense of Harvey (1967: 159), and results from the model may be 'right for the wrong reasons' (Favis-Mortlock *et al.*, 2001). To illustrate this point, Jakeman and Hornberger (1993) found that commonly used rainfall-runoff data contains only enough information to constrain a simple hydrological model with a maximum of four free parameters. An end-point to this problem is Beven's 'model equifinality',[4] whereby entirely different sets of input parameters can still produce similar model results (Beven, 1989).

The result of the above-described pressures upon model development is a nonvirtuous circle, whereby a 'better' environmental model inexorably has to describe more processes, or existing processes in more detail. Doing so:

- will probably increase the model's predictive power in a specific domain, but may cause it to fail unexpectedly elsewhere (because it is giving 'the right answer for the wrong reasons' in the original domain);
- requires more data, and so in practical terms narrows the circumstances in which the model may be used;
- may make the model less comprehensible.

Is there any way out of this vicious circle?

21.2 SELF-ORGANIZATION IN COMPLEX SYSTEMS

The foregoing suggests that, if an environmental model is to be considered 'better', it has to describe the environment in greater physically based detail. Does this necessarily imply a more complex model? Put more generally: does real-world complexity inevitably imply an equivalent complexity of the underlying generative process?

21.2.1 Deterministic chaos and fractals

Along with much other change, the 1960s brought a clue that this might not always be so. During this decade, deterministic chaos was first discovered by a number of workers such as the atmospheric physicist Edward Lorentz (Ruelle, 2001). However, the mathematical roots of chaos are a good deal older: notably in the work of mathematician Henri Poincaré around the beginning of the twentieth century (e.g. Jones, 1991). Deterministic chaos took some time to enter the scientific mainstream. This began in the 1970s, for example, with the work of biologist Robert May on chaotic population dynamics (May, 1976).

The tongue-in-cheek question[5] 'Does the flap of a butterfly's wings in Brazil set off a tornado in Texas?' summarizes one major attribute of deterministic chaos. In effect, the question asks if a large effect have a tiny cause. It can: such nonlinear 'extreme sensitivity to initial conditions' is a hallmark of chaotic systems. Small uncertainties in measurement or specification of initial conditions become exponentially larger, and the eventual state of the system cannot be predicted. Thus, for example, the chaotic component of the Earth's atmosphere means that weather forecasts rapidly diminish in reliability as one moves more than a few days into the future. It is important to note that whereas weather (i.e. the particular set of meteorological conditions on a specific day, at a specific location) cannot be predicted, climate (here, the range of meteorological conditions of a number of replicates of the meteorological conditions simulated for that day and location) *can* be predicted. This notion is at the heart of recently devised 'ensemble forecasting' techniques which are carried out using atmospheric models (Washington, 2000).

Deterministic chaos is also often associated with fractal (i.e. self-similar, scale-independent) patterns. Following seminal work by Benoit Mandelbrot (1975; see also Andrle, 1996), fractal patterns were acknowledged to be present in a wide variety of natural situations (but see e.g. Evans and McClean, 1995). This linkage between fractals and systems exhibiting deterministic chaos is suggestive of some deeper connection (cf. Cohen and Stewart, 1994).

But for environmental modellers, perhaps the most interesting insight from chaotic systems is that they do not have to be complicated to produce complex results. Lorentz's atmospheric model comprised only three non-linear equations, and May's population dynamics models were even simpler. In all such models, the results at the end of one iteration (in the case of a time-series model, at the end of one 'timestep') are fed back into the model and used to calculate results for the next iteration. This procedure produces a feedback loop. Some values of the model's parameters will cause it to eventually[6] settle down to a static equilibrium output value; with other values, the model's output will eventually settle down to cycle forever between a finite number of end-point values; but for others, the model will switch unpredictably between output values in an apparently random way. Thus in such chaotic systems, complex patterns can be the results of simple underlying relationships.

Both positive and negative implications follow from the discovery of such systems. For those of a deterministic cast of mind this is sobering since it represents the final death rattle of the notion of a predictable, 'clockwork', universe, even at the macroscale.[7] But there is also a strongly positive implication: complexity does not have to be the result of complexity!

Nonetheless, while this early work on deterministic chaos was intriguing and suggestive, it was not immediately 'useful' for most environmental modellers:

- While the output from chaotic functions is complex, it includes little in the way of immediately recognizable structure: at first glance it more resembles random noise. It is therefore qualitatively very different from the complex but highly structured patterns which we observe in many environmental systems.
- When analysing real-world measurements which plausibly possess a chaotic component, it has proved to be very difficult to go from the data to a description of the underlying dynamics (e.g. Wilcox *et al.*, 1991), even though simple models may be devised which produce simulated data closely resembling the measured values. But since there may be many such models, what certainty can we have that any one of them has captured the workings of the real-world system? (Again cf. Beven's 'model equifinality': Beven, 1993.)

For a while, it seemed that deterministic chaos might remain little more than an interesting diversion for environmental modellers.

21.2.2 Early work on self-organizing systems

Pioneering research carried out from the late 1980s, however, strongly supported the notion that structured complexity does not always require an underlying complexity of process. A major centre for this work was the Santa Fe Institute (Waldrop, 1994). Overviews of this diverse body of early research on self-organization and emergence in complex systems are presented by Coveney and Highfield (1995), Kauffman (1995) and Holland (1998), among others.

These early studies demonstrated that responses which are both complex and highly structured may result from relatively simple interactions (but very many of them) between the components of a system. Interactions between such components are governed by 'local' rules, but the whole-system ('global') response is to manifest some higher-level 'emergent' organization, following the formation of ordered structures within the system. The system thus moves from a more uniform ('symmetrical') state to a less uniform – but more structured – state: this is so-called 'symmetry-breaking'. Several aspects of the phenomenon of self-organization are summarized in Figure 21.2. Note the definition of emergence which is given: 'emergent responses cannot be simply inferred from the behaviour of the system's components'.

21.2.3 Attributes of self-organizing systems

Three key concepts which characterize self-organizing systems are feedback, complexity and emergence. All are interlinked.

21.2.3.1 Feedback

Feedback occurs when a system processes data about its own state. Thus, to use a humanly-engineered example, the governor of a steam engine is an arrangement whereby a valve is controlled by weighted arms which revolve at a rate that is controlled by the speed of the steam engine. As the arms rotate faster, increased centrifugal force makes them close the valve somewhat and so slow the engine. As the engine slows, the arms of the governor also rotate more slowly and so open the valve a little; the engine thus speeds up. It is the dynamic balance between the opening of the valve ('positive feedback') and the closing of the valve ('negative feedback') that enables the steam engine to maintain a constant speed. This interaction between positive and negative feedback permits the steam engine to process data about its own state.

In general, negative feedback in a system occurs when the system functions in such a way that the effects of a disturbance are counteracted over time, bringing the system back to its pre-disturbance state. In landscapes, an

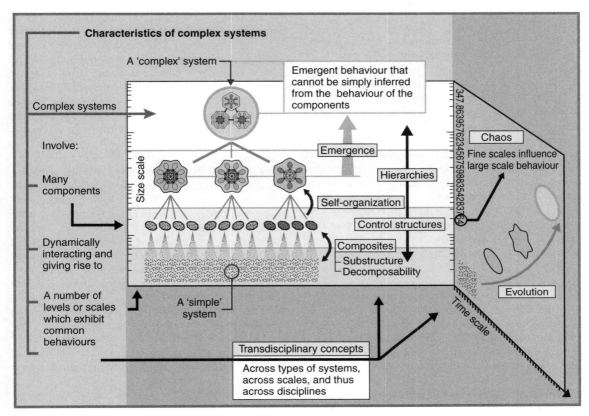

Figure 21.2 Schematic representation of some attributes of self-organizing complex systems. Graphic by Marshall Clemens (http://www.idiagram.com), from the 'Visualizing Complex Systems Science' project at http://necsi.org /projects/mclemens/viscss.html. Used by permission

example of negative feedback can occur when deposition of coarse sediment takes place in a channel section. The resulting increase in gradient at the downstream end of the deposit is reflected in an increase in flow velocity and bed shear strength, ultimately resulting in increased erosion of the channel bed and removal of the coarse sediment deposit, and a return to the earlier conditions.

Positive feedback takes place when a disturbance continues to force the system away from its earlier state. Whereas negative feedback counterbalances change and drives the system back to the pre-disturbance conditions, positive feedback reinforces change and may lead to an entirely new state of the system. An example of positive feedback can occur when soil erosion exposes an underlying soil horizon with a low permeability, which reduces the infiltration rate. Consequently, the rate of overland flow production increases and, in some cases, erosion increases as a result.[8]

In self-organizing systems, feedback plays a crucial role in the formation of spatial and temporal patterns. An example of the role of positive feedback in landscape evolution occurs when erosion locally lowers the surface, resulting in an increased concentration of water and still more erosion, until at some point a valley is formed which ultimately becomes part of a larger channel network. Viewed in this manner, the formation of a drainage network on a continental surface is the direct result of positive feedback that enhances initially minor differences in surface elevation. At the same time, however, the occurrence of negative feedback prevents the valleys from becoming overly deep, by the deposition of sediment in any low spots in the drainage network. Thus a balance is maintained: while 'successful' channels (see below) are deepened, no channel can become too deep too quickly. It is this nonlinear recursive interplay[9] which leads to the emergence of channel networks.

21.2.3.2 Complexity

In order to exhibit self-organization, a system must be complex i.e. must possess sufficient scope,[10] for component-level interactions (which can be characterized as positive and negative feedback) to give rise to system-wide, emergent responses. However, while all self-organizing systems are complex, not all complex systems are self-organizing (Çambel, 1993: 20). At present there is, though, no single, precise definition of complexity (Bar-Yam, 1997; Chaitin, 1999). Gallagher and Appenzeller (1999) loosely define a complex system as a system with properties that cannot be described fully in terms of the properties of its parts. Most authors do not provide a single definition of complexity, but instead describe various characteristics of complexity.

The notion of complexity is closely tied to our conceptualization of scale. Bar-Yam (1997: 258) points out that:

The physics of Newton and the related concepts of calculus, which have dominated scientific thinking for three hundred years, are based upon the understanding that at smaller and smaller scales – both in space and in time – physical systems become simple, smooth and without detail.

From this viewpoint, the assumption is that even the most complex of systems, when viewed at a 'component scale', somehow[11] becomes simpler, and thus more mathematically tractable. This assumption is at the heart of much of present-day mathematical modelling. The assumption is a reasonable one for systems in which self-organization does not take place,[12] but for those systems which self-organize, to ignore the complex within-system interactions which give rise to that self-organization is to throw the baby out with the bath water.

21.2.3.3 Emergence

The roots of the notion of emergence go back at least to *c*.330 BCE with Aristotle's description of synergy: 'The whole is more than the sum of its parts.' An emergent response is synergistic, but 'more so' (in a qualitative sense).

The continual flow of energy and matter through a thermodynamically dissipative system maintain it in a state far from equilibrium (Ruelle, 1993; Çambel, 1993). Ordered structures 'emerge' as a result of interactions between the system's subcomponents, such interactions being driven by the flow of matter and energy which

characterizes such systems. As these structures grow more common within the system, the system as a whole 'self-organizes'. This transition (see 'symmetry-breaking' above) often occurs quite rapidly and abruptly, in the manner of a phase change e.g. from water to ice. It is crucial to note that this increase in systemic organization is entirely a result of internal interactions, rather than resulting from some externally imposed controlling factor (although a flow of energy and matter through the system is essential).

Bar-Yam (1997: 10) points out that even in the scientific community there is still confusion regarding the nature of emergence. One 'fingerprint' of emergence is that the emergent properties of a system cannot easily be derived from the properties of the system components or subsystems (see Figure 21.2). Additionally, Bar-Yam (1997: 10–12) distinguishes between local and global emergence. In local emergence, the emergent response of the system is relatively resistant to local perturbations: for example, a standing wave on the lip of a waterfall will remain, despite innumerable small variations in flow conditions in its vicinity. In global emergence, the emergent response of the system is more susceptible to local perturbations: thus any living thing will be adversely affected by disturbance of even small parts of its body. Local and global emergence presumably form end-points in a continuum.

21.2.3.4 Self-organized criticality

Studies by Bak and co-workers (Bak *et al.*, 1988; Bak, 1996) on sand-pile models have provided a number of insights into other generic aspects of complex systems. While still controversial, this work on so-called 'self-organized criticality' (SOC) suggests that the presence of power law frequency-magnitude relationships, 1/f properties of time-series data, and spatial fractality form a kind of fingerprint for SOC, and hence for self-organization in the more general sense (Bak, 1996; Buchanan, 2000). This view suggests that self-organization can be manifested not only in emergent pattern formation, but also in terms of particular configurations of the internal dynamics of systems.

21.2.4 The counter-intuitive universality of self-organization

Self-organizing systems have been proposed or identified all around us:

- In purely physical systems e.g. crackling noise (Sethna *et al.*, 2001), chemical reactions (Tam, 1997) and sand-dune dynamics (Hansen *et al.*, 2001).

- In biological systems e.g. population dynamics (Solé *et al.*, 1999), bacterial patterns (Ben-Jacob and Levine, 2001), embryonic pattern formation (Goodwin, 1997), ecology (Laland *et al.*, 1999; Peterson, 2000), evolutionary dynamics (Lewin, 1997) and Gaia Theory (Lenton, 1998).
- In human systems e.g. urban structure (Portugali *et al.*, 1997), social networks (Watts, 1999; Winder, 2000), and the discipline of geography (Clifford, 2001).

Yet since our common experience is that bricks do not spontaneously organize themselves into houses, nor does a child's bedroom spontaneously tidy itself, at first acquaintance the notion of self-organization is distinctly counter-intuitive (Bar-Yam, 1997: 623).

This is probably to some extent a matter of preconceptions. An everyday observation that the hot cappuccino always cools to muddy uniformity. It is perhaps this kind of much-repeated experience which colours our expectations of 'the way things are'. But equally, we see living things – plants and animals – come into being, and maintain their organization for some time (i.e. as long as they are alive). Accustomed to this since infancy, we unquestioningly intuit that these 'kinds' of system are different in some fundamental way (although we may seldom ask ourselves what this difference is). So: why do these two kinds of systems behave so differently?

Distinctions between living and nonliving systems were much discussed at the end of the nineteenth century, when the laws of thermodynamics were being formulated by physicists such as Ludwig Boltzmann. These laws express some of the most profound scientific truths yet known, in particular the universal inexorability of dissolution and decay. To develop an understanding of self-organization, it is essential to comprehend the implications of the second law of thermodynamics. This states that entropy, or disorder, in a closed system can never decrease: it can be colloquially expressed as 'There is no such thing as a free lunch'. Thus in a 'closed' finite universe, the end result is a kind of cold-cappuccino uniformity, often described in studies of chemical equilibrium (Chorley, 1962). However, pioneering work by Ilya Prigogine and colleagues at the Free University of Brussels in the 1960s (e.g. Nicolis and Prigogine, 1989; Ruelle, 1993; Klimontovich, 2001) focused on systems which are 'open' from a thermodynamic perspective. In a thermodynamically open system (also called a 'dissipative system': Çambel, 1993: 56), there is a continuous flow of matter and energy through the system. This continuous flow of matter and energy permits the system to maintain itself in a state far from thermodynamic

equilibrium (Huggett, 1985) – at least while the flow continues. This is in contrast to the thermodynamically closed (or 'conservative') system, such as the cup of cappuccino together with its immediate surroundings: here there is no such flow of matter and energy into the system, only a movement of energy between the coffee and its surroundings. Thus, the coffee and its environment gradually equilibrate, with the coffee cooling and mixing, and the surrounding air warming slightly. Living systems, though, maintain their structure and do not equilibrate, as long as food and oxygen, etc. are available. So in a sense, the second law of thermodynamics defines the current against which all living things successfully swim while they are alive.

All self-organizing systems, living or nonliving, are alike in being thermodynamically open. This is not a new insight for many Earth scientists: for example, Leopold *et al.* (1964: 267) note that dynamic equilibrium of fluvial systems refers

to an open system in which there is a continuous inflow of materials, but within which the form or character of the system remains unchanged. A biological cell is such a system. The river channel at a particular location over a period of time similarly receives an inflow of sediment and water, which is discharged downstream, while the channel itself remains essentially unchanged.

Thus, each nonliving dissipative system must, as must any living system, maintain a flow of matter and energy through itself in order to retain its integrity. Without this, it becomes a thermodynamically closed system, with bland equilibration as its only future. While our intuition correctly distinguishes between closed and dissipative systems when the dissipative system is living, it does not reliably distinguish between closed and dissipative systems when the dissipative system is nonliving. Nonliving dissipative systems which organize themselves are therefore a surprise to us.[13]

21.3 CELLULAR AUTOMATON MODELS

If a real system is presumed to manifest self-organization and emergence, then one way to study this is by means of a model which is also capable of manifesting these phenomena. The main tool for modelling self-organization in spatial systems is the 'cellular automaton' (CA) model.

21.3.1 Self-organization on a cellular grid

CA models discretize continuous space[14] into a series of cells. These are usually part of a regular square or rectangular grid, but can also be e.g. hexagonal. This spatial discretization[15] is necessary if the model is to be computationally tractable. The rules and relationships which comprise the model are then applied at the scale of individual cells.

These rules and relationships may be viewed as positive and negative feedbacks, and thus each cell may be seen as an 'automaton' with its behaviour controlled by the positive and negative feedbacks to which it is subject. Interactions are usually (but not always: see e.g. Li, 1997) between adjacent or nearby cells: thus the model's interactions are all 'local'. If the CA model then self-organizes and gives rise to larger-scale responses, these will manifest as patterns on the cellular grid (e.g. Mahnke, 1999; Wooton, 2001; Wolfram, 2002).

Bar-Yam (1997: 490) suggests that 'The idea of a cellular automaton is to think about simulating the space rather than the objects that are in it', and (ibid.: 139) that 'Cellular automata are an alternative to differential equations for the modelling of physical systems.' While the same real-world system may be described either by a CA model, or by a more conventional approach (e.g. a fluid dynamics model for a fluvial system), it appears to be the case that the CA approach usually brings out different characteristics of the system when compared with the more conventional representation.[16]

21.3.2 Kinds of CA models

Watts (1999: 181ff.) gives a good overview of the history and development of CA models. They are now used in the study of self-organization in a wide range of fields (e.g. Bar-Yam, 1997). Wolfram (2002) gives a magisterial overview of CA models: however this book appeared after this chapter was drafted, so that a full treatment here is infeasible.

Several CA models are freely available on the Internet.[17] Note that whereas the rules in the 'Game of Life' CA model are deterministic, this is not necessarily the case particularly in more complicated CA models, such as the rill growth models described below. Similarly, the original configuration of the CA grid (i.e. at the beginning of the simulation) may be arrived by deterministic or stochastic means.[18]

A variant of the simple CA model which is suggested to be better suited to representing continuum systems such as fluvial flow is the 'lattice-gas model' (Wolfram, 1986; Garcia-Sanchez *et al.*, 1996; Pilotti and Menduni, 1997; see also http://poseidon.ulb.ac.be/lga_en.html).

21.3.3 Computational constraints to CA modelling

One constraint to the application of CA models is computational. Since the rules/relationships of the CA model are implemented on a per-cell basis rather than upon the entire grid, any calculations specified by the local rules may well need to be carried out a very large number of times during a lengthy simulation. Since the majority of present-day computers are fundamentally 'serial', i.e. processing user instructions on a strictly sequential basis, this can result in very long run times for large grids. An obvious solution is to use a parallel computer (Bar-Yam, 1997: 488), on which a number of instructions can be processed simultaneously. To implement a CA model on a parallel computer is not, however, as simple as it might appear in all cases.[19] This remains an area of active development.

21.3.4 Modelling self-organization: the problem of context and boundaries

'If everything in the universe depends on everything else in a fundamental way, it may be impossible to get close to a full solution by investigating parts of the problem in isolation' (Hawking, 1988). As Stephen Hawking acknowledges, it is increasingly recognized that reductionism, i.e. breaking apart many systems – including those of interest to geographers – into smaller, more tractable units, poses a risk for full understanding. The reductionist's concentration on the components of a self-organizing system, away from the context in which such components interact and give rise to emergent self-organization, will miss vital points about the way the system works. Even with a more holistic focus, the imposition of an artificial boundary between the system's components will constrain the interactions between components in the region of the boundary, with potentially strong effects on emergent responses of the system.[20] This is notably the case for CA models of self-organizing systems, where the 'problem of boundary conditions' may be severe. We must set boundaries, but doing so conceptually breaks some of the model's connections to the 'outside world', and so can result in a distorted model (Bar-Yam, 1997: 8). So in the final analysis, we must accept that even the best possible model of a self-organizing system remains incomplete.

21.3.5 Terminology: self-organization and cellular automata

The relationship between model and theory is the subject of much debate among philosophers of science and logicians. Harvey, during an extensive discussion of the model-theory dichotomy (1969: Chapters 10 to 12), suggests (ibid.: 145) that a model may be regarded as a formalized expression of a theory. Thus a CA model may be regarded as an expression of the notion of self-organization and emergence in complex systems.

In any developing area of science, it is inevitable that terminology is somewhat fluid and unsettled. Thus some workers (e.g. Coulthard and Macklin, 2002) eschew the phrase CA model altogether, preferring 'cellular model'. The distinction made appears to be the nature of the local rules applied to each automaton, with 'cellular model' used if these rules are relatively complicated, and 'CA model' used only if the local rules are relatively simple. Notwithstanding its imprecision, this is a usage which may or may not become generally accepted in time. Here, the two expressions are used interchangeably.

However, it is certainly undesirable to focus only on the methodology (CA or cellular model) to the neglect of the underlying theory (self-organization and emergence). To do so is tantamount to describing the operation of an internal combustion engine without considering the expanding gases which drive it. CA models, when used to represent some real-world system, are just tools for reproducing the self-organization and emergence which is assumed to also manifest in the real-world system. It is self-organization and emergence that are the deeper concepts, and which thus best deserve our attention.

21.3.6 Geomorphological applications of CA models

Geomorphological modellers have eagerly embraced self-organization. A number of studies have made use of CA models from the early 1990s.

21.3.6.1 Early geomorphological CA modelling studies

Chase (1992) used a cellular model to investigate the evolution of fluvially eroded landscapes over long periods of time and at large spatial scales. In this model, rainfall occurs on the simulated landscape in single increments called precipitons. After being dropped at a random position, a precipiton runs downslope and starts eroding and depositing sediment, depending on the conditions in the cell. Through time, as numerous precipitons modify the landscape, the simple rules give rise to a complex, fluvially sculpted landscape.

Werner and Hallet (1993) also used the CA approach to investigate the formation of sorted stone stripes by needle ice, and found that the formation of stone stripes and similar features, such as stone polygons and nets, reflects self-organization resulting from local feedback between stone concentration and needle ice growth rather than from an externally applied, large-scale template. Werner and Fink (1993) similarly simulated the formation of beach cusps as self-organized features: the CA model here was based on the interaction of water flow, sediment transport and morphological change. Werner (1995) again used the approach to study the formation of aeolian dunes, and Werner (1999) reviewed earlier work and pointed out research directions.

The CA model of Murray and Paola (1994, 1996, 1997) was developed to investigate the formation of braided channel patterns. Their model replicated the typical dynamics of braided rivers, with lateral channel migration, bar erosion and formation, and channel splitting, using local transport rules describing sediment transport between neighbouring cells. Thomas and Nicholas (2002) also tackled a similar problem.

Simulation of rill networks formed by soil erosion by water was the goal of the plot-scale RillGrow 1 model constructed by Favis-Mortlock (1996, 1998b). The model successfully reproduced realistic-appearing rill networks, and was also able to reproduce other 'global' responses such as the relationship between total soil loss and slope gradient.

In addition to focusing on the pattern-forming aspects of self-organization, there has also been a flurry of interest in modelling the more statistical aspects of self-organization such as SOC. Hergarten and Neugebauer (1998, 1999) investigated the magnitude-frequency distribution of landslides with a cellular model. Landslides were also the focus of a CA modelling study by Clerici and Perego (2000).

More recently, Hergarten *et al.* (2000) and Hergarten and Neugebauer (2001) have shifted the focus of their magnitude-frequency CA modelling work to drainage networks. Modelled and observed drainage basins and their channel networks were also the focus of the work of Rodríguez-Iturbe and co-workers (e.g.

Rodríguez-Iturbe and Rinaldo, 1997). This investigation concentrated on the fractal and multifractal properties of channel networks. The models developed by Rodríguez-Iturbe and co-workers are based on the continuity equations for flow and sediment transport which are solved on a two-dimensional, rectangular grid. A similar approach is used in the long-term landscape-evolution modelling of Coulthard *et al.* (2002), which focuses on small catchments, and the models of Tucker and Slingerland (1996, 1997) and De Boer (2001) which emphasise larger catchments.

21.3.6.2 The next generation of geomorphological CA models

All modelling approaches from this first generation of studies have been 'conceptual' in the sense that arbitrary units are used for cell size, elevation etc. In many cases (e.g. Murray and Paola, 1997) the scaling of their model is undefined, so that the size of the landscape area being simulated is unclear. Thus validation of these models by comparison of model results with observations is impossible.

While these models have clearly established the general validity and potential of the CA approach for landscape modelling, an obvious next step is to use more realistic units and to construct the models in such a way that their outputs can be more rigorously validated. This is the aim of the RillGrow 2 model, as described below.

21.4 CASE STUDY: MODELLING RILL INITIATION AND GROWTH

21.4.1 The RillGrow 1 model

Conventional process-oriented models of soil erosion by water represent flow in hillslope rills (channels cut by erosive flow) by applying commonly used hydraulics relationships, just as can be done for flow in any open channel. This is an empirically rooted, 'engineering' approach which works well in many circumstances. However, it also leads to a number of deficiencies in models which employ it. The first major deficiency is that it is a poor description of reality, since 'all rills are not created equal'. Hillslope erosion models represent rills as prismatic channels, equally spaced, and with a similar hydrological efficiency. This is not the case in reality: each rill has its own unique characteristics, although it probably shares a 'family resemblance' with others on the same hillslope. This assumed uniformity means that the 'patchiness' of hillslope erosion is not captured very well in models. Spatial variability

is under-estimated; in a classic study using multiply-replicated plots, Wendt *et al.* (1986) found a variability of both hydrology and erosion that was far greater than would be estimated by running a conventional erosion model with data for each plot. The second important deficiency takes the form of a logical inconsistency: on an uneroded surface (e.g. a freshly tilled field), rills do not yet exist. Since there are no rills, there are no channels, and the laws of hydraulic flow in channels cannot be applied. Thus, it is necessary to assume that, in some sense, rills 'pre-exist' when conventional erosion models are used. WEPP (Nearing *et al.*, 1989) assumes a 1-m rill spacing, whereas the user needs to specify initial rill spacing and dimensions for EUROSEM (Morgan *et al.*, 1998). Thus we have a 'what comes first: the chicken or the egg?' problem. To tackle these and other problems, we need to go beyond an engineering-type approach and focus on the initiation and temporal development of rill erosion.

Favis-Mortlock (1996, 1998b) constructed the Rill-Grow 1 model in order to test the hypothesis that the initiation and development of hillslope rills may be modelled using a self-organizing systems approach, i.e. driven by simple rules governing systemic interactions on a much smaller scale than that of the rills. The central idea was that some rills are more 'successful' than others, since they preferentially grow in size and sustain flow throughout a rainfall event. Thus they compete for runoff. From an initial population of many microrills, only a subset subsequently develops into larger rills as part of a connected network (Table 21.1).

Modification of the soil microtopography by erosion produces a positive feedback loop, with the most 'successful' rills (i.e. those conveying the most runoff) modifying the local microtopography to the greatest extent, and so most effectively amplifying their chances

Table 21.1 'Successful' and 'unsuccessful' rills

Category of rill	Rate of growth	Effectiveness during rainfall event
Successful	Higher	Becomes major carrier for runoff and eroded soil for part of hillslope; may 'capture' weaker rills
Unsuccessful	Lower	Becomes less and less important as a carrier for runoff and sediment; may eventually be 'captured' or become completely inactive

Source: Favis-Mortlock (1996)

of capturing and conveying future runoff. There is a limit to this growth, however (i.e. an associated negative feedback); each 'successful' rill's catchment cannot grow forever, since eventually the whole surface of the hillslope will be partitioned between the catchments of the 'successful' rills. The dynamics of this competitive process give rise to connected rill networks. Thus the hillslope erosional system is a dissipative system, with rainfall providing the essential input of matter and energy to the system, and runoff and sediment being the outputs.

The very simple RillGrow 1 CA model produced apparently realistic results (see e.g. Favis-Mortlock *et al.*, 1998) in terms of reproducing the observed characteristics of rill networks on plot-sized areas. However, it possessed some serious conceptual limitations which prevented the approach from being more rigorously validated. First, the algorithm used (broadly similar to the precipiton approach of Chase, 1992, described above) meant that the model did not operate within a true time domain, and so validation of relationships with a temporal aspect (e.g. the effects of rainfall intensity, or time-varying discharge) was impossible. In addition, the model assumed an infinite transport capacity, with erosion being entirely detachment-limited. Thus the model could only hope to reproduce correctly situations where deposition is minimal. Finally, the model possesses a rather weak physical basis.

In many respects, RillGrow 1 was a typical 'first generation' geomorphological CA model. In order to move beyond these limitations, RillGrow 2 was developed, the aim being to improve the process descriptions of the first version of the model, while (as far as possible) still retaining its simplicity. An early version of RillGrow 2 was described in Favis-Mortlock *et al.* (2000). The most recent version is summarized here. A detailed description of the model is forthcoming (Favis-Mortlock, in preparation).

21.4.2 The RillGrow 2 model

RillGrow 2, like RillGrow 1, operates upon an area of bare soil which is specified as a grid of microtopographic elevations (a DEM). Typically, cell size is a few millimetres, with elevation data either derived from real soil surfaces (Figure 21.3) by using a laser scanner (Huang and Bradford, 1992) or by means of photogrammetry (Lascelles *et al.*, 2002); or generated using some random function (cf. Favis-Mortlock, 1998b). Computational constraints mean that, for practical purposes, the microtopographic grid can be no larger than plot-sized. A gradient is usually imposed on this grid. The model has a variable timestep, which is typically of the order of 0.05 s. At each timestep, multiple raindrops are dropped at random locations on the grid, with the

Figure 21.3 Soil surface microtopography: the scale at which RillGrow 2's rules operate. The finger indicates where flow (From left to right) is just beginning to incise a microrill in a field experiment (see Lascelles *et al.*, 2000)
Source: Photograph © Martin Barfoot, 1997 (martin.barfoot@geog.ox.ac.uk), used by permission

number of drops depending on rainfall intensity. Run-on from upslope may also be added at an edge of the grid. Often, the soil is assumed to be saturated so that no infiltration occurs; however, a fraction of all surface water may be removed each timestep as a crude representation of infiltration losses.

Splash redistribution is simulated in RillGrow 2. Since this is a relatively slow process it is not normally calculated at every timestep. The relationship by Planchon *et al.* (2000) is used: this is essentially a diffusion equation based on the Laplacian of the continuity equation describing splash erosion, with a 'splash efficiency' term which is a function of rainfall intensity and water depth. Currently, the splash redistribution

and overland flow components of RillGrow 2 are only loosely coupled; while sediment which is redistributed by splash can be moved in or out of the store of flow-transported sediment, this is not done in an explicitly spatial manner.

Movement of overland flow between 'wet' cells occurs in discrete steps between cells of this grid. Conceptually, overland flow in RillGrow 2 is therefore a kind of discretized fluid rather like the 'sheepflow' illustrated in Figure 21.4.

For the duration of the simulation, each 'wet' cell is processed in a random sequence which varies at each timestep. The simple logic outlined in Figure 21.5 is used for the processing.

Figure 21.4 'Sheepflow': a visual analogy of RillGrow 2's discretized representation of overland flow. Which should we best focus on: an individual sheep or the 'flow' of the flock of sheep?
Source: Photograph © Martin Price, 1999 (martin.price@perth.uhi.ac.uk), used by permission

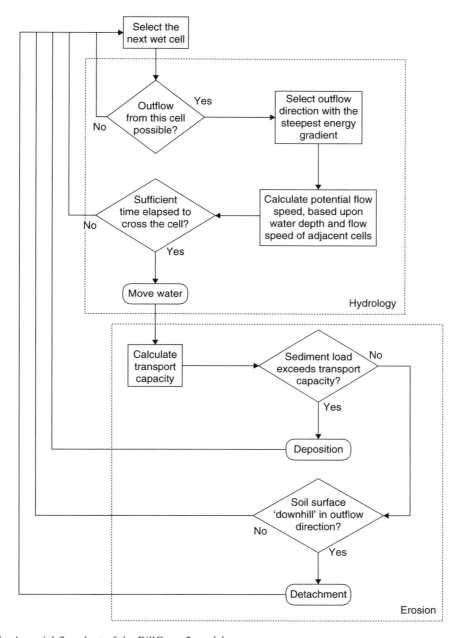

Figure 21.5 A partial flowchart of the RillGrow 2 model

Outflow may occur from a 'wet' cell to any of the eight adjacent cells. If outflow is possible, the direction with the steepest energy gradient (i.e. maximum difference in water-surface elevation) is chosen. The potential velocity of this outflow is calculated as a function of water depth and hydraulic radius. However, outflow only occurs if sufficient time has elapsed for the water to have crossed this cell. Thus outflow only occurs for a subset of 'wet' cells at each timestep.

When outflow does occur, the transport capacity of the flow is calculated using the previously calculated flow velocity with this S-curve relationship (Equation 5 in Nearing *et al.*, 1997):

$$\log_e(q_s) = \frac{\alpha + \beta \cdot e^{\gamma + \delta \cdot \log_e(\varpi)}}{1 + e^{\gamma + \delta \cdot \log_e(\varpi)}} \qquad (21.1)$$

where:

q_s is unit sediment load [$\text{kg m}^{-1}\text{s}^{-1}$]

$\alpha, \ \beta, \ \gamma, \ \delta$ are constants

and:

$$\varpi = \rho \cdot g \cdot S \cdot q$$

where:

ω = stream power [W m^{-2}]

ρ = density of water [kg m^{-3}]

g = gravitational acceleration [m s^{-2}]

S = energy slope [m m^{-1}]

q = unit discharge of water [$\text{m}^2 \text{s}^{-1}$].

If the sediment concentration of the water on the cell exceeds its transport capacity, deposition occurs. Deposition is calculated using a version of Equation 12 in Lei *et al.* (1998), which assumes deposition to be a linear function of the difference between sediment load and transport capacity. Note that no consideration is made of different settling times for each size fraction of the deposited sediment, nor of differences in properties (e.g. bulk density, erodibility) between re-entrained and newly eroded sediment.

Whereas outflow velocity and transport capacity are both determined by the energy gradient, it is assumed to be the soil-surface gradient which controls detachment. In RillGrow 2, detachment occurs only if both energy gradient and soil surface gradient are downhill. If the soil-surface gradient is uphill, then the energy gradient-driven outflow is assumed to be merely hydrostatic levelling. Where detachment does occur, its value is calculated using a probabilistic equation by Nearing (1991). This assumes that detachment is controlled by the probability of occurrence of random turbulent bursts and the difference between soil strength and shear stress generated by these bursts. The relationship used in RillGrow 2 is a reformulation of Equation 10 in Nearing (1991):

$$e = K \cdot S \cdot u \cdot P \qquad (21.2)$$

where:

e = detachment [kg s^{-1}]

K = a constant

u = outflow speed [m s^{-1}]

and:

$$P = 1 - \Phi\left(\frac{T - \tau_b}{\sqrt{S_T{}^2 + S_{\tau_b}{}^2}}\right) \qquad (21.3)$$

where:

Φ = the cumulative probability function of a standard normal deviate

T = a constant

S_T = the coefficient of variation of T (assumed constant)

S_{τ_b} = the coefficient of variation of τ_b (assumed constant)

and:

$$\tau_b = 150 \cdot \rho \cdot g \cdot h \cdot S$$

where:

$$h = \text{water depth [m]}.$$

While still relatively simple, RillGrow 2 is thus a development of the earlier RillGrow model in that it attempts to explicitly reproduce, in a true time domain, the effects of several processes involved in rill formation.

21.4.2.1 Results from RillGrow 2

During a series of 12 laboratory-based experiments at the University of Leicester (Lascelles *et al.*, 2000, 2002; Favis-Mortlock *et al.*, in preparation) simulated rainfall (i.e. from an overhead sprinkler system) was applied to a sandy soil in a 4×1.75 m flume. A range of slope angles and rainfall intensities were used (Table 21.2). Each experiment lasted for 30 minutes, during which time surface flow, discharge, sediment removal and flow velocities were measured. Prior to and following each

Table 21.2   Some characteristics of the Leicester flume experiments used to evaluate RillGrow 2

Experiment No.	Slope angle (degrees)	Rainfall (mm h^{-1})
X09	5	108
X10	10	129
X11	15	125
X12	5	117
X13	10	127
X14	15	126
X15	5	120
X16	10	126
X17	15	131
X18	5	121
X19	10	123
X20	15	125

Source: Adapted from Favis-Mortlock *et al.* (in preparation)

experimental run, digital photogrammetry was used to create Digital Elevation Models (DEMs) of the soil surface. The initial (pre-experiment) DEMs and other data were then used as inputs to RillGrow 2. For one experiment only (X11: Table 21.2), measured flow velocities, discharge and sediment yield were also used to calibrate the model with respect to soil roughness and erodibility. Once calibrated, the model's inputs were not further adjusted when simulating the other eleven experiments.

Full results are given in Favis-Mortlock *et al.* (in preparation), however, measured and simulated discharge and sediment delivery for all experiments are compared in Figure 21.6. While both total discharge and sediment loss were well simulated in all cases, sediment loss was consistently under-estimated for the low-gradient experiments. This result is thought to be because of the loose coupling between the splash and flow components of the model: the nonspatial representation of the exchange of sediment between splash and flow is most troublesome when flow detachment is low. While this under-estimation occurs at all gradients, it is most noticeable when flow detachment is low.

A more difficult test is for the model to reproduce the observed patterns of erosion at the end of the experiments. Results for six experiments are shown in Figure 21.7, with a 3D view in Figure 21.8.

For both experiments, the model was able to reproduce the main elements of the rill pattern; again more successfully for the high-gradient experiments than for the low-gradient ones. A similar result was seen for most of the other six experiments. Note, in Figure 21.8, that the rain-impacted surface of the soil is not reproduced by the model.

Thus, with microtopography as the only spatially explicit input, the CA model RillGrow 2 is able to predict the pattern of rills which will be formed. It does this by considering rill networks to be emergent, whole-system responses to interactions between flow, detachment and deposition at a scale of millimetres.

21.5 CONCLUSION

This chapter started by considering the apparently ever-increasing complexity of models which results from the desire to improve them. Following this, recent advances in understanding nonlinear systems were described, starting with deterministic chaos and followed by self-organization and emergence. These were drawn from a number of scientific areas. Next came a description of the CA modelling approach which builds upon these concepts, and a more detailed look at one CA model. Where has this led us?

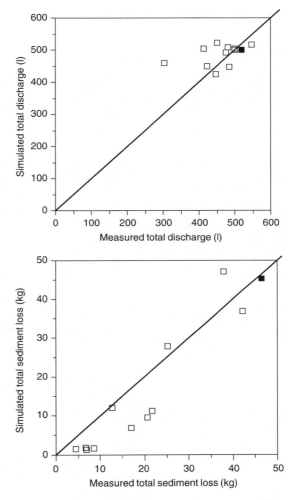

Figure 21.6 Total discharge (a) and sediment delivery (b) at the end of the Leicester flume experiments, measured values and as simulated by RillGrow 2. The black square indicates the value for X11; this was used to calibrate the model
Source: Favis-Mortlock *et al.* (in preparation)

The evidence does indeed seem to support the twin ideas that:

- self-organizing systems can give rise to complex patterns and relationships which are not necessarily the result of complex processes;
- CA approaches can be used to model such systems, and these models need not themselves be complex.

This is a most optimistic conclusion! CA models of systems in which self-organization is assumed to occur

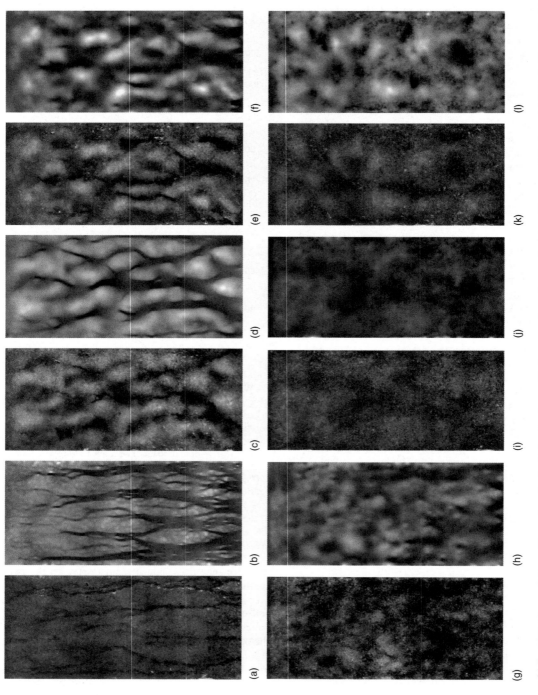

Figure 21.7 Plan views of the soil's surface in the Leicester flume. Darker areas have lower elevation. Top row: 15° slope (a) X11 (calibrated); (b) X11 simulated by RillGrow 2; (c) X14; (d) X14 simulated; Middle row: 10° slope; (e) X16; (h) X16 simulated. Lower row: 5° slope; (i) X15; (j) X15 simulated; (k) X12; (l) X12 simulated
Source: Favis-Mortlock *et al.* (in preparation)

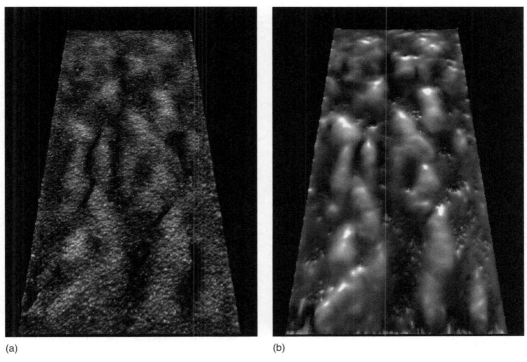

(a)   (b)

Figure 21.8   3D views of the soil surface in the Leicester flume, showing X13 with 10° slope: (a) experiment; (b) simulated by RillGrow 2
Source: Favis-Mortlock *et al.* (in press)

need only be simple at the 'component' level, yet are still capable of producing a great wealth of complex, emergent responses. When such models are designed to use realistic real-world values (e.g. RillGrow 2), it seems that the models, while simple, are nonetheless capable of reproducing realistic facsimiles of some of the complex features of real environments.

Finally, an important point re. discretization. Wolfram (2002, p. 327) notes that many systems in nature appear smooth or continuous, yet CA models involve only discrete elements. How can such models ever hope to reproduce what we see in nature? The crucial point here is that even though the individual elements of a system may be discrete, the average behaviour seen when looking at a large number of these components may well be continuous. The question then becomes, 'Are there a sufficient number of CA elements for their average behaviour to approximate the average behaviour of the real system?'

The mounting weight of model-based evidence, together with field- and laboratory-based studies not discussed here, tends to confirm the assumption that self-organization is a real feature of the real world. If this is

so, then one implication is a message of hope for environmental modellers. We are not doomed to ever more complex models!

While it may seem perverse that the study of complexity can lead us to simplicity, ground-breaking research on self-organization during the late 1980s and 1990s appears to have been a fresh wind, doing away with the tired idea of fitting the results of data-hungry models to sparse, observed data. Instead of the cumbersome drudgery of varying a large number of parameters and variables to obtain a better fit between the computed and observed data, models may again be used to generate new and exciting ideas, and to help solve the looming and difficult environmental problems which humanity must successfully tackle if it is to survive.

ACKNOWLEDGEMENTS

There is some overlap between this chapter and the rather longer paper of Favis-Mortlock and De Boer (in press). As in that paper, I would like to thank the following: John Boardman (University of

Oxford), Jack Cohen (University of Warwick), Anton Imeson (University of Amsterdam), Mark Nearing (USDA-ARS), John Thornes (King's College London), Brian Whalley (Queen's University Belfast); the 'self-organized lunchers' at Oxford (Susan Canney, Clive Hambler, Mike Packer and David Raubenheimer); and finally all those who attended the NATO-ASI 'Nonlinear Dynamics in Life and Social Sciences' (Moscow, 2000). Development of RillGrow 2 was partly funded by the UK Natural Environment Research Council (NERC GR3/11417: 'Rill initiation by overland flow and its role in modelling soil erosion').

NOTES

1. Cf. Chaitin's (1999) notion of 'algorithmic complexity'.
2. This is particularly the case in practical applications where the main aim is a successful replication of measured values, perhaps with the subsequent aim of estimating unmeasurable values, e.g. under some hypothetical conditions of climate or land use.
3. See section on terminology below.
4. The concept of equifinality is originally due to Von Bertalanffy (1968).
5. Originally the title of a 1970s lecture by Lorentz. There are now many variants.
6. The word 'eventually' is important here. The repetitive calculations which are often necessary are ideally suited to a computer, but not to a human. This is one reason why deterministic chaos had to wait for the widespread usage of computers for its discovery.
7. *'But what we've realized now is that unpredictability is very common, it's not just some special case. It's very common for dynamical systems to exhibit extreme unpredictability, in the sense that you can have perfectly definite equations, but the solutions can be unpredictable to a degree that makes it quite unreasonable to use the formal causality built into the equations as the basis for any intelligent philosophy of prediction'* (Berry 1988).
8. In some cases the exposed B horizon will have greater shear strength and so be better able to resist detachment by the extra flow. Here, an increase in runoff will not result in increased erosion.
9. Referred to as 'complicity' in Cohen and Stewart (1994). As Jack Cohen put it in an email to complex-sciences@necsi.org on 10 December 2000:

   a more exquisite problem for complex thinking is the problem of recursion. The system doesn't simply respond to the environment, the environment also responds to the system. The pattern of a river bed is the result of genuine interaction of this kind, and so is nearly every ongoing process in physics, biology or management.

   Another term for this is 'autocatalysis'.
10. In other words, a large number of independently varying degrees of freedom.
11. Because of the centre-tending effect of the law of large numbers: see e.g. Harvey (1969: 246).
12. Such as a idealized gas: see Bar-Yam (1997). It is also true (but in a different way) for almost all systems, even quite complicated ones, which have been constructed by some external agency e.g. by a human. A first question to ask when attempting to identify self-organization is, 'Is this system more than the sum of its parts?'
13. Yet from still another point of view, they certainly shouldn't be. For if nonliving systems do not organize themselves, how would they get organized in the first place? Unless we invoke an organizer (i.e. teleology), then from this perspective 'if self-organization did not exist, it would be necessary to invent it'.
14. Just as almost all models which operate in a true time dimension discretize continuous time into distinct 'timesteps'.
15. Discretization appears to be a fundamental operation for humans attempting to make sense of the world; see e.g. Chapter 1 of Bohm (1980).
16. This is related to the concepts of 'trivial duality' and 'nontrivial duality' in physics (Greene, 1999: 297 ff.). If each of two different descriptions of the same system tell us something that the other does not, this is 'nontrivial duality'.
17. For example:
    * http://psoup.math.wisc.edu/Life32.html
    * http://sero.org/sero/homepages/
    * http://www.automaton.ch/
    * http://www.exploratorium.edu/complexity/ CompLexicon/automaton.html
18. See http://www.santafe.edu/~hag/class/class.html for a hierarchical classification of cellular automata.
19. For example, what is the best way of splitting up the cellular grid between processors if cells must be processed in a random sequence (cf. Watts, 1999: 205)? What about local interactions between cells allocated to different processors?
20. 'Biologists cannot adopt a reductionist approach when working with a living organism, or it dies' (Jack Cohen, personal communication, 2000).

REFERENCES

Ackoff, R.L., Gupta, S.K. and Minas, J.S. (1962) *Scientific Method: Optimizing Applied Research Decisions*, John Wiley & Sons, New York.

Andrle, R. (1996) The west coast of Britain: statistical self-similarity vs. characteristic scales in the landscape, *Earth Surface Processes and Landforms* **21** (10), 955–962.

Aristotle (*c.* 330 BCE) *Metaphysica* 10f–1045a.

Bak, P. (1996) *How Nature Works*, Springer-Verlag, New York.

Bak, P., Tang, C. and Wiesenfeld, K. (1988) Self-organized criticality, *Physical Review A* **38** (1), 364–374.

Bar-Yam, Y. (1997) *Dynamics of Complex Systems*, Perseus, Reading, MA.

Ben-Jacob, E. and Levine, H. (2001) The artistry of nature, *Nature* **409**, 985–986.

Berry, M. (1988) The electron at the end of the universe, in L. Wolpert and A. Richards (eds) *A Passion for Science*, Oxford University Press, Oxford, 39–51.

Beven, K.J. (1993) Prophecy, reality and uncertainty in distributed hydrological modelling, *Advances in Water Resources* **16**, 41–51.

Beven, K.J. (1996) The limits of splitting: hydrology, *The Science of the Total Environment* **183**, 89–97.

Bohm, D. (1980) *Wholeness and the Implicate Order*, Routledge, London.

Buchanan, M. (2000) *Ubiquity*, Weidenfeld and Nicolson, London.

Çambel, A.B. (1993) *Applied Chaos Theory: A Paradigm for Complexity*, Academic Press, Boston.

Chaitin, G.J. (1999) *The Unknowable*, Springer-Verlag, Singapore.

Chase, C.G. (1992) Fluvial landsculpting and the fractal dimension of topography, *Geomorphology* **5**, 39–57.

Chorley, R.J. (1962) *Geomorphology and General Systems Theory*, US Geological Survey Professional Paper, No. 500-B.

Clerici, A. and Perego, S. (2000) Simulation of the Parma River blockage by the Corniglio landslide (Northern Italy), *Geomorphology* **33**, 1–23.

Clifford, N.J. (2001) Physical Geography – the naughty world revisited, *Transactions of the Institute of British Geographers* NS **26**, 387–389.

Cohen, J. and Stewart, I. (1994) *The Collapse of Chaos*, Penguin, London.

Coulthard, T.J., Macklin, M.G. and Kirkby, M.J. (2002) A cellular model of Holocene upland river basin and alluvial fan evolution, *Earth Surface Processes and Landforms* **27** (3), 269–288.

Coveney, P. and Highfield, R. (1995) *Frontiers of Complexity*, Faber and Faber, London.

De Boer, D.H. (2001) Self-organization in fluvial landscapes: sediment dynamics as an emergent property, *Computers and Geosciences* **27** (8), 995–1003.

Evans, I.S. and McClean, C.J. (1995) The land surface is not unifractal: variograms, cirque scale and allometry, *Zeitschrift für Geomorphologie N.F. Supplement* **101**, 127–147.

Favis-Mortlock, D.T. (1996) An evolutionary approach to the simulation of rill initiation and development, in R.J. Abrahart (ed.) *Proceedings of the First International Conference on GeoComputation* (Vol. 1), School of Geography, University of Leeds, Leeds, 248–281.

Favis-Mortlock, D.T. (1998a) Validation of field-scale soil erosion models using common datasets, in J. Boardman and D.T. Favis-Mortlock (eds) *Modelling Soil Erosion by Water*, Springer-Verlag NATO-ASI Series I-55, Berlin, 89–128.

Favis-Mortlock, D.T. (1998b) A self-organizing dynamic systems approach to the simulation of rill initiation and development on hillslopes, *Computers and Geosciences* **24** (4), 353–372.

Favis-Mortlock, D.T. (in preparation) The RillGrow 2 model.

Favis-Mortlock, D.T., Boardman, J. and MacMillan, V.J. (2001) The limits of erosion modeling: why we should proceed with care, in R.S. Harmon and W.W. Doe III (eds) *Landscape Erosion and Evolution Modeling*, Kluwer Academic/Plenum Publishing, New York, 477–516.

Favis-Mortlock, D.T., Boardman, J., Parsons, A.J. and Lascelles, B. (2000) Emergence and erosion: a model for rill initiation and development, *Hydrological Processes* **14** (11–12), 2173–2205.

Favis-Mortlock, D.T. and De Boer, D. (2003) Simple at heart? Landscape as a self-organizing complex system, in S.T. Trudgill and A. Roy (eds) *Contemporary Meanings In Physical Geography*, Edward Arnold, London, 127–171.

Favis-Mortlock, D.T., Guerra, A.J.T. and Boardman, J. (1998) A self-organising dynamic systems approach to hillslope rill initiation and growth: model development and validation, in, W. Summer, E. Klaghofer and W. Zhang (eds) *Modelling Soil Erosion, Sediment Transport and Closely Related Hydrological Processes*, IAHS Press Publication No. 249, Wallingford, 53–61.

Favis-Mortlock, D.T., Parsons, A.J, Boardman, J. and Lascelles, B. (in preparation) Evaluation of the RillGrow 2 model.

Favis-Mortlock, D.T., Quinton, J.N. and Dickinson, W.T. (1996) The GCTE validation of soil erosion models for global change studies, *Journal of Soil and Water Conservation* **51** (5), 397–403.

Gallagher, R. and Appenzeller, T. (1999) Beyond reductionism, *Science* **284**, 79.

Garcia-Sanchez, L., Di Pietro, L. and Germann, P.F. (1996) Lattice-gas approach to surface runoff after rain, *European Journal of Soil Science* **47** (4), 453–462.

Goodwin, B. (1997) *How the Leopard Changed its Spots: The Evolution of Complexity*, Phoenix, London.

Greene, B. (1999) *The Elegant Universe*, Jonathan Cape, London.

Hansen, J.L., Van Hecke, M., Haaning, A., Ellegaard, C., Andersen, K.H., Bohr, T. and Sams, T. (2001) Instabilities in sand ripples, *Nature* **410**, 324.

Harvey, D. (1969) *Explanation in Geography*, Edward Arnold, London.

Hawking, S. (1988) *A Brief History of Time*, Bantam Books, New York.

Hergarten, S. and Neugebauer, H.J. (1998) Self-organized criticality in a landslide model, *Geophysical Research Letters* **25** (6), 801–804.

Hergarten, S. and Neugebauer, H.J. (1999) Self-organized criticality in landsliding processes, in S. Hergarten and H.J. Neugebauer (eds) *Process Modelling and Landform Evolution*, Springer, Berlin, 231–249.

Hergarten, S. and Neugebauer, H.J. (2001) Self-organized critical drainage networks, *Physical Review Letters* **86** (12), 2689–2692.

Hergarten, S., Paul, G. and Neugebauer, H.J. (2000) Modeling surface runoff, in J. Schmidt (ed.) *Soil Erosion: Application of Physically-Based Models*, Springer-Verlag, Berlin, 295–306.

Holland, J.H. (1998) *Emergence: From Chaos to Order*, Perseus Books, Reading, MA.

Huang, C. and Bradford, J.M. (1992) Applications of a laser scanner to quantify soil microtopography, *Soil Science Society of America Journal* **56** (1), 14–21.

Huggett, R.J. (1985) *Earth Surface Systems*, Springer-Verlag, Berlin.

Jakeman, A.J. and Hornberger, G.M. (1993) How much complexity is warranted in a rainfall-runoff model? *Water Resources Research* **29**, 2637–2649.

Jetten, V., de Roo, A.P.J. and Favis-Mortlock, D.T. (1999) Evaluation of field-scale and catchment-scale soil erosion models, *Catena* **37** (3/4), 521–541.

Jones, H. (1991) Fractals before Mandelbrot: a selective history, in A.J. Crilly, R.A. Earnshaw and H. Jones (eds) *Fractals and Chaos*, Springer-Verlag, Berlin, 7–33.

Kauffman, S. (1995) *At Home in the Universe: The Search for Laws of Self-Organization and Complexity*, Oxford University Press, Oxford.

Kirchner, J.W., Hooper, R.P., Kendall, C., Neal, C. and Leavesley, G. (1996) Testing and validating environmental models, *The Science of the Total Environment* **183**, 33–47.

Kirkby, M.J., Naden, P.S., Burt, T.P. and Butcher, D.P. (1992) *Computer Simulation in Physical Geography*, 2nd edn, John Wiley & Sons, Chichester.

Klemeš, V. (1986) Dilettantism in hydrology: transition or destiny? *Water Resources Research* **22**, 177S–188S.

Klimontovich, Y.L. (2001) Entropy, information and ordering criteria in open systems, in W. Sulis and I. Trofimova (eds) *Nonlinear Dynamics in the Life and Social Sciences*, IOS Press, Amsterdam, 13–32.

Laland, K.N., Odling-Smee, F.J. and Feldman, M.W. (1999) Evolutionary consequences of niche construction and their implications for ecology, *Proceedings of the National Academy of Science* **96**, 10242–10247.

Lascelles, B., Favis-Mortlock, D.T., Parsons, A.J. and Boardman, J. (2002) Automated digital photogrammetry: a valuable tool for small-scale geomorphological research for the non-photogrammetrist? *Transactions in GIS* **6** (1), 5–15.

Lascelles, B., Favis-Mortlock, D.T., Parsons, A.J. and Guerra, A.J.T. (2000) Spatial and temporal variation in two rainfall simulators: implications for spatially explicit rainfall simulation experiments, *Earth Surface Processes and Landforms* **25** (7), 709–721.

Lei, T., Nearing, M.A., Haghighi, K. and Bralts, V.F. (1998) Rill erosion and morphological evolution: a simulation model, *Water Resources Research* **34** (11), 3157–3168.

Lenton, T.M. (1998) Gaia and natural selection, *Nature* **394**, 439–447.

Leonard, R.A., Knisel, W.G. and Still, D.A. (1987) GLEAMS: Groundwater Loading Effects of Agricultural Management Systems, *Transactions of the American Society of Agricultural Engineers* **30** (5), 1403–1418.

Leopold, L.B., Wolman, M.G. and Miller, J.P. (1964) *Fluvial Processes in Geomorphology*, Freeman, San Francisco.

Lewin, R. (1997) Critical mass: complexity theory may explain big extinctions better than asteroids, *New Scientist*, **155**, 23 August, 7.

Li, W. (1997) Nonlocal cellular automata, in H.F. Nijhout, L. Nadel, and D.S. Stein (eds) *Pattern Formation in the Physical and Biological Sciences*, Addison-Wesley, Reading, MA, 189–200.

Mahnke, R. (1999) Pattern formation in cellular automaton models, in J. Schmelzer, G. Röpke and R. Mahnke (eds) *Aggregation Phenomena in Complex Systems*, Wiley-VCH, Weinheim, 146–173

Mandelbrot, B.B. (1975) Stochastic models for the earth's relief, the shape and the fractal dimension of the coastlines, and the number-area rules for islands, *Proceedings of the National Academy of Sciences, USA* **72** (10), 3825–3828.

May, R.M. (1976) Simple mathematical models with very complicated dynamics, *Nature* **261**, 459–467.

Morgan, R.P.C., Quinton, J.N., Smith, R.E., Govers, G., Poesen, J.W.A., Chisci, G. and Torri, D. (1998) The EUROSEM model, in J. Boardman and D.T. Favis-Mortlock (eds) *Modelling Soil Erosion by Water*, Springer-Verlag NATO-ASI Series I-55, Berlin, 389–398.

Murray, A.B. and Paola, C. (1994) A cellular model of braided rivers, *Nature* **371**, 54–57.

Murray, A.B. and Paola, C. (1996) A new quantitative test of geomorphic models, applied to a model of braided streams, *Water Resources Research* **32** (8), 2579–2587.

Murray, A.B. and Paola, C. (1997) Properties of a cellular braided-stream model, *Earth Surface Processes and Landforms* **22** (11), 1001–1025.

Nearing, M.A. (1991) A probabilistic model of soil detachment by shallow turbulent flow, *Transactions of the American Society of Agricultural Engineers* **34** (1), 81–85.

Nearing, M.A., Foster, G.R., Lane, L.J. and Finkner, S.C. (1989) A process-based soil erosion model for USDA-Water Erosion Prediction Project technology, *Transactions of the American Society of Agricultural Engineers* **32** (5), 1587–1593.

Nearing, M.A., Norton, L.D., Bulgakov, D.A., Larionov, G.A., West, L.T. and Dontsova, K. (1997) Hydraulics and erosion in eroding rills, *Water Resources Research* **33** (4), 865–876.

Nicolis, G. and Prigogine, I. (1989) *Exploring Complexity*, Freeman, New York.

Peterson, G.D. (2000) Scaling ecological dynamics: self-organization, hierarchical structure, and ecological resilience, *Climatic Change* **44**, 291–309.

Pilotti, M. and Menduni, G. (1997) Application of lattice gas techniques to the study of sediment erosion and transport caused by laminar sheetflow, *Earth Surface Processes and Landforms* **22** (9), 885–893.

Planchon, O., Esteves, M., Silvera, N. and Lapetite, J.M. (2000) Raindrop erosion of tillage induced microrelief: Possible use of the diffusion equation, *Soil and Tillage Research* **56** (3–4), 131–144.

Portugali, J., Benenson, I. and Omer, I. (1997) Spatial cognitive dissonance and socio-spatial emergence in a self-organizing city, *Environment and Planning B* **24**, 263–285.

Rodríguez-Iturbe, I. and Rinaldo, A. (1997) *Fractal River Basins: Chance and Self-Organization*, Cambridge University Press, Cambridge.

Rosenblueth, A. and Wiener, N. (1945) The role of models in science, *Philosophy of Science* **12**.

Ruelle, D. (1993) *Chance and Chaos*, Penguin, London.

Ruelle, D. (2001) Applications of chaos, in W. Sulis and I. Trofimova (eds) *Nonlinear Dynamics in the Life and Social Sciences*, IOS Press, Amsterdam, 3–12.

Sethna, J.P., Dahmen, K.A. and Myers, C.R. (2001) Crackling noise, *Nature* **410**, 242–250.

Solé, R.V., Manrubia, S.C., Benton, M., Kauffman, S. and Bak, P. (1999) Criticality and scaling in evolutionary ecology, *Trends in Evolutionary Ecology* **14** (4), 156–160.

Tam, W.Y. (1997) Pattern formation in chemical systems: roles of open reactors, in H.F. Nijhout, L. Nadel and D.S. Stein (eds) *Pattern Formation in the Physical and Biological Sciences*, Addison-Wesley, Reading, MA, 323–347.

Thomas, R. and Nicholas, A.P. (2002) Simulation of braided river flow using a new cellular routing scheme, *Geomorphology* **43** (3–4), 179–195

Tucker, G.E. and Slingerland, R. (1996) Predicting sediment flux from fold and thrust belts, *Basin Research* **8**, 329–349.

Tucker, G.E. and Slingerland, R. (1997) Drainage basin responses to climate change, *Water Resources Research* **33** (8), 2031–2047.

Von Bertalanffy, L. (1968) *General Systems Theory: Foundations, Development, Applications*, George Braziller, New York.

Waldrop, M.M. (1994) *Complexity*, Penguin, London.

Washington, R.W. (2000) Quantifying chaos in the atmosphere, *Progress in Physical Geography* **24** (2), 499–514.

Watts, D.J. (1999) *Small Worlds: The Dynamics of Networks between Order and Randomness*, Princeton University Press, Princeton, NJ.

Wendt, R.C., Alberts, E.E. and Hjemfelt Jr, A.T. (1986) Variability of runoff and soil loss from fallow experimental plots, *Soil Science Society of America Journal* **50** (3), 730–736.

Werner, B.T. (1995) Eolian dunes: computer simulations and attractor interpretation, *Geology* **23**, 1107–1110.

Werner, B.T. (1999) Complexity in natural landform patterns, *Science* **284**, 102–104.

Werner, B.T. and Fink, T.M. (1993) Beach cusps as self-organized patterns, *Science* **260**, 968–971.

Werner, B.T. and Hallet, B. (1993) Numerical simulation of self-organized stone stripes, *Nature* **361**, 142–145.

Wilcox, B.P., Seyfried, M.S. and Matison, T.H. (1991) Searching for chaotic dynamics in snowmelt runoff, *Water Resources Research* **27** (6), 1005–1010.

Williams, J.R., Renard, K.E. and Dyke, P.T. (1983) EPIC – a new method for assessing erosion's effect on soil productivity, *Journal of Soil and Water Conservation* **38** (5), 381–383.

Winder, N. (2000) Contemporary human ecodynamics and the mathematics of history, in G. Bailey, R. Charles and N. Winder (eds) *Human Ecodynamics*, Oxbow, Oxford, 1–9.

Wischmeier, W.H. and Smith, D.D. (1978) *Predicting Rainfall Erosion Losses*, US Department of Agriculture, Agricultural Research Service Handbook 537, Washington, DC.

Wolfram, S. (1982) *Cellular Automata as Simple Self-Organising Systems*, Caltech preprint CALT-68-938, Pasudena, CA.

Wolfram, S. (1986) Cellular automaton fluids 1. Basic theory, *Journal of Statistical Physics* **45**, 471–526.

Wolfram, S. (2002) *A New Kind of Science*, Wolfram Media Inc., Champaign, IL, USA.

Wooton, J.T. (2001) Local interactions predict large-scale pattern in empirically derived cellular automata, *Nature* **413**, 841–844.

22

Data-Based Mechanistic Modelling and the Simplification of Environmental Systems[1]

PETER C. YOUNG, ARUN CHOTAI AND KEITH J. BEVEN

22.1 INTRODUCTION

The environment is a complex assemblage of interacting physical, chemical and biological processes, many of which are inherently nonlinear, with considerable uncertainty about both their nature and their interconnections. It is surprising, therefore, that stochastic dynamic models are the exception rather than the rule in environmental science research. One reason for this anomaly lies in the very successful history of physical science over the last century. Modelling in deterministic terms has permeated scientific endeavour over this period and has led to a pattern of scientific investigation which is heavily reductionist in nature. Such deterministic reductionism appears to be guided by a belief that physical systems can be described very well, if not exactly, by deterministic mathematical equations based on well known scientific laws, provided only that sufficient detail can be included to describe all the physical processes that are *perceived* to be important by the scientists involved. This leads inexorably to large, nonlinear models reflecting the scientist's perception of the environment as an exceedingly complex dynamic system.

Although deterministic reductionism still dominates environmental modelling, there are some signs that attitudes may be changing. There is a growing realization that, despite their superficially rigorous scientific appearance, simulation models of the environment based on deterministic concepts are more speculative extensions of our mental models and perceptions of the real world than necessarily accurate representations of the real world itself. The recent revived interest in the 'top-down' approach to modelling in the hydrological literature (e.g. Jothityangkoon *et al.*, 2000 and the references therein), for instance, is a response to the relative failure of the alternative reductionist ('bottom-up') philosophy in this area of study. But such scepticism is not new. It has its parallels in the environmental (e.g. Young, 1978, 1983; Beck, 1983) and ecosystems (e.g. see prior references cited in Silver, 1993) literature of the 1970s and early 1980s, where the present first author's contributions were set within the context of 'badly defined' environmental systems. To quote from Young (1983), which echoes earlier ideas (Young, 1978), for instance:

Although such reductionist analysis is perfectly respectable, it must be used very carefully; the dangers inherent in its application are manifold, but they are not, unfortunately, always acknowledged by its proponents. It is well known that a large and complex simulation model, of the kind that abounds in current ecological and environmental system analysis, has enormous explanatory potential and can usually be fitted easily to the meagre time-series data often used as the basis for such analysis. Yet even deterministic sensitivity analysis will reveal the limitation of the resulting model: many of the 'estimated' parameters are found to be ill-defined and only a comparatively small subset is important in explaining the observed system behaviour.

This paper goes on to point out that such overparameterization and the associated identifiability problems are quite often acknowledged, often implicitly, by the reductionist simulation model-builder. For

Environmental Modelling: Finding Simplicity in Complexity. Edited by J. Wainwright and M. Mulligan
© 2004 John Wiley & Sons, Ltd ISBNs: 0-471-49617-0 (HB); 0-471-49618-9 (PB)

example, the modelle sometimes constrains the values of certain 'better known' parameters and seeks to fit the model by optimizing the chosen cost function in relation to the remaining parameters, which are normally few in number. However, the model then has a degree of 'surplus content' not estimated from the available data, but based on a somewhat ad hoc *evaluation of all available prior knowledge of the system and coloured by the analyst's preconceived notions of its behavioural mechanisms. The paper concludes that:*

On the surface, this conventional simulation modeling approach seems quite sensible: for example, the statistician with a Bayesian turn of mind might welcome its tendency to make use of all a priori *information available about the system in order to derive the* a posteriori *model structure and parameters. On the other hand, he would probably be concerned that the chosen procedures could so easily be misused: whereas the constrained parameter optimization represents a quantitative and relatively objective approach, it is submerged rather arbitrarily within a more qualitative and subjective framework based on a mixture of academic judgment and intuition. Such a statistician would enquire, therefore, whether it is not possible to modify this framework so that the analyst cannot, unwittingly, put too much confidence in* a priori *perceptions of the system and so generate overconfidence in the resulting model.*

This and the other early papers then went on to present initial thoughts on such an objective, statistical approach to modelling poorly defined systems that tried to avoid the dangers of placing too much confidence in prior perceptions about the nature of the model. They also adumbrate very similar anti-reductionist arguments that have appeared recently in the hydrological literature and express some of these same views within a hydrological context (Jakeman and Hornberger, 1993; Beven, 2000, 2001). And quite similar anti-reductionist views are also appearing in other areas of science: for instance, in a recent lecture (Lawton, 2001), the current chief executive of the Natural Environment Research Council (NERC) recounted the virtues of the top-down approach to modelling ecological systems (although, for some reason, he did not appear to accept that such reasoning could also be applied to other natural systems, such as the physical environment).

In the subsequent period since the earlier papers were published, however, the first author has sought to develop this statistical approach within a more

rigorous systems setting that he has termed *Data-Based Mechanistic* (DBM) modelling. Prior to discussing the DBM approach, the present chapter will first outline the major concepts of statistical modelling that are important in any modelling process. Subsequently, two examples will be presented that illustrate the utility of DBM modelling in producing parametrically efficient (parsimonious) continuous or discrete-time models from environmental time series data. Finally, the chapter discusses briefly how this same methodology can be useful not only for the modelling of environmental and other systems directly from time series data, but also as an approach to the evaluation and simplification of large deterministic simulation models.

22.2 PHILOSOPHIES OF MODELLING

In considering questions of complexity and simplicity in mathematical modelling, it is important to note how the mathematical modelling of natural systems has developed over the past few centuries. In this regard, Young (2002a) points out that two main approaches to mathematical modelling can be discerned in the history of science; approaches which, not surprisingly, can be related to the more general deductive and inductive approaches to scientific inference that have been identified by philosophers of science from Francis Bacon (1620) to Karl Popper (1959) and Thomas Kuhn (1962):

- The *hypothetico-deductive approach*. Here, the *a priori* conceptual model structure is effectively a (normally simple) theory of behaviour based on the perception of the environmental scientist/modeller and is strongly conditioned by assumptions that derive from current environmental science paradigms.
- The *inductive approach*. Here, theoretical preconceptions are avoided as much as possible in the initial stages of the analysis. In particular, the model structure is not pre-specified by the modeller but, wherever possible, it is inferred directly from the observational data in relation to a more general class of models. Only then is the model interpreted in a physically meaningful manner, most often (but not always) within the context of the current scientific paradigms.

The DBM approach to modelling is of this latter inductive type and it forms the basis for the research described in the rest of this chapter. Previous publications (Young, 1978; Beck, 1983; Young *et al.*, 1996; Young, 1998 and the prior references therein) map the evolution of this DBM philosophy and its methodological underpinning in considerable detail. As these references demonstrate,

DBM models can be of various kinds depending upon the nature of the system under study. In the context of the present chapter, however, they take the form of linear and nonlinear, stochastic *Transfer Function* (TF) representations of the hydrological processes active in river catchments.

22.3 STATISTICAL IDENTIFICATION, ESTIMATION AND VALIDATION

The statistical approach to modelling assumes that the model is stochastic: in other words, no matter how good the model and how low the noise on the observational data happens to be, a certain level of uncertainty will remain after modelling has been completed. Consequently, full stochastic modelling requires that this uncertainty, which is associated with both the model parameters and the stochastic inputs, should be quantified in some manner as an inherent part of the modelling analysis.

In the statistical, time-series literature, such a stochastic modelling procedure is normally considered in two main stages: *identification* of an appropriate, identifiable model structure; and *estimation* (optimization, calibration) of the parameters that characterize this structure, using some form of estimation or optimization. Normally, if the data provision makes it possible, a further stage of *validation* (or *conditional validation*: see later) is defined, in which the ability of the model to explain the observed data is evaluated on data sets different to those used in the model identification and estimation stages. In this section, we outline these three stages in order to set the scene for the later analysis. This discussion is intentionally brief, however, since the topic is so large that a comprehensive review is not possible in the present context.

22.3.1 Structure and order identification

In the DBM approach to modelling, the identification stage is considered as a most important and essential prelude to the later stages of model building. It usually involves the identification of the most appropriate model order, as defined in dynamic system terms. However, the model structure itself can be the subject of the analysis if this is also considered to be ill-defined. In the DBM approach, for instance, the nature of linearity and nonlinearity in the model is not assumed *a priori* (unless there are good reasons for such assumptions based on previous data-based modelling studies). Rather it is identified from the data using nonparametric and parametric statistical estimation methods based on

a suitable, generic model class. Once a suitable model structure has been defined within this class, there are a variety of statistical methods for identifying model order, some of which are mentioned later. In general, however, they exploit some order identification statistics, such as the correlation-based statistics popularized by Box and Jenkins (1970), the well-known Akaike Information Criterion (AIC: Akaike, 1974), and the more heuristic YIC statistic (see e.g. Young *et al.*, 1996) which provides an alternative to the AIC in the case of transfer functions (where the AIC tends to identify over-parameterized models).

22.3.2 Estimation (optimization)

Once the model structure and order have been identified, the parameters that characterize this structure need to be estimated in some manner. There are many automatic methods of estimation or optimization available in this age of the digital computer. These range from the simplest, deterministic procedures, usually based on the minimization of least squares cost functions, to more complex numerical optimization methods based on statistical concepts, such as Maximum Likelihood (ML). In general, the latter are more restricted, because of their underlying statistical assumptions, but they provide a more thoughtful and reliable approach to statistical inference; an approach which, when used correctly, includes the associated statistical diagnostic tests that are considered so important in statistical inference. In the present DBM modelling context, the estimation methods are based on optimal, linear Instrumental Variable (IV) methods for transfer function models (e.g. Young, 1984, and the references therein) and nonlinear modifications of these methods (see later).

22.3.3 Conditional validation

Validation is a complex process and even its definition is controversial. Some academics (e.g. Konikow and Bredehoeft (1992), within a ground-water context; Oreskes *et al.* (1994), in relation to the whole of the earth sciences) question even the possibility of validating models. To some degree, however, these latter arguments are rather philosophical and linked, in part, to questions of semantics: what is the 'truth'; what is meant by terms such as validation, verification and confirmation? etc. Nevertheless, one specific, quantitative aspect of validation is widely accepted; namely 'predictive validation' (often referred to as just 'validation'), in which the predictive potential of the model is evaluated on data other than that used in the identification and estimation stages of the analysis. While Oreskes *et al.* (1994) dismiss this approach, which they term 'calibration and verification', their criticisms are rather weak and appear to be based on a perception that 'models almost invariably need additional tuning during the verification stage'. While some modellers may be unable to resist the temptation to carry out such additional tuning, so negating the objectivity of the validation exercise, it is a rather odd reason for calling the whole methodology into question. On the contrary, provided it proves practically feasible, there seems no doubt that validation, in the predictive sense it is used here, is an essential pre-requisite for any definition of model efficacy, if not validity in a wider sense.

It appears normal these days to follow the Popperian view of validation (Popper, 1959) and consider it as a continuing process of falsification. Here, it is assumed that scientific theories (models in the present context) can never be proven universally true; rather, they are not yet proven to be false. This yields a model that can be considered as 'conditionally valid', in the sense that it can be assumed to represent the best theory of behaviour currently available that has not yet been falsified. Thus, conditional validation means that the model has proven valid in this more narrow predictive sense. In the rainfall-flow context considered later, for example, it implies that, on the basis of the new measurements of the model input (rainfall) from the validation data set, the model produces flow predictions that are acceptable within the uncertainty bounds associated with the model.

Note this stress on the question of the inherent uncertainty in the estimated model: one advantage of statistical estimation, of the kind considered in this chapter, is that the level of uncertainty associated with the model parameters and the stochastic inputs is quantified in the time series analysis. Consequently, the modeller should not be looking for perfect predictability (which no-one expects anyway) but predictability which is consistent with the quantified uncertainty associated with the model.

It must be emphasized, of course, that conditional validation is simply a useful statistical diagnostic which ensures that the model has certain desirable properties. It is not a panacea and it certainly does not prove the complete validity of the model if, by this term, we mean the establishment of the 'truth' (Oreskes *et al.*, 1994). Models are, at best, approximations of reality designed for some specific objective; and conditional validation merely shows that this approximation is satisfactory in this limited predictive sense (see also Beven 2002). In many environmental applications, however, such validation is sufficient to establish the credibility of the model and to justify its use in operational control, management and planning studies (see also Chapter 1).

22.4 DATA-BASED MECHANISTIC (DBM) MODELLING

The term 'data-based mechanistic modelling' was first used in Young and Lees (1993) but the basic concepts of this approach to modelling dynamic systems have developed over many years. It was first applied within a hydrological context in the early 1970s, with application to modelling water quality in rivers (Beck and Young, 1975) and rainfall-flow processes (Young, 1974; Whitehead and Young, 1975). Indeed, the DBM water quality and rainfall-flow models discussed later in the present chapter are a direct development of these early models.

In DBM modelling, the most parametrically efficient (parsimonious) model structure is first inferred statistically from the available time series data in an *inductive* manner, based on a generic class of black-box models (normally linear or nonlinear differential equations or their difference equation equivalents). *After this initial black-box modelling stage is complete*, the model is interpreted in a physically meaningful, mechanistic manner based on the nature of the system under study and the physical, chemical, biological or socio-economic laws that are most likely to control its behaviour. By delaying the mechanistic interpretation of the model in this manner, the DBM modeller avoids the temptation to attach too much importance to prior, subjective judgement when formulating the model equations. This inductive approach can be contrasted with the alternative *hypothetico-deductive* 'grey-box' modelling, approach, where the physically meaningful but simple model structure is based on prior, physically based and possibly subjective assumptions, with the parameters that characterize this simplified structure estimated from data only *after* this structure has been specified by the modeller.

Other previous publications, as cited in Young (1998), map the evolution of the DBM philosophy and its methodological underpinning in considerable detail, and so it will suffice here to merely outline the main aspects of the approach:

1. The important first step is to define the objectives of the modelling exercise and to consider the type of model that is most appropriate to meeting these objectives. Since DBM modelling requires adequate data if it is to be completely successful, this stage also includes considerations of scale and the data availability at this scale, particularly as they relate to the defined modelling objectives. However, the

prior assumptions about the form and structure of this model are kept at a minimum in order to avoid the prejudicial imposition of untested perceptions about the nature and complexity of the model needed to meet the defined objectives.

2. Appropriate model structures are identified by a process of objective statistical inference applied directly to the time-series data and based initially on a given generic class of linear Transfer Function (TF) models whose parameters are allowed to vary over time, if this seems necessary to satisfactorily explain the data.

3. If the model is identified as predominantly linear or piece-wise linear, then the constant parameters that characterize the identified model structure in step 2 are estimated using advanced methods of statistical estimation for dynamic systems. The methods used in the present chapter are based on optimal Instrumental Variable (IV) estimation algorithms (see Young, 1984) that provide a robust approach to model identification and estimation and have been well tested in practical applications over many years. Here the important *identification* stage means the application of objective statistical methods to determine the dynamic model order and structure. Full details of these time series methods are provided in the above references and they are outlined more briefly in both Young and Beven (1994) and Young *et al.* (1996).

4. If significant parameter variation is detected over the observation interval, then the model parameters are estimated by the application of an approach to time dependent parameter estimation based on the application of recursive Fixed Interval Smoothing (FIS) algorithms (e.g. Bryson and Ho, 1969; Young, 1984; Norton, 1986). Such parameter variation will tend to reflect nonstationary and nonlinear aspects of the observed system behaviour. In effect, the FIS algorithm provides a method of nonparametric estimation, with the *Time Variable Parameter* (TVP) estimates defining the nonparametric relationship, which can often be interpreted in *State-Dependent Parameter* (SDP) terms (see section 22.6.2).

5. If nonlinear phenomena have been detected and identified in stage 4, the nonparametric state dependent relationships are normally parameterized in a finite form and the resulting nonlinear model is estimated using some form of numerical optimization, such as nonlinear least squares or Maximum Likelihood (ML) optimization.

6. Regardless of whether the model is identified and estimated in linear or nonlinear form, it is only accepted as a credible representation of the system if, in addition to explaining the data well, it also provides a description that has direct relevance to the physical reality of the system under study. This is a most important aspect of DBM modelling and differentiates it from more classical 'black-box' modelling methodologies, such as those associated with standard TF, nonlinear autoregressive-moving average-exogenous variables (NARMAX), neural network and neuro-fuzzy models.

7. Finally, the estimated model is tested in various ways to ensure that it is conditionally valid. This can involve standard statistical diagnostic tests for stochastic, dynamic models, including analysis which ensures that the nonlinear effects have been modelled adequately (e.g. Billings and Voon, 1986). It also involves validation exercises, as well as exercises in stochastic uncertainty and sensitivity analysis (see section 22.6).

Of course, while step 6 should ensure that the model equations have an acceptable physical interpretation, it does not guarantee that this interpretation will necessarily conform exactly with the current scientific paradigms. Indeed, one of the most exciting, albeit controversial, aspects of DBM models is that they can tend to question such paradigms. For example, DBM methods have been applied very successfully to the characterization of imperfect mixing in fluid flow processes and, in the case of pollutant transport in rivers, have led to the development of the *Aggregated Dead Zone* (ADZ) model (Beer and Young, 1983; Wallis *et al.*, 1989). Despite its initially unusual physical interpretation, the acceptance of this ADZ model (e.g. Davis and Atkinson, 2000 and the prior references therein) and its formulation in terms of physically meaningful parameters, seriously questions certain aspects of the ubiquitous Advection Dispersion Model (ADE) which preceded it as the most credible theory of pollutant transport in stream channels (see the comparative discussion in Young and Wallis, 1994). An example of ADZ modelling is described in section 22.6.1.

One aspect of the above DBM approach which differentiates it from alternative deterministic 'top-down' approaches (e.g. Jothityangkoon *et al.*, 2000) is its inherently stochastic nature. This means that the uncertainty in the estimated model is always quantified and this information can then be utilized in various ways. For instance, it allows for the application of uncertainty and sensitivity analysis based on *Monte Carlo Simulation* (MCS) analysis, as well as the use of the model in statistical forecasting and data assimilation algorithms, such as recursive parameter estimation and the Kalman filter.

The uncertainty analysis is particularly useful because it is able to evaluate how the covariance properties of the parameter estimates affect the probability distributions of physically meaningful, derived parameters, such as residence times and partition percentages in parallel hydrological pathways (see e.g. Young, 1992, 1999a and the examples below).

The DBM approach to modelling is widely applicable: It has been applied successfully to the characterization of numerous environmental systems including the development of the ADZ model for pollution transport and dispersion in rivers (e.g. Wallis *et al.*, 1989; Young, 1992); rainfall-flow modelling (Young and Beven, 1994; Young, 2001b and the prior references therein); adaptive flood forecasting and warning (Lees *et al.*, 1994; Young and Tomlin, 2000; Young, 2002a); and the modelling of ecological and biological systems (Young, 2000; Jarvis *et al.*, 1999); Other applications, in which the DBM models are subsequently utilized for control system design, include the modelling and control of climate in glasshouses (e.g. Lees *et al.*, 1996), forced ventilation in agricultural buildings (e.g. Price *et al.*, 1999), and inter-urban road traffic systems (Taylor *et al.*, 1998). They have also been applied in the context of macro-economic modelling (e.g. Young and Pedregal, 1999).

22.5 THE STATISTICAL TOOLS OF DBM MODELLING

The statistical and other tools that underpin DBM modelling are dominated by recursive methods of time series analysis, filtering and smoothing. These include optimal *Instrumental Variable* (IV) methods of identifying and estimating discrete and continuous-time TF models (e.g. Young, 1984, 2002b), including the *Simplified Refined Instrumental Variable* (SRIV) algorithm used in later practical examples; *Time Variable Parameter* (TVP) estimation and its use in the modelling and forecasting of nonstationary time series (e.g. Young, 1999b, and the prior references therein); and *State Dependent Parameter* (SDP) parameter estimation methods for modelling nonlinear stochastic systems (see Young, 1978, 1984, 1993a, 1998, 2000, 2001a; Young and Beven, 1994). Here, the TVP and SDP estimation is based on optimized *Fixed Interval Smoothing* (FIS) algorithms.

These recursive statistical methods can also be utilized for other environmental purposes. For example, as discussed in section 22.7, they can provide a rigorous approach to the evaluation and exploitation of large simulation models (e.g. Young *et al.*, 1996), where the analysis provides a means of simplifying the models.

Such *reduced order* representations can then provide a better understanding of the most important mechanisms within the model; or they can provide *dominant mode* models that can be used for control and operational management system design, adaptive forecasting, or data assimilation purposes (see e.g. Young *et al.*, 1996; Young, 1999a). The DBM data analysis tools can also be used for the statistical analysis of nonstationary data, of the kind encountered in many areas of environmental science (see Young, 1999b). For example, they have been applied to the analysis and forecasting of trends and seasonality in climate data (e.g. Young *et al.*, 1991); and the analysis of palaeoclimatic data (Young and Pedregal, 1998).

22.6 PRACTICAL EXAMPLES

Two practical examples will be considered here, both concerned with hydrological systems. The first will show how even a purely linear representation, in the form of the ADZ model for solute transport and dispersion, can provide a powerful approach to analysing experimental data. However, many environmental systems are nonlinear and so the second example will show how SDP modelling procedures can be exploited to handle such nonlinearity.

22.6.1 A linear example: modelling solute transport

Tracer experiments[2] are an excellent way of evaluating how a river transports and disperses a dissolved, conservative pollutant. Figure 22.1 shows a typical set of tracer data from the River Conder, near Lancaster in North West England. This river is fairly small, with a cobbled bed, and the experiment involved the injection of 199 mg of the dye tracer Rhodamine WT, with the measurement locations situated 400 metres apart, some way downstream of the injection location to allow for initial mixing. The river flow rate was measured at $1.3\,m^3.s^{-1}$.

Since the ADZ model was first introduced by Beer and Young (1983), it has become conventional to estimate the model in discrete-time TF form (since equivalent continuous-time estimation algorithms are not freely available) and then deduce the continuous-time (differential equation) model parameters from the estimated parameters of this discrete-time TF. In this example, however, discrete-time modelling is not very successful when applied to the data in Figure 22.1 using a relatively fast sampling interval of 0.25 minutes, both the discrete-time SRIV algorithm and the alternative Prediction Error Minimisation (PEM) algorithm in Matlab™ (Matlab is a well known and widely available numeric computation

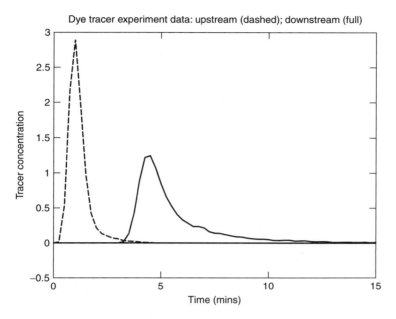

Figure 22.1 Dye-tracer experiment data for the River Conder, North West England: measured input, upstream concentrations (dashed line); measured output, downstream concentrations (full line)

and visualization software package) yield second-order models which do not explain the data very well. Moreover, while these algorithms produce well fitting third order models, these are clearly overparameterized and have complex roots, so that the models can be rejected on DBM grounds since they have no obvious physical interpretation.

Continuous-time SRIV modelling (e.g. Young, 1984, 2002b) of the tracer data is much more successful. This suggests strongly that the dynamic relationship between the measured concentrations at the input (upstream) and at the output (downstream) measurement sites is linear and second order, with the continuous-time TF identified by the continuous-time SRIV algorithm in the following form:

$$y(t) = \frac{b_o s + b_1}{s^2 + a_1 s + a_2} u(t-3) + \xi(t) \qquad (22.1)$$

where s is the Laplace Transform or derivative operator; i.e. $s^i y(t) = d^i y(t)/dt^i$. In ordinary differential equation terms, this model can be written as:

$$\frac{d^2 y(t)}{dt^2} + a_1 \frac{dy(t)}{dt} + a_2 y(t) = b_0 \frac{du(t-3)}{dt}$$
$$+ b_1 u(t-3) + \eta(t) \qquad (22.2)$$

where $\eta(t) = (s^2 + a_1 s + a_2)\xi(t)$. Here, time is measured in minutes and the pure time delay of 3 minutes on

the input variable, $u(t-3)$, is the purely advective plug flow effect. Although there is a little serial correlation but some heteroscedasticity in the estimated residuals $\xi(t)$, the variance is extremely low (0.0001), as reflected in the very high coefficient of determination based on these modelling errors of $R_T^2 = 0.9984$ (i.e. 99.84% of the $y(t)$ variance is explained by the input-output part of the model).[3] The noise model for $\xi(t)$ is estimated as a third order, continuous-time autoregressive moving average noise process with a zero mean, serially uncorrelated white noise source $e(t)$ of very low variance σ^2. In this situation, however, the noise on the output is so small that it can be ignored in the subsequent uncertainty and sensitivity analysis.

With such a high R_T^2, the model 22.1–22.2 obviously explains the data very well, as shown in Figure 22.2 which compares the deterministic (noise free) model output:

$$\hat{x}(t) = \frac{\hat{b}_o s + \hat{b}_1}{s^2 + \hat{a}_1 s + \hat{a}_2} u(t-3) \qquad (22.3)$$

with the measured tracer concentrations $y(t)$. The estimated parameters are as follows:

$$\hat{a}_1 = 2.051(0.073); \quad \hat{a}_2 = 0.603(0.055);$$
$$\hat{b}_0 = 1.194(0.014); \quad \hat{b}_1 = 0.642(0.056)$$

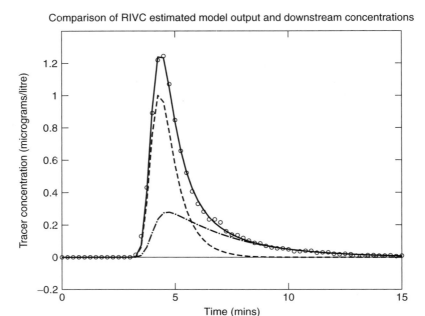

Comparison of RIVC estimated model output and downstream concentrations

Figure 22.2   Comparison of the SRIV identified and estimated ADZ model output $\hat{x}(t)$ and the measured concentration of tracer $y(t)$ at the downstream location. Also shown: inferred quick pathway (dashed) and slow pathway (dash-dot) concentration profiles

where the figures in parentheses are the estimated standard errors. The associated covariance matrix, which provides the basis for the subsequent MCS analysis, takes the form:

$$\mathbf{P}(\tilde{\mathbf{a}}) = \begin{bmatrix} 0.0054 & 0.0039 & -0.0008 & -0.0040 \\ 0.0039 & 0.0030 & -0.0005 & -0.0031 \\ -0.0008 & -0.0005 & 0.0002 & 0.0005 \\ -0.0040 & -0.0031 & 0.0005 & 0.0031 \end{bmatrix}$$

(22.4)

Introducing the estimated parameter values, the TF model 22.3 can be decomposed by partial fraction expansion into the following form:

$$\hat{x}(t) = \frac{0.6083}{1 + 0.5899s}u(t-3) + \frac{0.4753}{1 + 2.8118s}u(t-3)$$

(22.5)

which reveals that the model can be considered as a parallel connection of two first order processes which appear to characterize distinctive solute pathways in the system with quite different residence times: one 'quick', with a residence time $T_q = 0.5899$ minutes; and the other 'slow', with a residence time $T_s = 2.8188$ minutes. The associated steady state gains (e.g. Young, 1984), as obtained by setting $s = d/dt = 0$ in the two first-order TFs of 22.5, are $G_q = 0.6083$ and $G_s = 0.4753$,

respectively. These suggest a parallel partitioning of tracer with a partition percentage of $P_q = 57.1\%$ for the quick pathway, and $P_s = 42.9\%$ for the slow pathway.

The decomposition of the TF into the parallel pathway form 22.5, provides the information required to interpret the model in a physically meaningful manner, as required by DBM modelling. The first order model associated with each pathway can be considered as a differential equation describing mass conservation, as shown in Young (1999a). If it is assumed that the flow is partitioned in the same way as the dye, then the *Active Mixing Volume* (AMV: see Young and Less, 1993) of water associated with the dispersion of the solute in each pathway can be evaluated, by reference to this equation, the flow rate and the residence times. This yields a quick pathway AMV, $V_q = 26.3m^3$; and a slow pathway AMV, $V_s = 94.1m^3$, respectively. The associated *Dispersive Fraction* (DF), in each case, is calculated as the ratio of the AMV and the total volume of water in the reach, giving $DF_q = 0.12$ and $DF_s = 0.56$ (i.e. the acting mixing volumes are 12% and 56% of the total volume of water in each pathway, respectively). In other words, the slow pathway results in a considerably greater dispersion (and longer-term detention) of the dye than the quick pathway, as one might expect.

Given this quantitative analysis of the model 22.5, the most obvious physical interpretation of the parallel flow decomposition is a form of two-layer flow, with the slow pathway representing the dye in the water moving in and adjacent to the cobbled bed and banks of the river, which is being differentially delayed in relation to the quick pathway, which is associated with the more freely moving surface layers of water. The aggregated effect of each pathway is then an advective transportation delay of 3 minutes, associated with nondispersive plug flow; and an ADZ, defined by the associated AMVs and DFs in each case, which are the main mechanisms for dispersion of the dye in its passage down the river.

This parallel partitioning of the flow and solute also helps to explain the shape of the experimentally measured concentration profile. The individual concentration profiles for the quick and slow pathways, as inferred from the parallel partitioning, are shown as dashed and dash-dot curves, respectively, in Figure 22.2. It is clear from these plots that the quick pathway explains the initial sharp rise in the profile and its contribution is virtually complete after about 5 minutes. In contrast, the most significant contribution of the slow pathway is its almost total explanation of the long, raised tail on the profile, over the

last 10 minutes of the experiment. It is this ability of the ADZ model to explain the combination of the rapid initial rise and this long tail effect which most differentiates it from the classical advection-dispersion equation model (see Young and Wallis 1994; Lees *et al.*, 2000). This makes the ADZ explanation more appropriate for the modelling of the imperfect mixing phenomena that appear to dominate pollutant transport in *real* stream channels.

The above interpretation of the DBM modelling results is rather satisfying. But, as the first author has pointed out previously (Young, 1992, 1999a), calculations and inferences such as those above must be treated with care: not only is the physical interpretation conjectural until confirmed by more experiments, but also the effects of the estimated uncertainty in the stochastic model should always be taken into account. For example, MCS analysis, based on the estimated model parameter covariance matrix $\mathbf{P}(\hat{\mathbf{a}})$ in Equation 22.4, can be used to assess the sensitivity of the derived, physically meaningful model parameters to parametric uncertainty. For example, Figure 22.3 shows the normalized empirical probability distributions (histograms) of the AMVs obtained from MCS analysis based on 3000 random realizations (the complete MCS results are reported in Young, 1999a). The most obvious aspect

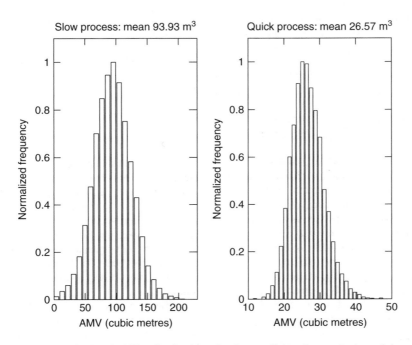

Figure 22.3 Normalized empirical probability distributions for the parallel pathway. Active mixing volumes obtained from 3000 MCS realizations

of these distributions is that, whereas the fast pathway AMV is quite tightly defined, despite the uncertainty on the estimated TF model parameters, the slow pathway AMV is rather poorly defined. In particular, while V_q is distributed quite tightly around the mean value of $26.56m^3$ (SRIV estimated value $26.27m^3$), with a standard deviation of $4.76m^3$, V_s has a much broader distribution about its mean value of $94.41m^3$ (SRIV estimated value $94.12m^3$), with a standard deviation of $32.29m^3$.

In the case of the other physically meaningful model parameters, the uncertainty levels on the partitioning percentages are quite similar for both pathways and there is some evidence of skewness in the distributions, with P_q skewed towards higher percentages and P_s towards lower. The standard deviation on the steady state gain for the slow pathway (0.2129) is over 10 times the standard deviation on the quick pathway (0.0209), showing that it is much less well defined; and it is interesting to note that the total steady state gain for the complete TF has a mean of 1.085 (SRIV estimate 1.066), with a standard deviation of 0.198, showing that it is insignificantly different from unity. This confirms that the tracer experiment was satisfactory, since the steady state gain is an estimate of the ratio of the area under the downstream concentration profile to that under the upstream profile; areas which should be the same if there are no dye losses (the computed ratio of areas is 1.0737). If this steady state gain was significantly different from unity, therefore, it would throw doubt on the efficacy of the experiment: for example, it would suggest that there had been significant adsorption on to organic material or that the pumps which supplied the river water to the fluorometers were poorly located and did not provide water which had a representative measure of the dye concentration at the measurement site.

It is clear that the kind of MCS analysis illustrated in Figure 22.3 provides valuable additional insight into the nature of parallel pathway models such as 22.5. This is also true for the related parallel pathway models used in rainfall-flow analysis (see next subsection), particularly when the parallel flow partitioning is to be used as a basis for exercises such as base-flow separation (see Jakeman *et al.*, 1990; Young 1992). But MCS analysis should not be limited to these kind of water resource applications: it is a generally useful and easy-to-use tool in data-based environmental systems analysis and so should be considered as a standard procedure that is essential for evaluating the efficacy of data-based environmental models and their sensitivity to uncertainty. It is certainly considered *sine qua non* for

success in DBM modelling. And, as shown in Young (1998, 1999a) and pointed out later in section 22.7, it is also valuable in the evaluation of large, deterministic simulation models.

22.6.2 A nonlinear example: rainfall-flow modelling

This example is concerned with the analysis of daily rainfall, flow and temperature data from the 'ephemeral' Canning River in Western Australia which stops flowing over Summer, as shown in Figure 22.4. These data have been analysed before and reported fully in Young *et al.* (1997). The results of this previous analysis are outlined briefly below but most attention is focused on more recent analysis that shows how the inductive DBM modelling can help to develop and enhance alternative conceptual ('grey-box') modelling that has been carried out previously in a more conventional hypothetico-deductive manner.

Young *et al.* (1997) show that, in this example, the most appropriate generic model form is the nonlinear SDP model (see section 22.5). Analysis of the rainfall-flow data in Figure 22.4, based on this type of model, is accomplished in two stages. First, nonparametric estimates of the SDPs are obtained using the *State Dependent parameter Auto-Regressive eXogenous Variable* (SDARX) model form (see Young, 2001a, b in which it is discussed at some length within a rainfall-flow context):

$$y_t = \mathbf{z}_t^T \mathbf{p}_t + e_t \qquad e_t = N(0, \sigma^2) \qquad (22.6)$$

where:

$$\mathbf{z}_t^T = [y_{t-1} \ y_{t-2} \ \dots \ y_{t-n} \ r_{t-\delta} \dots r_{t-\delta-m}]$$

$$\mathbf{p}_t = [a_1(z_t) \ a_2(z_t) \dots a_n(z_t) \ b_0(z_t) \dots b_m(z_t)]^T$$

where, in the present context, y_t and r_t are, respectively, the measured flow and rainfall and δ is a pure advective time delay. Here, the elements of the triad $[n \ m \ \delta]$ are identified as $n = 2$, $m = 3$, $\delta = 0$ and the parameters are all assumed initially to be dependent on a state variable z_t. In this case, the SDP analysis then shows that the state dependency is apparently in terms of the measured flow variable (i.e. $z_t = y_t$: see later explanation) and is limited to those parameters associated with rainfall r_t.

In the second stage of the analysis, the nonparametric estimate of the nonlinearity is parameterized in the simplest manner possible; in this case as a power law in y_t. The constant parameters of this parameterized nonlinear model are then estimated using a nonlinear

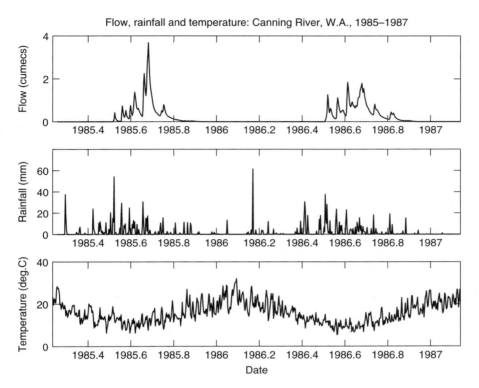

Figure 22.4 Daily rainfall-flow and temperature data for the ephemeral Canning River in Western Australia for the period 23 March 1985 to 26 February 1987

optimization procedure (see Young, 2001b). The resulting model is the following simplified version of the nonlinear, discrete-time, *SDP Transfer Function* (SDTF) model (e.g. Young, 2000):

$$y_t = \frac{\hat{B}(z^{-1})}{\hat{A}(z^{-1})} u_t + \xi_t \qquad (22.7)$$

where z^{-i} is the backward shift operator, i.e. $z^{-i} y_t = y_{t-i}$:

$$\hat{A}(z^{-1}) = 1 - 1.646 z^{-1} + 0.658 z^{-2}$$

$$\hat{B}(z^{-1}) = 0.0115 + 0.0185 z^{-1} - 0.0028 z^{-2}$$

$$u_t = c. y_t^{\hat{\beta}}. r_t \qquad (22.8)$$

with $\hat{\beta} = 0.85$. This shows that the input variable u_t is a nonlinear function in which the measured rainfall r_t is multiplied by the flow raised to a power $\hat{\beta}$, with the normalization parameter c simply chosen so that the steady state gain of the linear TF between u_t and y_t is unity.[4] In other words, the SDP analysis shows, in a

relatively objective manner, that the underlying *dynamics* are predominantly linear but the overall response is made nonlinear because of a very significant input nonlinearity.

This model not only explains the data well ($R_T^2 = 0.958$) it is also consistent with hydrological theory, as required by the tenets of DBM modelling. This suggests that the changing soil-water storage conditions in the catchment reduce the '*effective*' level of the rainfall and that the relationship between the measured rainfall and this effective rainfall (sometimes termed rainfall excess) u_t is quite nonlinear. For example, if the catchment is very dry because little rain has fallen for some time, then most new rainfall will be absorbed by the dry soil and little, if any, will be effective in promoting increases in river flow. Subsequently, however, if the soil-water storage increases because of further rainfall, so the '*runoff*' of excess water from the catchment rises and the flow increases because of this. In this manner, the effect of rainfall on flow depends upon the antecedent conditions in the catchment and a similar rainfall event occurring at different times and under different soil-water storage conditions can yield markedly different changes in river flow.

The linear TF part of the model conforms also with the classical 'unit hydrograph' theory of rainfall-flow dynamics: indeed, its unit impulse response at any time is, by definition, the unit hydrograph. And the TF model itself can be seen as a parametrically efficient method of quantifying this unit hydrograph. Additionally, as in the solute-transport example, the TF model can be decomposed by partial fraction expansion into a parallel pathway form which has a clear hydrological interpretation. In particular, it suggests that the effective rainfall is partitioned into three pathways: the instantaneous effect, arising from the [2 3 0] TF model form which, as might be expected, accounts for only a small 5.8% of the flow; a fast flow pathway with a residence time of 2.65 days which accounts for the largest 53.9% of the flow; and a slow flow pathway of 19.86 days residence time accounting for the remaining 40.3% of the flow. It is this latter pathway that leads to an extended tail on the associated hydrograph and can be associated with the slowly changing baseflow in the river (for a more detailed explanation and other examples, see Young, 1992, 1993a, 1998, 2001b; Young and Beven, 1994; Young *et al.*, 1997).

The most paradoxical and, at first sight, least interpretable model characteristic is that the effective rainfall nonlinearity is a function of flow. Although this is physically impossible, the analysis produces such a clearly defined relationship of this sort that it must have some physical connotations. The most hydrologically reasonable explanation is that the flow is acting as a surrogate for soil-water storage. Of course, it would be better to investigate this relationship directly by measuring the soil-water storage in some manner and incorporating these measurements into the SDP analysis. Unfortunately, it is much more difficult to obtain such 'soil moisture' measures and these were not available in the present example.

The temperature measurements are available, however, and this suggests that we should explore the model 22.7–22.8 further, with the object of enhancing its physical interpretation using these additional data.

Two interesting conceptual ('grey-box') models of rainfall-flow dynamics are the Bedford-Ouse River model (e.g. Whitehead and Young, 1975); and a development of this, the IHACRES model (Jakeman *et al.*, 1990). Both of these '*Hybrid-Metric-Conceptual*' (HCM) models (Wheater *et al.*, 1993) have the same basic form as 22.7–22.8, except that the nature of the effective rainfall nonlinearity is somewhat different. In the case of the IHACRES model, for instance, this nonlinearity is modelled by the following equations:

$$\tau_s(T_t) = \tau_s e^{\frac{\overline{T}_t - T_t}{f}} \tag{22.9a}$$

$$s_t = s_{t-1} + \frac{1}{\tau_s(T_t)}(r_t - s_{t-1}) \tag{22.9b}$$

$$u_t = c.s_t^\beta.r_t \tag{22.9c}$$

where T_t is the temperature; $\overline{T}_t$ is the mean temperature; s_t represents a conceptual soil-water storage variable; and c, τ_s, f and β are *a priori* unknown parameters. Comparing 22.9c with 22.8, we see that the main difference between the two models is that the measured y_t in 22.8, acting as a surrogate for soil-water storage, has been replaced by a modelled (or latent) soil-water storage variable s_t. The model 22.9b that generates this variable is a first order discrete-time storage equation with a residence time $\tau_s(T_t)$ defined as τ_s multiplied by an exponential function of the difference between the temperature T_t and its mean value $\overline{T}_t$, as defined in 22.9a.

In the original IHACRES model (e.g. Jakeman and Hornberger, 1993), $\overline{T}_t$ is normally set at 20°C, but the estimation results are not sensitive to this value. Also, s_t is not raised to a power, as in 22.9c. Some later versions of IHACRES have incorporated this parameter, but it has been added here so that the two nonlinearities in 22.8 and 22.9c can be compared. More importantly, its introduction is practically important in this particular example since, without modification, the IHACRES model (Figure 22.5) is not able to model the ephemeral Canning flow very well.

Using a constrained nonlinear optimization procedure similar to that in the previous example, the parameters in this modified IHACRES model are estimated as follows:

$$\hat{A}(z^{-1}) = 1 - 1.737z^{-1} + 0.745z^{-2}$$
$$\hat{B}(z^{-1}) = 0.0285 + 0.140z^{-1} - 0.160z^{-2} \tag{22.10}$$
$$\hat{\tau}_s = 65.5, \hat{f} = 30.1; \hat{\beta} = 6.1; \overline{T}_t = 15.9$$

These parameters are all statistically well defined and the model explains 96.9% of the flow y_t ($R_T^2 = 0.969$), marginally better than the DBM model. Moreover, as shown in Figure 22.5, it performs well in validation terms when applied, without re-estimation, to the data for the years 1977–78. The coefficient of determination in this case is 0.91 which is again better than that achieved by the DBM model (0.88). However, when validated against the 1978–79 data, the positions are reversed and the DBM model is superior. Overall, therefore, the two models are comparable in their ability to explain and predict the Canning River data.

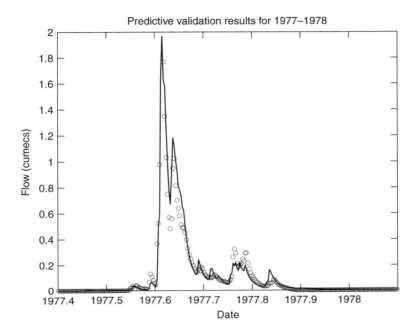

Figure 22.5   Modified IHACRES model validation results on 1977–78 data

Figure 22.6 compares the measured rainfall r_t (upper panel) with the effective rainfall (middle panel), as computed by Equation 22.9c. It is clear that, not surprisingly, the nonlinear transformation produced by Equations 22.9a–22.9c has a marked effect: in particular, as shown in Figure 22.7, the power law transformation in 22.9c, with $\beta = 6.1$ considerably modifies the soil-water storage s_t, effectively reducing it to zero, in relative terms, over the Summer period, as required. The reason why the modified IHACRES and DBM models perform similarly becomes clear if we compare the normalized (since they differ by a scale factor) effective rainfall variables for both models, as shown in the lower panel of Figure 22.6 (modified IHACRES, full line; DBM, dashed line). The similarity between these variables is obvious and the normalized impulse responses (unit hydrographs) of the models are also closely comparable.

Finally, it must be emphasized that this example is purely illustrative and it is not suggested that the modified IHACRES model identified here cannot be improved upon by the introduction of some alternative nonlinear mechanism. For instance, the estimation of a power law nonlinearity with such a large power of 6.1 seems a rather odd and dangerous way to handle this type of nonlinearity, although the reader will see that it is very effective in this case. Nevertheless, the example illustrates well how DBM modelling can, in a reasonably objective manner, reveal the *nature* of the nonlinearity required to model the data well and then seek out a physically meaningful parameterization that achieves this. In this example, it clearly demonstrates that the standard IHACRES model nonlinearity cannot model the data well enough unless it is modified in some manner. Of course, the power law nonlinearity is not the only, and definitely not the best, way of achieving this. For example, Ye *et al.* (1997) introduce a threshold-type nonlinearity on s_t, which makes good physical sense, and obtain reasonable results but with R_T^2 values significantly less than those obtained with the above model (around 0.88–0.89 for estimation and 0.82–0.88 for validation). Clearly more research is required on the characterization of the effective rainfall-flow nonlinearity in models such as these.

22.7 THE EVALUATION OF LARGE DETERMINISTIC SIMULATION MODELS

In this chapter, we have concentrated on data-based modelling and analysis. However, as we have pointed out, many environmental scientists and engineers, including ourselves, use more speculative simulation models of the deterministic-reductionist kind. Although we

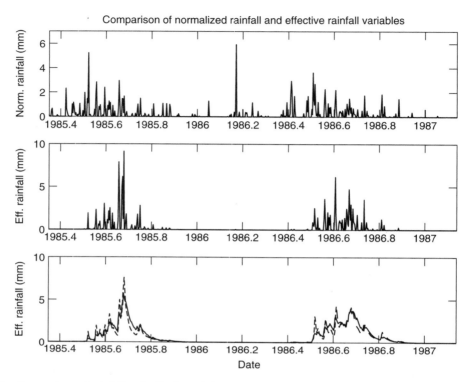

Figure 22.6   Comparison of rainfall and effective rainfall measures

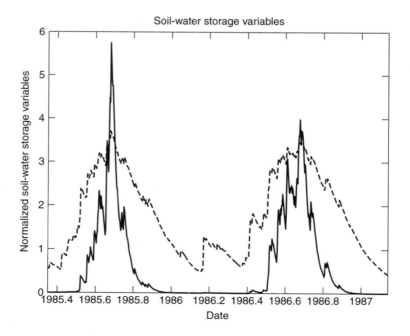

Figure 22.7   Comparison of the estimated soil-water storage variable s_t (dashed line) and s_t^β (full line)

would not advocate the use of such models if adequate experimental or monitored data are available, they can provide a very good method of extending our 'mental models' of environmental systems, often as a valuable prelude to the design of experimental and monitoring exercises or, more questionably, as an aid in operational control, management and planning exercises.

When we exploit speculative simulation models in these ways, however, we must ensure that their construction and use is preceded by considerable critical evaluation. One advantage of the DBM approach to modelling is that its methodological tools can also be exploited in such evaluation exercises. For instance, in Young *et al.* (1996), Young (1998, 1999a) and Young and Parkinson (2002), a technique known as *Dominant Mode Analysis* (DMA), which employs the same optimal IV methods of model identification and estimation used for DBM modelling, is able to identify simple, reduced order representations of a 26th order, nonlinear global carbon cycle dynamic model used in climate research. In particular, DMA shows that over the whole historical period, from the start of the Industrial Revolution to the end of the twentieth century, the output of this model can be reproduced to a remarkably accurate degree by a fourth-order, *linear* differential equation comprising the following third-order equation,

$$\frac{d^3y(t)}{dt^3} + 0.284\frac{d^2y(t)}{dt^2} + 0.0126\frac{dy(t)}{dt} + 0.0000255y(t)$$

$$= 0.255\frac{d^3u(t)}{dt^3} + 0.518\frac{d^2u(t)}{dt^2}$$

$$+ 0.0668\frac{du(t)}{dt} + 0.000815u(t), \qquad (22.11)$$

together with a mass conservation integrator. Furthermore, as in the examples of section 22.6 this third-order equation can be decomposed into a parallel connection of first-order models with residence times of 4.4, 19.1 and 467 years that makes reasonable physical sense. The shorter time constants are not dissimilar to the troposphere−stratosphere exchange time of 6 ± 2 years, and the air−sea exchange time of 11 ± 4 years, respectively, as assumed in the simulation model. And the larger time constant is almost certainly associated with the assumed value of 900 ± 250 years for the deep-ocean turnover time. Naturally, the complex interconnected nature of the simulation model modifies the 'effective' values of these time constants when the model is integrated, and so the identified dominant-mode residence times seem very reasonable.

Model reduction exercises of this kind demonstrate rather dramatically how the superficial complexity of simulation models can, as in the real data examples described in section 22.6, conceal underlying simple dynamics that are the main engine for the observed behaviour of the model in its response to input variations. Such reduced order models can function, therefore, as replacements for the model in those applications, such as forecasting and control, where the internal descriptive aspects of the complex model are of less importance. In other applications, such as 'what-if' studies, environmental planning and risk analysis, however, the reduced order model may not disclose clearly those *physically meaningful* parameters that control the dominant reduced order modes of behaviour and are, therefore, important in such applications. In these situations, it is possible to exploit other DBM tools that are able to identify the most important parameters in the complex simulation model and examine how these affect its dynamic behaviour. This is the domain of uncertainty and sensitivity analysis; analysis that has been revolutionized in recent years by our ability to apply MCS-based methods to complex simulation models (see e.g. Saltelli *et al.*, 2000; Beven *et al*, 2000; Thiemann *et al.*, 2001).

A typical example of such Monte Carlo analysis is described in Parkinson and Young (1998), where MCS and the related technique of *Generalized (or regional) Sensitivity Analysis* (GSA: see Spear and Hornberger, 1980) are used to assess the effect of input and parametric uncertainties (as defined by climate scientists) on the behaviour of the global carbon cycle simulation model mentioned above. This helps to reveal the most important physical parameters in the model and enhances the DMA-based model reduction exercises.

22.8 CONCLUSION

For too long in the environmental sciences, deterministic reductionism has reigned supreme and has had a dominating influence on mathematical modelling in almost all areas of the discipline. In general, such 'bottom-up', reductionist models are over-parameterized in relation to the information content of the experimental data, and their determinism sits uncomfortably with the acknowledged uncertainty that characterizes most environmental systems. This chapter has argued that parsimonious, 'top-down' models provide a more appropriate parameterization in most situations and that the uncertainty which pervades most environmental systems demands an alternative stochastic approach. In particular, stochastic models and statistical modelling procedures provide a means of acknowledging this uncertainty and quantifying its effects. Most often, however, the conventional statistical approach to stochastic model building is posed in a 'black-box' manner that fails to produce models that can be interpreted in physically meaningful terms. The Data-Based Mechanistic (DBM) approach to modelling discussed in this chapter tries to correct these deficiencies. It provides a modelling strategy that not only exploits powerful statistical techniques but also produces models that can be interpreted in physically meaningful terms and are normally more acceptable to environmental scientists and engineers.

Finally, it should be noted that most of the modelling and associated numerical analysis described in this chapter was carried out using the CAPTAIN Toolbox developed in CRES at Lancaster within the Matlab software environment (see http://www.es.lancs.ac.uk/cres/systems.html).

NOTES

1. This chapter is based, in part, on the papers by Young (1999a, 2001c).
2. The interested reader will find a more complex example of ADZ modelling in Young (2001c), where the same approach used here is applied to data from a tracer experiment conducted in a large Florida wetland area.
3. R_T^2 is defined as $R_T^2 = 1 - cov\{y(t) - \hat{x}(t)\}/cov\{y(t)\}$, where $\hat{x}(t)$ is the deterministic model output defined in Equation 22.3. Note that this is sometimes referred to as the 'Nash-Sutcliffe Efficiency' in the hydrological literature. It is not, however, the conventional coefficient of determination which, in classical statistics and time series analysis, is normally based on one step ahead prediction errors, rather than the modelling error, as here: see, e.g. Young (1993b).
4. This is an arbitrary decision in this case. However, if the rainfall and flow are in the same units, then this ensures that the total volume of effective rainfall is the same as the total flow volume.

REFERENCES

Akaike, H. (1974) A new look at statistical model identification, *IEEE Transactions on Automatic Control* **AC19**, 716–722.

Bacon, F. (1620) *Novum Organum*. [See: L. Jardine and M. Silverthorne (eds) (2000) *The New Organon*, Cambridge University Press, Cambridge.]

Beck, M.B. (1983) Uncertainty, system identification and the prediction of water quality, in M.B. Beck and G. Van Straten (eds), *Uncertainty and Forecasting of Water Quality*, Springer-Verlag, Berlin, 3–68.

Beck, M.B. and Young, P.C. (1975) A dynamic model for BOD-DO relationships in a non-tidal stream, *Water Research* **9**, 769–776.

Beer, T. and Young, P.C. (1983) Longitudinal dispersion in natural streams, *Journal of the Environmental Engineering Division, American Society of Civil Engineers* **102**, 1049–1067.

Beven, K.J. (2000) Uniqueness of place and process representations in hydrological modelling, *Hydrology and Earth System Sciences* **4**, 203–213.

Beven, K.J. (2001) How far can we go in distributed hydrological modelling? *Hydrology and Earth System Sciences* **5**, 1–12.

Beven, K.J. (2002) Towards a coherent philosophy for environmental modelling, *Proceedings of the Royal Society of London A* **458**, 2465–2484.

Beven, K.J., Freer, J., Hankin, B. and Schulz, K. (2000) The use of generalized likelihood measures for uncertainty estimation in high order models of dynmaic systems, in W.J. Fitzgerald, A. Walden, R. Smith and P.C. Young (eds) *Nonstationary and Nonlinear Signal Processing*, Cambridge University Press, Cambridge, 115–151.

Billings, S.A. and Voon, W.S.F. (1986) Correlation based model validity tests for nonlinear models, *International Journal of Control* **44**, 235–244.

Box, G.E.P. and Jenkins, G.M. (1970) *Time-Series Analysis: Forecasting and Control*, Holden-Day, San Francisco.

Bryson, A.E. and Ho, Y.-C. (1969) *Applied Optimal Control*, Blaisdell Publishing, Massachusetts.

Davis, P.M. and Atkinson, T.C. (2000) Longitudinal dispersion in natural channels: 3. An aggregated dead zone model applied to the River Severn, U.K., *Hydrology and Earth System Sciences* **4**, 373–381.

Jakeman, A.J. and Hornberger, G.M. (1993) How much complexity is warranted in a rainfall-runoff model? *Water Resources Research* **29**, 2637–2649.

Jakeman, A.J., Littlewood, I.G. and Whitehead, P.G. (1990) Computation of the instantaneous unit hydrograph and identifiable component flows with application to two small upland catchments, *Journal of Hydrology* **117**, 275–300.

Jarvis, A.J., Young, P.C., Taylor, C.J. and Davies, W.J. (1999) An analysis of the dynamic response of stomatal conductance to a reduction in humidity over leaves of *Cedrella odorata*, *Plant Cell and Environment* **22**, 913–924.

Jothityangkoon, C., Sivapalan, M. and Farmer, D.L. (2001) Process controls of water balance variability in a large semi-arid catchment: downward approach to hydrological model development, *Journal of Hydrology* **254**, 174–198.

Konikow, L.F. and Bredehoeft, J.D. (1992) Ground water models cannot be validated, *Advances in Water Resources* **15**, 75–83.

Kuhn, T. (1962) *The Structure of Scientific Revolutions*, University of Chicago Press, Chicago.

Lawton, J. (2001) Understanding and prediction in ecology, Institute of Environmental and Natural Sciences, Lancaster University, Distinguished Scientist Lecture.

Lees, M.J., Camacho, L.A. and Chapra, S. (2000) On the relationship of transient storage and aggregated dead zone models of longitudinal solute transport in streams, *Water Resources Research* **36** (1), 213–224.

Lees, M.J., Taylor, J., Chotai, A., Young, P.C. and Chalabi, Z.S. (1996) Design and implementation of a Proportional-Integral-Plus (PIP) control system for temperature, humidity and carbon dioxide in a glasshouse, *Acta Horticulturae* **406**, 115–123.

Lees, M., Young, P.C., Beven, K.J., Ferguson, S. and Burns, J. (1994) An adaptive flood warning system for the River Nith at Dumfries, in W.R. White and J. Watts (eds) *River Flood Hydraulics*, Institute of Hydrology, Wallingford, 65–75.

Norton, J.P. (1986) *An Introduction to Identification*, Academic Press, London.

Oreskes, N., Shrader-Frechette, K. and Bellitz, K. (1994) Verification, validation, and confirmation of numerical models in the earth sciences, *Science* **263**, 641–646.

Parkinson, S.D. and Young, P.C. (1998) Uncertainty and sensitivity in global carbon cycle modelling, *Climate Research* **9**, 157–174.

Popper, K. (1959) *The Logic of Scientific Discovery*, Hutchinson, London.

Price, L., Young, P.C., Berckmans, D., Janssens, K. and Taylor, J. (1999) Data-based mechanistic modelling and control of mass and energy transfer in agricultural buildings, *Annual Reviews in Control* **23**, 71–82.

Saltelli, A., Chan, K. and Scott, E.M. (2000) *Sensitivity Analysis*, John Wiley & Sons, Chichester.

Silvert, W. (1993) Top-down modelling in ecology, in P.C. Young (ed.) *Concise Encyclopedia of Environmental Systems*, Pergamon Press, Oxford, 605.

Spear, R.C. and Hornberger, G.M. (1980) Eutrophication in Peel Inlet – identification of critical uncertainties via generalized sensitivity analysis, *Water Research* **14**, 43–49.

Taylor, C.J., Young, P.C., Chotai, A. and Whittaker, J. (1998) Nonminimal state space approach to multivariable ramp metering control of motorway bottlenecks, *IEEE Proceedings, Control Theory and Applications* **145**, 568–574.

Thiemann, M., Trosset, M., Gupta, H. and Sorooshian, S. (2001) Bayesian recursive parameter estimation for hydrologic models, *Water Resources Research* **37**, 2521–2535.

Wallis, S.G., Young, P.C. and Beven, K.J. (1989) Experimental investigation of the Aggregated Dead Zone (ADZ) model for longitudinal solute transport in stream channels, *Proceedings of the Institute of Civil Engineers* Part 2, **87**, 1–22.

Wheater, H.S., Jakeman, A.J. and Beven, K.J. (1993) Progress and directions in rainfall-run-off modelling. Chapter 5 in A.J. Jakeman, M.B. Beck and M.J. McAleer (eds) *Modelling Change in Environmental Systems*, John Wiley & Sons, Chichester, 101–132.

Whitehead, P.G. and Young, P.C. (1975) A dynamic-stochastic model for water quality in part of the Bedford-Ouse River system, in G.C. Vansteenkiste (ed.) *Computer Simulation of Water Resources Systems*, North Holland, Amsterdam, 417–438.

Ye, W., Bates, B.C., Viney, N.R., Sivapalan, M. and Jakeman, A.J. (1997) Performance of conceptual rainfall-runoff models in low-yielding ephemeral catchments, *Water Resources Research* **33**, 153–166.

Young, P.C. (1974) Recursive approaches to time-series analysis, *Bulletin: Institute of Mathematics and its Applications* **10**, 209–224.

Young, P.C. (1978) A general theory of modeling for badly defined dynamic systems, in G.C. Vansteenkiste (ed.) *Modeling, Identification and Control in Environmental Systems*, North Holland, Amsterdam, 103–135.

Young, P.C. (1983) The validity and credibility of models for badly defined systems, in M.B. Beck and G. Van Straten (eds), *Uncertainty and Forecasting of Water Quality*, Springer-Verlag, Berlin, 69–100.

Young, P.C. (1984) *Recursive Estimation and Time-Series Analysis*, Springer-Verlag, Berlin.

Young, P.C. (1992) Parallel processes in hydrology and water quality: a unified time series approach, *Journal of the Institution of Water and Environmental Management* **6**, 598–612.

Young, P.C. (1993a) Time variable and state dependent modelling of nonstationary and nonlinear time series, in T. Subba Rao (ed.) *Developments in Time Series Analysis*, Chapman & Hall, London, 374–413.

Young, P.C. (ed.) (1993b) *Concise Encyclopaedia of Environmental Systems*, Pergamon Press, Oxford.

Young, P.C. (1998) Data-based mechanistic modelling of environmental, ecological, economic and engineering systems, *Environmental Modelling and Software* **13**, 105–122.

Young, P.C. (1999a) Data-based mechanistic modelling, generalized sensitivity and dominant mode analysis, *Computer Physics Communications* **115**, 1–17.

Young, P.C. (1999b) Nonstationary time series analysis and forecasting, *Progress in Environmental Science* **1**, 3–48.

Young, P.C. (2000) Stochastic, dynamic modelling and signal processing: time variable and state dependent parameter estimation, in W.J. Fitzgerald, A. Walden, R. Smith and P.C. Young (eds) *Nonstationary and Nonlinear Signal Processing*, Cambridge University Press, Cambridge, 74–114.

Young, P.C. (2001a) The identification and estimation of nonlinear stochastic systems, in A.I. Mees (ed.) *Nonlinear Dynamics and Statistics*, Birkhauser, Boston.

Young, P.C. (2001b) Data-based mechanistic modelling and validation of rainfall-flow processes, in M.G. Anderson and P.D Bates (eds) *Model Validation in Hydrological Science*, John Wiley & Sons, Chichester, 117–161.

Young, P.C. (2001c) Data-based mechanistic modelling of environmental systems, *Keynote paper presented at the first plenary session of the International Federation on Automatic Control (IFAC) Workshop on Environmental Systems, Tokyo, Japan, August 21st, 2001.*

Young, P.C. (2002a) Advances in real-time flood forecasting, *Philosophical Transactions of the Royal Society: Mathematical, Physical and Engineering Sciences* **360**, 1433–1450.

Young, P.C. (2002b) Optimal IV identification and estimation of continuous-time TF models, *Proceedings International Federation on Automatic Control (IFAC) Congress, Barcelona.*

Young, P.C. and Beven, K.J. (1994) Data-based mechanistic modelling and the rainfall-flow nonlinearity, *Environmetrics* **5**, 335–363.

Young, P.C. and Lees, M.J. (1993) The active mixing volume: a new concept in modelling environmental systems, in V. Barnett and K.F. Turkman (eds) *Statistics for the Environment*, John Wiley & Sons, Chichester, 3–43.

Young, P.C. and Parkinson, S.D. (2002) Simplicity out of complexity, in M.B. Beck (ed.) *Environmental Foresight and Models*, Elsevier, Amsterdam, 251–301.

Young, P.C. and Pedregal, D.J. (1998) Recursive and en-bloc approaches to signal extraction, *Journal of Applied Statistics* **26**, 103–128.

Young, P.C. and Pedregal, D.J. (1999) Macro-economic relativity: government spending, private investment and unemployment in the USA 1948–1998, *Journal of Structural Change and Economic Dynamics* **10**, 359–380.

Young, P.C. and Tomlin, C.M. (2000) Data-based mechanistic modelling and adaptive flow forecasting, in M.J. Lees and P. Walsh (eds) *Flood Forecasting: What Does Current Research Offer the Practitioner?*, BHS Occasional Paper No. 12, produced by the Centre for Ecology and Hydrology on behalf of the British Hydrological Society, 26–40.

Young, P.C. and Wallis, S.G. (1994) Solute transport and dispersion in channels, in K.J. Beven and M.J. Kirkby (eds) *Channel Networks*, John Wiley & Sons, Chichester, 129–173.

Young, P.C., Jakeman, A.J. and Post, D.A. (1997) Recent advances in data-based modelling and analysis of hydrological systems, *Water Science Technology* **36**, 99–116.

Young, P.C., Parkinson, S.D. and Lees, M. (1996) Simplicity out of complexity in environmental systems: Occam's Razor revisited, *Journal of Applied Statistics* **23**, 165–210.

Young, P.C., Ng, C.N., Lane, K. and Parker, D. (1991) Recursive forecasting, smoothing and seasonal adjustment of nonstationary environmental data, *Journal of Forecasting* **10**, 57–89.

23

Pointers for the Future

JOHN WAINWRIGHT AND MARK MULLIGAN

23.1 WHAT HAVE WE LEARNED?

The task now returns to us to highlight the simplicity in the complexity that has gone before. Are there ways in which the complexity of environmental systems can be understood, and, if so, what are the tools that are used to evaluate them? As suggested by the title of one of the most commonly used texts on numerical computing (Press *et al.*, 1992; see also Cross and Moscardini, 1985), modelling is as much of an art as a science. A number of discussions in this book (particularly the model-comparison exercises discussed in Chapters 20 and 21) suggest that this is especially the case for environmental models. Models test our conceptualizations of our environment, so it is not surprising perhaps that models do not always (ever?) agree. What we are looking at is how best to *represent* the environment, and 'best' will of course depend on why it is we want to represent the environment at all. In the same ways that artistic representations of the environment may modify the way it looks to tell us more about it (and ourselves) than a simple photographic reproduction could do (Figure 23.1), so too do our models attempt to abstract meaning from the complexity we observe. Many environmental scientists will be used to making schematic sketches of their observations in the field. In a lot of respects, the development of models attempts to take this schematization further, within a more formal framework that provides some means of testing our ideas. Only by an iterative testing of our models – confronting them with as wide a range of different datasets and simulation contexts as possible – can we hope to learn more and provide syntheses of our understanding in which we can have a high degree of confidence. In this section, we provide an overview of some of the ways in which this process might happen and related limitations, drawn on the preceding chapters of the book.

23.1.1 Explanation

As noted by Baird in Chapter 3, there are a number of situations where different explanations of the same phenomena are available. Favis-Mortlock related this situation to the debate on equifinality as defined by Beven (1996; see also Cooke and Reeves, 1976, for an earlier debate based on qualitative modelling). In a sense, these debates relate to ideas of using multiple working hypotheses (Chamberlain, 1890) to evaluate competing ideas. Both Baird and Haraldsson and Sverdrup see modelling as part of a methodology employing Popperian falsification to test between competing ideas. Yet we have seen that in a number of cases, our data are insufficiently strong to allow us to use such an approach (see below). Explanation comes as part of an iterative process where we question both our models and our data (see Wainwright *et al.*, 2000, for examples). In a number of places, Baird and others suggest that there is a 'Nature' or a 'real world' that we are trying to model. In this sense, a lot of ongoing modelling work employs a critical realist methodology (cf. Richards, 1990; Bhaskar, 1997). There is an underlying assumption that there are real features that we attempt to reproduce, structured by processes which we can only observe via their effects. Modelling allows us to close the explanatory loop by providing the link between cause and effect. Some would see this process as tinkering to produce the correct answer (Oreskes *et al.*, 1994). While the naïve calibration of models to produce the 'right' answer

Environmental Modelling: Finding Simplicity in Complexity.   Edited by J. Wainwright and M. Mulligan
© 2004 John Wiley & Sons, Ltd   ISBNs: 0-471-49617-0 (HB); 0-471-49618-9 (PB)

(a)

(b)

Figure 23.1   Comparison of (a) an 'absolute' and (b) an abstract representation of a landscape

against some measured value is certainly problematic in this respect, the use of this criticism is itself questionable, as pointed out by Young, Chotai and Beven. As noted by Haraldsson and Sverdrup, we can learn more when the result is incorrect than when it is correct. Models are always an approximation, and always limited in terms of their 'truth' to the extent that their use does not go beyond the assumptions made in making that approximation. The truth is out there!

In disciplines such as ecology (Osborne) and geomorphology, this explanatory loop enables us to tackle difficult questions relating to the link between process and form (see below). Without this link, most empirical approaches to these disciplines possess very poor explanatory power. But not all models provide the same level of explanatory power – as is suggested here by Lambin. While different models are perhaps more suited to different methodological contexts, we should beware of placing too high a burden on some types of model. This issue poses something of a problem, though, when the results of our modelling pass out of the hands of the modeller and into the policy domain (Engelen, Mulligan). Oreskes *et al.* (1994) have suggested that this problem means that we should place strict limits on the ways models are employed. Models are an important way of communicating our results (see below), but we should be careful to consider that science is a socially situated activity. As we saw in Chapter 12, there are complex levels of behaviour and interaction that control decisions relating to environmental questions. Ultimately, this process becomes a recursive one, where the model results are employed within a wider framework that then controls what sort of research is carried out, and thus the sorts of future models that are produced. The history of climate modelling, as discussed by Harvey, is a very clear example here as is that of catchment hydrological modelling (Mulligan).

But we should remember that the social situation is not simply an external issue. It occurs within science itself and the practice of modelling too. Baird points out issues of scientific fashions, where certain explanations tend to be preferred over others, despite the fact that there is no clear rationale for making one choice over another. Future discoveries and methodologies may mean that either choice is ultimately incorrect, so we should beware of becoming too dogmatic about our explanations, and continue to question current orthodoxies. Major advances in science have tended to develop in this way. Preconceptions, as we pointed out in the Introduction (see also Favis-Mortlock) are always with us; we accept them at our peril! They may relate to disciplinary boundaries that prevent the rapid advancement of

our understanding (Michaelides and Wainwright). There is sufficient commonality across the modelling methodology that is carried out within traditional disciplinary boundaries for us to be able to discuss issues and overcome the limitations posed by such myopia and after all, models can help us to communicate across these very same boundaries, because the language of modelling is common to them all. In environmental modelling there is no such thing as 'not my field', as the criticism of increased specialization as a function of reductionist perspectives in a number of the chapters has illustrated. Communication between different modellers is important, as the discussion by Zhang, Drake and Wainwright of different definitions of the term 'scale' has illustrated.

There is no reason to prefer one form of working over another. Young, Chotai and Beven illustrate the benefits of both hypothetic-deductive and inductive approaches, while Thornes highlights the use of a heuristic framework. All can provide powerful means of reaching conclusions in different contexts. If one mode of working leaves us at an impasse, we should consider whether an alternative might provide a way out.

Most explanations in environmental science are based on a tension between parsimony and generality as noted for fluvial systems by Michaelides and Wainwright and for catchments by Mulligan. As we pointed out in the Introduction, complex systems theory is essentially a rewording of Ockham's razor (there is nothing new under the Sun, as Newton might have said!). Although we might often talk about 'laws of nature', environmental science deals with a higher level of aggregation where fundamental laws are not appropriate. Thus, it is difficult to produce models with a sufficient level of generality to be suitable for all applications (even assuming sufficient computer power were available). In this vein, Michaelides and Wainwright question how easy it is to interpret holistic results. Such a question relates to perceptual problems related to scale as pointed out by Osborne, Perry and Bond, and Zhang, Drake and Wainwright, among others. It is often assumed that we simply need to find the right model components and link them together to tackle this problem. But, as noted by Young, Beven and Chotai, there are different ways of linking them together too. To reach our ultimate explanatory goals, we thus need to provide the means of finding optimal model structures.

23.1.2 Qualitative issues

As we noted above, modelling can often be considered to be as much of an art as a science (consider

Penrose's [1989] discussion of 'inspirational flashes'). Integrated models are considered to be a fundamental way forward of improving our understanding, but as noted by Engelen, their development remains a relatively subjective process (see discussion in Chapter 12 also). The production of such models often throws up a number of ambiguities and inconsistencies. Thus, their development provides another means of furthering our understanding of environmental systems, following an iterative approach, as discussed above. The level to which we can represent the environment depends on our understanding and computational power. But as noted by Engelen, the very process of producing models in this way forces us to refine our ways of thinking about problems and produce tools that assist our thought process. There is a tendency not to question the use of calculators in everyday life (for example, in shop tills or indeed in sophisticated laboratory equipment) – why should there be so much resistance to using models as *appropriate tools* to solving questions of environmental understanding? Without such tools, our explanations are reduced to the level of analogies, as pointed out by Favis-Mortlock. The limitations of such approaches are well understood by every archaeologist, and considered in relation to environmental problems in Meyer *et al.* (1998).

The model-building process often provides a means of collecting information from 'nonscientific' sources about the ways in which specific systems operate. Engelen points to an intermediate stage in integrated analysis where qualitative models can be built up from knowledge acquired from a wide range of sources. Twery also demonstrates how rule-based approaches to modelling can allow the codification of institutional knowledge. Such knowledge is often lost as individuals retire or die (or goes out of disciplinary fashion). The loss of such information often leads to the reinvention of modelling wheels.

In a related sense we should beware of assuming that models provide a totally objective means of tackling problems. Often, there are hidden assumptions in the ways different people approach the modelling process. Wright and Baker discuss this problem in relation to a comparison of different applications of the same model to the same problem (see also the comparison of erosion models in Favis-Mortlock, 1998). Models are sensitive to boundary conditions, discretizations and parameterizations as discussed in Chapter 1, so we should not be surprised at this result. Such comparisons allow us to investigate more robust approaches and the extent to which knowledge and interpretations are embedded within our models.

23.1.3 Reductionism, holism and self-organized systems

Favis-Mortlock notes that 'We are not doomed to ever more complex models!' The coupling together of simple models can provide significant insights, even in modelling the global climate, as noted by Harvey. Reductionist approaches come about because of our difficulties in conceptualizing processes beyond scales relating to human experience, as noted above. But they frequently defeat the object, in that we are interested in explaining the whole. Simple models that can illustrate emergent behaviour are useful exploratory tools, and can illustrate the dominant controls on different parts of the environmental system. They can help explain chaotic and complex behaviour in a way that linearized models cannot, as pointed out by Favis-Mortlock. Perry and Bond, Osborne and Mazzoleni *et al.* demonstrate how the use of individual-based models can provide explanations of emergent behaviour in this respect (see also the discussion in Chapter 12).

Michaelides and Wainwright question whether there are different levels of emergent behaviour and whether or not they form hierarchies that might allow us to simplify the modelling process between different scales. Where different forms of complex system come together, we may need to deal with one as a stochastic variable to assist analysis, as demonstrated by Thornes.

23.1.4 How should we model?

Sverdrup and Haraldsson suggest that if modelling is being carried out to develop our understanding of the environmental system, we should build our own models rather than simply apply a readily available model. The ready-made model may be inappropriate to our specific application, and it may be difficult *a priori* to assess the extent to which this may be so. The opposite viewpoint is expounded by Wright and Baker, who suggest that (at least for complex CFD code), we are better off applying a tried-and-trusted code, perhaps even from a commercial source. Even so, it is important for the underlying concepts to be thoroughly understood to avoid the occurrence of problems further down the line of the modelling process.

An appropriate answer to this question is, as is often the case, something of a compromise between these two extremes. In particular, the purpose of the models is an important consideration, as noted by Perry and Bond. Similarly, Twery notes that in policy applications as well as in others, different questions may be most appropriately answered by different forms of model.

This is clearly a case of horses for courses! While the development of more efficient modelling frameworks and toolkits means that it is increasingly easy to develop our own models, we need to beware of reinventing the wheel. Thornes and Haraldsson and Sverdrup question the wider usefulness of models other than to modellers. This viewpoint is essentially derived from the 'understanding' perspective. As illustrated by Engelen, some models may be designed specifically for use, even if they might lack the most powerful levels of explanation. Haraldsson and Sverdrup define 'survivor' models as those which remain in use after some time by the wider community. It is not necessarily the case that such models are always the best – they may simply be the easiest to apply, or the cheapest, or the ones that fit a particular pattern of explanatory fashion, or the ones that are most prolifically written about – but it may be appropriate to investigate what models others are using before embarking on a new study.

Lambin defines a number of more advanced models as 'game-playing tools' in order to develop further understanding. In some senses, this approach is similar to the heuristic method put forward by Thornes. It is important to retain a sense of fun in our investigations, not least so that our ideas do not become dulled and we fail to see alternatives.

23.1.5 Modelling methodology

We have discussed modelling methodology in detail in Chapter 1, so only provide a few brief points here. Harvey notes that parameterization is a function of the scales at which environmental models operate (see also the discussion above). In one sense, it can be considered as an emergent property of the way a system operates at a smaller spatial and/or temporal scale. Yet parameterization is often paid too scant a regard in the application of models. The sophistication of our models is often much greater than the capacity of our data collection efforts to parameterize them. Similarly, Haraldsson and Sverdrup note that calibration is too often used to force a result that tells us nothing about the system (and everything about the modeller's preconceptions – see also Young, Chotai and Beven). If calibration is employed, there should always be an attempt to assess whether it is reasonable, otherwise the whole exercise is virtually pointless. In their assessment of traditional approaches to slope-stability analysis, Collison and Griffiths emphasize the need for user input into the model, which meant that the model was often simply used to confirm the investigator's initial hunch. Self-finding solutions of the newer approaches reduce the potential for operator-induced error, as discussed above.

Visualization can be an important role of the modelling process, as illustrated by Engelen, Burke, Twery, Mulligan and Wright and Baker. But we should not be misled by the presentation of graphical results that may hide underlying weaknesses in the approach. (Keith Beven often warns to be wary of modellers presenting 3D graphics, because it means they have generally spent more time in producing them than the model results!). However, the communication and understanding of model results are often as important as the results themselves and output sophistication has to grow in line with model sophistication or we will not learn enough about model (and system) behaviour from the process of modelling. As with every aspect of our scientific approach, there should be a transparency in what we do and how we present our methods and results (see Haraldsson and Sverdrup).

23.1.6 Process

Models provide a means of addressing the link between the observable and the underlying cause. The underlying process-form debate is a critical one in ecology (Osborne) and geomorphology. One of the main advantages of the modelling approach is that it allows us to understand the limitations of traditional forms of explanation. Interactions of simple processes lead to the emergence of form. Difficulties in interpretation arise because of inherent nonlinearities due to scale effects in both process and form (e.g. Zhang, Drake and Wainwright). Both process and form possess elements of complexity (Michaelides and Wainwright), and it is not necessarily the case that we need complexity in one to explain complexity in the other.

23.1.7 Modelling in an integrated methodology

Despite isolationist claims (unfortunately from both sides), modelling is not an activity that exists in isolation. Baird notes that field or laboratory work is often seen simply as a means of testing model output – it is not surprising then that modellers are often perceived as aloof and ignorant! As we have noted already, we always take our preconceptions into the modelling process, so we should at least try to make these informed preconceptions. In reality, there is a strong loop between field work that suggests new models, that require new data for testing, that suggest further model developments, and so on. Michaelides and Wainwright also demonstrate that there is an important interaction of physical and numerical models in the same way. Modelling is often promoted because of its relatively low cost (e.g.

Collison and Griffiths, Wright and Baker), and indeed this is a major strength, allowing detailed analysis. But it can only provide a partial solution on its own (e.g. Klemeš, 1986, 1997).

23.1.8 Data issues

The distinction of 'modellers' and 'field workers' can lead to problems with the use of field data for model testing (Baird). Data collected in particular ways may contain hidden effective parameterizations that generally lead to unnecessary calibrations and the propagation of further error through the model system. Error enters the problem throughout of the modelling process, where Quinton notes that our ability to measure properties accurately means problems in terms of parameterization and model testing. Although there is general acceptance that parameters contain uncertainty, it is generally assumed, albeit implicitly, that the data used to test models is without error (but see Young, Chotai and Beven). This assumption is completely illogical! Most measurements themselves are models: a mercury thermometer represents temperature change as the change in volume of a thin tube of mercury, a pyranometer uses the increase in temperature of an assumed perfect absorber (black body) as a surrogate for the incident radiation load. Nearing demonstrates that natural variability in measured properties has significant implications for the best-case scenarios of model tests. We cannot hope to produce more accurate results than the properties we measure or their spatio-temporal variability (see also Quinton). The baby should remain with its bath water!

Complex models have significant data requirements, as illustrated by Nearing for process-based soil-erosion models. Yet there is often a reluctance to fund the work required to collect the necessary data. The low-cost conundrum strikes again! But without detailed spatial measurements and long time series, we will be unable to evaluate model performance beyond simply trivial levels. This lack makes the modelling no more than an inexpensive research pastime that may lead to a better system conceptualization, but is held short of its potential use in contributing to the solution of serious environmental problems. Large datasets are becoming available via remote sensing and GIS integration of existing databases, but as Zhang, Drake and Wainwright point out, there is a need to interpret field and remote sensing measurements appropriately when using them to parameterize or evaluate a model.

Models may require unmeasurable properties, such as the complex three-dimensional structure of the subsurface (see Mulligan and Collison and Griffiths) or

properties which are measurable but not at the 'effective' scale required by the models. All robust models require some form of observational paradox, in that we perturb the system in order to collect data for parameters or testing. The integration of the modelling and field-work programmes can help to reduce the impacts of these perturbations. It could be questioned (Engelen and Chapter 12) whether the development of databases from local informants for integrated models incorporating human behaviour might cause behavioural modifications, and such models always to be one step behind reality. It is also important to support models with sufficient observations. Quinton illustrates the case that some relationships built into models may only be based on a small number of measurements. Without an understanding of this limitation, it is impossible to know where to focus the research effort to provide improvements when models fail.

23.1.8.1 Data and empirical models

Empirically based approaches may be found even in process-based models, as illustrated by Haraldsson and Sverdrup and by Nearing. In a sense, this statement is the same as Harvey's definition of parameterization. Very few environmental models contain no empirical content (e.g. some CFD applications: Wright and Baker), and it is important to be aware of this limitation. Given that process-based modelling is designed to provide an improvement in terms of portability issues (cf. Grayson *et al.*, 1992), this limitation is significant. There will always be some limit to portability, and unless we remember this, we may end up unnecessarily rejecting the process model (rather than the hidden parameterization). Inductive approaches may mean that empirical approaches are very useful for defining the appropriate model structure from data (Young, Beven and Chotai), at least at a particular level of aggregation. Twery also notes that empirical models may provide an adequate level of process representation for certain applications.

23.1.8.2 Data and scale

Zhang, Drake and Wainwright illustrate how 'scale' can be a thorny issue in that different researchers perceive it to be a completely different question. Certainly, dealing with scale is a nontrivial process that requires quite sophisticated analysis, as illustrated by Harvey and Perry and Bond. Scale is implicitly built into all our model representations and into our field measurements. Applications of data measured at one scale to a model that is integrated at another may lead to completely misleading results. Further work is required on how to make measurements at scales that are appropriate, both for parameterization and model evaluation, and how to make the scales of modelling converge with those of measurement.

23.1.9 Modelling and policy

As noted above, Twery points out that models used in an applied context can often be relatively simple because relatively straightforward predictions are required. Supporting this argument, Nearing suggests that different approaches are appropriate in modelling for conservation issues, while Thornes demonstrates how simple concepts can give relative directions of change that can be used in management decisions. Yet there is a contradiction here in that Quinton suggests that environmental managers require 'accuracy' in prediction and Osborne notes that 'policy-makers require definitive answers'. However, Twery also notes that there are problems of dealing with uncertainty in an applied context (as with all other contexts!). Visualization may be an appropriate means of dealing with the latter problem (Engelen), but there are serious issues of how uncertainty is conceptualized by different groups. Is an incorrect but definit(iv)e result better than a result that will be perceived as vague (or evasive) when it is presented with a large margin of error? Certainly, there needs to be more communication about what is possible in terms of prediction (see above), even if there is no clear answer to this question at present.

Engelen demonstrates another need for simplicity in policy-based approaches, in that speed in producing results can have a significant impact on the uptake of a particular model. Simplicity in individual components may lead to models being better able to deal with integrated analysis. However, Quinton cautions against the hidden complexity of many models, in that a GUI may hide a vast database of hidden parameterizations and assumptions. Schneider (1997) highlights the same issue within the context of integrated assessment models (IAMs) for the impact of global climate change. In the multidisciplinary issues of climate change and land use change impacts that have been the mainstay of research in environmental science research of late, it is very difficult to get away from hidden complexity because indicating the whole complexity would disable any capacity for understanding by any other than the model architect.

23.1.10 What is modelling for?

We have elaborated in detail above how modelling is an important component of explanation. It is a means of evaluating our lack of understanding (Haraldsson and Sverdrup, Michaelides and Wainwright). But as noted by Kirkby (1992), models can serve a whole range of different purposes. Among other things, modelling can be used for integration of information from different disciplines (Michaelides and Wainwright), interpolation (Wright and Baker), knowledge storage and retrieval (Engelen), communication (Haraldsson and Sverdrup) and learning (Engelen) and as a digital laboratory. It is a virtual Swiss army knife to complement the one we take into the field!

23.1.11 Moral and ethical questions

Perhaps as a subset of the last question, Osborne points out that modelling can be important for addressing questions of environmental change where direct experimentation would be morally unacceptable. As well as moral grounds, Wright and Baker suggest we can also address the practically unfeasible. Since environmental issues incorporate economic, social, health, welfare, cultural, natural and geographical factors, there will always be important moral questions at stake, and it is thus vital that we adopt an appropriate ethical stance towards the results we produce, and the interpretations we allow to be placed upon them.

23.2 RESEARCH DIRECTIONS

There is a tendency to assume that all modellers follow the same methodology. What should be apparent from the different contributions in this book is that there are many different approaches, and a lot of healthy debate about the ways forward. We provide here a brief overview of developments that may be productive in this light. This list should be read in conjunction with the text of Chapter 1.

- There is a need to define frameworks for the robust analysis of environmental systems that deal with their complex, open character.
- Complex systems analysis may provide one way forward in this respect, but there is a clear need to integrate bottom-up approaches with top-down approaches, at least while computer power is limited.
- Nevertheless, appropriate levels of aggregation of different processes need to be defined, not least because they represent the progress in our level of understanding.
- As part of this question, parameterization may best be understood as an emergent characteristic of a system at different hierarchical levels.

- Techniques must be developed to deal with errors at all stages of the modelling (including data-collection) process. We know that error propagation is important, but lack the means of overcoming it.
- There is a need to encourage re-use of good modelling concepts while ensuring that outdated concepts are not incentivized to remain in circulation when they have 'passed their sell-by date'. Tools need to be developed which open up modelling to a wider audience through the removal of the technical overheads which currently exist in the development of models. This will help break down the barriers between 'modellers' and 'field workers' and ensure a wider understanding of the role and purpose of modelling in environmental science. Generic modelling languages and graphical model-building tools are the first step in this direction.
- Models are a potentially important research output since they can summarize research findings in a much more interactive (and sometimes more useful) way than scientific publications. They will never be the mainstream scientific publications until ways are found to peer review and quality control them. Indeed, it is worrying that many models are so complex that their construction is rarely transparent in the publications which use them. This lack of transparency is not acceptable for laboratory or field methodologies and should not be acceptable for modelling experiments either. The WWW provides the means to distribute even complex model descriptions to the audience of academic journals.

23.3 IS IT POSSIBLE TO FIND SIMPLICITY IN COMPLEXITY?

The straightforward answer to this question is yes. Whether we will ever be happy with the particular answer is another question (probably answered by a resounding no!). Fundamentally, do the possibilities of complex system theory offer us progress in terms of finding more than the sum of the parts of our environmental system? Much work on General Systems Theory from the 1960s and 1970s made similar claims as to providing the 'ultimate answer'. As people became progressively more bogged down in increasingly complicated (if not complex) models, disillusion set in as it was realized that increasing amounts of computer power would only tend to compound problems. As people await sufficient computer power to run cellular or individual-based models with $10^{EXTREMELY\ LARGE\ NUMBER}$ of cells/individuals, will we see history repeating itself? You will be aware from the Introduction that we do not possess a crystal ball, so this question we will leave unanswered for the present . . .

REFERENCES

Beven, K.J. (1996) The limits of splitting: hydrology, *The Science of the Total Environment* **183**, 89–97.

Bhaskar, R. (1997) *A Realist Theory of Science*, 2nd edn, Verso, London.

Chamberlain, T.C. (1890) The method of multiple working hypotheses, *Science* **15** [It may be easier to find the reprinted versions in *Science* **148**, 754–759 (1965) or in R. Hilborn and M. Mangel (1997) *The Ecological Detective*, Princeton University Press, Princeton, NJ.]

Cooke, R.U. and Reeves, R. (1976) *Arroyos and Environmental Change in the American South-West*, Clarendon Press, Oxford.

Cross, M. and Moscardini, A.O. (1985) *Learning the Art of Mathematical Modelling*, John Wiley & Sons, Chichester.

Favis-Mortlock, D.T. (1998) Validation of field-scale soil erosion models using common datasets, in J. Boardman and D.T. Favis-Mortlock (eds) *Modelling Soil Erosion by Water*, Springer-Verlag NATO-ASI Series I-55, Berlin, 89–128.

Grayson, R.B., Moore, I.D. and McMahon, T.A. (1992) Physically based hydrologic modelling: II. Is the concept realistic? *Water Resources Research* **28**, 2659–2666.

Kirkby, M.J. (1992) Models, in M.J. Clark, K.J. Gregory and A.M Gurnell (eds) *Horizons in Physical Geography*, Macmillan, London.

Klemeš, V. (1986) Operational testing of hydrologic simulation models, *Hydrological Sciences Journal* **31**, 13–24.

Klemeš, V. (1997) Of carts and horses in hydrologic modelling, *Journal of Hydrologic Engineering* **1**, 43–49.

Meyer, W.B., Butzer, K.W., Downing, T.E., Turner II, B.L., Wenzel, G.W. and Wescoata, J.L. (1998) Reasoning by analogy, in S. Rayner and E.L. Malone (eds) *Human Choice and Climate Change Vol. 3: Tools for Policy Analysis*, Battelle Press, Columbus, OH, 217–289.

Oreskes, N., Shrader-Frechette, K. and Bellitz, K. (1994) Verification, validation and confirmation of numerical models in the Earth Sciences, *Science* **263**, 641–646.

Penrose, R. (1989) *The Emperor's New Mind*, Oxford University Press, Oxford.

Press, W.H., Teukolsky, S.A., Vetterling, W.T. and Flannery, B.P. (1992) *Numerical Recipes in FORTRAN: The Art of Scientific Computing*, Cambridge University Press, Cambridge.

Richards, K.S. (1990) 'Real' geomorphology, *Earth Surface Processes and Landforms* **15**, 195–197 [and discussion in vol. **19**, 269f.].

Schneider, S.H. (1997) Integrated assessment modelling of global climate change: transparent rational tool for policy making or opaque screen hiding value-laden assumptions? *Environmental Modelling and Assessment* **2**, 229–249.

Wainwright, J., Parsons, A.J. and Abrahams, A.D. (2000) Plot-scale studies of vegetation, overland flow and erosion interactions: case studies from Arizona and New Mexico, *Hydrological Processes* **14**, 2921–2943.

Index